D1476321

Yong-Cheng Ning

Structural Identification of Organic Compounds with Spectroscopic Techniques

Chemistry Library

Related Titles from Wiley-VCH

Ernö Pretsch, Gábór Tóth, Morton E. Munk, Martin Badertscher

Computer-Aided Structure Elucidation: Spectra Interpretation and Structure Generation

ISBN 3-527-30640-4, 2003

Stefan Berger, Siegmar Braun

200 and More NMR Experiments: A Practical Course

ISBN 3-527-31067-3, 2004

Horst Friebolin

Basic One- and Two-Dimensional NMR Spectroscopy

ISBN 3-527-31233-1, 2004

Robert M. Silverstein, Francis X. Webster, David Kiemle

Spectrometric Identification of Organic Compounds
7th Edition

ISBN 0-471-39362-2, 2004

Helmut Günzler, Hans-Ulrich Gremlich

IR Spectroscopy
An Introduction

ISBN 3-527-28896-1, 2002

Structural Identification of Organic Compounds with Spectroscopic Techniques

With a Foreword by Richard R. Ernst,
Nobel Prize Winner 1991

Yong-Cheng Ning

WILEY-VCH Verlag GmbH & Co. KGaA

Prof. Dr. Yong-Cheng Ning
Department of Chemistry
Tsinghua University
100084 Beijing
PR China

All books published by Wiley-VCH are carefully produced. Nevertheless, authors, editors, and publisher do not warrant the information contained in these books, including this book, to be free of errors. Readers are advised to keep in mind that statements, data, illustrations, procedural details or other items may inadvertently be inaccurate.

Library of Congress Card No.: applied for

British Library Cataloguing-in-Publication Data:
A catalogue record for this book is available from the British Library.

Bibliographic information published by Die Deutsche Bibliothek
Die Deutsche Bibliothek lists this publication in the Deutsche Nationalbibliografie; detailed bibliographic data is available in the Internet at <http://dnb.ddb.de>.

© 2005 WILEY-VCH Verlag GmbH & Co. KGaA, Weinheim

All rights reserved (including those of translation in other languages). No part of this book may be reproduced in any form – by photoprinting, microfilm, or any other means – nor transmitted or translated into machine language without written permission from the publishers. Registered names, trademarks, etc. used in this book, even when not specifically marked as such, are not to be considered unprotected by law.

Printed in the Federal Republic of Germany
Printed on acid-free paper

Composition K+V Fotosatz GmbH, Beerfelden
Printing betz-druck GmbH, Darmstadt
Bookbinding Litges & Dopf Buchbinderei GmbH, Heppenheim

ISBN 3-527-31240-4

Foreword

The progress made in instrumental analytical chemistry over the past two decades has been staggering. It is thus timely to present another comprehensive treatise on some of the most powerful tools from the chemist's arsenal that can be used to answer questions on the molecular properties of matter. Professor Yong-cheng Ning, one of the leading spectroscopists in China, has made a wise selection of the techniques to be included: nuclear magnetic resonance, mass spectrometry and infrared and Raman spectroscopy. All are treated in great depth and provide a sound basis of understanding, but the discussion is still comprehensible to chemists interested in their own immediate application. Many instructive examples are presented, illustrating detailed spectral analysis of organic compounds. The book is well balanced and will prove to be extremely useful to a wide range of experimental scientists.

Indeed, chemical analysis is widespread in almost all sciences. It has certainly become indispensable in materials science for establishing structure-property relationships. The deeper the understanding of biochemical and life processes, the more chemical concepts are required for explanations, and it has become apparent that biochemistry represents just an advanced form of the chemistry of complex systems. Instrumental analysis tools have profoundly changed molecular biology over recent decades. Even in biomedicine, the same techniques have proved to be very insightful. Instrumental chemical analysis has truly become one of the basic pillars of modern science.

Not without good reason, four chapters of the book deal with nuclear magnetic resonance (NMR). Indeed, NMR has proved to be one of the most versatile analytical tools to be developed so far. Its adaptability to a wide range of applications is quite remarkable, and even after almost 60 years of development, it is still being utilized in novel and more sophisticated applications, particularly in the understanding of the function of larger and larger biopolymers. Certainly, mastering advanced pulse sequences is one of the secrets of the progress of NMR. These aspects are treated well and extensively in this book.

Organic mass spectrometry has also made remarkable progress over recent decades. It has become an invaluable tool for determining the primary structure of from small up to very large molecules. In addition, it is a wonderful tool for exploring ion reactions in the gas phase. With the development of powerful ion-

Structural Identification of Organic Compounds with Spectroscopic Techniques. Yong-Cheng Ning
Copyright © 2005 WILEY-VCH Verlag GmbH & Co. KGaA, Weinheim
ISBN: 3-527-31240-4

ization techniques of molecules, almost without any size limitations, mass spectrometry has become one of the most important tools in organic chemistry and in molecular biochemistry.

Infrared (IR) and Raman spectroscopy are certainly the most traditional of the techniques mentioned, developed during the first half of the twentieth century. Although some of the traditional fields of application have been challenged by NMR and mass spectrometry, IR and Raman have maintained their importance as fast and convenient tools of analysis. In addition, progress in the optical domain over recent decades has continued and even accelerated, and today no chemist can operate successfully without access to optical spectroscopy.

Even more important than the individual tools is their wise combination. Today, too many specialists only know their own work horse and try to apply it to all conceivable problems. Breaking the barriers between the various fields and applying the entire arsenal of tools will be essential to further progress. In this respect, the treatise by Professor Yong-cheng Ning is very welcome. It can open the so far closed windows between the various analytical techniques and perhaps further fertilize better coordinated progress in instrumental analysis.

Winterthur, March 9, 2003 Prof. Dr. Richard R. Ernst

Preface

Structural identification of organic compounds using a combination of various spectroscopic data is undoubtedly a very important area of work for chemists. Organic spectroscopy is now a highly developed and still rapidly evolving discipline. Chemists should know of any up-to-date developments so that they can apply advanced techniques in their own research. Of course they hope to understand the principles, advantages and application limits of any new techniques. Furthermore, they need to know that spectra are dependent on measuring conditions. For example: the peaks of quaternary carbon atoms could be lost in their ^{13}C spectrum when the delay between two pulses is not long enough; there are artifacts in 2D NMR spectra; the molecular ion peak of a measured compound could be missed in its mass spectrum when using electron ionization, etc. Chemists can correctly interpret spectra only on the basis of having essential knowledge of the organic spectroscopy.

Organic spectroscopy is an inter-discipline between physics and chemistry. It is based on physics. At present many monographs on nuclear magnetic resonance, mass spectrometry and infrared spectroscopy written by physicists are difficult for chemists to understand. In addition, these books may be far removed from the chemical application of the spectroscopic techniques. On the other hand, monographs written by chemists emphasize the interpretation of spectra, but they do not discuss the physical basis of organic spectroscopy in any great detail. There is thus a gap between these two types of monographs.

This monograph presents the principles of organic spectroscopy in great depth in simple terms, and it builds a bridge between these two types of monographs. Chemists will gain a deeper understanding of the principles of organic spectroscopy with this monograph. A typical example is the discussion on pulse sequence units that construct pulse sequences for 2D NMR. The principles of the pulse sequence units are presented in a way that is easy for readers to understand. Once the functions of the units have been mastered, the principles of related pulse sequences, which consist of these units, can be better appreciated. Further discussion of this is presented in Appendix 1, "Product Operator Formalism for Pulse Sequences".

Up-to-date developments in NMR, MS, IR and Raman are covered in the book, for example, the pulsed-field gradient technique, LC-NMR, DOSY, the interpretation of mass spectra produced by soft ionization techniques, 2D IR, which can be verified from Appendix 3, "Subject Index (Spectroscopic Methods

Structural Identification of Organic Compounds with Spectroscopic Techniques. Yong-Cheng Ning
Copyright © 2005 WILEY-VCH Verlag GmbH & Co. KGaA, Weinheim
ISBN: 3-527-31240-4

and Theories)". The material has been prepared from the latest literature and information provided by manufacturers of spectroscopic instruments.

Although considerable space in the book is devoted to the discussion of spectroscopic theories, key tricks for the interpretation of spectra, including some of the author's original ideas, are treated thoroughly. For instance, the analysis of split patterns of peaks in the ^1H spectrum is more important and reliable than that of chemical shifts, which is illustrated by many examples. In addition, the book summarizes the rules for the interpretation of spectra and the spectroscopic patterns of common functional groups. The interpretation of spectra is always given in detail, especially for ^1H spectra. The monograph provides abundant spectral data so that it is convenient for the reader to deduce an unknown structure from the spectra of the compound.

A whole chapter deals with the determination of configurations and conformations of organic compounds and some biological molecules. It should be stressed that the discussions are presented from the point of view of spectroscopic methodology.

This book will be very useful for chemists who are engaged in research on organic structures as well as graduate students of related specialities. In fact, the book is an updated translation of the Chinese version. The First Edition won a second-class prize for "Excellent Teaching Materials" from the State Education Commission of China in 1992. The Second Edition was selected as one of the excellent textbooks for graduate students by the Ministry of Education of China in June 2003. Only two books from the field of chemistry were selected at the time. More than 16 500 copies have been sold-out in the mainland of China. It has also been printed twice in complex Chinese characters in Taiwan.

The author wishes to express his deepest gratitude to Prof. Dr. Richard R. Ernst, the single Nobel Prize winner for chemistry in 1991, who wrote the Foreword to this book. The success of the Chinese version of the book can, to a considerable extent, be attributed to his first Foreword. It is certain that his second Foreword to the present version will continue to play an important role.

The author would like to record his sincere thanks to Prof. Di-hui Qing of the Department of Foreign Languages, Xidian University, and his graduate students, Miss Wen-ning Tong, Miss Yan-ping Xu, Miss Yan-ping Wang, Miss Zhihong Cao, Miss Yuan Yuan and Mr. Chang-an Li, who have checked and refined the manuscript. The English version is closely related to their work.

My gratitude is also extended to Professor Chun-tao Che of the Chinese University of Hong Kong, who provided about 30 spectra for the book and some examples in Sections 8.4 and 3.5.

The work of Mr. Hai-jun Yang, who was in charge of producing the figures electronically, is also greatly appreciated.

Last but not least, the author would like to express his sincere thanks to Bruker, Varian, JEOL, Micromass (VG), Finnigan-Mat, Applied Biosystems, Hewlett Packard, Bio-rad, Nicolet, PerkinElmer, Shimadzu, Hitachi, Dilor, Ionspect and Bear. The book would not have been completed without their contributions to the latest developments in organic spectroscopy.

Beijing, 2004 Yong-Cheng Ning

Contents

Structural Identification of Organic Compounds with Spectroscopic Techniques. Yong-Cheng Ning
Copyright © 2005 WILEY-VCH Verlag GmbH & Co. KGaA, Weinheim
ISBN: 3-527-31240-4

1
Introduction to Nuclear Magnetic Resonance

Two groups lead, respectively, by Bloch and Purcell discovered almost simultaneously the phenomenon of nuclear magnetic resonance (NMR). Bloch and Purcell shared the Nobel Prize for physics in 1952.

NMR is the most powerful method for the identification of organic compounds, and is widely applied in many fields.

In this chapter, the basic principles and concepts of NMR spectroscopy are described. Discussions on ^{1}H spectra, ^{13}C spectra and 2D NMR spectroscopy will be given in Chapters 2, 3 and 4, respectively.

References 1, 2 are provided for the reader who would like to have a further understanding of the basic concepts of NMR spectroscopy.

1.1
Basic Principle of NMR

1.1.1
Nuclear Magnetic Momentum

Magnetic nuclei are the objects studied by NMR. The atomic nucleus consists of neutrons and positively charged protons so that a nucleus can possess magnetic momentum when it "spins" about the nuclear axis. The spinning motion of a nucleus is determined by spin quantum number I. There are three different cases:

1. A nucleus with an even number of neutrons and an even number of protons has a zero spin quantum number, for example, ^{12}C, ^{16}O, ^{32}S and so forth.
2. A nucleus with an odd number of neutrons and an even number of protons, or with an even number of neutrons and an odd number of protons, has a half-integral spin quantum number, for example:
   $I=1/2$, such as ^{1}H, ^{13}C, ^{15}N, ^{19}F, ^{31}P, ^{77}Se, ^{113}Cd, ^{119}Sn, ^{195}Pt, ^{199}Hg, and so forth.
   $I=3/2$, such as ^{7}Li, ^{9}Be, ^{11}B, ^{23}Na, ^{33}S, ^{35}Cl, ^{37}Cl, ^{39}K, ^{63}Cu, ^{65}Cu, ^{79}Br, ^{81}Br and so forth.
   $I=5/2$, such as ^{17}O, ^{25}Mg, ^{27}Al, ^{55}Mn, ^{67}Zn and so forth.
   $I=7/2$, 9/2 and so forth.

Structural Identification of Organic Compounds with Spectroscopic Techniques. Yong-Cheng Ning
Copyright © 2005 WILEY-VCH Verlag GmbH & Co. KGaA, Weinheim
ISBN: 3-527-31240-4

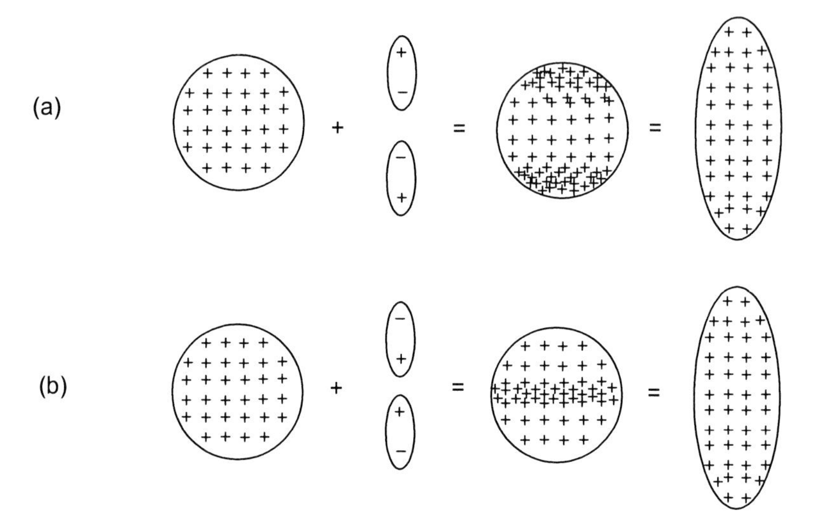

Fig. 1.1 Non-uniform distributions of nuclear charges and electric quadrupole moments.

3. A nucleus with an odd number of neutrons and an odd number of protons has an integral spin quantum number, for example: $I=1$ for ^{2}H, ^{6}L, ^{14}N; $I=2$ for ^{58}Co; $I=3$ for ^{10}B.

From the above, it follows that only nuclei belonging to cases 2 and 3 are the objects that can be studied by NMR. Those nuclei which have a non-zero spin quantum number are called magnetic nuclei. Furthermore, only the nuclei with $I=1/2$ are suitable for NMR measurement because they have a uniform charge distribution over the nuclear surface. As a result, they have no electric quadrupole moment (see below) so they can be recorded as narrow peaks in their NMR spectra. On the other hand, all other magnetic nuclei (with $I>1/2$) have a non-uniform charge distribution over the nuclear surface, as shown in Fig. 1.1, which leads to broadened peaks in their NMR spectra.

The distribution shown in Fig. 1.1a can be considered as the addition of a uniform charge distribution and a pair of electric dipoles, whose negative "poles" are towards the nuclear "equator", which leads to the concentrated positive charge distribution at two nuclear "poles." If the nucleus extends longitudinally to give a uniform charge distribution, it has a positive electric quadrupole moment according to the following equation:

$$Q = (2/5)Z(b^2 - a^2) \tag{1-1}$$

where Q is the electric quadrupole moment of the spheroid; b and a are the half longitudinal axis and the half transverse axis, respectively; Z is the charge carried by the spheroid.

Similarly, the distribution shown in Fig. 1.1b possesses a negative electric quadrupole moment.

All nuclei with an electric quadrupole moment (positive or negative) have a specific relaxation mechanism, which leads to a rapid relaxation so as to broaden their peaks. This is why only the nuclei with $I=1/2$ are suitable for NMR measurement.

The nucleus with a non-zero spin quantum number has an angular momentum, $\boldsymbol{P}$, the magnitude of which is given by Eq. (1-2):

$$P=\sqrt{I(I+1)}\frac{h}{2\pi}=\sqrt{I(I+1)}\hbar \tag{1-2}$$

where h is Planck's constant:

$$\hbar=\frac{h}{2\pi}$$

The nucleus with an angular momentum has a magnetic moment, given by Eq. (1-3):

$$\boldsymbol{\mu}=\gamma\boldsymbol{P} \tag{1-3}$$

where γ is the constant of proportionality relating $\boldsymbol{\mu}$ and $\boldsymbol{P}$, known as the magnetogyric ratio, which is an important property of the nucleus.

1.1.2
Quantization of Angular Momentum and Magnetic Moment

According to quantum mechanics, when a magnetic nucleus is placed in a static magnetic field B_0, which is along the z direction, the angular momentum of the nucleus is quantized and it will adopt one of $(2I+1)$ orientations with respect to the external magnetic field. The allowed projections of the angular momentum on the z axis, P_z, are confined to several discrete values, which are given by Eq. (1-4):

$$P_z=m\hbar P_z=m\hbar \tag{1-4}$$

where m is the magnetic quantum number of the nucleus. It has $2I+1$ values, $m=I$, $I–1$, $I–2...I$.

The quantization of angular momenta in a static magnetic field is shown in Fig. 1.2.

The projections of magnetic moments of the nucleus on the z axis, μ_z, are given by Eq. (1-5):

$$\mu_z=\gamma P_z=\gamma m\hbar \tag{1-5}$$

When a nucleus is placed in a magnetic field B_0, which is along the z axis, the energy of the nucleus is given by

$$E=-\mu\cdot\boldsymbol{B}_0=-\mu_z B_0 \tag{1-6}$$

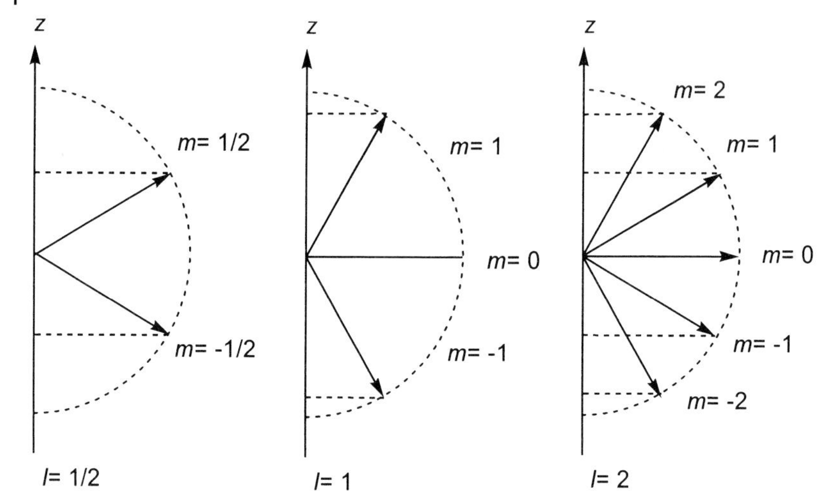

Fig. 1.2 The possible orientations for angular momenta.

Substituting Eq. (1-5) into (1-6) leads to

$$E = -\gamma m\hbar B_0 \tag{1-7}$$

Thus, the energy differences between various energy levels are

$$\Delta E = -\gamma\Delta m\hbar B_0 \tag{1-8}$$

According to the selection rule of quantum mechanics, only transitions with $\Delta m = \pm 1$ are allowable, so that the energy difference for allowed transitions is

$$\Delta E = \gamma\hbar B_0 \tag{1-9}$$

On the other hand, $m = I, I\text{–}1, \ldots, -I$, that is, the magnetic moment has $2I+1$ orientations. Using Eq. (1-6), we obtain the energy difference for allowed transitions:

$$\Delta E = \frac{2\mu_z B_0}{2I} = \frac{\mu_z B_0}{I} \tag{1-10}$$

where μ_z is the maximum projection of $\boldsymbol{\mu}$ on the z axis.

1.1.3
Nuclear Magnetic Resonance

There are various energy levels for the nucleus with a magnetic moment in a static magnetic field. The nucleus will undergo a transition if an electromagnetic wave with a frequency given by Eq. (1-9) is used. This transition is nuclear magnetic resonance (NMR). Thus the fundamental equation can be derived as follows:

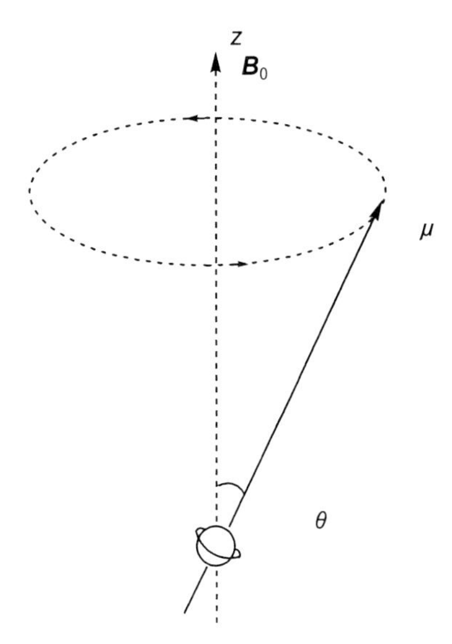

Fig. 1.3 A magnetic nucleus precesses in a static magnetic field.

$$h\nu = \gamma\hbar B_0$$

$$\nu = \frac{\gamma B_0}{2\pi} \tag{1-11}$$

where ν is the frequency of the electromagnetic wave. Its relevant circular frequency ω is

$$\omega = 2\pi\nu = \gamma B_0 \tag{1-12}$$

NMR can be discussed from another point of view. In a static magnetic field B_0 (along the z axis), a precession motion (also known as the Lamor precession) of a magnetic nucleus takes place because the nuclear spinning axis makes an angle θ with respect to the z axis, which is shown in Fig. 1.3. This situation is similar to that where a spinning gyroscopic top precesses when it is tilted with respect to the gravitational field.

The Lamor precession frequency ω_L is given by Eq. (1-13):

$$\omega_L = \gamma B_0 \tag{1-13}$$

Precession directions are determined by the sign of the γ.

Suppose that a linear polarized oscillating magnetic field, that is, an electromagnetic radiation, with the Lamor angular frequency ω_L is applied to the plane perpendicular to the static magnetic field B_0. This magnetic field can be split into two components rotating in opposite directions. One of these components

Tab. 1.1 NMR properties of some magnetic isotopes

Isotope	*Resonance frequency* (MHz)*	*Natural abundance (%)*	*Relative sensitivity (with respect to the proton at a fixed B_0)*	*Magnetic moment μ (multiply by $eh/4\pi mC$)*	*Spin quantum number I*	*Electric quadrupole moment ($e \times 10^{-24}$ cm)*
^{1}H	42.577	99.9844	1.000	2.79270	1/2	–
^{2}H	6.536	1.56×10^{-2}	9.64×10^{-3}	0.85738	1	2.77×10^{-3}
^{13}C	10.705	1.108	1.59×10^{-2}	0.70216	1/2	–
^{14}N	3.076	99.635	1.01×10^{-3}	0.40357	1	2×10^{-2}
^{15}N	4.315	0.365	1.04×10^{-3}	–0.28304	1/2	–
^{17}O	5.722	3.7×10^{-2}	2.91×10^{-2}	1.8930	5/2	-4×10^{-3}
^{19}F	40.055	100	0.834	2.6273	1/2	–
^{31}P	17.235	100	6.64×10^{-2}	1.1305	1/2	–
^{33}S	3.266	0.74	2.26×10^{-3}	0.64274	3/2	-6.4×10^{-2}
^{35}Cl	4.172	75.4	4.71×10^{-3}	0.82089	3/2	-7.97×10^{-2}
^{37}Cl	3.472	24.6	2.72×10^{-3}	0.68329	3/2	-6.21×10^{-2}
^{79}Br	10.667	50.57	7.86×10^{-2}	2.0990	3/2	0.33
^{81}Br	11.498	49.43	9.84×10^{-2}	2.2626	3/2	0.28
^{127}I	8.519	100	9.35×10^{-2}	2.7939	5/2	–0.75

* Resonance frequency at 1 Tesla.

rotating in the opposite direction to that of nuclear precession is ineffective. The other component excites the nuclear precession because of the same direction of rotation and the same frequency. Some energy is transferred from the electromagnetic wave to the magnetic nuclei, which is NMR.

The NMR properties of some magnetic isotopes, which exist in organic compounds, are given in Tab. 1.1.

1.2 Chemical Shift

In 1950 Proctor and Yu found two NMR signals for a solution of ammonium nitrate. Clearly these two signals belong to ammonium ions and nitrate ions, respectively, that is, NMR signals can distinguish nuclei in different chemical surroundings for a given isotope.

1.2.1 Shielding Constant

The above-mentioned phenomenon can be explained by the shielding effect of the electrons surrounding the nuclei. The magnitude of the magnetic field actually experienced by the nuclei is slightly less than that of the applied field. Therefore, Eq. (1-11) should be modified as

$$\nu = \frac{\gamma}{2\pi} B_0(1 - \sigma) \tag{1-14}$$

where σ is known as the shielding constant or the screening constant. It is a field-independent factor and is related to the chemical surroundings of the nuclei. Different isotopes possess various σ values covering maybe several orders of magnitude, but σ values of all isotopes are much less than 1.

σ can be shown to be

$$\sigma = \sigma_d + \sigma_p + \sigma_a + \sigma_s \tag{1-15}$$

σ_d is the diamagnetic term, which is contributed by s electrons. The induced electron circulation produces a diamagnetic field. The term "diamagnetic shielding" arises from the fact that the induced field is opposed to the applied field. The greater the density of s electrons around the nucleus, the less field strength the nucleus experiences. Acted on by the shielding effect, the resonance peak of the nucleus will shift towards the right.

σ_p is the paramagnetic term, which is contributed by p and d electrons that are distributed unsymmetrically about the nucleus. Because of the hindrance by other chemical bonds, the direction of the induced field coincides with that of the applied field, hence the term "paramagnetic shielding." σ_a indicates the anisotropic influences from neighboring groups and σ_s shows the solvent (or medium) effects.

The effects from σ_d and σ_p are much greater than those from σ_a and σ_s. σ_p is much more important than σ_d for all isotopes except ^1H.

1.2.2 Chemical Shift δ

For a given isotope, nuclei in various functional groups will show their NMR signals with different abscissas because of their different σ values. A specific substance is selected as an internal standard, the peak for which is set as the origin of an NMR spectral abscissa. All peak positions of functional groups of a compound are defined as

$$\delta = \frac{\nu_{sample} - \nu_{standard}}{\nu_{standard}} \times 10^6 \tag{1-16}$$

where ν_{sample} and $\nu_{standard}$ are the resonant frequencies of the functional group and the standard, respectively.

δ, known as the chemical shift, shows the peak position for a particular functional group and is in ppm (parts per million), which is dimensionless. Its sign is negative when a signal is positioned on the right side of the standard and positive when on the left side.

Since the numerator of Eq. (1-16) is far less than the denominator by several orders of magnitude, and $\nu_{standard}$ is very close to the nominal frequency of the NMR spectrometer, ν_0, $\nu_{standard}$ is replaced by ν_0.

Tetramethylsilane is usually applied as the standard because it has only a single peak and it can be easily removed from the sample (bp = 27 °C). An additional advantage is that common functional groups have positive δ values.

1.3
Spin–spin Coupling

In 1952, Gutowsky et al. reported that two ^{19}F peaks existed for a $POCl_2F$ solution, which led to the discovery of spin–spin couplings.

1.3.1
Spin–spin Coupling Produces NMR Signal Splitting

The splitting of the ^{19}F signal is produced by ^{31}P being connected to ^{19}F. Because ^{31}P possesses a spin quantum number of 1/2, it has two orientations with almost equal probabilities: either approximately parallel or anti-parallel to the applied magnetic field, which leads either to an increase or to a decrease in the magnetic field experienced by the ^{19}F. Consequently, the singlet corresponding to Eq. (1-14) is replaced by a doublet with a roughly identical intensity. This effect of producing the spin–spin splitting is known as spin–spin coupling, the magnitude of which, known as the coupling constant J and measured in Hertz, is measured by the space between adjacent split peaks in a multiplet. If more than one magnetic nucleus (with a spin quantum number of 1/2) couples with another magnetic nucleus, the NMR signals of the nucleus to be coupled will show further splitting. The splitting pattern can be described by Pascal's triangle (Tab. 1.2), where n is the number of nuclei that participate in coupling. Multiplicity indicates the split signal of the nuclei to be coupled. The relative peak intensities of a multiplet are shown by the developed binomial coefficients in Pascal's triangle, from which the $n+1$ rule can be understood: $n+1$ peaks will be produced by the coupling of n nuclei.

Tab. 1.2 Pascal's triangle

n	*Relative peak intensities Developed binomial coefficients*	*Multiplicity*
0	1	singlet
1	1 1	doublet
2	1 2 1	triplet
3	1 3 3 1	quartet

1.3.2
Energy Level Diagram

The energy level diagram is very useful for theoretical discussions on NMR. An AX system, which consists of two coupled magnetic nuclei, will be used as an example. The details of spin systems will be described in Section 2.3.

Consider that both of the two nuclei have a spin quantum number of 1/2, so each has two orientations with respect to the applied magnetic field. We use α to denote the orientation corresponding to the magnetic quantum number m of 1/2 and β to that of −1/2. Consequently, four energy levels exist, which are A(α) X(α), A(β) X(α), A(α) X(β) and A(β) X(β), simplified as $\alpha\alpha$, $\beta\alpha$, $\alpha\beta$ and $\beta\beta$, respectively.

1. The energy level diagram without coupling between A and X.
   The energy level diagram, which contains four energy levels, 1, 2, 3 and 4, according to the order of increasing energy, is shown in Fig. 1.4.
   The transition from level 1 to level 2 changes the α state of the A nucleus to the β state without changing the X spin state, so the signal belongs to A. The same is true of the transition from level 3 to level 4. On the other hand, both the transition from level 1 to level 3 and that from level 2 to level 4 produce the X signal.
   Because of the equal energy differences, there is only a peak for A and a peak for X.
   Fig. 1.4 b is usually used to give a clear demonstration.
2. The energy level diagram with coupling between A and X.
   Because of the mutual actions between magnetic nuclei, the energy levels of the AX system will be changed when a coupling between A and X exists. Suppose that levels 1 and 4 are raised by a magnitude of $J/4$, and correspondingly, levels 2 and 3 will be decreased by the same magnitude. Consequently, on the basis of Fig. 1.4, a new energy level diagram, Fig. 1.5, is drawn, in which levels 1′, 2′, 3′ and 4′ are transferred from levels 1, 2, 3 and 4 of Fig. 1.4, respectively.
   Related calculations lead to Fig. 1.6.

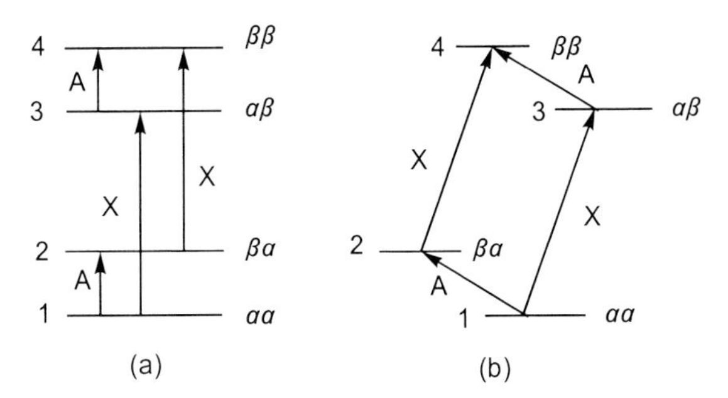

Fig. 1.4 The energy level diagram of an AX system without the coupling between A and X.

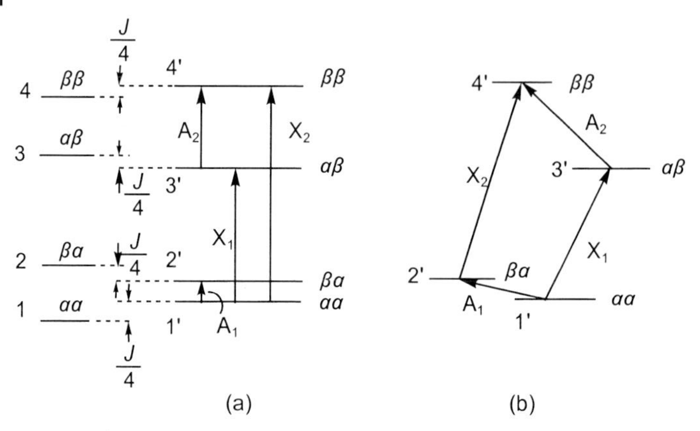

Fig. 1.5 The energy level diagram of an AX system with the coupling between A and X.

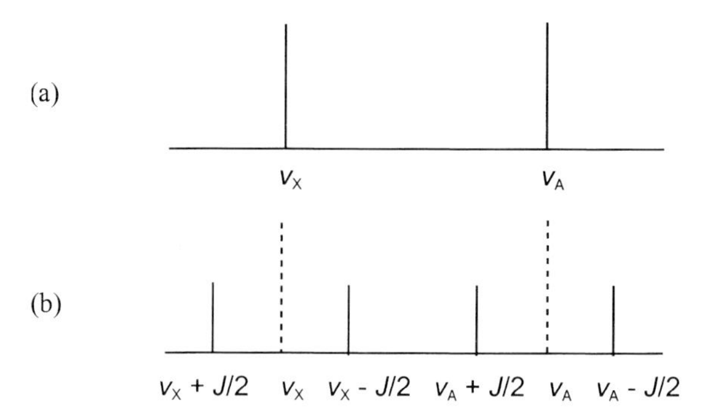

Fig. 1.6 The NMR spectrum of an AX system (a) without coupling between A and X (b) with coupling between A and X.

When a coupling between A and X exists, there are two sets of split peaks for A and X, respectively. Each set of peaks keeps the same δ value (measured at the middle point between the two peaks) because the coupling does not affect the chemical shift. Two peaks, which belong to one nucleus, are separated by a spacing, J. The two peaks have the same intensity. The coupling constant J, measured in Hertz, is independent of the applied magnetic field or the spectrometer frequency.

1.3.3
Coupling Constant *J*

The magnitude of J is closely associated with the number of chemical bonds across the related coupled nuclei. Consequently, the number is denoted as an upper left superscript to J, for example, ^1J, ^2J, and so forth. As the coupling is conducted through binding electrons, the magnitude of J decreases rapidly with an increase in the number. Thus, the term long-range spin–spin coupling is obtained when the number is greater than a particular integer n. In the ^{1}H spectra, n is 4 because a coupling across three bonds is common, and n is 2 for the ^{13}C spectra.

The J possesses an algebraic value, which is distinguished by a positive or negative sign. If two coupled nuclei with the same orientation have a higher energy level than that without the coupling, or two coupled nuclei with opposite orientations have a lower energy level than that without the coupling, their J has a positive value. On the other hand, J has a negative value. The absolute value of J can be read or calculated from the related spectrum, but the sign of J is difficult to obtain.

1.4 Magnetization

1.4.1 Magnetization Concept

Since NMR is a phenomenon of an ensemble of magnetic nuclei, it can be discussed from a macroscopic point of view.

In a static magnetic field B_0, magnetic moments of a particular type of nuclei precess about B_0 along several cones, the number and orientations of which are determined by the spin quantum number I. If $I=1/2$, there are two cones, on which the magnetic moments precess with the same Lamor frequency. According to the Boltzmann's distribution, the population of magnetic moments precessing about B_0 is slightly larger than that of magnetic moments precessing about $-B_0$. Therefore, it could be considered that only the magnetic moments that represent the difference of the two populations precess about B_0 to form a resultant magnetization, as shown in Fig. 1.7.

Suppose that there are N nuclei in a unit volume, we define the magnetization $\boldsymbol{M}$ as

$$\boldsymbol{M} = \sum_{i=1}^{N} \mu_i \tag{1-17}$$

where μ_i is the i-th magnetic moment of a nucleus in a unit volume.

It is clear that

$$\boldsymbol{M} = \boldsymbol{M}_{||} + \boldsymbol{M}_{\perp} \tag{1-18}$$

and

$$\boldsymbol{M}_{||} = \boldsymbol{M}_{+} + \boldsymbol{M}_{-} \tag{1-19}$$

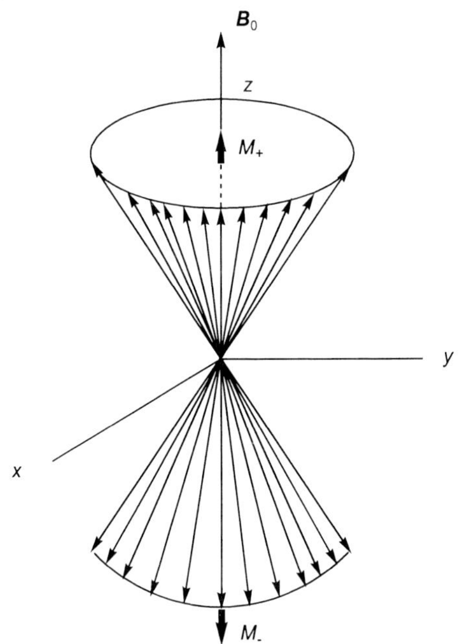

Fig. 1.7 Magnetic moments of nuclei with $I=1/2$ precess on two cones.

where $\boldsymbol{M}_{||}$ and $\boldsymbol{M}_{\perp}$ are the components along $\boldsymbol{B}_0$ and perpendicular to $\boldsymbol{B}_0$, respectively.

If only a static magnetic field exists then

$$\boldsymbol{M}_{||} = \boldsymbol{M}_z = \boldsymbol{M}_+ - \boldsymbol{M}_- \tag{1-20}$$

$$\boldsymbol{M}_{\perp} = 0 \tag{1-21}$$

1.4.2 Rotating Frame

The effect of an oscillating magnetic field (a linear polarized alternating magnetic field), that is, a radiofrequency wave, will now be discussed. A linear polarized alternating magnetic field is composed of two counter-rotating magnetic fields, of which one has the same rotating direction as that of the nuclear magnetic moments. Therefore, the rotating magnetic field with the Lamor frequency can affect the nuclear magnetic moments, and the other rotating magnetic field has no function in NMR.

A laboratory framework consists of three mutually perpendicular axes: the x, the y and the z axes. The last has the same direction as $\boldsymbol{B}_0$. Now we will define a rotating frame, in which the z' axis has the same direction as $\boldsymbol{B}_0$ or the z axis while the x' and the y' axes rotate about the z' axis at the Lamor frequency in the same rotat-

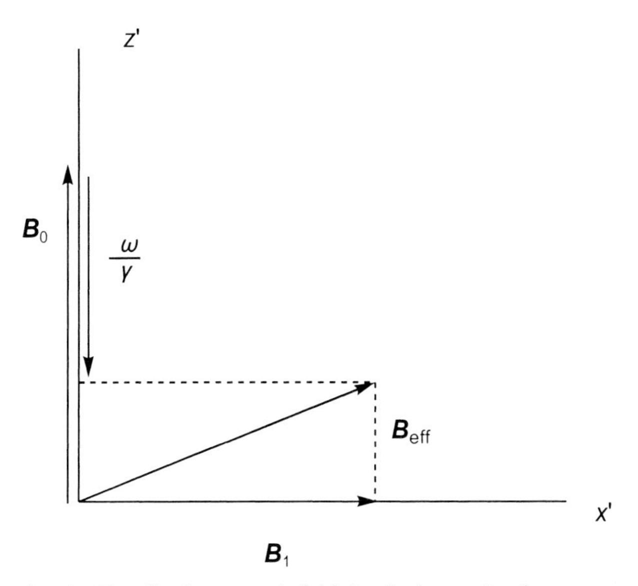

Fig. 1.8 The effective magnetic field, $\boldsymbol{B}_{eff}$, in the rotating frame.

ing direction as that of the precession of nuclear magnetic moments. In the rotating frame, $\boldsymbol{B}_1$ is relatively static so that its direction can be set along the x' axis.

By using the rotating frame, the magnetization $\boldsymbol{M}$ can be discussed more easily. $\boldsymbol{M}$ is affected by $\boldsymbol{B}_0$ along the z' axis and by $\boldsymbol{B}_1$ along the x' axis. However, since the frame is rotating, a term $\boldsymbol{B}_v$, known as a virtual field, has to be added to $\boldsymbol{B}_0$ and $\boldsymbol{B}_1$. $\boldsymbol{B}_v$ has the opposite direction to $\boldsymbol{B}_0$ and its magnitude is given by the following equation:

$$B_v = \omega/\gamma \tag{1-22}$$

where γ is the magnetogyric ratio of the isotope and ω is the rotating frequency of the rotating frame.

$\boldsymbol{M}$ is affected by the resultant magnetic field from $\boldsymbol{B}_0$, $\boldsymbol{B}_1$ and $\boldsymbol{B}_v$, known as $\boldsymbol{B}_{eff}$, as shown in Fig. 1.8.

Fig. 1.8 can be expressed by the following equation:

$$\boldsymbol{B}_{eff} = \boldsymbol{B}_0 + \boldsymbol{B}_v + \boldsymbol{B}_1 \tag{1-23}$$

Since at resonance

$$\omega_0 = \gamma B_0 \tag{1-12}$$

hence

$$B_0 = \omega_0/\gamma$$

where ω_0 denotes the resonant frequency.

Substituting Eq. (1-12) into Eq. (1-23) yields

$$\boldsymbol{B}_{\text{eff}} = \boldsymbol{B}_0 + \boldsymbol{B}_{\text{v}} + \boldsymbol{B}_1 = [(\omega_0 - \omega)/\gamma]\boldsymbol{k} + B_1\boldsymbol{i} \tag{1-24}$$

where $\boldsymbol{k}$ and $\boldsymbol{i}$ are the unit vectors of the rotating frame, respectively.

When $\omega=\omega_0$, Eq. (1-24) becomes

$$\boldsymbol{B}_{\text{eff}} = \boldsymbol{B}_1 \tag{1-25}$$

This equation means that the magnetization $\boldsymbol{M}$ is affected only by $\boldsymbol{B}_1$ under the resonant conditions. Therefore, $\boldsymbol{M}$ will rotate about the x' axis to produce a transverse component, $\boldsymbol{M}_\perp$, which is related to an NMR signal. This can be understood as follows: $\boldsymbol{M}_\perp$ is rotating within the laboratory framework, in which a detection coil is set. The rotating $\boldsymbol{M}_\perp$ will continuously cut the detection coil to produce a potential, that is, an NMR signal.

If $\omega \neq \omega_0$, the magnetization rotates about the $\boldsymbol{B}_{\text{eff}}$ as given by Eq. (1-24). When $\omega_0 - \omega > B_1$, the magnetization rotates about the z' axis, so that there is no transverse component, that is, no NMR signal.

1.5 Relaxation Process

1.5.1 What is a Relaxation Process?

All types of absorption spectroscopy follow the same principle: sample molecules undergo transitions from a lower energy level to a higher energy level, where the molecules absorb electromagnetic wave quantum with the appropriate energy, which is equal to the difference between the two energy levels. At the same time, some molecules at the higher energy level return to the lower energy level under the same conditions. Because the probability of the transition from the lower energy level to the higher energy level is equal to that of the transition from the higher energy level to the lower energy level, and the population at the lower energy level is slightly greater than that at the higher energy level, according to Boltzmann's distribution law, an absorption signal can be produced. If the absorption signal is to be maintained, some molecules at the higher energy level must be able to return continuously to the lower energy level. This process is known as relaxation. The relaxation in absorption spectroscopy can take place spontaneously. However, its rate is determined by the energy difference between the two energy levels. The smaller the energy difference between the two energy levels, the slower the rate of relaxation. Since NMR transitions have the smallest energy differences in absorption spectroscopy, the relaxation in NMR has to be considered. If the relaxation is not effective, the system will reach saturation, which means the population at the higher energy level is equal to that at the lower energy level, resulting in no NMR signal.

1.5.2
Longitudinal and Transverse Relaxation

Relaxation can be understood more thoroughly through the concept of the magnetization vector $\boldsymbol{M}$. When the NMR condition is satisfied, $\boldsymbol{M}$ (or the cone in Fig. 1.7) tips towards the y' axis of the rotating frame, and correspondingly, the two components of $\boldsymbol{M}$ change: $M_{||}$ decreases from M_0 to a specific value; $M_{\perp}$ increases from zero to a specific value. As soon as the declination of the magnetization vector begins, the relaxation returns the two components of the magnetization vector to the original (equilibrium) state ($M_{||}=M_0$, $M_{\perp}=0$). Although there are some relationships between the relaxation of $M_{||}$ and that of $M_{\perp}$, the two relaxations, known as the longitudinal relaxation and the transverse relaxation, respectively, possess different mechanisms and physical significance.

The longitudinal relaxation is the relaxation of the longitudinal component (along the z axis) of the magnetization vector. Before the declination of $\boldsymbol{M}$,

$$M_{||} = M_0 = M_+ - M_-$$

which means $M_{||}$ is associated with the population difference between two related energy levels. From the point of view of energy, nuclear magnetic resonance is a process during which a spin system absorbs some energy from its environment to increase the population of the higher energy level. On the other hand, the longitudinal relaxation returns $M_{||}$ to M_0 so that the population of the higher energy level decreases. This is the process during which the spin system releases some of its energy to its environment. In brief, the longitudinal relaxation is associated with energy exchange between a spin system and its environment. The longitudinal relaxation is also called spin–lattice relaxation, where "lattice" implies the environment.

The transverse relaxation is the relaxation of the transverse component (in the $x'y'$ plane) of the magnetization vector. Prior to nuclear magnetic resonance, magnetic moments of spins precess randomly on the surface of the cone, so that their projections on the $x'y'$ plane have a uniform distribution. There is neither coherence among the phases of the precessing magnetic moments nor a resultant $\boldsymbol{M}_{\perp}$. When the resonance takes place, the magnetization vector is declined to create a transverse component on the y' axis (or in the $x'y'$ plane). On the basis of the consideration of the declination of the cone in Fig. 1.7, the projections of precessing magnetic moments form a distribution with a fan-shaped symmetry about the y' axis, which means that the projections of the magnetic moments possess some coherence. The transverse relaxation means that this coherence returns to a uniform distribution. Therefore, the transverse relaxation is an entropy effect.

The inhomogeneity of an applied magnetic field affects the spins of various volume elements with differing magnetic field strengths. Therefore, the inhomogeneity also contributes to the transverse relaxation.

The transverse relaxation is associated with the interactions between the precessing nuclear magnetic moments. It decentralizes their distribution about the

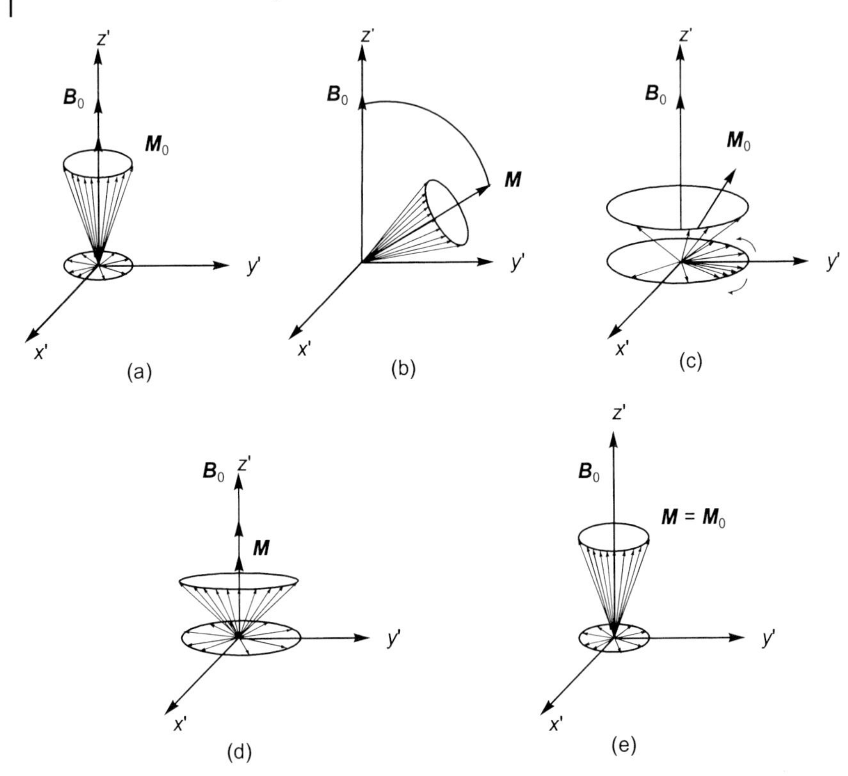

Fig. 1.9 The relaxation process of the magnetization ***M*** (a) under equilibrium conditions. (b) ***M*** has been declined from the z' axis by the action of $\boldsymbol{B}_1$. (c) Both longitudinal relaxation and transverse relaxation take place after the action of $\boldsymbol{B}_1$; the transverse relaxation is clear. (d) The transverse relaxation has finished and the longitudinal relaxation continues. (e) The longitudinal relaxation has finished; ***M*** comes back to the equilibrium condition.

y' axis to attain a uniform distribution about the origin. Consequently, the transverse relaxation is known as spin–spin relaxation.

The rates of the processes of the longitudinal and transverse relaxations are characterized by $1/T_1$ and $1/T_2$, respectively. T_1 and T_2 are similar to the time constant for a first-order chemical reaction. T_1 is the longitudinal relaxation time, and T_2 the transverse relaxation time. Both are measured in seconds.

The two following equations hold:

$$\mathrm{d}M_{||}/\mathrm{d}t = \mathrm{d}M_z/\mathrm{d}t = -(M_z - M_0)/T_1 \tag{1-26}$$

$$\mathrm{d}M_{\perp}/\mathrm{d}t = -(M_{\perp} - 0)/T_2 = -M_{\perp}/T_2 \tag{1-27}$$

where M_0 is the value of $M_{||}$ under the original (equilibrium) conditions; and zero is the value of $M_{\perp}$ under the original (equilibrium) conditions.

When $\boldsymbol{M}$ is declined from the z' axis towards the y' axis, $M_{||}$ decreases while $M_{\perp}$ is formed. After the action of $\boldsymbol{B}_1$, $M_{||}$ and $M_{\perp}$ relax with their time constants T_1 and T_2, respectively, according to Eq. (1-26) or Eq. (1-27). This process is shown in Fig. 1.9, from which it is seen that just at the last moment, $M_{||}$ approaches M_0, so that $T_1 \geq T_2$.

1.5.3
Width of an NMR Signal

An NMR signal possesses some width because of the uncertainty principle of quantum mechanics.

$$\Delta E \cdot \Delta t \approx h \tag{1-28}$$

where Δt is the duration a particle stays in an energy level, and h is Planck's constant.

In an NMR process, Δt is determined by spin–spin interactions with time constant T_2. Thus

$$\Delta E \cdot T_2 \approx h \tag{1-29}$$

Since

$$\Delta E = h\Delta \upsilon$$

one obtains

$$\Delta \upsilon \approx \frac{1}{T_2} \tag{1-30}$$

The width calculated using Eq. (1-30) is the so-called natural width. Because the inhomogeneity of an applied magnetic field also contributes to the transverse relaxation, apparent transverse relaxation time T_2' is shorter than T_2. Therefore, Eq. (1-30) becomes

$$\upsilon \approx \frac{1}{T_2'} \tag{1-31}$$

Consequently, the real line widths measured in NMR experiments are larger than the natural widths.

1.6
Pulse-Fourier Transform NMR Spectrometer

In the past, NMR spectrometers were continuous wave spectrometers, through which an NMR spectrum was recorded by the continuously changing magnetic field strength or radiofrequency (RF). At most only the nuclei of one functional group were resonated at any particular moment in this way. All the other nuclei were waiting for the resonance to occur, so that the sensitivity of the measurement was low.

When there is only a small amount of sample available, it is necessary to repeat the scans to improve the signal-to-noise ratio. The accumulated signal-to-noise ratio is proportional to the square root of the number of scans, $\sqrt{n}$, because the signal intensity is proportional to n while the noise intensity is proportional to the square root of the number of scans. Since the value of n is rather large and the time for one scan is rather long, the accumulation of data by a continuous wave spectrometer is time-consuming.

To overcome the above-mentioned shortcomings, a new type of NMR spectrometer is required, that is, the pulse-Fourier transform NMR spectrometer.

1.6.1
Application of Strong and Short RF Pulses

In a rotating frame, the effective field of the magnetic nuclei is given by Eq. (1-24), that is:

$$\boldsymbol{B}_{\text{eff}} = (\omega_0/\gamma - \omega/\gamma)\boldsymbol{k} + B_1\boldsymbol{i} \tag{1-24}$$

This equation can be rewritten as:

$$B_{\text{eff}} = \left[\left(\frac{\omega_0}{\gamma} - \frac{\omega}{\gamma}\right)^2 + B_1^2\right]^{1/2} = \frac{1}{\gamma}[(\omega_0 - \omega)^2 - (\gamma B_1)^2]^{1/2} \tag{1-32}$$

where ω_0 is the resonant frequency of the nuclei; ω is the rotating circular frequency of the frame; and B_1 is the magnetic induction strength of the rotating polarized magnetic field.

The general equation for nuclei with various chemical shifts and coupling constants is

$$B_{\text{eff}} = \frac{1}{\gamma}[(\omega_{0i} - \omega)^2 + (\gamma B_1)^2]^{1/2} \tag{1-33}$$

where B_{eff} is the effective field of the i-th nucleus; and ω_{0i} is the resonant frequency of the i-th nucleus.

Suppose that the spectral width of an NMR spectrum is ΔF (measured in Hertz) and that B_1 is sufficiently strong, so that

$$\gamma B_1 \gg 2\pi\Delta F \tag{1-34}$$

Under these conditions, the second term in the square brackets of Eq. (1-33) can be neglected. Consequently, one has

$$B_{\text{eff}} \approx B_1 \tag{1-35}$$

Eq. (1-35) leads to a very important conclusion: if B_1 is sufficiently strong, all nuclei with different δ and J values have an approximate effective field, B_1. In other words, all magnetization vectors corresponding to all signals in an NMR spectrum are rotated towards the y' axis under the action of $\boldsymbol{B}_1$ along the x' axis. Therefore, all nuclei resonate simultaneously although they possess different δ and J values.

As with Eq. (1-13), the rotation of all magnetization vectors about the x' axis can be described by the following equation:

$$\Omega = \gamma B_1 \tag{1-36}$$

where Ω is the angular velocity of $\boldsymbol{M}$ about the x' axis.

Suppose that the duration of the action of B_1 is t_p, during which the rotated angle of $\boldsymbol{M}$ is α, the rotation can be expressed by the following equation:

$$\alpha = \Omega t_p \tag{1-37}$$

Substituting Eq. (1-36) into Eq. (1-37), one obtains:

$$\alpha = \gamma B_1 t_p \tag{1-38}$$

Suppose $\alpha = 90^\circ$ (correspondingly, the pulse is known as the $\pi/2$ pulse), Eq. (1-37) becomes

$$\pi/2 = \gamma B_1 t_p$$

$$t_p = \frac{\pi}{2\gamma B_1} \tag{1-39}$$

The substitution of Eq. (1-34) into Eq. (1-39) yields

$$t_p \ll \frac{1}{4\Delta F} \tag{1-40}$$

From Eqs. (1-34) and (1-40), it can be seen that all nuclei can be excited at the same time by using a strong, short pulse.

If Eq. (1-34) can not be satisfied, $\boldsymbol{M}$ will rotate not about $\boldsymbol{B}_1$ but about $\boldsymbol{B}_{\text{eff}}$, as can be seen from Fig. 1.8.

1.6.2
Time Domain Signal and Frequency Domain Spectrum, and their Fourier Transform

Under the action of a strong, short pulse, all magnetization vectors decline from the x' axis towards the y' axis so their transverse components, the measurable $M_{\perp i}$, are produced. After the action, we have

$$M_{y'_i} = M_{y'_i}(0)e^{-\frac{t}{T_{2i}}}\cos(\omega_{0_i} - \omega)t \tag{1-41}$$

where $M_{y'_i}$ is the measured signal of the i-th nucleus at time t; t is the time recorded at the end of the pulse; $M_{y'_i}(0)$ is the measured signal of the i-th nucleus at $t=0$; T_{2i} is the transverse relaxation time of the i-th nucleus; ω_{0i} is the resonant frequency of the i-th nucleus; and ω is the rotating frequency of the rotating frame.

The term $e^{-t/T_{2i}}$ in Eq. (1-41) originates from the transverse relaxation described by Eq. (1-27). The term $\cos(\omega_{0i}-\omega)t$ arises from the fact that $M_{\perp i}$ rotates in the $x'y'$ plane when $\omega_{0i} \neq \omega$. The signals, $M_{y'_i}$, are termed “free induction decay”, FID, because they are freely precessing transverse components that are induced after the action of pulses and they are being reduced according to their transverse relaxation times. Since each $M_{y'_i}$ has its own ω_{0i} and T_{2i}, the measured signal is an interferogram resulting from the addition of all FIDs. This is a time domain signal since the variable is time. The Fourier transform changes the time domain signal into a frequency spectrum, which we can recognize as follows:

$$F(\omega) = \frac{1}{2\pi}\int_{-\infty}^{\infty} f(t)e^{i\omega t}dt \tag{1-42}$$

Alternatively, the Fourier transform can change a frequency domain spectrum into a time domain signal, that is,

$$\text{FID} = f(t) = \int_{-\infty}^{\infty} F(\omega)e^{i\omega t}d\omega \tag{1-43}$$

In the two equations above, the sign of “i” is the imaginary unit.

Because of the different units in Eqs. (1-42) and (1-43), a coefficient, $1/2\pi$, is present in Eq. (1-42).

An intuitive explanation of the Fourier transform is useful to the understanding of abstract calculation. According to the trigonometric expression of a complex number, that is,

$$e^{-i\omega t} = \cos\omega t - i\sin\omega t \tag{1-44}$$

Eq. (1-42) can be rewritten as

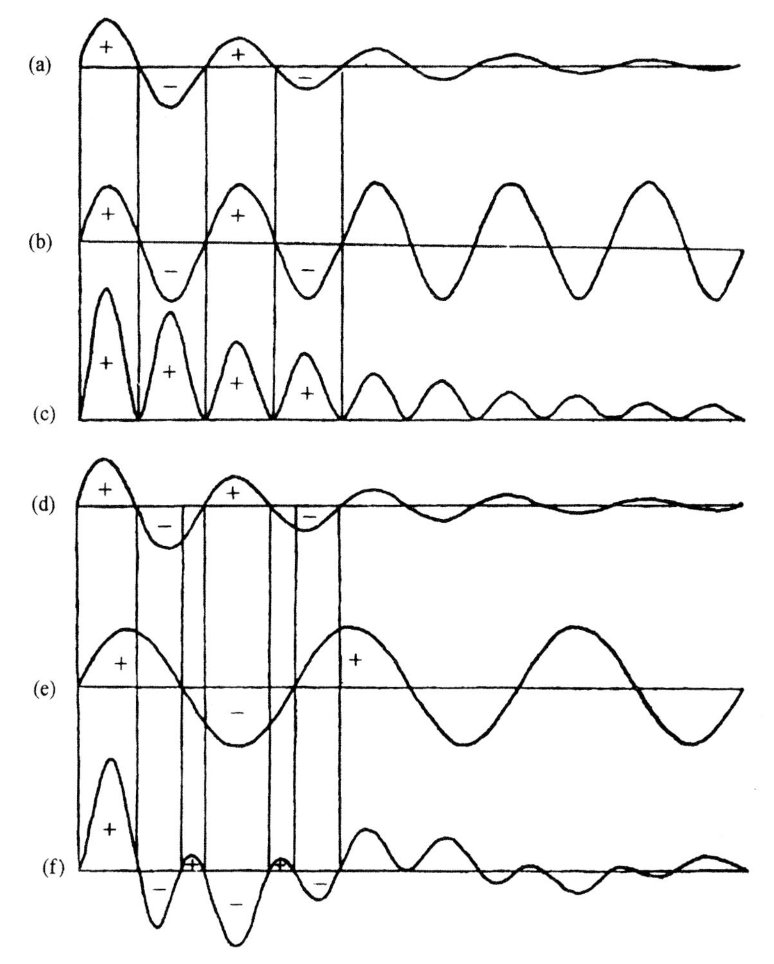

Fig. 1.10 An intuitive explanation of the Fourier transform. (a) $f(t) = \sin\omega_1' t \times e^{-t/T}$; (b) $\sin\omega_1 t$, $\omega_1 = \omega_1'$; (c) $\sin\omega_1' t \times e^{-t/T} \times \sin\omega_1 t$; (d) $f(t) = \sin\omega_2' t \times e^{-t/T}$; (e) $\sin\omega_2 t$, $\omega_2 = (3/5)\omega_2'$; (f) $\sin\omega_2' t \times e^{-t/T} \times \sin\omega_2 t$.

$$F(\omega) = \frac{1}{2\pi} \int_{-\infty}^{\infty} f(t)[\cos\omega t - i\sin\omega t]\mathrm{d}t \tag{1-45}$$

A real and an imaginary component can be obtained from Eq. (1-45).

Now $\int_{-\infty}^{\infty} f(t)\sin\omega t dt$ will be used as an example. Similarly, we can calculate the other component since $\cos\omega t = \sin(\omega t + \pi/2)$.

The calculation of $\int_{-\infty}^{0} f(t)\sin\omega t dt$ can be understood from Fig. 1.10.

If ω' in $f(t) = \sin\omega' \times e^{-t/T}$ is equal to ω in $\sin\omega t$ and both $f(t)$ and $\sin\omega t$ have the same phase, the product of every component is positive. Therefore, the integration, which is the addition of all products, is positive, that is, it has a non-zero value.

On the other hand, if $\omega' \neq \omega$, then some products are positive and some are negative leading to an integration of zero, which can be proved strictly by mathematics.

$f(t)$ is an addition of FIDs with many frequencies. A spectral width will be set for a spectral recording. An ω value, which is varied gradually and discontinuously from one end to another of the spectral width, is taken for the calculation of the Fourier transform. If this ω is equal to an ω' in FIDs, a non-zero integral value is obtained. Consequently, a peak is present in an NMR spectrum. On the other hand, if any ω' of FIDs is not equal to the ω, the integral value is zero, that is, there is no peak at the position of the abscissa, ω. An NMR spectrum is obtained after the discontinuous calculation for the spectral width to be set.

1.6.3
FT-NMR with Respect to the Fourier Decomposition

FT-NMR can be understood from another point of view, that is, from the Fourier decomposition.

A square waveform can be decomposed into a series of harmonic components, as shown in Fig. 1.11.

From Fig. 1.11 it can be seen that the more harmonic components are added, the nearer the result of the addition approaches the square waveform. If the number of components increases to infinity, the addition forms the square waveform. It can be assumed that a sample can "perceive" infinite frequencies under the action of a pulse with a square waveform, which means that nuclei with different resonant frequencies can resonate simultaneously through a single pulse.

However, the discussion above is only an explanation in principle. If an NMR experiment uses a pulse as shown in Fig. 1.11, amplitudes of effective components are too weak.

The NMR experiment applies an electromagnetic wave modulated by rectangular pulses, which is shown in Fig. 1.12.

The modulation means that one function is multiplied by another function. Through the physical concepts, a sample can perceive many discrete frequencies positioned within a wide range of frequencies, as shown in (c) of Fig. 1.11.

The envelope curve of the amplitudes of frequency components is expressed by the following equation:

$$H(f - f_0) = \frac{At_p \sin[\pi(f - f_0)t_p]}{PD\pi(f - f_0)t_p} \tag{1-46}$$

where $H(f - f_0)$ is the intensity of the component with a frequency of $(f - f_0)$; A is the intensity of the rectangular pulse; PD is the cycle of the rectangular pulse; and t_p is the duration of the rectangular pulse.

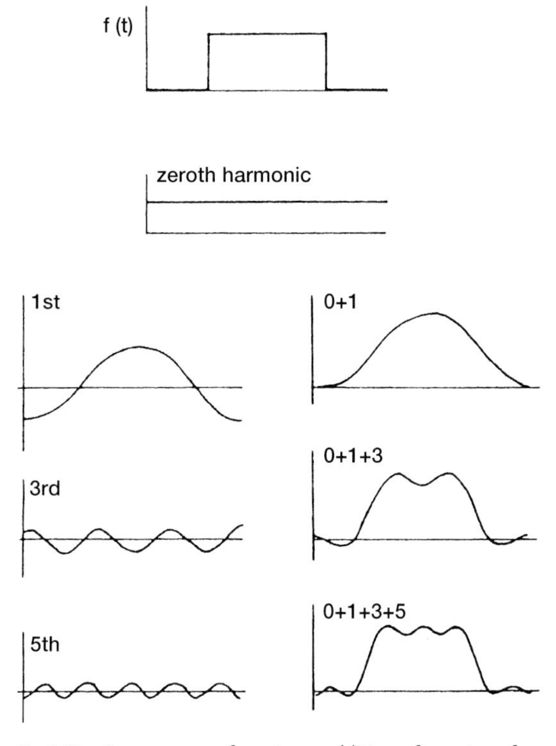

Fig. 1.11 A square waveform is an addition of a series of harmonic components. Reprinted from reference [2], with permission from Elsevier.

From Fig. 1.12 d or Eq. (1-46), it is clear that $H(f - f_0) = 0$ when $(f - f_0) = 1/t_{p0}$. Because t_p is very short (several microseconds), $(f - f_0)$ with an amplitude greater than zero has a very large range. If $(f - f_0) \ll 1/t_p$ (which is the case in practice), the $H(f - f_0)$ values are very close to $H(f_0)$, which means that the NMR signals have perfect quantitative relationships in spite of their different resonant frequencies.

The frequency interval between every two components is equal to $1/PD$, which is generally less than the width of an NMR signal, so that the NMR signals cannot be missed.

FT-NMR has been discussed in the two previous sections. In fact, these two arguments are related, but each stresses one of the two sides accordingly.

The first argument, covering the equations from (1-32) to (1-40), gives a clear explanation of the necessity of a pulse with strong power and of a short duration. Under the conditions with $(\gamma B_1)^2 \gg (\omega_{0i} - \omega)^2$, all nuclei in different functional groups resonate simultaneously and they have approximately quantitative signals.

The second argument, discussed in Section 1.6.3, explicates intuitively the frequency distribution of a pulse. As a short t_p, during which the energy transmission for the nuclei is achieved, is necessary to obtain a good quantitative result, it could be understood that the power of the pulse must be strong.

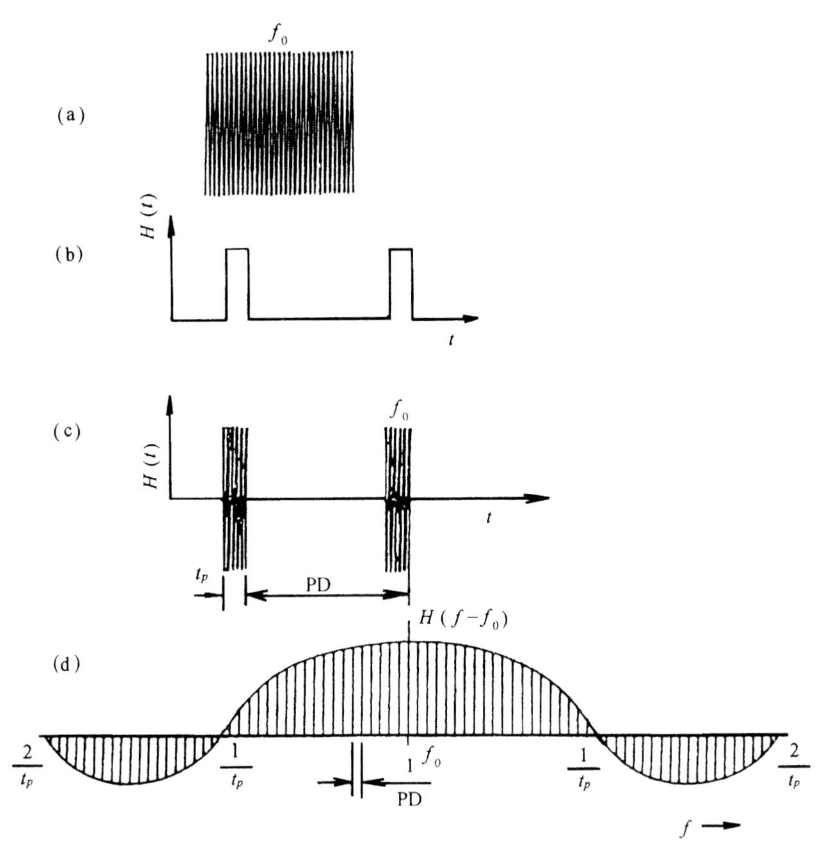

Fig. 1.12 The principle of pulse-FT-NMR. (a) A continuous RF wave with a frequency f_0 of equi-amplitude. (b) Periodic square pulses. (c) The result of (a) modulated by (b). (d) Discrete frequency components obtained by (c).

Now return to the first argument, which is based on the rotating frame with a rotating frequency f_0, which is just the center of the frequency distribution in Fig. 1.12d. In addition, ΔF in Eq. (1-34) corresponds to the chosen frequency width in the broad frequency distribution. Therefore, these two arguments are actually related.

1.6.4
Advantages of an FT-NMR Spectrometer

The change from continuous wave (CW) NMR spectrometers to FT-NMR spectrometers was a milestone for NMR instruments. FT-NMR spectrometers have many advantages over CW NMR spectrometers. The main advantages are as follows:

1. All nuclei in different functional groups are simultaneously in resonance, which greatly improves the efficiency of NMR measurements.

2. Since the repeat time of pulses is short (less than several seconds in general), an accumulation of data acquisition can be accomplished in a time much shorter than that by CW NMR spectrometers.
3. For the FT–NMR spectrometer, its pulse emission and data acquisition are carried out according to a time-sharing system so that the energy leakage from the emitter into the receptor, which occurs in CW spectrometers, is removed.
4. Pulse sequences, by which 2D NMR experiments are created, can only be used in FT–NMR spectrometers.

On the basis of 1, 2 and 3, NMR measurements with FT-NMR spectrometers can be accomplished with a much smaller amount of sample and/or in a much shorter time compared with CW spectrometers. The application of pulse sequences has lead to a significant evolutionary stage in NMR spectroscopy: from one-dimensional NMR spectra into multi-dimensional NMR spectra.

1.7 Recent Developments in NMR Spectroscopy

The main progress in NMR spectroscopy has been associated with the application of pulse sequences. This will be described in Chapter 4.

NMR measurements with micro-quantities of samples will be discussed in Section 9.3.3.

LC-NMR, the latest hyphenated technique between a separation system and an analytical instrument, will now be presented. The combination of the superior separation capabilities of HPLC with the exceptional structural elucidation capabilities of NMR forms a powerful and versatile tool for examining the chemical structures of a complex mixture. For example, both the identification of several known alkaloids and the characterization of new chemical structures are possible without previous isolation and purification [3].

The LC-NMR instrument consists of an HPLC system and an NMR spectrometer with a flow cell, which has a very small volume (e.g. 60 μL). LC-NMR is more difficult than LC-MS, in which the solvent is removed before MS measurements. However, in LC-NMR experiments, the solvents, the peak intensities of which are much stronger than those of samples, by several orders of magnitude, cannot be removed. Even the ^{13}C satellites are much stronger than the signals of the solutes. In addition, because of the application of gradient elution, the solvent peaks will shift in the course of the elution, which means that researchers are confronted with moving targets. Therefore, the suppression of the solvent peaks is the most critical task in LC-NMR. The low sensitivity of NMR is another difficulty for LC-NMR. However, the following developments have lead to the practical application of LC-NMR.

1. The frequency used in NMR spectrometers has been raised to 920 MHz, which considerably improves the sensitivity. In practice, an NMR spectrometer with a frequency of higher than 500 MHz is suitable for LC-NMR.

2. On the basis of shaped pulses (see Section 4.1.10) and pulsed field gradients (see Section 4.1.9), effective techniques for solvent suppression have been developed. Some of these are termed "excitation sculpting" [4], from which their function can be understood. LC-NMR experiments can now be carried out without deuterated organic solvents, which are very expensive.
3. The progress in automation and software has made the process of LC-NMR easier.

Our attention will now focus on the suppression of solvent peaks.

There are two types of LC-NMR experiment: continuous-flow (on-flow) monitoring and stop-flow monitoring (which includes using loops to store samples). The monitoring of LC-NMR in real time can be achieved in the on-flow mode but 2D NMR spectra of LC-NMR can only be recorded in the stop-flow mode.

From the point of view of hyphenated instruments, the continuous-flow (on-flow) mode is ideal. However, it is restricted to measuring ^{1}H spectra because of its low sensitivity.

WET (water suppression enhanced through the T_1 effect) [5, 6] is usually applied in LC-NMR. The WET sequence consists of several shaped RF pulses, which are aimed at the various solvent frequencies that need to be suppressed. Each shaped RF pulse follows a pulsed field gradient and a short delay to dephase a transverse magnetization vector of solvents. In this way, five or more of the solvent peaks can be suppressed. In addition, selective ^{13}C decoupling suppresses the ^{13}C satellites.

When LC-NMR is manipulated in the on-flow mode, scout scans are applied. A scout scan is a one-transient, one-pulse experiment to locate the positions of the solvent peaks. Special software then adjusts the transmitter so as to point to the largest solvent peak and keeps this position at a constant frequency. Once the solvent peak position is determined by the scout scan, an adjusted solvent peak suppression is executed and acquisitions are accumulated. The process is then repeated.

In addition to LC-NMR, there is also LC-NMR-MS, SFC (supercritical fluid chromatography)-NMR, and so forth.

The analysis of a mixture can be achieved without separation by using DOSY (diffusion ordered spectroscopy) which will be described in Section 4.13.

1.8 References

1 T.C. Farrar, E.D. Becker, *Pulse and Fourier Transform NMR, Introduction to Theory and Methods*, Academic Press, 1971.

2 D. Show, *Fourier Transform N.M.R. Spectroscopy* (Second Edition), Elsevier, 1984.

3 G. Bringmann, C. Gunther et al., *Anal. Chem.* **1998**, 70, 2805–2811.

4 T.L. Hwang, A.J. Shake, *J. Magn. Reson. A* **1995**, 112, 275–279.

5 H.S. Smallcombe, S.L. Patt et al., *J. Magn. Reson. A* **1995**, 117, 295–303.

6 R.J. Ogg, P.B. Kingsley et al., *J. Magn. Reson. B* **1994**, 104, 1–10.

2
^{1}H NMR Spectroscopy

Of all magnetic nuclei ^{1}H NMR occupies the most important position because it has the highest sensitivity amongst all magnetic isotopes, with an isotopic abundance close to 100%, and it is present in all organic compounds.

A typical ^{1}H NMR spectrum is shown in Fig. 2.1.

The abscissa of the NMR spectrum is the chemical shift δ. The direction from left to right represents an increase in field strength or a decrease in frequency. The ordinate of the NMR spectrum is peak intensity. Integral values, noted under the corresponding peaks, show the areas of related peaks. The peaks at about 1.19, 2.60 and 7.16 ppm are assigned to the CH_3–, –CH_2– and the substi-

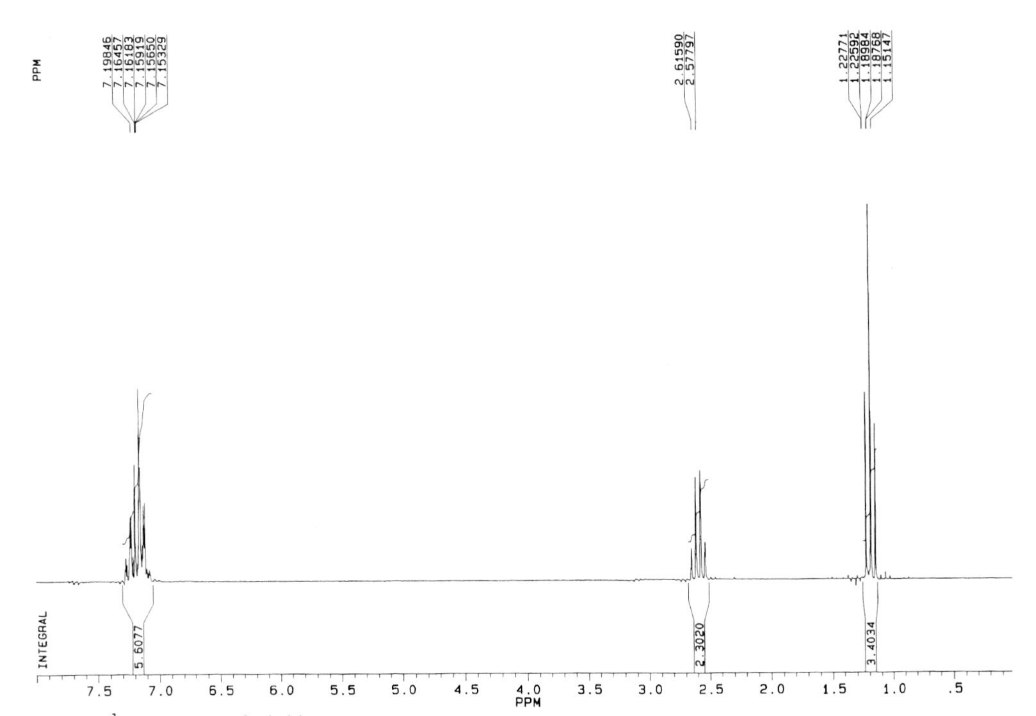

Fig. 2.1 ^{1}H spectrum of ethyl benzene.

Structural Identification of Organic Compounds with Spectroscopic Techniques. Yong-Cheng Ning
Copyright © 2005 WILEY-VCH Verlag GmbH & Co. KGaA, Weinheim
ISBN: 3-527-31240-4

tuted benzene ring, respectively. Three integral values of 3, 2 and 5 show the related ratios of the hydrogen atoms of the three functional groups.

The information obtained from a ^{1}H NMR spectrum is chemical shifts, coupling constants (and splitting patterns) and peak areas. The first two items will be discussed in detail in this chapter. The importance of the quantitative data obtained from NMR, which are very difficult to obtain by MS and IR, will be emphasized. The determination of the number of hydrogen atoms of each functional group in an unknown molecule is very useful when deducing the structure. Some specific cases, for example, the determination of the average number of ethylene oxides, n, in a non-ionic surfactant, can be carried out easily using ^{1}H NMR rather than by other spectroscopic methods.

Many conclusions obtained from this chapter, such as chemical equivalence, magnetic equivalence, dynamic nuclear magnetic resonance, etc., can be applied to the spectral interpretation of other magnetic isotopes.

2.1 Chemical Shift

2.1.1 Reference for Chemical Shift

As described in Section 1.2.2, TMS is usually used as a reference for the chemical shift in ^{1}H and ^{13}C NMR spectra. When TMS is added to the solution of a deuterated organic solvent of the sample to be measured, it experiences the same field strength as the sample, so it is called the internal reference. When a capillary tube with an organic solution of TMS is inserted into an NMR tube containing a D_2O solution of a sample, TMS experiences a field strength different to that in the D_2O solution, because of the different bulk susceptibilities between D_2O and the organic solvent. In this instance TMS is known as the external reference. Correspondingly, a precise δ value should be corrected.

2.1.2 Factors Affecting Chemical Shifts

Chemical shifts are determined by shielding constants σ expressed by Eq. (1-15). Because a hydrogen atom has only one s electron, its chemical shift is determined mainly by its σ_d. It can be predicted that a peak will be shifted towards the left (deshielding shift) if its extranuclear electron density decreases through structural changes or through the influence of the solvent. On the contrary, the shielding effect shifts the peak towards the right.

The main factors affecting chemical shifts can be summarized as follows.

Tab. 2.1 Effect on shift of substitutions into a methyl group

Compound	CH_3F	CH_3OCH_3	CH_3Cl	CH_3I	CH_3CH_3	$Si(CH_3)_4$	CH_3Li
δ (ppm)	4.26	3.24	3.05	2.16	0.88	0	–1.95

1. Electronegativity of substituents

The stronger electronegativity the substituent possesses, the more the peak of the hydrogen atom connected to the substituted carbon atom shifts towards the left. Substitution into a methyl group is an example (Tab. 2.1).

The induction effect of substituents can extend along the carbon chain. The hydrogen atoms at the α-carbon atom could have an obvious shift. The β-hydrogen atoms could have a discernible shift. However, the γ- and subsequent hydrogen atoms hardly have any shifts.

Since the electronegativity of common functional groups is greater than that of the hydrogen atom, it follows that $\delta(CH) > \delta(CH_2) > \delta(CH_3)$.

The substitution effect on an unsaturated hydrocarbon is complicated and both the induction effect and the conjugation effect have to be considered.

2. s–p Hybridization of the connected carbon atom

The δ value of hydrogen atoms attached to an unsaturated carbon atom is greater than that of hydrogen atoms attached to a saturated carbon atom, which can be explained by the percentages of s electrons in the carbon atom. The increment of the percentage from 25% (sp^3) to 33% (sp^2) leads to bond electrons approaching the carbon atom, which produces a deshielding effect on the hydrogen atoms attached to the unsaturated carbon atom.

3. Ring current effect

Although carbon atoms in both ethylene and benzene are sp^2 hybridized, the former has a δ value of 5.23 ppm while the latter, 7.3 ppm. The deviation can be explained by the ring current effect.

Suppose that a benzene molecule is perpendicular to the applied magnetic field $\boldsymbol{B}_0$, a ring current produced from the delocalized electrons of the benzene is induced. It produces an induced magnetic field, which opposes $\boldsymbol{B}_0$ in the middle of the molecule but reinforces $\boldsymbol{B}_0$ at the periphery. Therefore, the protons of benzene experience a deshielding shift because they are situated at the periphery of the benzene molecule. As the benzene molecule tumbles through its solution, its NMR signal is that averaged from all its orientations. If the benzene molecule is parallel to the applied magnetic field B_0, there is no effect from B_0. On average, the hydrogen atoms of a benzene molecule still have a larger δ value. The ring current effect of a benzene molecule is shown in Fig. 2.2.

The ring current effect functions not only in benzene molecules but also in all annular conjugated molecules with $4n+2$ delocalized electrons. The protons situ-

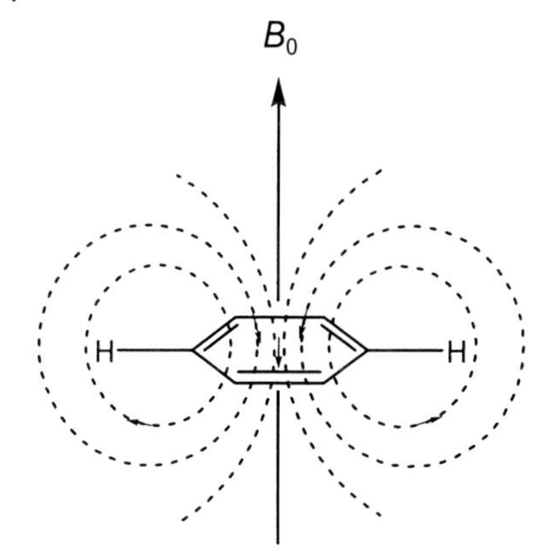

Fig. 2.2 Ring current effect of a benzene molecule.

H_a
H_b

$\delta\ H_a = 9.28$ppm
$\delta\ H_b = -2.99$ppm

C 2-1

CH_3
CH_3

$\delta_{CH_3} = -4.25$ppm
$\delta_{CH} = 8.14\text{-}8.67$ppm

C 2-2

ated above or below these molecules can show strong shielding shifts (their δ values can be less than zero) and the protons at the periphery of these molecules show strong deshielding shifts. Some examples are given in **C2-1** and **C2-2**.

4. Anisotropic shielding from neighboring atoms and bonds

For simplicity, let us consider a diatomic molecule AB. Influenced by the applied field $\boldsymbol{B}_0$, a magnetic moment $\boldsymbol{\mu}_A$ is induced at atom A.

$$\boldsymbol{\mu}_A = \chi_A \boldsymbol{B}_0 \tag{2-1}$$

where χ_A is the magnetic susceptibility of atom A.

$\boldsymbol{\mu}_A$ can be split into three components: $\mu_A(x)$, $\mu_A(y)$ and $\mu_A(z)$ in which $\mu_A(x)$ is along the A–B bond. The AB molecule can be oriented in any direction, with three particular directions being chosen, as shown in Fig. 2.3 [1].

The contribution of the three components to the shielding of atom B is given by

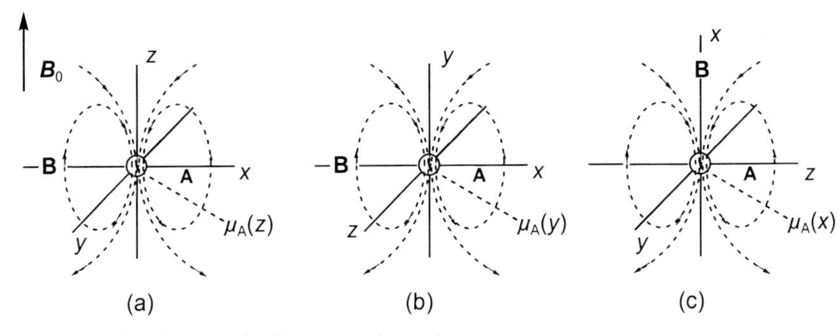

Fig. 2.3 Induced μ_A can be decomposed into three components.

$$\Delta\sigma = \frac{1}{12\pi} \sum_{i=x,y,z} \chi_A^i (1 - 3\cos^2\theta_i)/R^3 \tag{2-2}$$

where χ_A^i is the magnetic susceptibility of atom A in the i direction; θ_i is the angle formed between $\mu_A(i)$ and the A–B bond; and R is the distance between atoms A and B.

In Fig. 2.3 a and b, the induced field at atom B is parallel to $\boldsymbol{B}_0$ and thus atom B is deshielded. However, in Fig. 2.3 c, the induced field at atom B is opposed to $\boldsymbol{B}_0$ so that atom B is shielded. Since the AB molecule tumbles rapidly in solution, $\Delta\sigma$ for atom B is an averaged value. If atom A possesses magnetic isotropy, which means that $\chi_A(x)=\chi_A(y)=\chi_A(z)$, then $\Delta\sigma=0$. On the contrary, if atom A possesses magnetic anisotropy, that is, $\chi_A(x) \neq \chi_A(y) \neq \chi_A(z)$, the following equations hold:

$$\mu_A(x) = \chi_A(x) B_0 = \chi_{||} B_0 \tag{2-3}$$

where $\chi_{||}$ is the magnetic susceptibility of atom A along the bond:

$$\mu_A(y) = \chi_A(y) B_0 = \chi'_{\perp} B_0 \tag{2-4}$$

$$\mu_A(z) = \chi_A(z) B_0 = \chi''_{\perp} B_0 \tag{2-5}$$

where $\chi'_{\perp}$ and $\chi''_{\perp}$ are the magnetic susceptibilities of atom A along the y axis and the z axis, respectively.

When the electron cloud is distributed symmetrically on the section perpendicular to the bond (e.g., C–C, C≡C), $\chi'_{\perp}=\chi''_{\perp}$. Let

$$\Delta\chi = \chi_{||} - \chi_{\perp} \tag{2-6}$$

and Eq. (2-3) becomes

$$\Delta\sigma = \Delta\chi(1 - 3\cos^2\theta)/12\pi R^3 \tag{2-7}$$

Therefore, because of the magnetic anisotropy of atom A, $\Delta\chi \neq 0$, which means the shielding constant of atom B will be influenced by A.

This consideration can now be extended from a diatomic molecule to a polyatomic molecule. The equations from (2-1) to (2-7) are still effective provided the parameters of atom A are replaced by those of related bonds: χ is the magnetic susceptibility of the bond; $\Delta\sigma$ is the shielding or deshielding effect produced from the bond; R is the distance between point B and the middle point of the bond; and θ is the angle between the bond and the line connecting the middle point of the bond to point B. When the electron cloud of the bond is symmetrical about the bond, Eq. (2-7) is still valid. Because of the anisotropy of chemical bonds, all single bonds, double bonds and triple bonds show anisotropic shielding effects. Depending on the values of $\chi'_{\perp}$ and $\chi'_{\perp}$ (that is, depending on which one of them is greater), each bond forms two shielding or deshielding cones along its bond axis. When $\chi'_{\perp} \neq \chi''_{\perp}$, the shielding (or deshielding) cone is elliptical. The functional groups of –N=O, –C=O, and so forth belong to this situation. Some examples are shown in Figures 2.4 and 2.5, in which "+" implies the shielding effect and "–" the deshielding effect.

Alkyne hydrogen atoms are affected by a strong shielding effect, which is related to the rotation of their π electrons around the bond axis, so that they clearly have δ values smaller than those of alkene hydrogen atoms.

The fact that two *geminal* hydrogen atoms attached to the same carbon atom in a six-membered ring have different δ values can easily be explained by the anisotropic shielding effect. H_{1eq} is situated in the deshielding cone of bonds

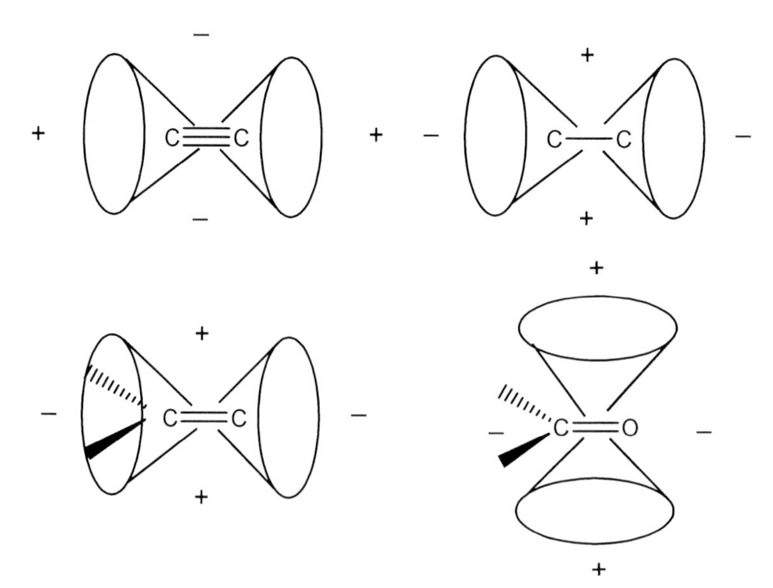

Fig. 2.4 Shielding or deshielding cones of chemical bonds.

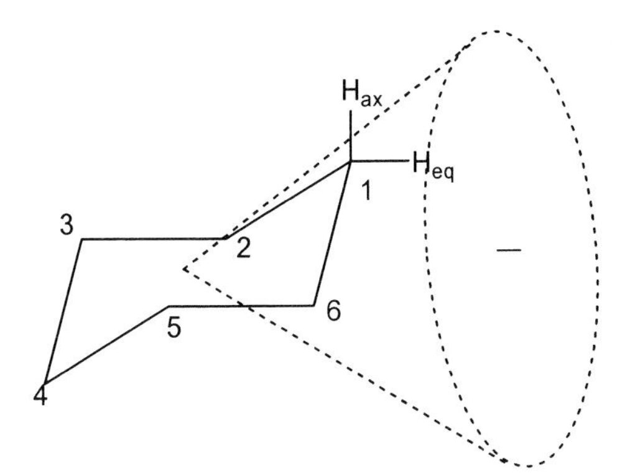

Fig. 2.5 Two *geminal* hydrogen atoms attached to the same carbon atom in a six-membered ring experience magnetic anisotropic effects.

2–3 and 5–6 so that it has a greater δ value than H_{1ax}. Their difference in δ is about 0.5 ppm.

The above-mentioned ring current effect possesses the anisotropic shielding effect. However, it is mainly characterized by $4n+2$ delocalized electrons and its effect is stronger than that of the anisotropy of chemical bonds.

5. Electric dipole of neighboring groups

A strong polar group in an organic molecule affects other groups through its electric field, which changes the electronic cloud densities of other functional groups and therefore the related shielding constants.

6. Van der Waals force

If a hydrogen atom is close to another atom with a distance shorter than the sum of the Van der Waals radii of the two atoms, the extranuclear electron of the hydrogen atom is repelled so that σ_d decreases as does H_b in **C2-3**.

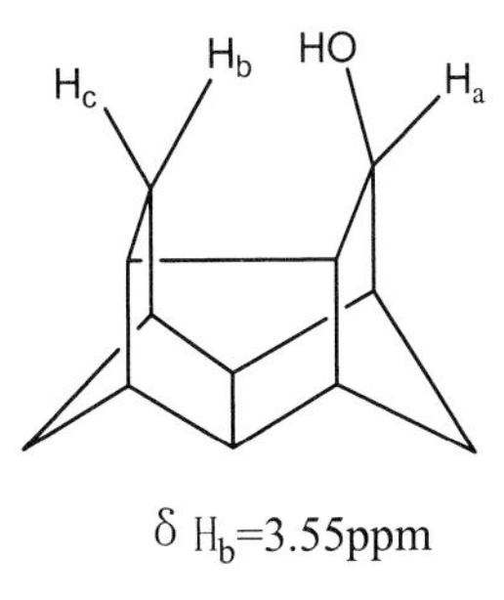

δ H_b=3.55ppm

δ H_c=0.88ppm

C2-3

7. Effect of the medium

The medium can affect the chemical shift values of the sample to be measured in the following ways:

- The medium changes the bulk magnetic susceptibility of the solution being measured. As a result, the magnetic induction field of the solution and the δ values of the sample are changed.
- The extranuclear electron cloud shape of the analyte will be changed when medium molecules with magnetic anisotropy approach analyte molecules. Consequently, the various parts of the sample molecule will experience different shielding or deshielding effects.
- Polar groups of a sample molecule induce an electric field in addition to the medium. This induced field in the medium affects in turn the other functional groups of the sample.

Anisotropy of the medium can be applied in NMR measurements. For example, crowded peaks can sometimes be separated by adding a small amount of deuterated benzene.

8. Hydrogen bond

Both intermolecular and intramolecular hydrogen bonds strongly deshieled related hydrogen atoms. Therefore, the carboxylic acid has a δ value greater than 10 ppm in general. An example is **C2-4**.

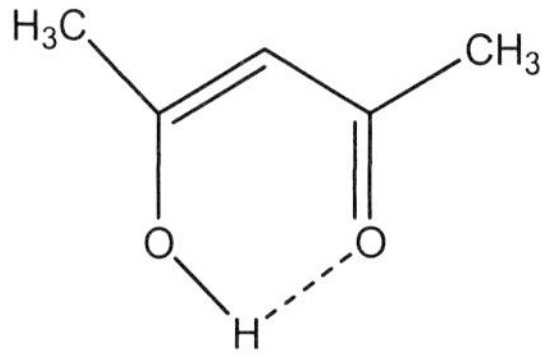

C2-4 δ H=15.4ppm

Because of the effect of hydrogen bonds, the chemical shift values of amines, hydroxyl, etc., can cover a range of δ values, which are determined by the concentration of samples, the temperature of solution to be measured, and so forth.

2.1.3
Chemical Shift Values of Common Functional Groups

The chemical shift values of common functional groups containing hydrogen atoms are shown in Tab. 2.2, from which an overall idea of the δ values of ^{1}H NMR can be obtained.

The δ value of the methyl group can be estimated from that of the related methylene group. Its δ value is smaller than that of the latter.

Tab. 2.2 Chemical shift values of functional groups containing hydrogen atoms

Functional group	δ_H *(ppm)*	*Functional group*	δ_H *(ppm)*
$-(CH_2)_n-CH_3^*$	0.87	$-HC=CH-$	4.5–8.0
$-C=C-CH_3^*$	1.7–2.0	(benzene ring)	6.5–8.0
(benzene ring)$-CH_3^*$	2.1–2.4	(pyridine ring, N, H)	8.0–8.8
$-C(=O)-CH_3^*$	2.1–2.6	(pyridine ring, N, H, H)	6.5–7.3
$-N-CH_3^*$	2.2–3.1	$R-NH_2$ $R-NH-$	0.5–3.0
$-O-CH_3^*$	3.5–4.0	$Ar-NH_2$ $Ar-NH-$	3.0–4.8
$-C-CH_2-C-$	1.2–1.4	$R-OH$	0.5–5.0
$-C-CH_2-N-$	2.3–3.5	$Ar-OH$	4.0–10.0
$-C-CH_2-O-$	3.5–4.5	$-C(=O)-H$	9.5–10.0
$-C\equiv CH$	2.2–3.0	$-C(=O)-OH$	9.0–12.0
$>C=CH_2$	4.5–6.0		

Tab. 2.3 Empirical screening parameters for Shoolery's formula

Functional group	*Screening parameter*	*Functional group*	*Screening parameter*
–R	0.0	–OCOPh	2.9
–CC–	0.8	$–NH_2$	1.0
–C≡C–	0.9	$–NR_2$	1.0
–Ph	1.3	$–NO_2$	3.0
–Cl	2.0	–SR	1.0
–Br	1.9	–CHO	1.2
–I	1.4	–COR	1.2
–OH	1.7	–COOH	0.8
–OR	1.5	–COOR	0.7
–OPh	2.3	–CN	1.2
–OCOR	2.7		

Tab. 2.4 Empirical parameters for calculations with Eq. (2-9) [2]

–H	0	0	0
–Alkyl	0.45	–0.22	–0.28
$–CH_2–Ar$	1.05	–0.29	–0.32
–X,X = F, Cl, Br	0.70	0.11	–0.04
$–CH_2O$	0.64	–0.01	–0.02
$–CH_2N$	0.58	–0.10	–0.08
–C=C	1.00	–0.09	–0.23
–C≡C	0.47	0.38	0.12
–Ar	1.38	0.36	–0.07
–F	1.54	–0.40	–1.02
–Cl	1.08	0.18	0.13
–Br	1.07	0.45	0.55
–I	1.14	0.81	0.88
–OR,R (saturated)	1.22	–1.07	–1.21
–OR,R (unsaturated)	1.21	–0.60	–1.00
–OCOR	2.11	–0.35	–0.64
–NR,R (saturated)	0.80	–1.26	–1.21
–NR,R (unsaturated)	1.17	–0.53	–0.99
–CHO	1.02	0.95	1.17
–CO	1.10	1.12	0.87
–CO*	1.06	0.91	0.74
–COOH	0.97	1.41	0.71
–COOH*	0.80	0.98	0.32
–COOR	0.80	1.18	0.55
–COOR*	0.78	1.01	0.46
$–CONR_2$	1.37	0.98	0.46
–COCl	1.11	1.46	1.01
–CN	0.27	0.75	0.55

* Conjugated.

The chemical shift values of the methylene group can be estimated by Shoolery's empirical formula:

$$\delta = 1.25 + \Sigma\sigma \qquad (2\text{-}8)$$

where σ is the empirical screening parameter of the functional groups substituting the methylene, which are listed in Tab. 2.3.

Because only α-substitution effects are considered, the values calculated by Eq. (2-8) may deviate considerably.

The chemical shift values of the methine group can also be estimated by Eq. (2-8) with the same empirical parameters of Tab. 2.3. However, the constant of 1.25 in Eq. (2-8) is replaced by 1.50.

The chemical shift values of the alkene hydrogen atoms can be estimated by Eq. (2-9) with corresponding parameters (Tab. 2.4) [2].

$$\delta_{C=C-H} = 5.25 + Z_{gem} + Z_{cis} + Z_{trans} \qquad (2\text{-}9)$$

where Z_{gem}, Z_{cis} and Z_{trans} are the empirical parameters of *geminal*, *cis*- and *trans*-substituting functional groups, respectively.

The chemical shift values of the residual hydrogen atoms in the substituted benzene can be estimated by Eq. (2-10) and the corresponding parameters (Tab. 2.5) [2].

Tab. 2.5 Empirical parameters for calculations with Eq. (2-10) [2]

Substituent	Z_2	Z_3	Z_4
–H	0	0	0
$-CH_3$	–0.20	–0.12	–0.22
$-CH_2CH_3$	–0.14	–0.06	–0.17
$-CH(CH_3)_2$	–0.13	–0.08	–0.18
$-CH{=}CH_2$	0.06	–0.03	–0.10
$-C{\equiv}CH$	0.15	–0.02	–0.01
–F	–0.26	0.00	–0.20
–Cl	0.03	–0.02	–0.09
–Br	0.18	–0.08	–0.04
–I	0.39	–0.21	0.00
–OH	–0.56	–0.12	–0.45
$-OCH_3$	–0.48	–0.09	–0.44
–OCO–Ph	–0.09	0.09	–0.08
$-NH_2$	–0.75	–0.25	–0.65
$-NO_2$	0.95	0.26	0.38
–CHO	0.56	0.22	0.29
$-COCH_3$	0.62	0.14	0.21
–COOH	0.85	0.18	0.27
$-COOCH_3$	0.71	0.11	0.21

$$\delta = 7.26 + \sum_{i=2}^{4} Z_i \tag{2-10}$$

where Z_2, Z_3 and Z_4 are the empirical parameters for the *ortho-*, *meta-* and *para-*substituting functional groups, respectively.

Because of the influence of hydrogen bonds, the δ value of the hydrogen atom connected to the heteroatom covers a certain range. The δ value is related to the concentration of the sample to be measured, solvent and temperature.

2.2
Coupling Constant *J*

Coupling constants reveal structural information, especially stereochemical information on organic compounds. Their absolute values can be determined from ^{1}H spectra. However, their relative signs can be determined only under specific circumstances.

2.2.1
Vector Model for Couplings

The vector model describes a simple physical mechanism for couplings even though it is an approximate and oversimplified model. The model is based on the following three rules. Firstly, because of Fermi's contact effect, a nucleus has a spin orientation opposite to those of the spins of the electrons around the nucleus (strictly speaking, this holds on condition that the γ value of the nucleus is positive). Secondly, according to Pauling's principle, a pair of electrons situated in an orbit has opposite spin orientations. Thirdly, according to Hund's

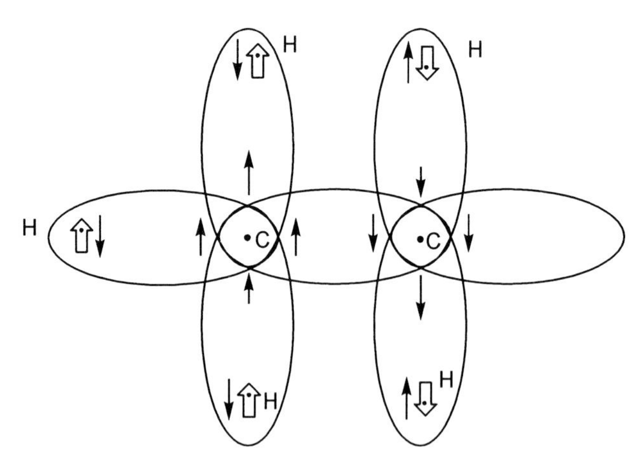

Fig. 2.6 Vector model for the couplings in a saturated chain.

rule, electrons around the same nucleus but occupying different orbits have the same spin orientation. The descriptions given above are illustrated in Fig. 2.6.

From Fig. 2.6, it follows that two hydrogen nuclei separated by an odd number of bonds have opposite spin orientations. Consequently, the sign of the coupling constant between them is positive (see Section 1.3.3). On the other hand, if the two nuclei are separated by an even number of bonds, the sign of the coupling constant is negative.

Also from Fig. 2.6, it follows that the couplings are transferred through chemical bonds. Thus the coupling effect decreases rapidly with the number of chemical bonds through which the coupling is transferred.

2.2.2 ^1J and ^2J

The ^1J of hydrogen nuclei can be illustrated in ^{1}H spectra only for the direct connections between hydrogen nuclei and other magnetic nuclei. The most important ^1J for ^{1}H is $^1J_{13C-1H}$, which will be discussed in Section 3.3.1.

The most frequently encountered ^2J of hydrogen nuclei is that of two hydrogen nuclei connected to the same carbon atom. This ^2J is denoted as J_{gem} since the related coupling is called the *geminal* coupling.

Spin couplings always exist. However, the splitting of couplings only appears if the two coupled nuclei have different chemical shifts. The two hydrogen atoms situated at the end of an alkene bond usually show their split pattern because the two nuclei are generally chemically non-equivalent due to the anisotropic effect of the alkene bond. Owing to the rapid rotation of a methyl group, its three hydrogen atoms are chemically equivalent so that there is no coupling split by ^2J. For a CH_2 in a saturated ring, its two hydrogen atoms have different chemical shift values (see Section 2.1.2) so that a split by the ^2J between the two hydrogen atoms will appear. For a CH_2 in a saturated chain, the two hydrogen atoms always have approximately the same δ values because of the rapid rotation. Under these conditions, the split by the ^2J may be indiscernible.

The factors affecting the algebraic value of ^2J are as follows:

1. s–p Hybridization
   The ^2J in methane is –12.4 Hz while that in an alkene is +2.3 Hz. The difference of 14.7 Hz results from their different hybridizations.
2. Substituents
   The algebraic value of ^2J will be increased (in the positive direction, with its absolute value decreased) by the substitution of an electron-attracting functional group and will be decreased by the substitution of an electron-donating functional group, which is shown in Tab. 2.6.
3. Conformation
   Compounds **(C2-5)**$_{eq}$ and **(C2-5)**$_{ax}$ will be taken as examples. In the former, the non-bonded electron pair of the S atom is parallel to the axial bond of the hydrogen atom at the vicinal carbon atom so that the ^2J value in **(C2-5)**$_{eq}$ is more positive than that in **(C2-5)**$_{ax}$.

Tab. 2.6 Changes in ^2J with substitutions

Compound	CH_4	CH_3Cl	CH_2Cl
^2J (Hz)	–12.4	–10.8	–7.5

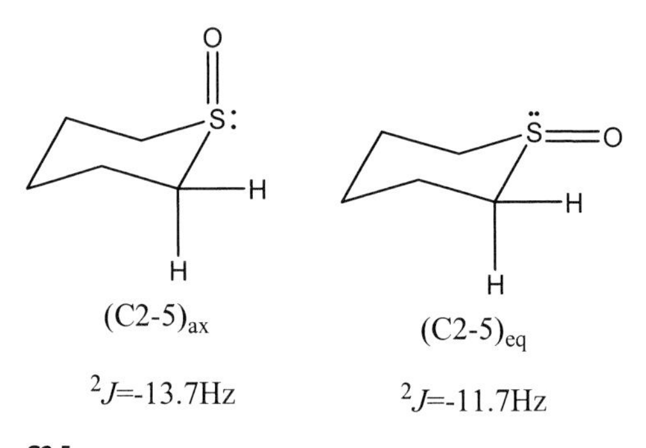

C2-5

4. The effect from a vicinal π bond
   A vicinal π bond (including >C=C<, phenyl, >C=O, –C≡C–, etc.) makes the saturated $^2J_{H\text{–}C\text{–}H}$ shift in the negative direction.
5. The size of the membered ring
   The ^2J in a three-membered ring obviously has a value that shifts in the positive direction (its absolute value is decreased), which is differentiated from those of other rings. This particular feature of the three-membered ring is also shown in other types of spectra, such as fairly high absorption frequencies in an IR spectrum and rather small δ values in an ^{1}H spectrum.

2.2.3
^3J

Since the splits by ^2J are always absent because two *geminal* hydrogen atoms frequently have approximately the same δ values and the splits by long-range couplings are not obvious, ^3J plays the most important role in ^{1}H spectra.

$^3J_{H\text{–}C\text{–}C\text{–}H}$ is also denoted by J_{vic} as the related coupling is called the *vicinal* coupling. Because in general two *vicinal* hydrogen atoms have different δ values, the splits by ^3J are always visible.

The factors affecting ^3J are as follows:

1. Dihedral angle Φ
   The value of ^3J depends on the dihedral angle that is formed by the related H–C–C–H, as shown in Fig. 2.7. This relationship is expressed by the well known Karplus equation:

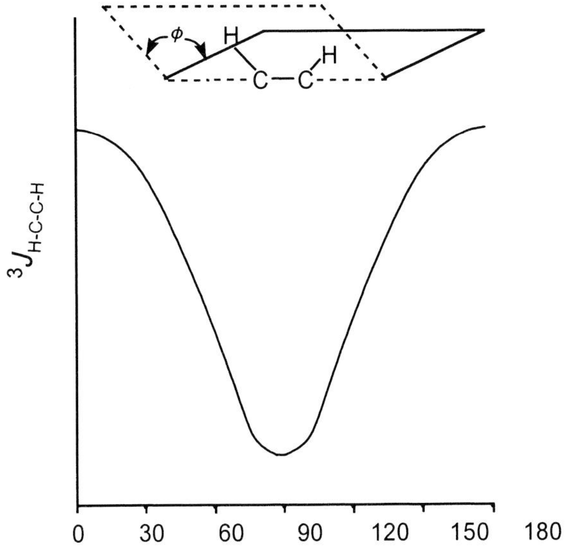

Fig. 2.7 ^3J is the function of related dihedral angle.

$$^3J = J_0 \cos^2 \Phi + C(\Phi = 0 - 90^\circ)$$

$$^3J = J_{180} \cos^2 \Phi + C(\Phi = 90 - 180^\circ) \qquad (2\text{-}11)$$

where J_0 is the value of ^3J when $\Phi = 0^\circ$; J_{180} is the value of ^3J when $\Phi = 180^\circ$; and C is a constant.

The Karplus equation is illustrated in Fig. 2.7.

Because $J_{180} > J_0$, the equation above can be rewritten as

$$^3J = A + B \cos \Phi + C \cos^2 \Phi \qquad (2\text{-}12)$$

where A, B and C are three constants. Some people take $A=7$, $B=-1$ and $C=5$.

On the basis of Eq. (2-11) or Eq. (2-12), the following phenomena can be interpreted.

i. $J_{trans} > J_{cis}$ in H–C=C–H
   Because J_{trans} is J_{180} while J_{cis} is J_0, $J_{trans} > J_{cis}$.
ii. $J_{aa} > J_{ae} \geq J_{ee}$ in a six-membered ring
   If both *vicinal* hydrogen atoms in a six-membered ring are situated at axial bonds, their related dihedral angle is approximately 180° so that their *vicinal* coupling constant is almost J_{180}. Suppose that a hydrogen atom is situated at an axial bond and its *vicinal* hydrogen atom is situated at an equatorial bond, their related dihedral angle is φ_{ae}. If two *vicinal* hydrogen atoms in a six-membered ring are situated at two equatorial bonds,

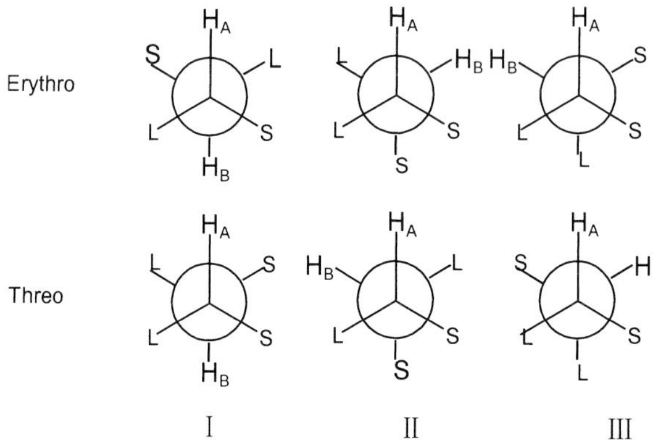

L:large substituent, S:small substituent.

Fig. 2.8 Newman projections of the *erythro*- and *threo*-forms.

their related dihedral angle is φ_{ee}. As $\varphi_{ae} \approx \varphi_{ee} \approx 60^\circ$, both J_{ae} and J_{ee} are smaller than J_{aa}.

iii. The *erythro*-form and the *threo*-form have different J values
The Newman projections of the *erythro*- and *threo*-forms are shown in Fig. 2.8.

The ^3J value read from a related spectrum is an averaged value of three conformers of the *erythro*- or *threo*-form. When the dihedral angle formed by H_A and H_B is 180° (I of the *erythro*-form or I of the *threo*-form in Fig. 2.8), the related ^3J is large and when the dihedral angle formed by H_A and H_B is 60°, the related ^3J is small. Because two large functional groups repel each other in conformer I of the *threo*-form, the existence probability of this conformer is less than 1/3. Similarly, it could be that the existence probability of conformer I of the *erythro*-form is greater than 1/3. As a result, the averaged ^3J value of the *threo*-form is less than that of the *erythro*-form. Hence stereochemical information can be obtained from the measurement of ^3J.

2. Electronegativity of substituents
^3J values will decrease with the enhancement of the electronegativities of the substituents. The ^3J values of alkenes decrease more rapidly than those of alkanes.
3. Bond lengths
^3J values will increase with a decrease in bond lengths in unsaturated six-membered rings, in which the related dihedral angles are zero.
4. Bond angles
^3J will increase with a decrease in bond angles.

2.2.4 Coupling Constants of Long-range Couplings

In ^{1}H NMR, the couplings across four or more bonds are known as long-range couplings. The coupling constants in a saturated compound decrease rapidly with an increase in the number of bonds between two coupled nuclei. Only particular structural units, such as H/\/\H or H/\/\/H, have a small long-range coupling constant (generally less than 2 Hz).

In an unsaturated system, long-range couplings can be transferred to bonds further away through π electrons. Some examples are listed as follows:

1. H–C=C–C–H (an allylic system) and H–C–C=C–C–H (a homoallylic system).
2. Conjugated systems.
3. The systems containing accumulated unsaturated bonds.
4. The couplings between hydrogen atoms in the benzene ring and those in the side chain.

In some particular structures, such as in compound **C2-6**, the coupling constant of the homoallylic system has a value as high as 9.63 Hz. The *trans*-homoallylic system has a coupling constant of 8.04 Hz.

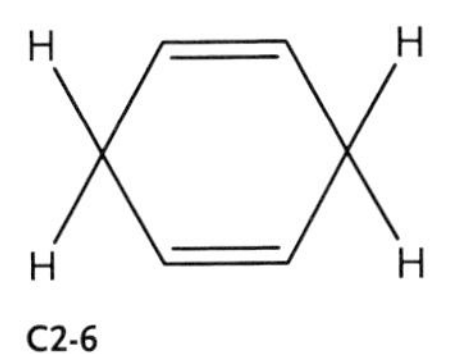

C2-6

In the ordinary (one-dimensional) ^{1}H spectrum, the existence of long-range couplings is known from the enhancement of the width at the half height of peaks and the splitting of related peaks is hardly noticeable.

2.2.5 Couplings in a Phenyl Ring or in a Heteroaromatic Ring

The ^3J values in a phenyl ring are 6–9 Hz.

Because a phenyl ring is a large conjugated system, its ^4J=1–3 Hz and ^5J= 0–1 Hz.

The ^3J values in a heteroaromatic ring are related to the positions of the two coupled hydrogen atoms, whether they are close to or far from the heteroatom. When the two hydrogen atoms are close to the heteroatom, their coupling constant ^3J is smaller while if the two hydrogen atoms are far from the heteroatom, their coupling constant ^3J is greater.

The J values in common functional groups are listed in Tab. 2.7.

Tab. 2.7 *J* values in common functional groups

Structure unit		*J_{AB} typical value (Hz)*
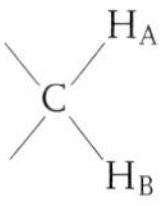 H_A–C–H_B (geminal)		–10 to –15
CH_ACH_B		7
Cyclohexane H_A, H_B	*ax-ax*	8–11
	ax-eq	2–3
	eq-eq	2–3
H_A–C=C–H_B (trans)		15–17
C=C(H_A)(H_B)		0–2
H_A–C=C–H_B (cis)		10–11
Benzene H_A, H_B	^3J	8
	^4J	2
	^5J	0.3
Pyridine (positions 2–6)	$J(2–3)$	5
	$J(3–4)$	8
	$J(2–4)$	1.5
	$J(3–5)$	1
	$J(2–5)$	0.8
	$J(2–6)$	0

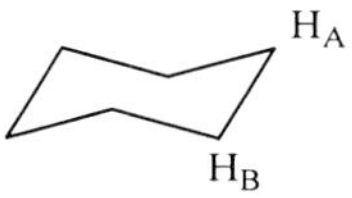

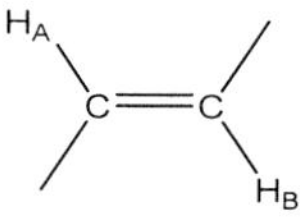

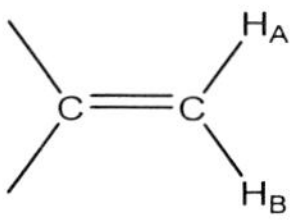

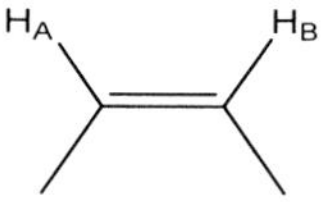

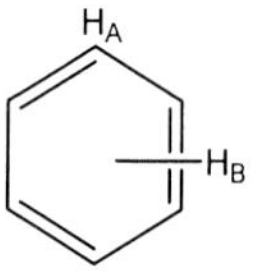

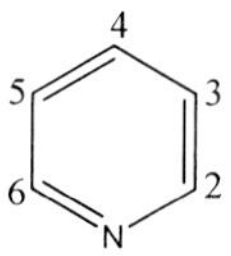

2.3 Spin–spin Coupling System and Classification of NMR Spectra

The content presented in this section can be applied not only to the proton but also to other magnetic nuclei.

2.3.1 Chemical Equivalence

Chemical equivalence is an important concept of stereochemistry. If two atoms (or two identical functional groups) have the same chemical environment, they are chemically equivalent. Two chemically non-equivalent functional groups have different properties: different rates of reaction, different spectroscopic parameters, etc., and therefore, chemical equivalence can be studied by spectroscopic methods.

Citric acid (**C2-7**) will be used as an example for this topic.

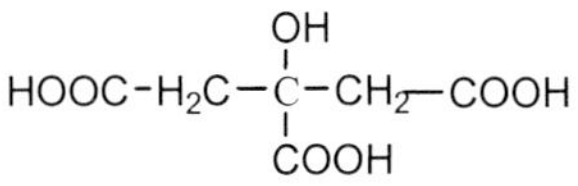

OH
HOOC-H$_2$C–C–CH$_2$–COOH
COOH

C2-7

From the structural formula it looks as if the two carboxyl groups are chemically equivalent. However, in zymolytic reactions they have different rates, which means that the two carboxyl groups are chemically non-equivalent.

The structure of δ-vitamin E is shown in **C2-8.**

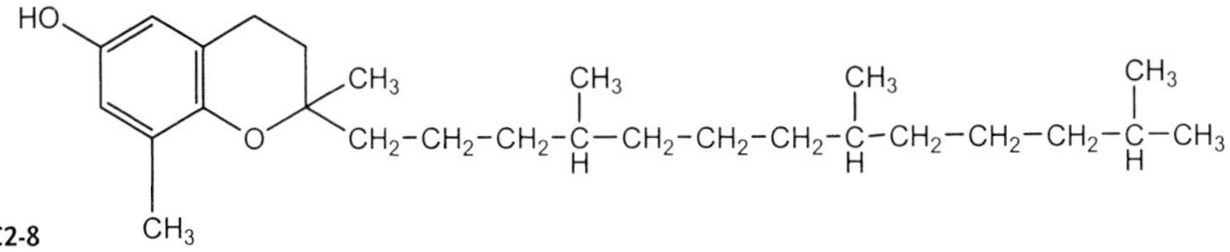

HO
O
CH$_3$
CH$_3$ CH$_3$ CH$_3$
CH$_2$–CH$_2$–CH$_2$–C–CH$_2$–CH$_2$–CH$_2$–C–CH$_2$–CH$_2$–CH$_2$–C–CH$_3$
H H H
CH$_3$

C2-8

The two methyl groups connected to the same carbon atom show two peaks in their ^{13}C spectrum, which illustrates their chemical non-equivalence [3].

From the discussion above, it is known that the appearance of a ^{1}H NMR is closely associated with the chemical equivalence of the sample to be recorded. Therefore, it is necessary to discuss chemical equivalence in detail. This topic will be discussed for two situations.

1. All nuclei in a molecule are relatively motionless

When all nuclei in a molecule are relatively motionless, the chemical equivalence of two identical functional groups can be judged by symmetrical operations.

If two functional groups are interchangeable through some symmetrical operation, such as a rotation around a C_2 axis, they are homotopic. In any solvent, these two functional groups will have the same chemical shift value, which means that they have the same resonant frequency. The term "isochronous" is used to describe this situation.

If two functional groups are interchangeable through a reflection in a symmetrical plane, they are enantiotopic, which means that they are mirror images. In an achiral solvent containing no chiral atom, two enantiotopic functional groups are isochronous as well as chemically equivalent. However, in a chiral solvent they are anisochronous so that they are chemically non-equivalent. Compound **C2-9** will be taken as an example, where R′ and R″ are two identical functional groups.

R'=R"

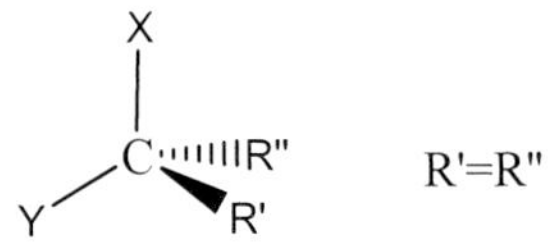

C2-9

For the purpose of the discussion, R′ and R″ are denoted differently. If compound **C2-9** is observed from R′, the three other functional groups have the order of X–R″–Y in a clockwise rotation. On the other hand, if it is observed from R″, the three other functional groups have the order of X–Y–R″ in a clockwise rotation. Therefore, R′ and R″ are anisochronous in a chiral solvent. The difference in their resonant frequencies is determined by the properties of X and Y, the solvent and the spectrometer.

If two functional groups are not interchangeable through any symmetrical operation, they are diastereotopic, which means that they are anisochronous and thus chemically non-equivalent. Under some particular conditions, they may coincidentally have the same resonant frequency, and are accidentally isochronous.

2. There are rapid intramolecular motions

Common intramolecular motions include the rotation around the chain, ring reversal, etc.

Because of the existence of rapid intramolecular motions, some functional groups, which are not interchangeable through symmetrical operations, may become chemically equivalent. The effect of the rotation of two identical functional groups (or two identical atoms) around the chain can be taken as an example. The Newman projection (Fig. 2.9) will now be used for a discussion of the intramolecular rotation.

The molecule used as an example is RCH_2–CXYZ, where X, Y and Z are three different functional groups (Fig. 2.10).

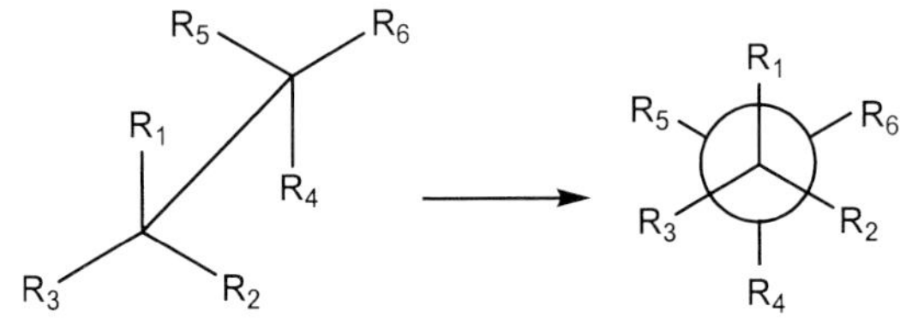

Fig. 2.9 Newman projection.

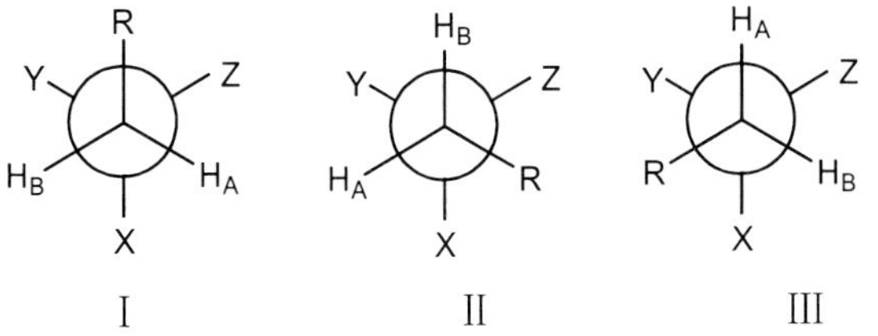

Fig. 2.10 Three conformers of RCH_2–CXYZ.

With the Newman projection, when H_A, H_B and R are overlapped by X, Y and Z, respectively, the molecular potential energy is maximum. Hence the time aspect of this situation can be neglected. Therefore, this molecule is always situated in one of the three conformations shown in Fig. 2.10. Because these functional groups have different sizes and shapes, the existence times of the three conformers as well as the percentages of the three conformers are different. The chemical shift values of H_A and H_B can be expressed by the weighted δ values of the three conformers, that is:

$$\delta(H_A) = P_I\ \delta_{XZ} + P_{II}\ \delta_{XY} + P_{III}\ \delta_{YZ} \qquad (2\text{-}13)$$

$$\delta(H_B) = P_I\ \delta_{XY} + P_{II}\ \delta_{YZ} + P_{III}\ \delta_{XZ} \qquad (2\text{-}14)$$

where P_I, P_{II} and P_{III} are the probabilities of the existence of the conformers I, II, and III respectively; δ_{XY} is the chemical shift value of the hydrogen atom when it is situated between X and Y, and similarly we have δ_{YZ} and δ_{XZ}.

Because $P_I \neq P_{II} \neq P_{III}$, $\delta(H_A) \neq \delta(H_B)$.

When the temperature of the sample is increased, the rotation around the chain will be more rapid so that P_I, P_{II} and P_{III} will be close to each other. However, they can not be equalized exactly under experimental conditions. Even when $P_I = P_{II} = P_{III}$, it is still true that $\delta(H_A) \neq \delta(H_B)$, because the chemical environments of H_A and H_B are not exactly identical. For example, the chemical environment of H_A in conformer I is not identical to that of H_B in conformer III. When H_A is in conformer I, X and H are situated at one of its sides while Z and R are situated at its other side. However, when H_B is in conformer III, X and R are situated at one of its sides while Z and H are situated at its other side. Therefore, H_A and H_B are still chemically non-equivalent.

From Fig. 2.10, it is seen that H_A and H_B will be chemically equivalent if R is replaced by H. Therefore, three hydrogen atoms in a methyl group are chemically equivalent and they always have the same δ value.

The discussion of ring reversal is similar to that of the rotation around the chain. When a six-membered ring reverses rapidly, the axial and equatorial hydrogen atoms interchange with each other so that they have a common signal.

3. Prochirality

If a carbon atom is connected to four different functional groups, this carbon atom forms a chiral center and it is a chiral carbon atom. If a carbon atom is connected to two identical functional groups, this carbon atom forms a prochiral center and it is a prochiral carbon atom. In fact, prochiralilty has been encountered in compounds **C2-7** and **C2-9**.

A substitution test can be used to discuss the relationship between two functional groups, which are connected by a prochiral carbon atom. Before the substitution, the prochiral compound is shown as X/X. After the substitution, the prochiral compound becomes X/T or T/X. There are three possibilities:

i. If $X/T \equiv T/X$, which means X/T and T/X are superposable, they are homotopic.
ii. If X/T and T/X are mirror images, they are enantiotopic.
iii. If X/T and T/X are diastereotopic, they are diastereotopic.

Because the substitution test is tedious, a simple method has been proposed [4].

If the molecule to be discussed has a symmetrical plane and the symmetrical plane bisects the angle of XCX, where two X groups are the two functional groups attached to the prochiral carbon atom, these two X groups are enantiotopic (if rapid intramolecular motions exists, this symmetrical plane should bisect the angle for every conformer). If this condition is not totally satisfied, these two X groups are diastereotopic. Compound **C2-7** is an example of diastereomers. Another example is **C2-10**, in which the two methyl groups at the quaternary carbon atom are diastereotopic, so are the two methyl groups in the isopropyl group.

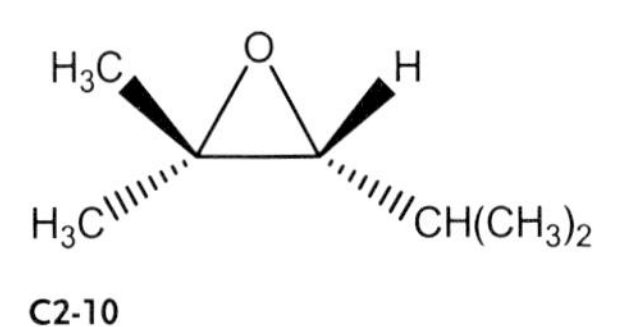

C2-10

4. Consideration of the chemical equivalence of two hydrogen atoms of a CH_2 (or two identical functional groups attached to the same carbon atom)

Here are some strong criteria for judging the chemical equivalence:

i. Two hydrogen atoms of a CH_2 in a ring are chemically non-equivalent, which occurs frequently.
ii. When a single bond cannot rotate rapidly, two identical functional groups are chemically non-equivalent, a typical example of which is compound **C2-11**.
iii. The two hydrogen atoms of a CH_2 attached to a chiral carbon atom are always chemically non-equivalent, of which Fig. 2.10 is an example.
iv. The chemical equivalence of the two hydrogen atoms of a CH_2 can be discussed using the rule of the symmetrical plane, which is a general rule. If a prochiral center is connected to a chiral center, this molecule has no symmetrical plane so that the two identical functional groups attached to the prochiral center are certainly chemically non-equivalent. On the other hand, if a prochiral center is not connected to a chiral center, the two identical functional groups attached to the prochiral center may be still diastereotopic, such as the two CH_2 groups in compound **C2-12** because the molecular symmetrical plane does not bisect the angle formed by H–C–H of the CH_2 group.

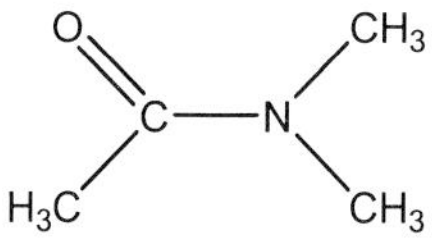

C2-11 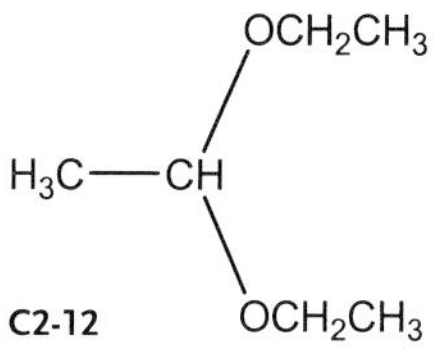 **C2-12**

Because the two hydrogen atoms are chemically non-equivalent, they form an AB system to produce four lines (see Section 2.4.1). Through the splitting from their adjacent methyl groups, the CH_2 group produces 16 lines. However, because the two CH_2 groups are symmetrical in the molecule, their lines are strictly overlapped in the same positions.

2.3.2 Magnetic Equivalence

h>If the following conditions are satisfied, the two nuclei are magnetically equivalent.

1. They are chemically equivalent.
2. They have the same coupling constant (in both magnitude and sign) for any other nucleus.

An example for magnetic non-equivalence is **C2-13**.

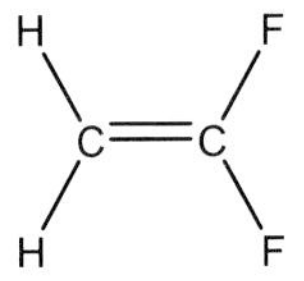

C2-13

From the symmetry of **C2-13**, the two hydrogen atoms as well as the two fluorine atoms are chemically equivalent. However, one hydrogen atom couples with a fluorine atom through *cis*-coupling, and the other hydrogen atom, through *trans*-coupling. Therefore, they do not satisfy the second condition above, neither do the two fluorine atoms (**C2-14** and **C2-15**). Because the two hydrogen atoms are chemically non-equivalent, they have more than ten lines in their ^{1}H spectrum.

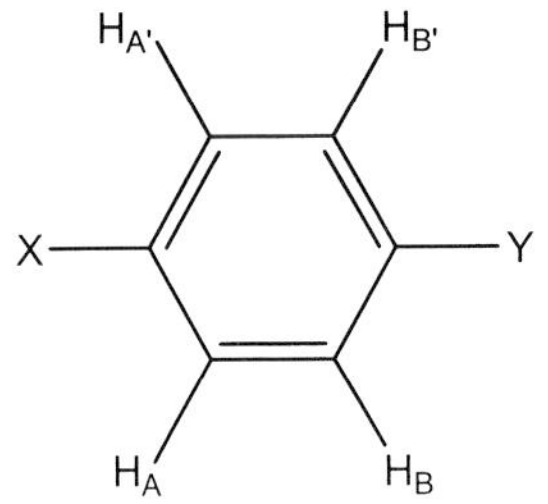

C2-14

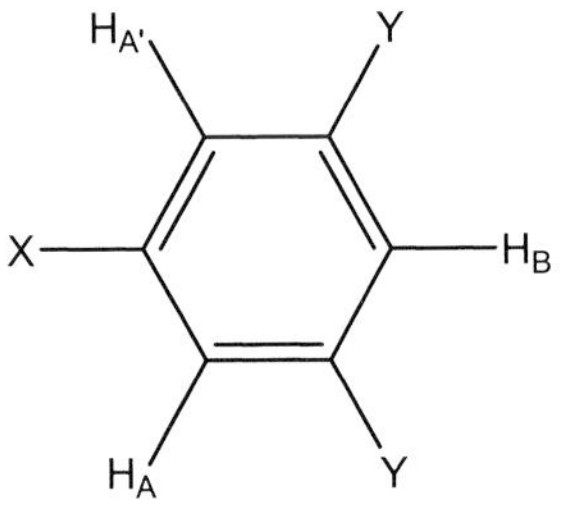

C2-15

In **C2-14**, H_A and $H_{A'}$ can interchange through the rotation around their C_2 axis so that they are chemically equivalent. However, the coupling between H_A and H_B is an *ortho*-coupling with coupling constant ^3J while the coupling between $H_{A'}$ and H_B is a *para*-coupling with coupling constant ^5J. Therefore, H_A and $H_{A'}$ are magnetically non-equivalent. In **C2-15**, H_A and H'_A are not only chemically equivalent but also magnetically equivalent because H_A or $H_{A'}$ has a ^4J coupling with the other nucleus, H_B (if neither X nor Y is a magnetic nucleus).

2.3.3
Spin System

1. Definition

Coupled nuclei form a spin system. Every nucleus couples with other nuclei of the system but does not couple with any other nucleus outside the system. It is not required that a nucleus couples with all other nuclei of the system. Spin systems are separated by a quaternary carbon atom or a heteroatom.

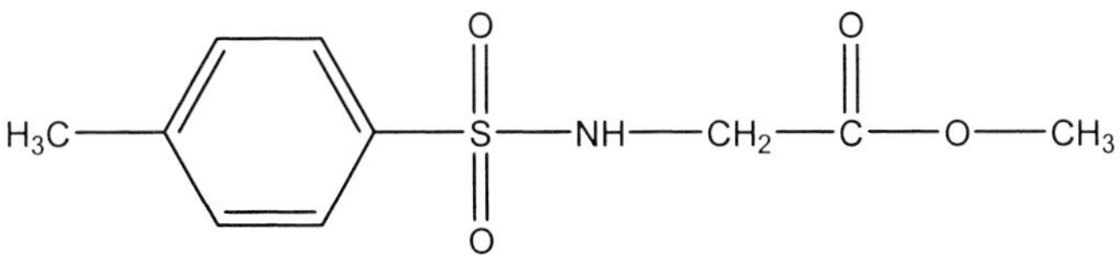

C2-16

In compound **C2-16**, $H_3C–C_6H_4–$ is a spin system. The two other spin systems of the molecule are $–NH–CH_2–$ and $–CH_3$.

2. Nomenclature

i. The nuclei with the same chemical shift values are denoted by a capital letter.
ii. Two spin groups with two close chemical shift values are denoted by two letters that are close to each other in the alphabet, such as A and B, or X and Y. In contrast, two spin groups with two significantly separated chemical shift values are denoted by two letters that are far away from each other in the alphabet, such as A and X.
iii. In a nucleus group, if all nuclei are magnetically equivalent, the number of nuclei is denoted as a subscript at the lower-right of the capital letter.
iv. If all nuclei of a nucleus group are not magnetically equivalent, they are differentiated by superscript "′", such as MM′M″ for three nuclei which are not magnetically equivalent but have the same δ value.

The above-mentioned spin system can be denoted as $A_3MM'XX'$.

The items i, iii and iv are clear. Now some supplementary discussion will be given for point 2.

The intensity of a coupling between two nuclei of a spin system is related closely to the difference between their chemical shift values. When $\Delta\delta$ and J are expressed in the same units (in Hz or ppm), the intensity is determined by $\Delta\delta/J$ or $\Delta\nu/J$, where either $\Delta\delta$ or $\Delta\nu$ is the difference in their chemical shift values. When $\Delta\nu \gg J$, the coupling is weak. Under these conditions, some terms can be omitted in the related theoretical calculation. In fact, its spectrum is very simple. On the contrary, when $J \approx \Delta\nu$ or $J > \Delta\nu$, the coupling is strong. Under these conditions, the related theoretical calculation cannot be simplified. In fact, its spectrum is complex. Two nuclei with a strong coupling are denoted as A and B while two nuclei with a weak coupling are denoted as A and X. When $\Delta\delta/J > 6$, the coupling is weak.

2.3.4 Classification of NMR Spectra

NMR spectra can be classified into the first-order and the second-order spectra.

The first-order spectra can be analyzed by the $n+1$ rule (or the $2nI+1$ rule for a nucleus with $I \neq 1/2$). The conditions for producing the first-order spectra are as follows:

1. The ratio of $\Delta\delta/J$ is large, at least 6.
2. The nuclei in a group are magnetically equivalent. This condition is even more important than 1. The two hydrogen atoms in **C2-13** are not magnetically equivalent so that they form a second-order spectrum and its ^{1}H spectrum is complex.

The characteristics of the first-order spectra are as follows:

1. The number of peaks is illustrated by the $n+1$ rule, where n is the number of adjacent hydrogen atoms that are magnetically equivalent. If the nuclei to be considered are coupled by n_1 nuclei with J_1 and n_2 nuclei with J_2 with $n_1+n_2=n$, they have $(n_1+1)(n_2+1)$ peaks.
2. The intensities of peaks within a peak set can be expressed approximately as the coefficients of the developed binomial (see Section 1.3).
3. Both δ and J can be read directly from the spectrum. δ is the center position of the peak set while J is the interval between two adjacent peaks (in Hz).

If the above-mentioned conditions cannot be satisfied at the same time, the second-order spectra will result, which are differentiated from the first-order spectra by the following facts:

1. In general, the number of peaks is greater than that calculated by the $n+1$ rule.
2. The intensities of peaks in a peak set show complicated ratios.
3. Generally neither δ nor J can be read out directly from the spectrum. Therefore, it is necessary to interpret the second-order spectra.

2.4 Common Second-order Spectra

Several common second-order spectra will be discussed in this section with respect to theoretical analysis. The spectral patterns of a number of common functional groups will be described in the next section from the practical application point of view.

The analysis of the second-order spectra is simply the application of quantum mechanisms to NMR spectroscopy. Several monographs published in the mid-1960s deal with this subject systematically [5, 6]. In the present book we will analyze several common second-order spectra in order to emphasize the physical concepts.

2.4.1 AB System

The AB system is frequently encountered, such as an isolated CH_2 in a ring, a benzene ring substituted by four functional groups, etc.

The energy level diagram of the AB system is similar to that of the AX system (Fig. 1.3) but the wave functions of the energy levels 2 and 3 are a linear combination of $\alpha\beta$ and $\beta\alpha$.

The spectral pattern of the AB system is shown in Fig. 2.11. Its related parameters satisfy the following equations.

$$J_{AB} = \nu_1 - \nu_2 = \nu_3 - \nu_4 \tag{2-15}$$

$$\begin{aligned}\Delta = \delta_{AB} = \delta_A - \delta_B &= \sqrt{D^2 - J^2} \\ &= \sqrt{(\nu_1 - \nu_3)^2 - (\nu_1 - \nu_2)^2} \\ &= \sqrt{(\nu_1 - \nu_4)(\nu_2 - \nu_3)}\end{aligned} \tag{2-16}$$

$$\frac{I_2}{I_1} = \frac{I_3}{I_4} = \frac{D+J}{D-J} = \frac{\nu_1 - \nu_4}{\nu_2 - \nu_3} \tag{2-17}$$

where I is the intensity of the peak.

From Eq. (2-17), it can be seen that the two inner peaks (2 and 3) of an AB system are always stronger than the two outer peaks (1 and 4). Only under the conditions where $\Delta_{AB} \gg J_{AB}$, can the intensities of the two outer peaks approach those of the two inner peaks. Therefore, it is very difficult for two hydrogen atoms to form an AX system.

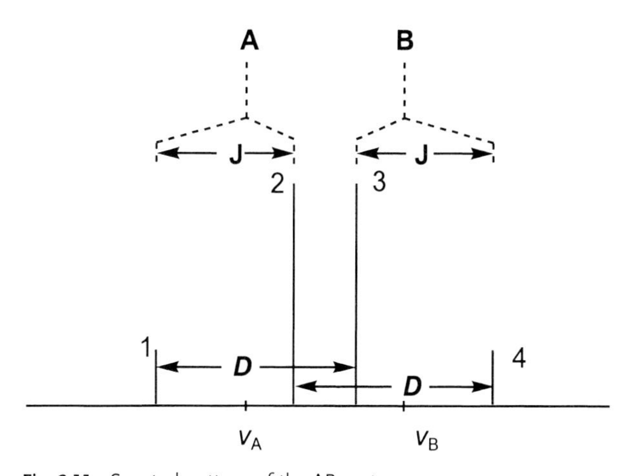

Fig. 2.11 Spectral pattern of the AB system.

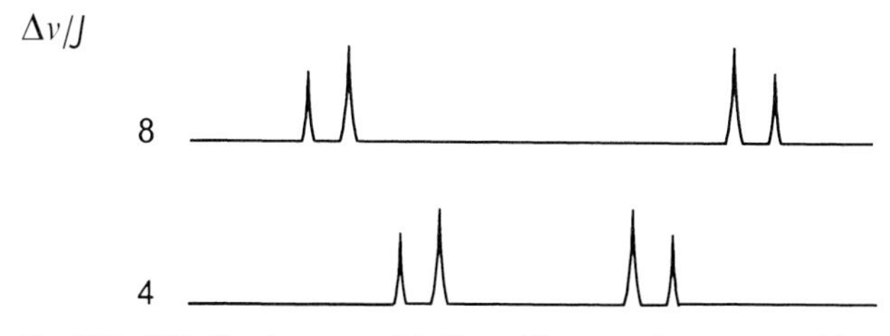

Fig. 2.12 With the decrease of $\Delta\nu/J$, an AX system becomes an AB system, or even an A_2 system.

From Eq. (2-16), it can be found that $D^2 > J^2$. Thus J is the interval between two up-field peaks (3 and 4) or two down-field peaks (1 and 2). This conclusion can also be drawn from Fig. 2.12.

From Fig. 2.12, it is obvious that the spectrum of a two-spin system is changed from a first-order spectrum to a second-order spectrum when the ratio of $\Delta\nu/J$ decreases gradually. When $\Delta\nu = 0$, all peaks are concentrated into one line. These two tendencies are also true for the other second-order spectra.

2.4.2 AB_2 System

The AB_2 system exists in a phenyl ring with three symmetrical substituents, etc.

The energy level diagram of an AB_2 system is shown in Fig. 2.13.

The selection rules for NMR transitions are: (1) $\Delta m = \pm 1$ and (2) the symmetry of the wave function is unchanged after the transition.

The transitions of an AB_2 system are shown in Fig. 2.13 by lines connecting two related energy levels. There are nine transitions, in which the transition $4 \rightarrow 3$ corresponds to a combination line with a weak intensity. Therefore, an AB_2 system has nine peaks.

The spectral pattern of an AB_2 system is shown in Fig. 2.14, in which $\nu_A > \nu_B$. If $\nu_A < \nu_B$, the pattern will rotate 180° around its center point. Under these conditions, the marks will be changed correspondingly.

The positions of the peaks of an AB_2 system satisfy the following equations:

$$\nu_1 - \nu_2 = \nu_3 - \nu_4 = \nu_6 - \nu_7 \tag{2-18}$$

$$\nu_1 - \nu_3 = \nu_2 - \nu_4 = \nu_5 - \nu_8 \tag{2-19}$$

$$\nu_3 - \nu_6 = \nu_4 - \nu_7 = \nu_8 - \nu_9 \tag{2-20}$$

In the pattern of an AB_2 system, lines 5 and 6 are close together, which is also shown in Fig. 2.14.

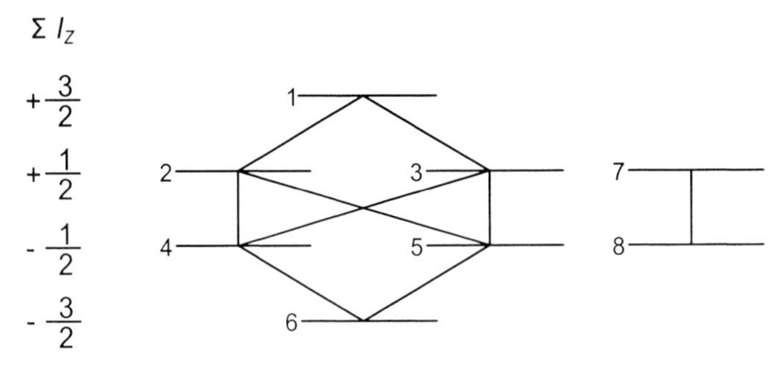

Fig. 2.13 Energy level diagram of an AB_2 system.

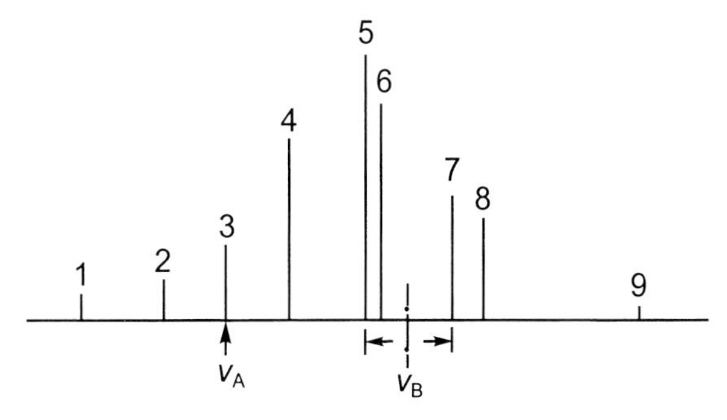

Fig. 2.14 Spectral pattern of an AB_2 system.

δ_A can be read directly from the pattern, as

$$\delta_A = \nu_3 \tag{2-21}$$

This corresponds to the transition between energy levels 7 and 8, both of which have anti-symmetrical wave functions.

The following equations hold for the AB_2 system:

$$\delta_B = \frac{1}{2}(\nu_5 + \nu_7) \tag{2-22}$$

$$J_{AB} = \frac{1}{3}[(\nu_1 - \nu_4) + (\nu_6 - \nu_8)] \tag{2-23}$$

Eq. (2-23) can easily be remembered in the following way. The term $(\nu_1 – \nu_4)$ illustrates the split width of nucleus A while the term $(\nu_6 – \nu_8)$ represents, approximately, the split width of nucleus B because line 6 is near line 5 and line 9 is weak as it is a combination line. Because the system contains three nuclei, the J value should be calculated from the total split width of the spin system divided by 3.

2.4.3 AMX System

The spectrum of the AMX system is a first-order spectrum, as shown in Fig. 2.15. Three δ values and three J values can be read directly from the spectrum.

2.4.4 ABX System

There are several ways of interpreting the spectrum of the ABX system. The procedure presented here has the advantage of emphasizing the physical concepts.

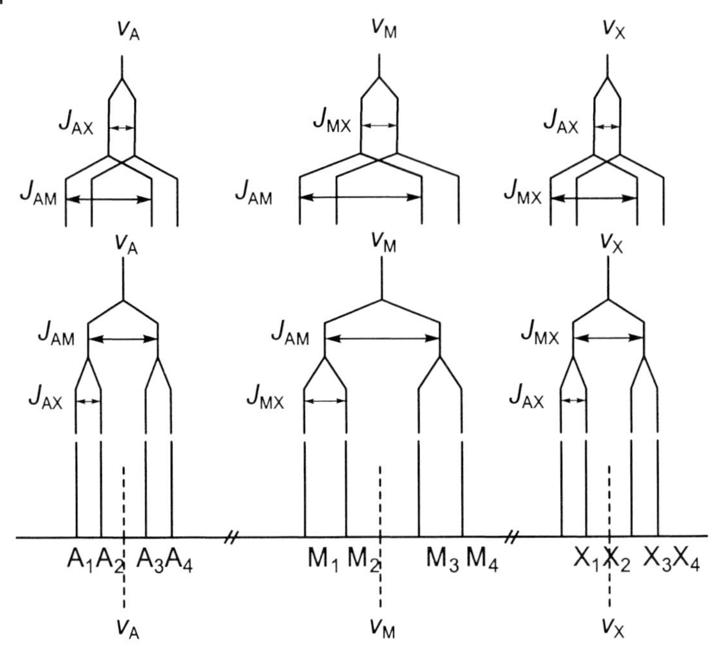

Fig. 2.15 Spectrum of the AMX system.

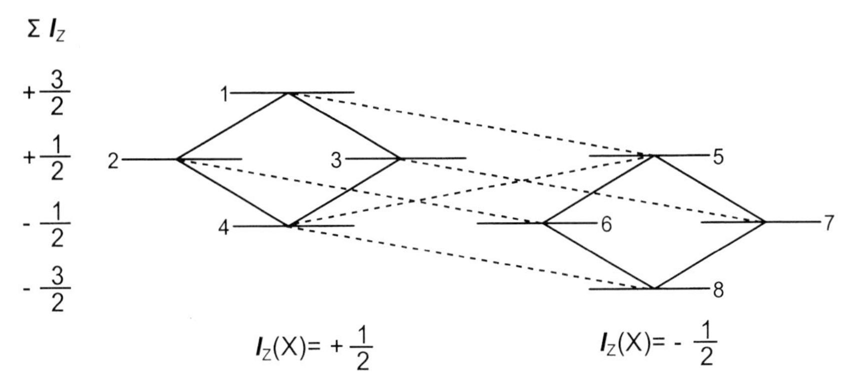

Fig. 2.16 Energy level diagram of an ABX system.

In an ABX system, the coupling between A and B is strong since they have close δ values. On the other hand, the couplings between X and A or between X and B are weak. Hence, the molecules of the ABX system can be classified into two groups according to the spin orientations of X, α and β. Therefore, the energy level diagram of the ABX system can be divided into two sub-systems: 1, 2, 3 and 4; 5, 6, 7 and 8, as shown in Fig. 2.16.

Energy levels 1, 2, 3 and 4 correspond to the X nuclei with $I_Z = 1/2$ while energy levels 5, 6, 7, and 8 correspond to the X nuclei with $I_Z = -1/2$. Transitions

1–5, 2–6, 3–7 and 4–8 belong to the transitions of X. Transitions 4–5 and 3–6 produce combination lines which are, in general, weak.

From the analysis above, it can be shown that the spectrum of an ABX system consists of two parts. The AB part has eight peaks, which are two sets of sub-spectra of ab systems. The X part has four strong peaks and two weak peaks.

2.4.5
AA′BB′ System

A phenyl ring with two symmetrical substituents (*para*-substitution or *ortho*-substitution by two identical functional groups), or $-CH_2-CH_2-$, etc., can form the AA′BB′ system.

A characteristic of the AA′BB′ system is that the spectrum is symmetrical around the middle point.

2.5
Spectra of Common Functional Groups

In this section, the spectral patterns of some common functional groups will be dealt with from the point of view of the interpretation of ^1H spectra, which is very useful for those workers wishing to improve their skills in interpreting ^1H spectra.

2.5.1
Substituted Phenyl Ring

1. Mono-substituted phenyl ring

When five hydrogen atoms are found in the aromatic region from the integral curve, the existence of a mono-substituted phenyl ring can be established.

In Section 2.3, it was shown that the complexity of a ^1H spectrum is determined by $\Delta\nu/J$. As the values of J do not vary considerably, the complexity is mainly determined by the substituents, which change the δ values of the hydrogen atoms of the substituted phenyl ring. To analyze its ^1H spectra, Chamberlain divided the substituents of a phenyl ring into two types [7]. In the present monograph, a concept will be presented that can be used to analyze the ^1H spectra through the combined consideration of the peak shapes and the chemical shifts of the remaining hydrogen atoms in a substituted phenyl ring. This is based on the electron effects of the substituents, which are classified into three types. Using this concept of the three types of substituents, the reader holds the key to interpreting the spectra of the substituted phenyl ring.

1. The first type of substituents

These are the functional groups that do not change the δ values of the remaining hydrogen atoms of the substituted phenyl ring to any considerable extent; $-CH_3$, $-CH_2$, $-CH-$, $-CH{=}CHR$, $-C{\equiv}CR$, $-Cl$, $-Br$, etc., belong to this type.

As the *ortho-*, *meta-* and *para*-hydrogen atoms have approximate δ values, the five remaining hydrogen atoms form a peak set within a narrow δ region with the appearance that the peaks in the middle are strong and the peaks at two edges are weak.

2. The second type of substituents

The second type of substituents is the functional groups that contain the saturated heteroatom. Because of the p–π conjugation between the non-bonding electrons of the heteroatom and the delocalized electrons of the phenyl ring, the electron density of the substituted phenyl ring is increased, especially at the *ortho-* and *para-* positions. As a result, the substituted phenyl ring is subjected to activation so that the electrophilic reactions can take place easily at the *ortho-* and *para*-positions. From the point of view of NMR, the *ortho-* and *para*-hydrogen atoms have upfield shifts after the substitution. The *meta*-hydrogen atoms also have an upfield shift but the magnitude is less than that of the *ortho-* and *para*-hydrogen atoms. Therefore, the peaks of the substituted phenyl ring fall into two groups. Three hydrogen atoms (*ortho-* and *para*-hydrogen atoms) have peaks at the far right-hand side. The two *meta*-hydrogen atoms have peaks at the far left-hand side. As each *meta*-hydrogen atom has two adjacent hydrogen atoms, its peaks are illustrated as a triplet with respect to ^3J, which is split further by ^4J and ^5J. However, the split by ^3J is the most obvious, a characteristic of the second type of substituents.

$-OH, -OR, -NH_2, -NHR, -NR'R''$, and so forth belong to this type.

3. The third type of substituents

The third type of substituent is the functional groups which contain the unsaturated heteroatom. Because of the electronegativity of the heteroatom, the electron density of the substituted phenyl ring is decreased, especially at the *ortho-* positions. As a result, the substituted phenyl ring is subjected to passivation. From the viewpoint of NMR, all the five hydrogen atoms, especially the two *ortho*-hydrogen atoms, have downfield shifts after substitution. Since each *ortho-* hydrogen atom has one adjacent hydrogen atom, its peaks are illustrated as a doublet with respect to ^3J, which is split further by ^4J and ^5J. Therefore, the characteristic of the third type of substituents is the doublet with respect to ^3J, which is situated at a fairly downfield position.

$-CHO, -COR, -COOR, -COOH, -CONHR, -NO_2, -N{=}NR$,
etc., belong to this type.

To interpret the spectra of the substituted phenyl ring, the following points should be noted:

i. Attention should focus on the analysis of the peak shapes split by ^3J.
ii. The analysis of peak shapes should be combined with the analysis of δ values.

2. *Para*-substituted phenyl ring

As the *para*-substituted phenyl ring has a C_2 symmetry axis, the four remaining hydrogen atoms form an AA′BB′ system, which has a spectrum symmetrical about the center point.

Because the four remaining hydrogen atoms are two pairs of adjacent hydrogen atoms, their spectrum is much simpler than that of an ordinary AA′BB′ system. The spectrum of the *para*-substituted phenyl ring is the simplest one among all spectra of substituted phenyl rings, with its characteristic being two sets of doublets split by ^3J. The two inner peaks are strong while the two outer peaks are weak. All these peaks are split further by ^4J and ^5J.

If a substituent has long-range couplings with its *ortho*-hydrogen atoms, such as structure **1**, the doublet formed by the *ortho*-hydrogen atoms will decrease in height while the width at the half height will increase.

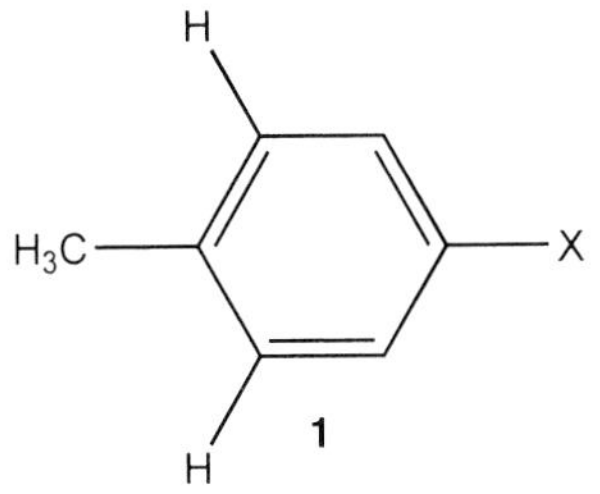

3. *Ortho*-substituted phenyl ring

i. The *ortho*-substitution into a phenyl ring by two identical functional groups forms an AA′BB′ system whose spectrum is symmetrical about the center point.

ii. The *ortho*-substitution into a phenyl ring by two different functional groups forms the most complex spectra among all spectra of substituted phenyl rings.

4. *Meta*-substituted phenyl ring

In general, the spectrum of the *meta*-substituted phenyl ring is complex. However, an isolated hydrogen atom, which is situated between the two substituents, always shows a singlet without the split by ^3J. From the singlet, it is possible to determine the *meta*-substituted phenyl ring if this singlet is not significantly overlapped by other peak sets.

5. Multi-substituted phenyl ring

A tri-substituted phenyl ring forms a system of AMX, ABC or AB_2.

A tetra-substituted phenyl ring forms an AB system.

A penta-substituted phenyl ring leaves only an isolated hydrogen atom, which has a singlet.

In general, the more different the δ values of the remaining hydrogen atoms, the more similar the spectrum is to a first-order spectrum.

2.5.2
Substituted Heteroaromatic Ring

Because of the existence of the heteroatom, the hydrogen atoms in a heteroaromatic ring have different δ values. The substitution increases these differences in δ further. Therefore, the spectra of a substituted heteroaromatic ring can be analyzed as a first-order spectrum. It should be noted that the coupling constants change with position with respect to the heteroatom (see Tab. 2.7).

2.5.3
Mono-substituted Ethylene

In a mono-substituted ethylene, several couplings: *cis*-, *trans*- and *geminal*-couplings co-exist. If the substituted ethylene is connected to an alkyl group, the coupling from ^3J and ^4J will split the spectrum further. Therefore, its spectrum can be very complex.

The structural unit of **2** will be taken as an example.

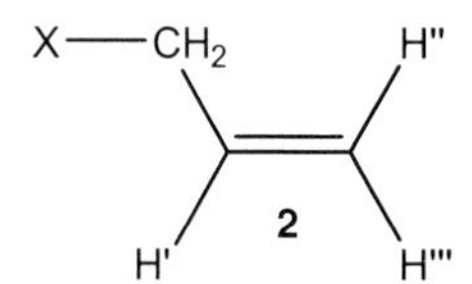

In general, H′ has 12 lines, which can be determined to be a first-order spectrum. The 12 lines are denoted as d×d×t or (d, d, t), in which one "d" is the doublet produced by the *trans*-coupling of H″, the other "d" represents the doublet produced by the *cis*-coupling of H‴ and "t" describes the triplet produced by the ^3J coupling of the CH_2. If the central lines of the four triplets are determined, J_{trans} and J_{cis} can be found easily.

Both H″ and H‴ have complex patterns because they are coupled with several coupling constants. Their couplings can be determined as follows: (1) H″ and H‴ are split into a doublet, respectively, by H′ with a considerable value of the coupling constant J_{trans} or J_{cis}. (2) H″ and H‴ couple each other with a small coupling constant ^2J. (3) H″ or H‴ couple CH_2 with a small coupling constant ^4J. Since J_{trans} and J_{cis} are prominent among the above-mentioned coupling constants, they determine the main distribution of their peaks.

2.5.4
Normal Long-chain Alkyl

A normal long-chain alkyl is a common structural unit with the formula X–$(CH_2)_n$–CH_3. Because other functional groups are generally electrophilic with respect to alkyl, the α-CH_2 has a larger δ value and the β-CH_2 has a slightly larger δ value than other CH_2 moieties, which have peaks around 1.25 ppm. Because these CH_2 have approximate δ values and $^3J \approx 7$ Hz, they form a strong coupling system with numerous peaks but are concentrated in a narrow region.

2.6 Methods for Assisting the Spectrum Analysis

2.6.1 Using a Spectrometer with a High Frequency

As discussed in Section 2.3, the ratio of $\Delta\nu/J$ determines the complexity of a ^{1}H spectrum. The value of J is determined by the structure so that it is not related to the spectrometer. However, the value of $\Delta\nu$ or $\Delta\delta$ changes when in units of Hz (but it is constant in ppm). Therefore, the use of an instrument with a high frequency will increase the ratio of $\Delta\nu/J$ so that the spectrum will be simplified.

In the simplification of spectra using a spectrometer with a high frequency, the ratio of S/N will be greatly improved. Since $\Delta E \propto \nu$, ΔE will increase with the enhancement of ν. According to Eq. (1-21), the difference between the populations, Δn, will consequently be proportional to ν. On the other hand, the intensity of the signals is proportional to ν. As a result, theoretically the improvement of S/N will be proportional to the second power of ν, but in practice, it will be proportional to the 1.5-th power of ν.

For the two above-mentioned reasons, the development of NMR spectrometers is directed towards an increase in the frequency of the instruments.

2.6.2 Deuterium Exchange

D_2O is used most frequently for deuterium exchange.

The hydrogen atom connected to a heteroatom (O, N or S) has a weak chemical bond, which leads to chemical exchange of the hydrogen atom. The rate of the chemical exchange reaction is in the order OH > NH > SH. If a sample contains these functional groups, a deuterium exchange experiment can be carried out after recording its ^{1}H spectrum. Several drops of deuteroxide (D_2O) are added to the sample solution, which is then shaken thoroughly. The hydrogen atom connected to a heteroatom will be replaced by a deuterium atom from D_2O. As a result, the peak of the hydrogen atom will disappear in the ^{1}H spectrum recorded the second time, which confirms the existence of the hydrogen atom.

Although the δ values of OH, NH and SH vary with experimental conditions, the ranges of their variations differ. What is more, they have different peak shapes, so OH, NH and SH can be differentiated. Their peak shapes will be discussed in detail in Section 2.8. OH usually has a sharp peak while NH and SH frequently have a blunt peak.

2.6.3
Medium Effect

The molecules of benzene, acetonitrile, etc., have anisotropy. When these molecules exist in the sample solution, they have various shielding–deshielding influences on different parts of the sample molecule. Therefore, some overlapped peaks of the sample may be separated after the addition of several drops of one of these deuterated compounds.

2.6.4
Shift Reagents

Shift reagents are complexes of a β-diketone with Eu or Pr, but in particular with the former.

The molecule of a shift reagent and the molecule of the sample to be measured form a new complex. The non-bonding electrons of the metal have a paramagnetic moment, which affects the magnetic nuclei of the sample through space. This effect is known as pseudo-contact interaction, which is proportional to $(3\cos^2\theta - 1)/r^3$, where r is the distance between the metal ion and the magnetic nucleus and θ is the angle between the paramagnetic moment of the electrons and $\mathbf{r}$. The interaction decreases rapidly with an increase in r and it is also related to θ. The functional group, where the shift reagent combines with the sample molecule, shows the greatest shift in its δ value. Overlapped peaks can be separated because they have different r and θ values.

2.6.5
Spectral Simulation by Computer

As an FT-NMR spectrometer is equipped with a computer, spectral simulations can be easily performed.

A set of NMR parameters (chemical shifts and coupling constants) is fed into the spectrometer, which can then calculate a spectrum according to these parameters. These parameters are adjusted to make the calculated spectrum approach that of the measured spectrum. If two spectra are identical, the adjusted parameters may be the answers to the chemical shifts and coupling constants of the measured spectrum. However, it should be noted that it is possible that there could be more than one set of solutions for a given spectrum.

2.7
Double Resonance

Double resonance is also known as double irradiation.

In an experiment on double resonance, two alternating magnetic fields B_1 and B_2 are used simultaneously. B_1 is used to produce nuclear magnetic resonance while B_2 is used to irradiate one type of nuclei in a certain functional group, which could be the same isotope as that being measured or a different isotope. The isotope irradiated by B_2 is denoted in parentheses { } and the isotope irradiated by B_1 is denoted before the parentheses, for example, $^{13}C\{^1H\}$, $^1H\{^1H\}$. If more isotopes are irradiated, the experiment is known as multiple resonance.

Similarly to Eq. (1-12), $\omega = \gamma B_0$, we can write

$$\omega_2 = \gamma B_2$$

or

$$\nu_2 = \frac{\gamma}{2\pi} B_2 \tag{2-24}$$

where γ is the magnetogyric ratio of the isotope irradiated by B_2; B_2 is the intensity of the irradiating magnetic field.

The ν_2 calculated according to Eq. (2-24) describes the spectral width affected by the perturbation. The more intense B_2 is, the greater width of ν_2. Therefore, the experiment of double resonance is closely associated with the intensity of B_2. In this section, only two experiments, spin decoupling and NOE (nuclear Overhauser effect), which are the most important of the double resonance experiments, will be described.

2.7.1 Spin Decoupling

Spin splits can provide structural information. However, if the splits are very complicated, the information is difficult to analyze. Spin decoupling can simplify a complicated spectrum, give information about the coupling systems, and so forth.

Take an AX system as the example of spin decoupling. The peak of A is split into a doublet by X. When A is irradiated by B_1 at frequency ν_1, which is necessary to produce the signal of A, X is irradiated by B_2 at another frequency ν_2. If B_2 is strong enough, the nuclei of X are in resonance and they are saturated. All nuclei of X transit rapidly between the two energy levels so that the supplementary magnetic fields at A averages out to zero. Therefore, the coupling of A by X is removed, which means that the nucleus of A has only a singlet.

If both A and X are hydrogen nuclei, the experiment is denoted as $^1H\{^1H\}$, which is known as a homonuclear decoupling. The experiment in which a ^{13}C spectrum is recorded while ^1H is irradiated, is known as a heteronuclear decoupling, which is denoted as $^{13}C\{^1H\}$.

O—CH_2—CH_2—CH_2—CH_3

O=P—O—CH_2—CH_2—CH_2—CH_3

O—CH_2—CH_2—CH_2—CH_3

C2-17

The CH_2 connected to the oxygen atom in **C2-17** couples its adjacent CH_2 and the ^{31}P through the chemical bonds of P–O–C–H, the coupling constant of which has a value close to that of $J_{H-C-C-H}$. As a result, the CH_2 has an approximate quartet. If the β-CH_2 and ^{31}P are irradiated by B_2 and B_3, respectively, the α-CH_2 will show a singlet. This experiment is a triple resonance.

In addition to the variation of the peak shape of the decoupled nuclei, its peak position may change if the difference between ν_A and ν_X is small. The change in resonant frequency is called the Bloch-Siegert shift.

Spin decoupling has the following functions

1. Determination of coupling relationships in a coupled system

The downfield part of the ^1H spectrum of an alkaloid is shown in Fig. 2.17a. The upfield part of the ^1H spectrum consists of a doublet (δ=1.17 ppm, three hydrogen atoms) and a quartet (δ=5.56 ppm, one hydrogen atom).

Because the spectrum was recorded by an instrument with 400 MHz, the majority of the peak sets in the spectrum are shown as first-order spectral patterns. With the integral curve (not shown in the figure), the hydrogen atoms and the occupied region of every peak set can be determined, such as a doublet (δ=7.28 ppm, one hydrogen atom); a triplet (δ=7.24 ppm, one hydrogen atom), and so forth. These analyzed results are listed in Tab. 2.8, in which the peak sets are denoted as the Roman numerals in the order of the increasing field strength.

The peak sets are irradiated successively from I to IX.

I is a singlet (denoted as s). Obviously, there is no coupling relationship between I and other peak sets so that it will not be irradiated.

II and III correspond to two hydrogen atoms. When they are irradiated, the doublets of IV and VI become two singlets, respectively, which is denoted as d→ s, as shown in Fig. 2.17b and is listed in Tab. 2.9. The result means that a coupling relationship exists between II and IV and between III and VI.

When IV is irradiated, II becomes a singlet with a shift towards the right. This singlet is superimposed on the doublet of III to produce the spectrum shown in Fig. 2.17c. Because V is close to IV and the power level for the irradiation of IV is too high, the peak set of V is weakened significantly and it probably leads to a change in shape in the peak set of IX.

The coupling relationship between V and IX is confirmed clearly when V is irradiated. The peak set of IX changes from a triplet to a doublet. Similar to the irradiation of IV, because IV is close to V and the power level for the irradiation

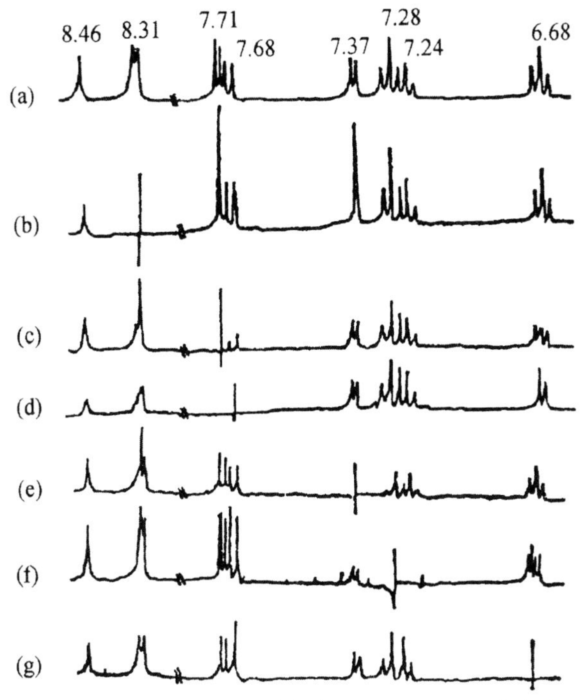

Fig. 2.17 Downfield portion of the ^{1}H spectrum of an alkaloid. (a) Without decoupling. (b)–(g) With decoupling.

Tab. 2.8 Peak sets of the ^{1}H spectrum of an alkaloid

No.	*I*	*II, III*	*IV*	*V*	*VI*	*VII*	*VIII*	*IX*
δ (ppm)	8.46	8.31	7.71	7.68	7.37	7.28	7.24	6.68
Number of H	1	2	1	1	1	1	1	1
Splitting	s	d	d	d	d	d	t	t*

* approximately

Tab. 2.9 Decoupling result of the ^{1}H spectrum of an alkaloid

Indicium in Fig. 2.17	*Irradiated peak set*	*Corresponding spectral changes*
(b)	II, III	IV: d → s, VI: d → s
(c)	IV	II: d → s, IX: with the shape changed
(d)	V	XI: t → d, II and II:: with the shapes changed
(e)	VI	III; d → s
(f)	VII	IX: with the shape changed
(g)	VIII	VIII: t → d, V: d → s

of V is too high, the peak set of IV is removed. The change in the peak set at 8.31 ppm can be analyzed as in Fig. 2.17 (c).

The e, f and g of Fig. 2.17 can be analyzed similarly. The decoupling result is listed in Tab. 2.9.

From Tab. 2.9 it can be established that the following four spin systems exist.

i. One isolated hydrogen atom (I).
ii. Two AX systems: II–IV and III–VI.
iii. One four spin system: V–IX–VIII–VII.

Based on other spectral data and by comparing its homologues, the structure of the alkaloid can be determined to be **C2-18**.

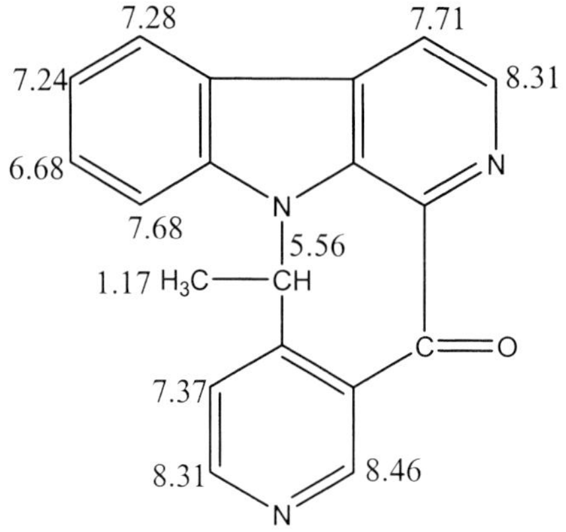

7.28 7.71 7.24 8.31 6.68 N N 7.68 5.56 1.17 H_3C—CH C=O 7.37 8.31 8.46 N

C2-18

2. Simplification of a complex spectrum

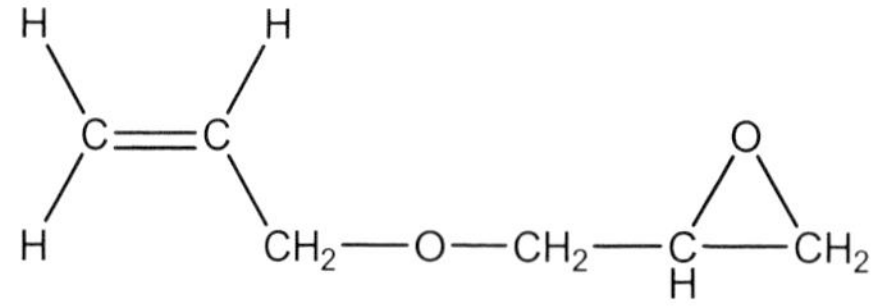

$H_2C=CH-CH_2-O-CH_2-CH(O)CH_2$

C2-19

The =CH– is illustrated as four triplets, which are not shown clearly in Fig. 2.18 because the triplets are overlapped partially. When the peak set at $\delta \approx 4.03$ ppm (corresponding to the peak of –C=C–CH_2–O–) is irradiated, these four triplets become four singlets, which are shown clearly at the top of Fig. 2.18. Also under this irradiation, the peak set of =CH_2 is simplified because the long-range coupling between =CH_2 and the –CH_2– is removed.

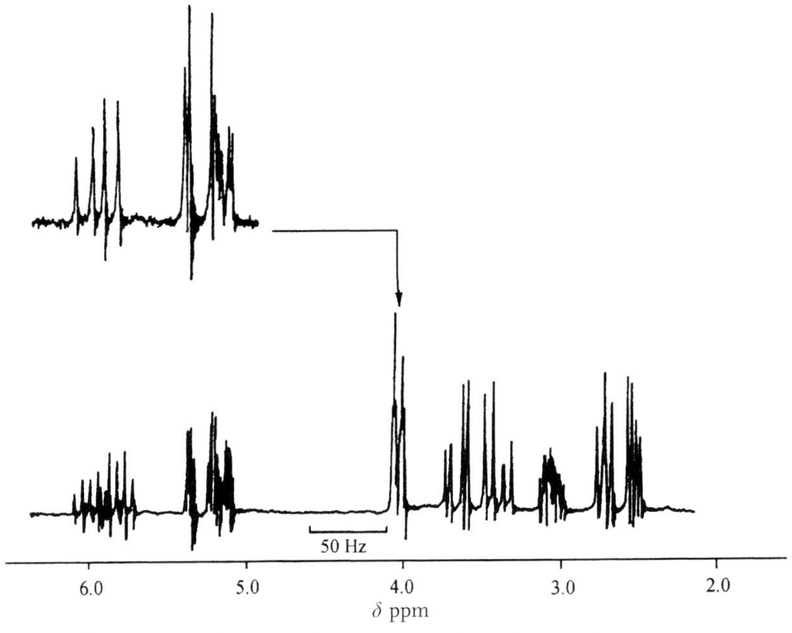

Fig. 2.18 ^{1}H spectrum of compound **C2-19** and its decoupling result.

Spin decoupling has other functions, such as the determination of the peak set position in hidden signals and the removal of the quadrupole moment effect.

2.7.2
Nuclear Overhauser Effect

The nuclear Overhauser effect discussed in this section is in one sense narrow.

In 1953, Overhauser studied the solution of metal sodium in liquid ammonia, which is paramagnetic. When the electron of the sodium was resonant and saturated by a field with a high frequency, the equilibrium of the nuclear populations of the sodium between energy levels was disturbed, which led to an enhancement of the sodium signal. This is called the Overhauser effect.

It was found that the Overhauser effect also exists in a spin system. If two nuclei are close (their distance is <5 Å) and one of them is irradiated so as to be saturated, the peak area of another nucleus will be changed. This is known as the nuclear Overhauser effect, which is not related to the number of the chemical bonds separating the two nuclei.

NOE is very effective for the configuration and conformation determinations of organic compounds. This topic will be discussed in detail in the last chapter.

The mechanism for producing NOE is dipole coupling between magnetic nuclei. The coupling discussed earlier is spin–spin coupling, also known as scalar coupling. The mechanism of the dipole coupling is different from that of the spin–spin coupling. Because magnetic nuclei have a magnetic moment, they can interact with each other through space if they are in close proximity. The dipole coupling is only related to the distance between the two related nuclei.

On the basis of the energy level diagram, the principle of NOE can be discussed [8, 9].

For two nuclei, I and S, related calculations lead to a so-called Solomon equation. It is used for the quantitative treatment of NOE. The equation is as follows:

$$\eta = \frac{\gamma_S}{\gamma_I} \cdot \frac{W_2 - W_0}{2W_{1I} + W_2 + W_0} \tag{2-25}$$

where η is the enhancement of NOE, namely the increased multiple of the signal intensity of I when S is irradiated; γ_S and γ_I are the magnetogyric ratios of S and I, respectively; W_2 is the probability of the double quantum transition ($\alpha\alpha \rightarrow \beta\beta$ or $\beta\beta \rightarrow \alpha\alpha$); W_0 is the probability of the zero quantum transition ($\alpha\beta \rightarrow \beta\alpha$ or $\beta\alpha \rightarrow \alpha\beta$); and W_{1I} the probability of the single quantum transition of I.

From Eq. (2-25), it can be seen that NOE results from $W_2 - W_0$. If $W_2 > W_0$, the signal of I will increase by the irradiation of S, which is a positive NOE. If $W_2 < W_0$, the signal of I will decrease by the irradiation of S, which is a negative NOE.

It is necessary to find W_2, W_0 and W_{1I} for the determination of η. From related theoretical calculations, the expressions for W_2, W_0 and W_{1I} are as follows:

$$W_2 = \frac{3}{5}\kappa^2 \frac{\tau_c}{1 + (\omega_I + \omega_S)^2 \tau_c^2} \tag{2-26}$$

$$W_0 = \frac{1}{10}\kappa^2 \frac{\tau_c}{1 + (\omega_I - \omega_S)^2 \tau_c^2} \tag{2-27}$$

$$W_{1I} = \frac{3}{20}\kappa^2 \frac{\tau_c}{1 + \omega_I^2 \tau_c^2} \tag{2-28}$$

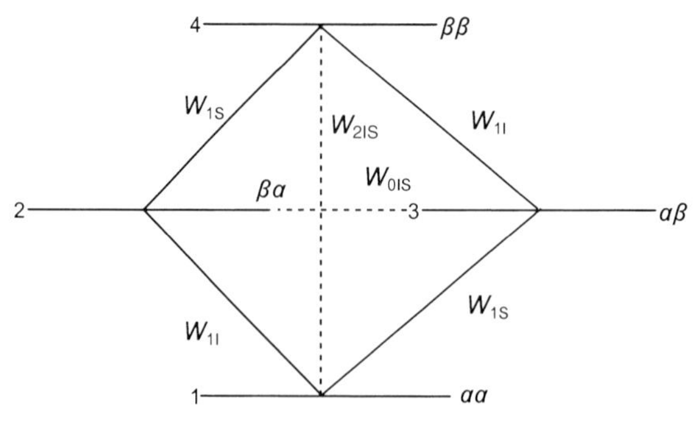

Fig. 2.19 Principle of NOE.

$$\kappa = \frac{\mu_0}{4\pi} \hbar \gamma_I \gamma_S r_{IS} \tag{2-29}$$

where ω_I and ω_S are the resonant frequencies of I and S, respectively; τ_c is the correlation time, a time constant during which the sample molecule loses phase coherence in its solution; and r_{IS} is the distance between I and S.

Substitution of Eqs. (2-26)–(2-28) into Eq. (2-25) and simplification yields

$$\eta = \frac{\gamma_S}{\gamma_I} \cdot \frac{\dfrac{6}{1+(\omega_I+\omega_S)^2\tau_c^2} - \dfrac{1}{1+(\omega_I-\omega_S)^2\tau_c^2}}{\dfrac{6}{1+(\omega_I+\omega_S)^2\tau_c^2} + \dfrac{1}{1+(\omega_I-\omega_S)^2\tau_c^2} + \dfrac{3}{1+\omega_I^2\tau_c^2}} \tag{2-30}$$

If I and S are homonuclear, $\gamma_I=\gamma_S$, $(\omega_I - \omega_S)\tau_c \ll 1$, and $\omega_I \approx \omega_S \approx \omega$. Under these conditions, Eq. (2-30) can be simplified as:

$$\eta = \frac{5 + \omega^2\tau_c^2 - 4\omega^4\tau_c^4}{10 + 23\omega^2\tau_c^2 + 4\omega^4\tau_c^4} \tag{2-31}$$

A small molecule tumbles rapidly so that it has a small τ_c. As a result, $\omega\tau_c \ll 1$. Under these conditions, Eq. (2-31) becomes approximately

$$\eta = 50\% \tag{2-32}$$

For a large molecule, it has a significant τ_c. As a result, $\omega\tau_c \gg 1$. Under these conditions, Eq. (2-31) becomes approximately

$$\eta = -1 \tag{2-33}$$

Therefore, from a small molecule to a large molecule, η will decrease from 0.5 to –1. From Eq. (2-31), the condition for $\eta=0$ can be obtained as follows:

$$5 + \omega^2\tau_c^2 - 4\omega^4\tau_c^4 = 0$$

$$\omega\tau_c \approx 1.12 \tag{2-34}$$

If $\eta \neq 0$ (regardless of a positive or a negative value), NOE can be observed. On the contrary, if $\eta=0$, NOE can not be observed. Hence it is necessary to avoid $\omega\tau_c \approx 1.12$ in the experiment.

From Eq. (2-31), it follows that the measurement of NOE is related to the spectrometer, which determines ω. Therefore, if a sample is measured with another spectrometer, the conditions for the measurement of NOE have to be re-established.

With the development of pulse sequences, NOE can be measured with a very high accuracy (see Section 4.1.10).

The discussion above is for a homonuclear case. If I and S are heteronuclear nuclei and small molecules tumble rapidly, Eq. (2-30) can be simplified as

$$\eta = \frac{\gamma_S}{2\gamma_I} \tag{2-35}$$

If γ_S and γ_I are different in sign, $\eta < 0$, as is the case for ^{15}N and ^{1}H.

In ^{1}H spectroscopy the most important use of NOE is the application in stereochemistry. The dipole–dipole interaction is proportional to $1/r^6$, where r is the distance between the two nuclei. Therefore, if two protons are close in space (regardless of any connection through chemical bonds) and one proton is irradiated, the signal of another proton will be enhanced. On the contrary, if two protons are a long way from each other in space, there is no NOE.

2.8 Dynamic Nuclear Magnetic Resonance

2.8.1 Description of Dynamic Nuclear Magnetic Resonance

DNMR (dynamic nuclear magnetic resonance) is an independent branch of nuclear magnetic resonance spectroscopy, which is used to study particular dynamic processes by NMR to obtain kinetic and thermodynamic parameters. Two theoretical treatments can be applied to this topic. The first is a classical treatment based on the Bloch equation [5, 10]. The second is a density matrix method based on quantum mechanics [11]. In the present monograph only the basic concepts of DNMR will be presented.

Every analytical instrument has its own time scale, which corresponds to the shutter rate for a camera. When a natural process occurs at a rate much higher than the time scale of the instrument, an average result of the process is recorded by the instrument. When a natural process takes place at a rate much lower than the time scale of the instrument, an instantaneous situation of the process is recorded. The dimension of the time scale is the second while the dimension of frequency is s^{-1}. Therefore, the time scale and frequency are reciprocals of each other. The frequency of infrared spectroscopy lies in the region of 10^{13}–10^{12} Hz while the frequency of NMR spectrometers is 10^{8}–10^{7} Hz. Therefore, the time scale of NMR is lower than that of IR by several orders of magnitude. In fact, the time scale of NMR is less than $1/10^8$ s. In dynamic processes, the difference between the δ values of two functional groups, which can be removed in DNMR, is measured by the difference between the resonant frequencies, $\Delta\nu$. Under these conditions, the corresponding time scale is $1/\Delta\nu$, which is of the order of milliseconds. Therefore, the extent of variation of some dynamic processes can cover a range from a slow process to a rapid process with respect to the time scale of NMR. As a result, dynamic processes can be studied satisfactorily by NMR.

One more word about the time scale of NMR. According to the author's opinion, the time scale for 2D NMR is frequently $1/(2J)$ where J is the associated coupling constant.

If a rapid process is measured by NMR, the measured parameters (chemical shifts and coupling constants) are average values for the period of the measurement.

Dynamic processes include intramolecular hindered rotations, chemical exchange reactions, etc. The intramolecular hindered rotation of **C2-20** will now be taken as an example.

C2-20

Because the C–N bond has some of the properties of a double bond, the rotation about the C–N bond is hindered. Therefore, the two methyl groups can not be chemically equivalent through the rotation around the C–N bond. At room temperature, these two methyl groups show two separated singlets. However, the potential barrier, which corresponds to the activation energy in a chemical reaction, will gradually decrease with an increase in the sample temperature. As a result, the intramolecular rotation around the C–N bond will accelerate. If the sample temperature is high enough, these two methyl groups will give only one peak. Therefore, if the sample temperature changes from a low temperature to a high temperature, these two methyl groups will have the spectra shown in Fig. 2.20 a–h.

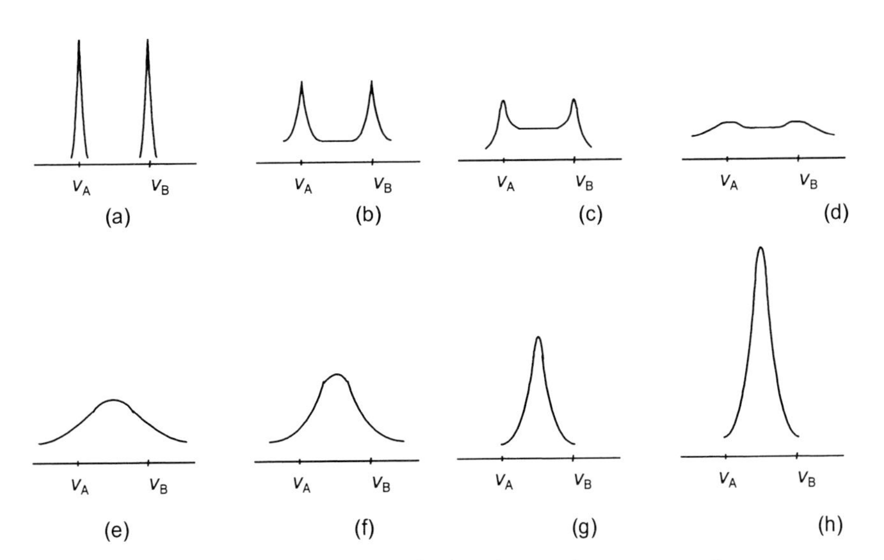

Fig. 2.20 Line shape changes of NMR signals when the sample temperature changes.

From Fig. 2.20 it can be seen that at a very low or at a very high temperature, the peak shape of the two methyl groups is sharp. At moderate temperatures they show two blunt peaks. At a particular temperature (Fig. 2.20e), two blunt peaks just join to form a blunt peak. This temperature is called the coalescence temperature, T_C, which is an important parameter in a DNMR experiment. Thermodynamic and kinetic parameters can be calculated from T_C.

The following processes are similar to that for the intramolecular hindered rotation.

1. Conformation exchanges, such as the reversal of a six-membered ring, which leads to an interchange between an equatorial hydrogen atom and an axial one.
2. Isomerization reactions.
3. Chemical exchange reactions and others [12].

2.8.2 Spectral Peak of Reactive Hydrogen Atom (OH, NH and SH)

The discussion above does not belong to the category of identification of organic compounds. However, the related concepts are useful for the interpretation of ^1H spectra. When an interchange reaction has a rapid rate, such as

$$RCOOH_{(a)} + HOH_{(b)} \leftrightarrow RCOOH_{(b)} + HOH_{(a)}$$

the following equation holds:

$$\delta_{observed} = N_a \times \delta_a + N_b \times \delta_b \qquad (2\text{-}36)$$

where $\delta_{observed}$ is the observed (averaged) chemical shift of the reactive hydrogen atoms in the system; N_a and N_b are the molar fractions of (a) and (b), respectively; δ_a and δ_b are the chemical shifts of (a) and (b), respectively.

From Eq. (2-36) it can be seen that the aqueous solution of the acetic acid only gives a peak of reactive hydrogen atoms although there are water and acetic acid molecules.

If there are several types of reactive hydrogen atoms, such as hydroxyl, amino and carboxyl, in a system, and rapid interchange reactions take place, all types of reactive hydrogen atoms also only give one peak. Equation (2-36) can be written in a general form:

$$\delta_{observed} = \sum N_i \times \delta_i \qquad (2\text{-}37)$$

where N_i is the molar fraction of the i-th reactive hydrogen atom; δ_i is the chemical shift of the i-th reactive hydrogen atom.

Examples of common functional groups are –OH, –NH and –SH. The order of their rates of chemical interchange reactions is OH > NH > SH. The hydrosulfide group, SH, does not show a rapid interchange reaction under commonly used experimental conditions. When only a rapid interchange reaction takes place, all reactive hydrogen atoms show just one peak, the chemical shift of which is calculated according to Eq. (2-37). In addition, the coupling between the reactive hydrogen atom and its adjacent functional group is removed. As a result, both the reactive hydrogen atom and its adjacent functional group do not show a coupling split through the coupling between them, which should be noticed when interpreting ^{1}H spectra.

Since hydroxyl and amino groups exist in many organic compounds, their peaks will be discussed in detail as follows:

1. Hydroxyl group OH

The rates of the interchange reactions of alcohol, phenol and carboxyl functional groups are high. Therefore, these functional groups show a non-blunt peak under the typical conditions used. Because these functional groups can associate with each other due to the hydrogen bond, the δ values of these functional groups can vary over a large range, which is connected with the experimental conditions, such as the concentration of the sample to be measured, temperature, solvent, etc. If the sample to be measured is very pure and not contaminated by the trace acid or base that is present as the catalyst for the interchange reactions, a slow interchange reaction will occur for the hydroxyl group. Both the hydroxyl group and its adjacent functional group will show coupling splits by the ^{3}J coupling between them.

After an NMR measurement, several drops of D_2O are added to the sample solution, which is then shaken and measured again. If the sample contains a hydroxyl group, its peak will be removed because the hydrogen atoms of the hydroxyl group are replaced by deuterium atoms. This is the most reliable method of confirming the existence of a hydroxyl group (alcohol, phenol and carboxylic acid), and it is better than the measurements by IR or MS. Besides D_2O, deuterated acetic acid can be used in this test. In this case, the signal of the hydroxyl group will shift in the downfield direction.

When deuterated dimethyl sulfoxide is used as the solvent, the hydroxyl group of the sample will associate strongly with the solvent to reduce greatly the rate of the interchange reaction. As a result, the measured ^{1}H spectrum has the following advantages:

i. Two separated signals, that of the hydroxyl group and that of water (existing in the sample or in the solvent), can be recorded.
ii. If the sample is a polyol, every hydroxyl group will be recorded separately.
iii. If the solution is neutral, the peak of the hydroxyl group will be split by its adjacent functional group, so that primary, secondary and tertiary alcohols can be differentiated.

2. Amino group NH

The peak shape of NH is affected by two factors: the rate of the interchange reaction and that of quadrupole moment relaxation.

A primary amine, $-NH_2$, will be taken as an example. Firstly, the interchange reaction will be considered. If the reaction occurs rapidly, the amino group will show a sharp peak (the effect of the quadrupole moment relaxation is not considered at this time). If the reaction occurs slowly, the amino group will show three peaks with an intensity ratio of 1:1:1 as the spin quantum number I of ^{14}N (its natural abundance is 99.6%) is 1 so that $2nI+1=3$.

The effect of the quadrupole moment relaxation is considered next. If a nucleus has a quadrupole moment, the nucleus has an uneven charge distribution on its surface. As a result, it will have a specific relaxation mechanism.

The sample molecule tumbles rapidly in the sample solution. If the distribution of extranuclear electrons of the sample molecule is not spherically symmetrical, the tumbling of the molecule will produce a fluctuating electric field, which produces a moment of force that acts on the nucleus with a quadrupole moment so as to change the orientation of the nucleus. As a result, the nucleus is relaxed. Namely, if the distribution of extranuclear electrons of the sample molecule is not spherically symmetrical and the charge distribution on the surface of the nucleus is not symmetrical either, a specific relaxation, quadrupole moment relaxation, will take place.

If this mechanism is very effective, the nucleus relaxes very rapidly so that the nucleus does not produce a coupling effect on its adjacent nuclei, as if it were not a magnetic nucleus. ^{35}Cl, ^{37}Cl, ^{79}Br, ^{81}Br and ^{127}I belong to this case. They do not split their adjacent functional groups, behaving as if they were not magnetic nuclei.

If this mechanism is not effective, the nucleus will split its adjacent functional group.

Therefore, if the ^{14}N nucleus has a rapid quadrupole moment relaxation, it does not split the peak of the hydrogen atom of NH, which should show a sharp peak (with the effect of an interchange reaction not being considered at this time). If the ^{14}N nucleus has a very slow quadrupole moment relaxation, the peak of the hydrogen atom of NH should show three sharp peaks. If the ^{14}N nucleus is in the middle position between the above two limits, the peak of the hydrogen atom of NH should show a blunt peak.

If two factors, the rate of the interchange reaction and that of the quadrupole moment relaxation, are considered at the same time, the peak shape of NH could be sharp or blunt. From the point of view of statistics, NH frequently shows a blunt peak. No matter what the peak shape is, the peak of NH can be removed by isotopic exchange with D_2O.

2.9 Interpreting ^{1}H NMR Spectra

The structure of an unknown compound can be deduced mainly on the basis of NMR data. If an unknown structure is not complex, it is possible to deduce the structure by a combination of its H spectrum, ^{13}C spectrum and molecular formula. An ^{1}H spectrum can provide abundant structural information so that it plays an important role in deducing an unknown structure or selecting the most likely structure from several possible structures. The peak shapes resulting from spin split should be emphasized in the interpretation of ^{1}H spectra. This will be discussed in this section in detail.

2.9.1 Sampling and Measurement

1. Prepare a sample solution with an appropriate solvent

Deuterated solvents are used for the preparation of the sample solutions to be measured. Deuterium nuclei do not produce signals in ^{1}H NMR spectra and they are also used for locking the magnetic field during the measurement.

The principal consideration for solvent selection is the solubility of the measured sample in the chosen solvent. $CDCl_3$ is used most frequently because it is cheap and it can dissolve many samples, except for highly polar compounds for which D_2O, acetone-d_6, etc. are appropriate.

Some particular samples require such solvents as benzene-d_6, dimethyl sulfoxide-d_6, dimethyl formamide-d_7, pyridine-d_5, etc.

The prepared solution should be low in viscosity, otherwise the resolution of the obtained ^{1}H spectrum will deteriorate.

2. Measurement

Recording a ^{1}H spectrum is a single pulse experiment in which two parameters, frequency width and scan number, are important. If the frequency width that is set is not sufficient, some signals (e.g., those of the carboxylic acid, associated phenol, etc.) situated beyond the spectral width will fold into the spectrum to give signals with the wrong chemical shift values. A sufficient scan number guarantees an appropriate S/N ratio.

An integration curve (or a set of integrated values for every peak set) gives the proportion of hydrogen atoms in every peak set.

If the sample to be measured is considered to contain the hydrogen atoms connected to a heteroatom, by shaking the solution with drops of D_2O and measuring it again, one can verify the existence of the hydrogen atoms.

2.9.2 Steps for ^{1}H Spectrum Interpretation

1. Find impurity peaks, solvent peak and spinning side-bands

The amounts of impurities should be much less than that of the sample to be measured. Therefore, the peaks of impurities can be discerned from those of the sample.

Deuterated solvents can not be 100% deuterated isotopically (e.g., only from 99 to 99.8%). The remaining trace hydrogen atoms will produce corresponding peaks. They can be found from their particular δ values, for example, the solvent peak of $CDCl_3$ appears at about 7.26 ppm.

If the NMR spectrometer is not well shimmed, spinning side-bands can appear at both sides of strong peaks. Spinning side-bands can be discerned easily by changing the spinning rate.

2. Calculate the unsaturation number (the index of hydrogen deficiency)

The unsaturation number is the number of double bonds and rings in a molecular structure. When a compound contains C, H, O, X (halogen) and N, its unsaturation number, Ω, can be calculated from its molecular formula and the valence value of the N atoms.

When the N atoms are trivalent, we have

$$\Omega = \mathrm{C} + 1 - \frac{\mathrm{H}}{2} - \frac{\mathrm{X}}{2} + \frac{\mathrm{N}}{2} \qquad (2\text{-}38)$$

where C, H, X, and N are the numbers of carbon, hydrogen, halogen and nitrogen atoms in the molecule, respectively.

When the N atoms are pentavalent, we have

$$\Omega = \mathrm{C} + 1 - \frac{\mathrm{H}}{2} - \frac{\mathrm{X}}{2} + \frac{3}{2}\mathrm{N} \qquad (2\text{-}39)$$

3. Determine the number of hydrogen atoms in every peak set in the ^{1}H spectrum

According to the molecular formula and the integrated value of each peak set in the ^{1}H spectrum, the distribution of the hydrogen atoms in each peak set can be calculated. If the molecular formula has not yet been determined, it is possible to determine the numbers of hydrogen atoms corresponding to the peaks on the basis of particular peaks, for example, those of the methoxy group or hydroxyl group, and so forth.

If the ^{1}H spectrum is complex, it is difficult to determine the distribution of hydrogen atoms just by integration values. In this case, other NMR data (e.g., DEPT, H,C-COSY) should be used.

4. Consideration of molecular symmetry

The number of peak sets in a ^{1}H spectrum decreases when the molecule to be measured has a symmetrical structure. The hydrogen atoms, which are symmetrical in the structure, have superimposed peak sets at the same chemical shift value.

5. Inspection of every peak set about its δ and J values

If the molecule has no symmetrical plane or if it has a symmetrical plane that does not bisect the related bond angle, its ^{1}H spectrum may be much more complex than has been estimated.

Inspecting every peak set in an ^{1}H spectrum is very important either for postulating an unknown structure or for the assignment from the spectrum.

From the experience of this author, analyzing coupling splitting patterns of every peak set in an ^{1}H spectrum is the most important thing for ^{1}H spectrum interpretation because the information obtained from the coupling is more important and more reliable than that from the δ value, which may have some exceptions.

Because of the wide application of high-frequency NMR spectrometers, it is common that peak sets in ^1H spectra can be analyzed as first-order spectra, in which case a peak set consists of split lines separated by coupling constants. Therefore, in the peak set, some pairs, all of which are separated by a particular space, can be found. This space, the distance between two related lines, corresponds to a coupling constant. Several such distances can be found in a peak set, which means that the peak set is split by these coupling constants. As a coupling exists in two coupled peak sets, the distance that corresponds to this coupling constant can be found in both related peak sets. Therefore, the key to analyzing a ^1H spectrum is to establish which are the pairs of equally spaced peaks both in a peak set and in two coupled peak sets. The fact that two peak sets contain the same distances between two related peaks means that the two peak sets are coupled to each other with the coupling constant equal to the distance in hertz. If there are several distances in a peak set, the author suggests finding these distances by going from a shorter to a longer one. Some examples will be used to illustrate this later.

6. Construct possible structures from deduced structural units

According to chemical shift values and coupling relations between peak sets (split by several coupling constants) or multiplets (split just by one coupling constants), the deduced structural units of an unknown compound can be assembled into several possible structures, which should agree with the ^1H spectrum.

7. Assign the ^1H spectrum on the basis of the deduced structures

Assignment of a ^1H spectrum implies analyzing every peak set or multiplet which should have an appropriate δ value and reasonable coupling splittings, as compared with the deduced structures. From the assignment, the most reasonable structure can be selected from the possible structures. The assignment is a necessary step for postulating a structure from a ^1H spectrum. Sometimes, the assignment is used to confirm a postulated structure. Some examples will be given later.

It is worth emphasizing the analysis of coupled splittings. If there are some discrepancies between the two postulations, the result obtained from the coupled splitting analysis deserves to be considered first. This is because a δ value can not be estimated precisely from any calculation or found from the same structural unit in a similar chemical environment. What is more, there are exceptions to δ values, but there are few exceptions to coupling splittings.

In addition to the examples presented in Section 2.9.3, some examples showing the importance of the analysis of coupled splittings are given below:

i. If there are three peak sets in the same vicinity of an ^1H spectrum for a substituted benzene: *d*, $J \approx 8$ Hz; *d*, $J \approx 2$ Hz; and $d \times d$, $J \approx 8$, 2 Hz, the structural unit of a 1-, 2-, 4-substituted benzene can be deduced. These three peak sets are ascribed to the 6-, 3- and 5-hydrogen atoms, respectively.

ii. Three structural units, a pyridine ring with two substituents, $-NH_2$ and $-CH_3$, have been postulated for an unknown structure. On the basis of the split pattern of the substituted pyridine, the three remaining hydrogen atoms are determined to be at the 4-, 5- and 6-positions of the pyridine. Because long-range coupling is shown in the ^{1}H spectrum, it can be concluded that the methyl is substituted at the 3-position. If the methyl is substituted at the 2-position, there is no splitting from long-range coupling in the ^{1}H spectrum since the splitting from ^5J can not be found.

iii. In an ^{1}H spectrum there are two singlets that correspond to two structural units: –C–CH–C– and –OH. On the basis of the analysis of the peak shapes, the sharp singlet is assigned to the peak of the former and the blunt one to that of the latter.

2.9.3 Examples of ^{1}H Spectrum Interpretation

Example 1

An unknown compound has a molecular formula of $C_6H_4OCl_2$. The downfield region of its ^{1}H spectrum is shown in Fig. 2.21. There is a singlet at 5.89 ppm in the upfield region. The singlet disappears after the sample solution is shaken with D_2O. Try to postulate what its structure is.

Solution

As the molecule contains an oxygen atom, the singlet at 5.89 ppm should be the peak of a hydroxyl group.

The unsaturation number of the unknown compound, 4, is calculated from the molecular formula, which implies the existence of a benzene ring or a heteroaromatic ring. As the molecule contains no heteroatom except for the hydroxy group, the unknown compound should contain a benzene ring. Thus its structure is **3**:

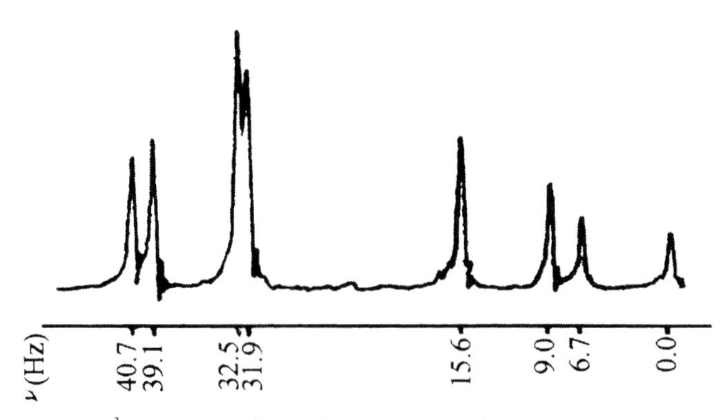

Fig. 2.21 ^{1}H spectrum of an unknown compound.

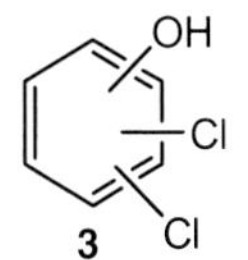

All that remains is to determine the positions of the three substituents on the benzene ring.

Fig. 2.21 shows a typical pattern of an AB_2 system, which indicates the symmetry of the substituted benzene ring. Therefore, only two structures are possible: 2,6-dichlorophenol and 3,5-dichlorophenol. The calculation of J_{AB} can help in the correct structure being chosen from the two possible structures.

Recall Eq. (2.23):

$$J_{AB} = \frac{1}{3}[(\nu_1 - \nu_4) + (\nu_6 - \nu_8)]$$

Substituting the values shown in Fig. 2.22 into the above equation, we obtain

$$J_{AB} = \frac{1}{3}[(15.6 - 0) + (40.7 - 32.5)] \text{ Hz}$$
$$= 7.93 \text{ Hz}$$

Therefore, the unknown compound is 2.6-dichlorophenol, structure 4:

Example 2

An unknown compound has a molecular formula of $C_{10}H_{12}O_3$, with the data of its ^{1}H spectrum shown in Tab. 2.10. Try to postulate what its structure is.

Tab. 2.10 Data from the ^{1}H spectrum of an unknown compound

δ (ppm)	*11.35*	*8.08*	*7.85*	*6.32*	*4.31*	*2.22*	*1.39*
Splitting	s (blunt)	d	d, d	d	q	s	t
J (Hz)		1.29	8.60, 1.29	8.60	7.14		7.14
Integrated value	0.83	1.00	1.03	1.04	2.06	3.28	3.17

Solution

Firstly consider the integrated value. In a ^{1}H spectrum the ratio of a peak area to its related number of hydrogen atoms is often not a constant. The ratio calculated from a peak in the aliphatic hydrogen region is always greater than that in the aromatic hydrogen region. In addition, the peak area of reactive hydrogen atoms is always smaller than its expected value. Therefore, the number of hydrogen atoms of the functional groups is 1, 1, 1, 1, 2, 3 and 3, respectively, from the left to the right in Tab. 2.10. The result agrees with the formula.

The calculated unsaturation number is 5, which means that the unknown compound contains an aromatic ring and a double bond.

From the chemical shift and the peak shape of the peak at 11.35 ppm and the presence of oxygen atoms in the compound, the peak can be ascribed to a hydroxy group.

On the basis of the peak pattern in the aromatic region of the ^{1}H spectrum, d (^3J), d (^4J) and d × d (^3J and ^4J), the structural unit of a 1-, 2-, 4-substituted benzene can be concluded (see point 7 in Section 2.9.2).

The peak with three hydrogen atoms at 2.22 ppm should be a methyl group. From its chemical shift value and its peak shape we come to the conclusion that it is one of the three substituents on the benzene ring.

The peaks at 4.31 and 1.39 ppm couple with each other. They can be deduced to be an ethyl group. As the CH_2 group has a large chemical shift value of 4.31 ppm, it should be connected to a strongly electronegative functional group. From the elemental composition of the unknown compound and its unsaturation number, it is reasonable to assume that the ethyl group should be part of the structural unit of $-COOCH_2CH_3$.

So we now know all structural units of the unknown compound. The only thing necessary to complete the unknown structure is to assemble them, that is, to determine the positions of the three substituents on the benzene ring. In such a case, analysis of the δ values of the remaining hydrogen atoms of the benzene ring is very important. Note that only one hydrogen atom has an upfield shift, which means that only one hydrogen atom is adjacent to the secondary functional group, OH (see Section 2.5.1). On the other hand, two hydrogen atoms have a downfield shift, which means that two hydrogen atoms are adja-

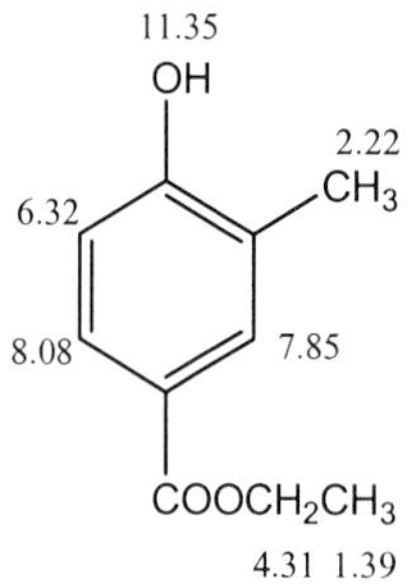

11.35
OH
2.22
CH_3
6.32
8.08
7.85
$COOCH_2CH_3$
4.31 1.39

5

cent to the tertiary functional group, $-COOCH_2CH_3$. Therefore, the unknown structure can be deduced to be **5**.

Example 3

An unknown compound has a molecular formula of $C_8H_{18}O_2$. Its ^1H spectrum is shown in Fig. 2.22, with a peak area ratio of 2:3:1:12 from left to right. Try to postulate what its structure is .

Solution

To begin with, the peaks from 3.30 to 3.62 ppm will be interpreted. At first glance it appears that there is a quartet in this region. However, this interpretation can be ruled out for the following two reasons. Firstly, the intensities of the "quartet" do not have the ratio of 1:3:3:1 as binomial coefficients. Secondly, the peak interval of the quartet does not appear in any other region. It is known that a peak interval corresponding to a coupling constant exists in two coupled peak sets. Therefore, these four peaks should be an AB system pattern with J_{AB}=11 Hz in a ^1H spectrum recorded with a 100 MHz spectrometer. This AB system results from an isolated non-chemically equivalent CH_2 group. The doublet with a small peak separated by 2.5 Hz at 3.36 ppm corresponds to a hydrogen atom. From the peak interval of 2.5 Hz, a multiplet at 1.81 ppm coupled to the doublet can be found. Another peak interval of 7 Hz in the multiplet at

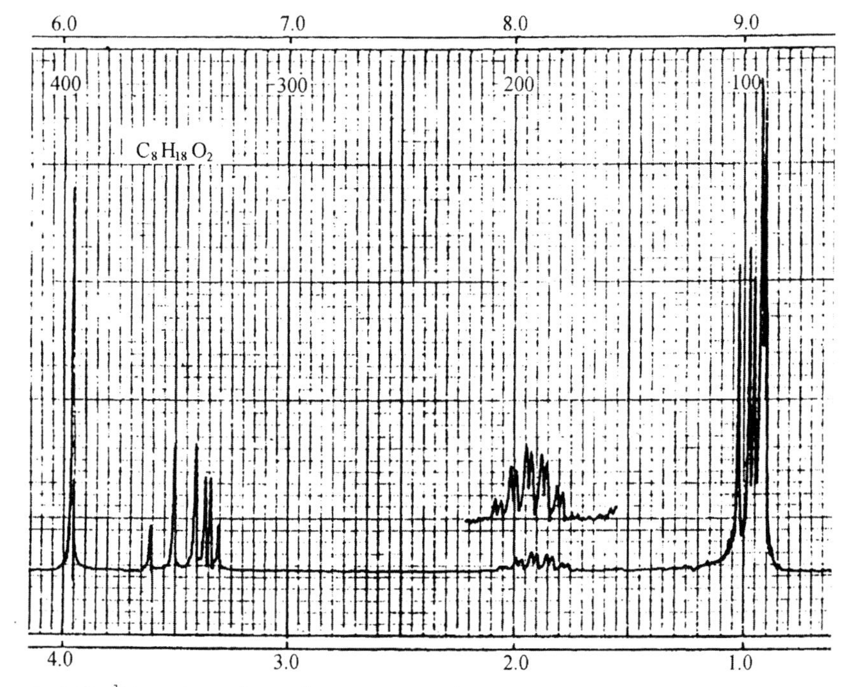

Fig. 2.22 ^1H spectrum of an unknown compound.

1.81 ppm corresponds to another coupling constant, which appears in the peak sets at about 0.95 ppm, consisting of five peaks and corresponding to 12 hydrogen atoms. These five peaks are denoted as 1 to 5 in the direction of from left to right. The above-mentioned 7 Hz corresponds to the distance from 1 to 3 or from 2 to 5. From the heights of peaks 4 and 5, we know that these two peaks are two singlets resulting from two isolated methyl groups.

On the basis of the interpretation above, the structure of the unknown compound can be postulated as **6**.

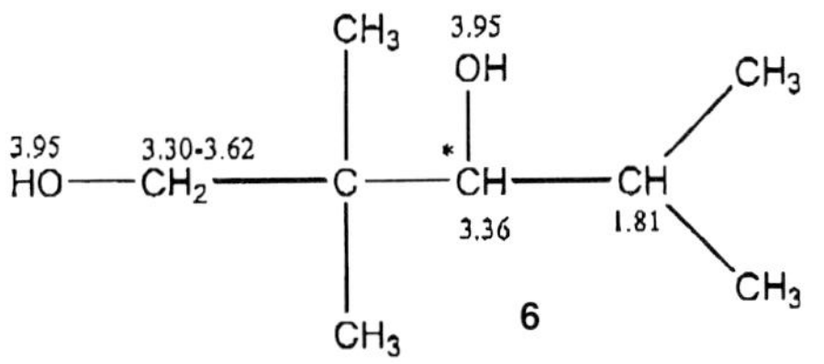

CH_3 3.95 OH CH_3

3.95 3.30-3.62 *

$HO—CH_2—C—CH—CH$

3.36 1.81

CH_3 6 CH_3

Because the molecule contains a chiral carbon atom (noted by "*") and the molecular structure does not have a symmetry plane, two methyl groups connected to the quaternary carbon atom are not chemically equivalent, neither are the two hydrogen atoms of the methylene group.

Two hydroxyl groups have the same δ value and a narrow line shape because of their rapid exchange reaction.

Example 4

Compound **C2-19** has the ^1H spectrum, as shown in Fig. 2.18 in Section 2.7.1. The enlarged peak sets from Fig. 2.18 are shown in Fig. 2.23. Try to assign these peaks.

Solution

In this example, only coupling splitting patterns are interpreted.

1. CH_2=CH*–CH–, (H_a)
   H_a has the splitting pattern of d × d × t, which corresponds to three coupling constants: J_{trans}, J_{cis} and ^3J. As described above, the smallest coupling constant, ^3J, will be differentiated from the two other coupling constants first. The four central lines of triplets are denoted with a point, ".". The peak interval in every triplet corresponds to ^3J. The distance between the first and second points or between the third and fourth points corresponds to J_{cis}. The distance between the first and third points or between the second and fourth points corresponds to J_{trans}.
2. H*$_2$C=CH–, (H_b and H_c)
   There are three coupling relationships for H_b: between H_b and H_a with J_{trans}; between H_b and H_c with ^2J; and between H_b and H_d, H_e with ^4J. Thus the coupling splitting pattern of H_b is d×d×t. The ^4J is the smallest coupling

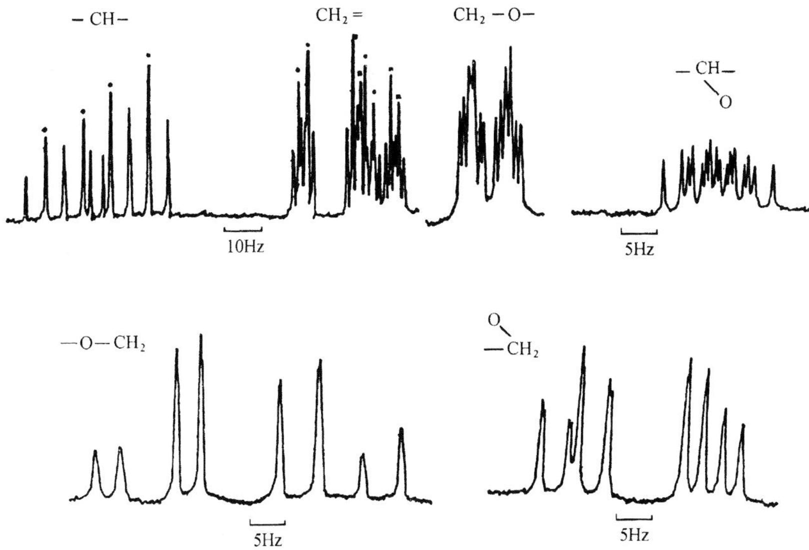

Fig. 2.23 Enlarged ^{1}H spectrum of **C2-19**.

constant among the three coupling constants. Therefore, the central lines of triplets from ^{4}J are marked first.

The interpretation of H_c is similar to that of H_b. Its central lines of triplets from ^{4}J should be marked at the same time.

H_b and H_a have a common coupling constant J_{trans}, which is read from the peak set of H_a. With this distance, the peak set of H_b can be found, which is denoted as ".". Similarly, the peak set of H_c can be found through the coupling constant J_{cis}, which is denoted as " *".

The distance between two adjacent "." or two adjacent " *" gives the coupling constant for ^{2}J.

3. H_2C=CH–CH*$_2$–O–, (H_d and H_e)
These are split into a doublet by H_a with a coupling constant ^{3}J, and then further split into a triplet by H_b and H_c with a small coupling constant ^{4}J. Because H_d and H_e are chemically non-equivalent, the peak set (d × t) has a still smaller splitting.
4. –O–CH*$_2$–CH (H_f and H_g)
Because the methylene group is connected to the chiral carbon atom, the two hydrogen atoms are chemically non-equivalent. Their peak sets can be approximately interpreted as an AB system, which is split further by H_h with two coupling constants $^{3}J_{fh}$ and $^{3}J_{gh}$, respectively. The effects of long range couplings can also be noted from a tiny splitting at the peak tops.
5. H_h
H_h couples four vicinal hydrogen atoms with four different ^{3}J values, respectively. Therefore, the coupling splitting pattern of H_h is d × d × d × d. The 16

peaks are denoted as 1 to 16 from right to left. The four coupling constants corresponding to four particular peak intervals are as follows:

1–2, 3–6, 4–7, 5–9, 8–12, 10–13, 11–14, 15–16;
1–3, 2–6, 4–8, 5–10, 7–12, 9–13, 11–15, 14–16;
1–4, 2–7, 3–8, 5–11, 6–12, 9–14, 10–15, 13–16;
1–5, 2–9, 3–10, 4–11, 6–13, 7–14, 8–15, 12–16.

6. H_i and H_j
These are the two hydrogen atoms of a CH_2 moiety in the three-membered ring. Therefore, they are chemically non-equivalent. Their interpretation is similar to that of H_f and H_g.

Example 5

A synthesized compound has the postulated structure of **C2-21**. Its ^{1}H spectrum is shown in Fig. 2.24. The Figs. 2.25 and 2.26 are enlarged specific areas spectra of Fig. 2.24. Try to confirm the structure.

OH
10
6
5
8
7
9
2
Cl
Cl

C2-21

Solution

Firstly let us interpret Fig. 2.24. The multiplet at about 2.05 ppm is the peak for the solvent acetone-d_6. The peaks at 2.88 and 3.72 ppm are impurity peaks. A 1,2,4-trisubstituted benzene ring can be easily confirmed by the following three

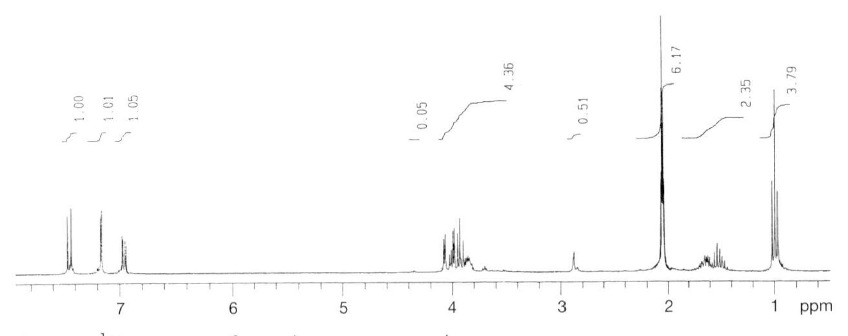

Fig. 2.24 ^{1}H spectrum of an unknown compound.

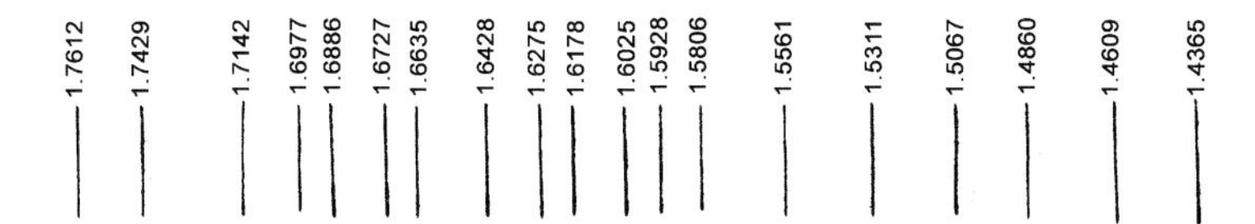

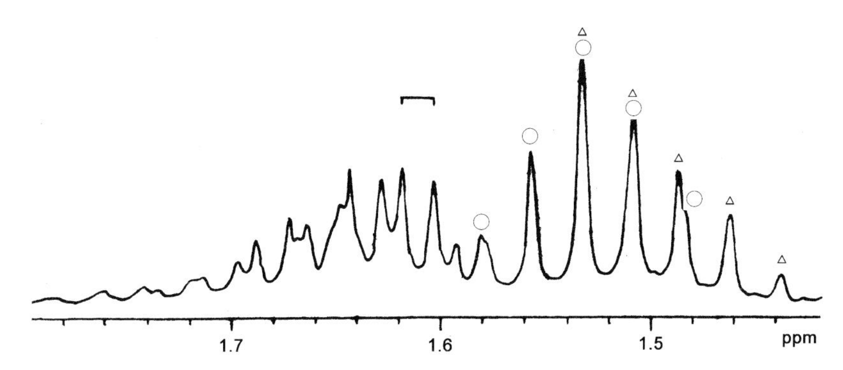

Fig. 2.25 Enlarged spectra of a specific section of Fig. 2.24.

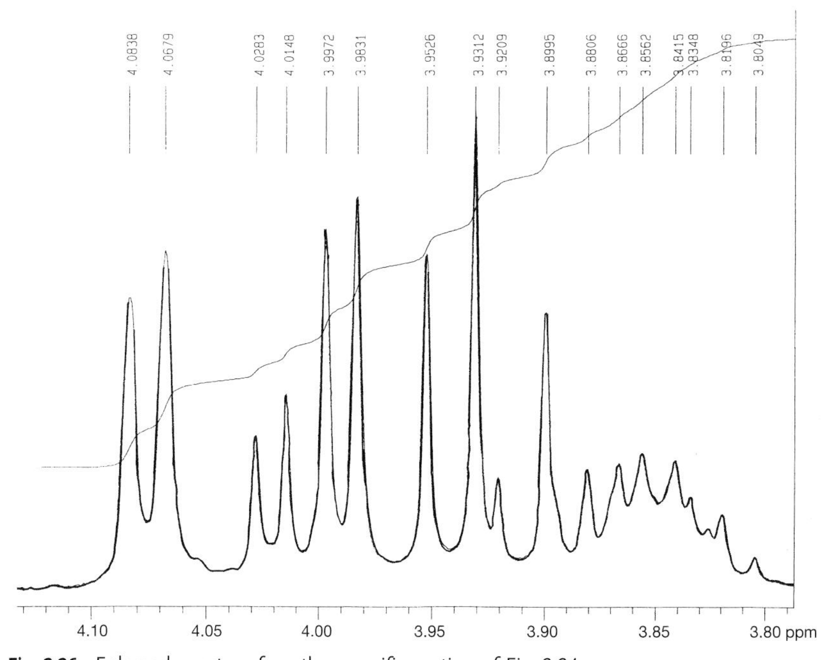

Fig. 2.26 Enlarged spectra of another specific section of Fig. 2.24.

peak sets: the doublet (d, 8.97 Hz, 7.45 ppm), the doublet (d, 2.94 Hz, 7.17 ppm) and the peak set (d × d, 8.79 Hz, 2.94Hz, 6.96 ppm). The three peak sets correspond to 5-, 2-, and 6-hydrogen atoms on the benzene ring, respectively. Their chemical shift values are reasonable. The two *ortho*-hydrogen atoms of the substituent, OR, have smaller δ values. The triplet at 0.99 ppm can be assigned to the terminal methyl group. The assignments of two peak sets at about 1.6 and 3.9 ppm are rather difficult.

Next let us interpret Fig. 2.25, which is the enlarged spectrum from the peak sets at about 1.6 ppm in Fig. 2.24. From the integration value it is known that all of these peaks correspond to two hydrogen atoms. Two partially overlapped quintets can be discerned. One quintet, denoted by "o", covers 1.58 to 1.48 ppm. Another quintet, denoted by "Δ", covers 1.53 to 1.44 ppm. The coupling constant for each quintet is about 7.32 Hz, which corresponds to a normal ^3J in a chain alkyl group.

The coupling constant measured between the two central lines of the two quintets is about 13.53 Hz, which is the typical value for two *geminal* hydrogen atoms. To summarize, the two quintets can be assigned to a hydrogen atom of the methylene group at the 9-position. These two hydrogen atoms are chemically non-equivalent because they are connected to the chiral carbon atom. Therefore, one of the two hydrogen atoms forms half of an AB system (with the ^2J about 13 Hz). Then these two peaks are further split into two quintets when the ^3J between CH and the H is equal to the ^3J between the methyl and the H.

Now the peaks from 1.58 to 1.71 ppm are analyzed. Firstly, seven peak intervals with the same distance denoted by "┌──┐" can be found, and, then, two quartets.

Another H of the methylene group which forms another half of the AB system is coupled by the methyl group and the methane group. Because the two coupling constants are different, another H of the methylene group shows a q × d pattern. The ^2J value, 13.71 Hz, can be found from the two quartets.

Fig. 2.26, the enlarged spectrum of the peak sets at about 3.9 ppm in Fig. 2.24, will now be interpreted. From the integration values it is known that these peaks result from four hydrogen atoms. The left ten peaks are denoted as 1–10. There are four equal distances: 3–5, 4–6, 7–9 and 8–10. This distance corresponds to a rather smaller ^2J of 9.5 Hz. Thus these eight peaks should be those of a methylene group. It should be the methylene group at the 7-position. The two hydrogen atoms have different coupling constants with H-8. Therefore, we can find two peak intervals: 3–4 (or 5–6) and 7–8 (or 9–10). Because the δ value of H-8 is close to those of H-7 and H-7′, the coupling splitting pattern of H-7 is not like the X part in an ABX system, which covers from 3.89 to 3.80 ppm. Peaks 1 and 2 correspond to one hydrogen atom. They are assigned to the hydroxyl group. This argument is confirmed by D_2O exchange.

To sum up, it can be confirmed that the synthesized compound has the expected structure. From this example, it is seen that compounds without a complicated structure may have a complicated ^1H spectrum.

Example 6

This example is one part of the identification of an unknown compound. It has been chosen as an example of the interpretation of ^1H spectra because the importance of analyzing peak shape can be understood further from this example.

From related spectra, structural units **7**, **8** and **9** of an unknown compound have been determined. Try to assemble them to complete the unknown structure from the aromatic region of its enlarged ^1H spectrum.

7

$-C(=O)-N(CH_2CH_2CH_2CH_3)-C(=O)-$

8

$-OCH_3$

9

Solution

To assemble the unknown structure the substitution positions of the two substituents with respect to the naphthalene group have to be determined. There are many possibilities for the substitutions. However, the most reasonable structures (maybe only one) can be found by analysis of the peak shapes and chemical shift values of the remaining hydrogen atoms of the naphthalene group.

Related data of Fig. 2.28 can be summarized in Tab. 2.11.

From Tab. 2.11, two spin systems I–IV–III and II–V can be determined. It is clear that I, IV and III correspond to three adjacent hydrogen atoms of one ring of the naphthalene group and II and V correspond to two adjacent hydrogen atoms of the other ring of the naphthalene group.

Because the two hydrogen atoms that are situated in the two rings of the naphthalene group, respectively, have rather large δ values, it is reasonable that both rings of the naphthalene group should be substituted by a carbonyl group, respectively (see Section 2.5.1). In addition to the fact that only one hydrogen atom has a fairly small δ value, the substitution positions of the two substituents with respect to the naphthalene group can be determined as in structure **10**.

Tab. 2.11 Data of the ^1H spectrum of the aromatic hydrogen atoms of an unknown compound

No.	*I*	*II*	*III*	*IV*	*V*
δ (ppm)	8.6014	8.5560	8.5227	7.7009	7.0428
Splitting	$d(^3J) \times d(^4J)$	$d(^3J)$	$d(^3J) \times d(^4J)$	$d(^3J) \times d(^3J)$	$d(^3J)$

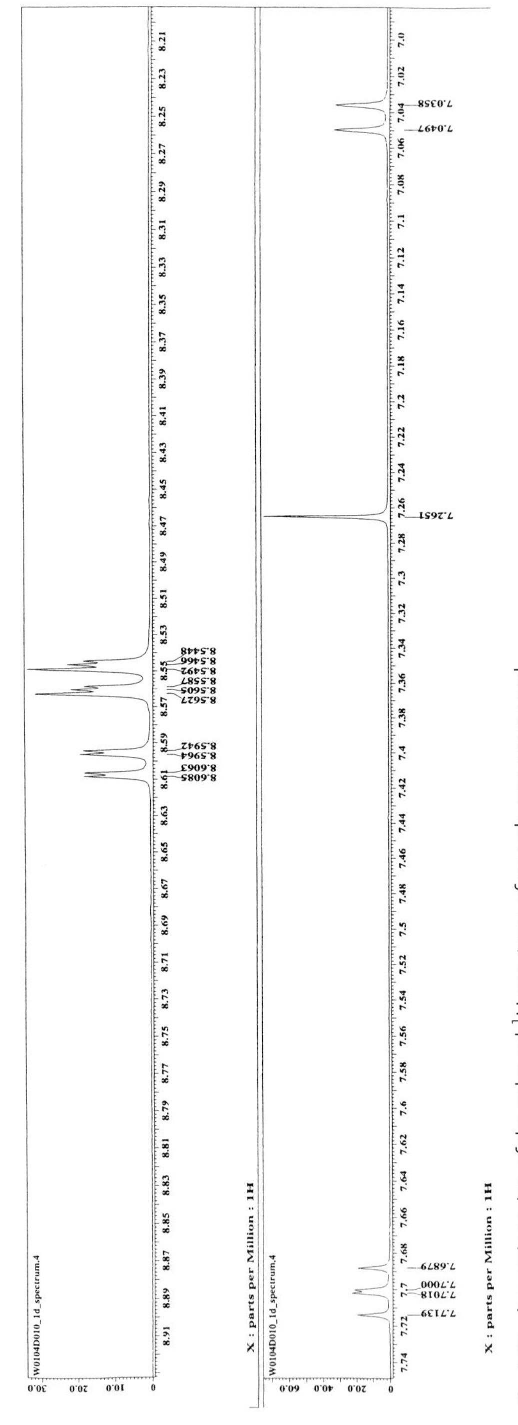

Fig. 2.27 Aromatic region of the enlarged ^{1}H spectrum of an unknown compound.

OCH_3

$O{=}C$ $C{=}O$

N

$CH_2CH_2CH_2CH_3$

10

In Example 8 of Section 8.4, the whole process for the identification of this compound will be given.

2.10 References

1 H. Günther, *NMR Spectroscopy, An Introduction*, Wiley, New York, 1980.

2 J.T. Clerc et al., *Structural Analysis of Organic Compounds*, Budapest, Akademiai kiado, 1981.

3 Yuxin Hua, Yongcheng Ning, *J. Spectrosc.* (Chinese version), **1989**, *6*, 294–299.

4 W.B. Jennings, *Chem. Rev.*, **1975**, *75*, 307–322.

5 J.A. Pople, W.G. Schneider, H.J. Bernstein, *High Resolution Nuclear Magnetic Resonance*, McGraw-Hill, New York, 1959.

6 J.W. Emsley, J. Feeney, L.H. Sutcliffe, *High Resolution Nuclear Magnetic Resonance Spectroscopy*, Pergamon Press, Oxford, 1965.

7 N.F. Chamberlain, *The Practice of NMR Spectroscopy, with Spectra–Structure Correlation for Hydrogen-1*, Plenum, New York, 1974.

8 J.H. Noggle, R.E. Schirmer, *The Nuclear Overhauser Effect*, Chemical Application, Academic Press, New York, 1971.

9 D. Newhaus, M. Williamson, *The Nuclear Overhauser Effect in Structural and Conformational Analysis*, VCH, Weinheim, 1989.

10 H.S. Gutowsky et al., *J. Chem. Phys.*, **1956**, *25*, 1228–1234.

11 J.I. Kaplan, G. Fraenkl, *NMR of Chemically Exchanging System*, Academic Press, New York, 1980.

12 J.B. Lambert, H.F. Shurvell, L. Verbit, R.G. Cooks, G.H. Stout, *Organic Structural Analysis*, MacMillan, London, 1976.

3 ^{13}C NMR Spectroscopy

3.1 Introduction

3.1.1 Advantages of ^{13}C NMR Spectra

1. Carbon atoms form the skeleton of an organic molecule, hence information about them is very useful for the identification of the structure of an unknown compound.

   A benzene ring substituted by six substituents, a saturated carbon atom connecting four functional groups, etc., contains no hydrogen atoms so that it has no ^{1}H signals. However, a ^{13}C NMR spectrum can provide structural information. Similarly, the presence of some functional groups, that contain no hydrogen atoms but do contain carbon atoms, such as carbonyl groups, can be identified from their ^{13}C NMR spectra.
2. The chemical shift region of ^{1}H spectra for common functional groups is about 10 ppm while that of ^{13}C NMR spectra is about 200 ppm, and can be even up to about 600 ppm for some synthetic compounds. As a result, a slight variation in the structure of a sample can be seen in its ^{13}C NMR spectrum.
3. In ^{13}C NMR spectroscopy, several decoupling methods and several pulse sequences for the determination of the orders of carbon atoms have been developed. Therefore, abundant structural information can be obtained from ^{13}C NMR spectra.

The relaxation times of carbon atoms can be determined precisely since they are fairly long. Relaxation times can be used to help in the assignment of peaks in a ^{13}C spectrum, which is useful for deducing an unknown structure.

References [1–4] are listed for the reader who wishes to study the related topics further.

Structural Identification of Organic Compounds with Spectroscopic Techniques. Yong-Cheng Ning
Copyright © 2005 WILEY-VCH Verlag GmbH & Co. KGaA, Weinheim
ISBN: 3-527-31240-4

3.1.2
Difficulties in the Measurement of ^{13}C NMR Spectra

Although the advantages of ^{13}C NMR spectra have been known for a long time, the development of ^{13}C NMR spectroscopy lagged behind that of ^{1}H NMR spectroscopy for about 20 years, as a result of the following factors. Firstly, the magnetogyric ratio, γ, of ^{13}C nuclei is a quarter of that of ^{1}H nuclei. As a result, the former has a much lower sensitivity than the latter. Secondly, the natural abundance of ^{13}C is about 1% while that of ^{1}H is about 100%. Not until commercial pulsed-FT NMR spectrometers were available was ^{13}C NMR spectroscopy applied in routine analysis.

Because the sensitivity of a ^{13}C NMR spectrum is much lower than that of a ^{1}H NMR spectrum, the acquisition of the data from the former is still much slower than that of the latter.

3.1.3
^{13}C NMR Spectra

For ^{13}C NMR measurements, the first spectrum to be recorded is a complete decoupling spectrum (see Section 3.3.2). Peak heights in the ^{13}C spectrum do not quantitatively show carbon atom numbers. If a particular pulse sequence is used, quantitative information about carbon atom numbers can be obtained.

In practical applications the ^{13}C and ^{1}H spectra of a compound are complementary.

Many conclusions, such as chemical equivalence, magnetic equivalence, and so forth, which were discussed in Chapter 2, and the relevant descriptions can be applied in ^{13}C spectroscopy.

3.2
Chemical Shift

As carbon atoms form the skeleton of an organic molecule, any structural variation of an organic molecule will be evident in its ^{13}C spectrum, and this is much more evident than in its ^{1}H spectrum. The range in the variation of chemical shifts of a substituted benzene ring is only about 1.5 ppm in ^{1}H spectra but about 60 ppm in ^{13}C spectra.

Before a description of the chemical shifts of common functional groups is given, a general discussion about δ_C will be introduced.

3.2.1
Paramagnetic Shielding is the Decisive Factor for Chemical Shifts

The general expression for the screening constant σ was given in Eq. (1-15). σ_d plays an important role in the ^{1}H spectrum. However, it is σ_p that plays a key role in the ^{13}C spectrum.

Pople and Karplus calculated paramagnetic deshielding [5, 6]. Their results have been quoted in the literature:

$$\sigma_p = -\frac{e^2h^2}{2m^2C^2}(\Delta E)^{-1}\langle r^{-3}\rangle_{2p}[Q_{AA} + \sum_B Q_{AB}] \tag{3-1}$$

where the minus sign indicates that the effect of σ_p is opposite to that of σ_d: the larger the value of $|\sigma_p|$, the stronger the deshielding of the carbon atom; $(\Delta E)^{-1}$ is the reciprocal of the mean electronic excitation energy; $\langle r^{-3}\rangle_{2p}$ is the expectation value of the inverse cube of the distance between the electrons in the 2p orbital and the nucleus; Q is the bond order in the molecular orbital formalism; Q_{AA} is the contribution of the density of the electrons in the 2p orbital; and Q_{AB} is the bond order between the nucleus and its adjacent nuclei.

It should be noted that the substitution by an electronegative functional group will increase both δ_H and δ_C although the two mechanisms are totally different: diamagnetic shielding for δ_H and paramagnetic shielding for δ_C.

A general consideration of δ_C has been given in this section. The δ_C for several common functional groups will now be discussed with respect to structural and steric factors.

3.2.2
Alkanes and their Derivatives

Factors affecting δ_C

1. The electronegativity of substituents

   The electronegativity of substituents is the main feature for the δ values of chain aliphatic alkyl groups. In the case of the substitution by an electronegative functional group, the α-carbon atom will have a considerable downfield shift. The β-carbon atom will have a slight downfield shift. If a carbon atom is not connected to a heteroatom, its chemical shift will generally be less than 55 ppm. If a carbon atom is connected to heteroatoms, its chemical shift can reach 80 ppm or even more.

2. Steric effect

   i. The sizes and number of substituents

      If the hydrogen atoms of an alkane chain are substituted by alkyl groups, the chemical shift of the substituted carbon atom will increase. A set of data is listed in Tab. 3.1.

Tab. 3.1 Chemical shift of substituted carbon atoms in alkane chains

$R=CH_3$	CH_3R	CH_2R_2	CHR_3	CR_4
δ (ppm)	5.7	15.4	24.3	31.4

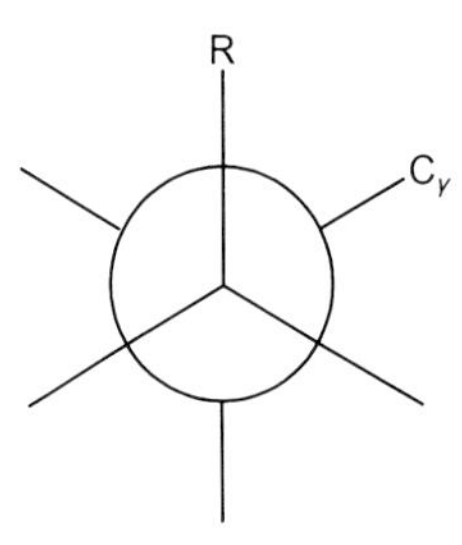

Fig. 3.1 γ-Gauche effect.

The larger the alkyl group, the greater the increase in the chemical shift of the carbon atom. The more branches the substituted carbon atom has, the greater the chemical shift.

ii. γ-Gauche effect
Any substitution decreases the δ value of the carbon atom, which is at the γ-position to the substitution. It is known that the substitution of an electronegative functional group increases the δ values of α- and β-carbon atoms; however, on the contrary, the chemical shift of the γ-carbon atom is decreased, a phenomenon called the γ-gauche effect, which can be explained by the steric effect. The alkyl chain can rotate. γ-Gauche conformation occupies about one third of the time. Under this conformation, extranuclear electrons of the γ-carbon atom are "pressed" by the substituent R so that they move towards the carbon atom, the latter having a fairly small δ value. After averaging all three conformations, the γ-gauche effect still exists (Fig. 3.1).

From the explanation above, it can be seen that δ_C is sensitive to stereochemistry and the measurement of δ_C is an important part of the study of stereochemistry.

3. Hyperconjugation effect
The substitution by a heteroatom from the second row of the Period Tab. (N, O or F) will produce a greater upfield shift for the γ-carbon atom than that by an alkyl group, which can be explained by hyperconjugation (Fig. 3.2). The chemical bond of C–X is short, where X represents one of these heteroatoms. The non-bonding electrons of the heteroatom are superimposed partially with the electrons of the γ-carbon atom, which leads to an increase in the electron density of the carbon atom so that its δ value decreases.

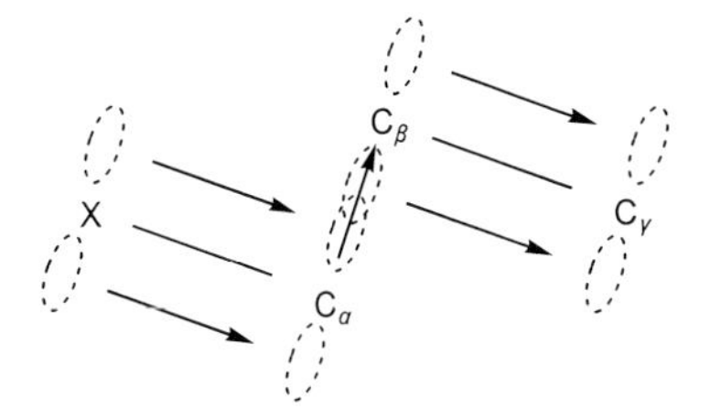

Fig. 3.2 Hyperconjugation effect.

4. Heavy atom effect
   The heavy atom effect is also known as the heavy halogen effect. After the substitution of a hydrogen atom by a halogen atom (Br or I), the substituted carbon atom will have a smaller δ value. This phenomenon can be explained as follows. The numerous electrons of the halogen atom increase the shielding effect of the substituted carbon atom so that the peak of the latter shifts upfield.

The approximate calculation of δ_C

1. Alkanes
   Two calculation methods, the Lindeman-Adams method and the Grant-Paul method [2], are used for the calculation of the chemical shifts of alkanes.
2. Substituted alkanes
   Shifts for alkanes can be calculated using empirical equations.
   The calculation of δ_C of an alkyl group is carried out in two steps. Firstly, the chemical shifts of an alkane (without substitution) are calculated, then the chemical shifts of an alkyl group.

3.2.3 Cycloalkanes and their Derivatives

The factors affecting the chemical shift of the alkyl group are still applicable to the substituted cycloalkanes. Additional information is that the γ-gauche effect is stronger than for substituted alkanes when the ring of the cycloalkane is rigid.

The chemical shifts of cycloalkanes vary only slightly from a four-membered to a 17-membered ring.

The chemical shifts of a substituted cyclohexane can be calculated.

The calculation of the δ_C value of a polycyclic alkane is difficult. Its chemical shifts are related to the geometric parameters (bond lengths and bond angles) and to its skeleton. Therefore, the chemical shifts of a substituted polycyclic alkane can only be roughly estimated on the basis of the chemical shifts of the unsubstituted polycyclic alkane.

The chemical shifts of some polycyclic alkanes are shown in structures **11**, **12**, **13** and **14**.

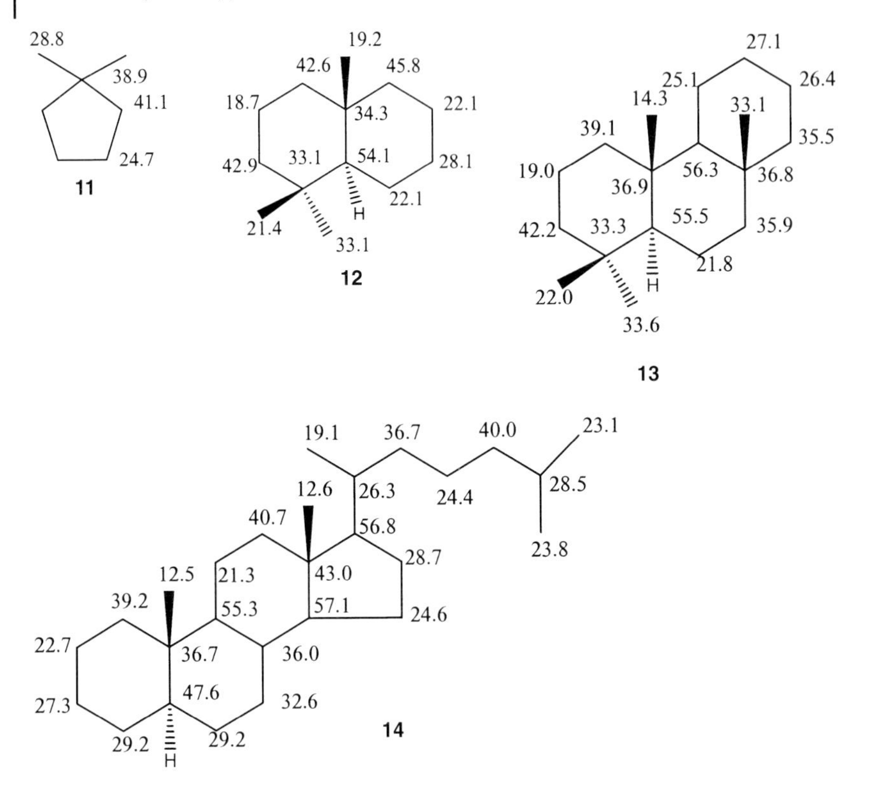

3.2.4
Alkenes and their Derivatives

Chemical shift range of alkenes and their derivatives and the factors affecting their chemical shifts

1. The chemical shift of ethylene is 123.3 ppm while those of its derivatives are between 100 and 150 ppm.

   In ^{1}H spectra, the ring current effect of a benzene ring increases considerably the chemical shifts of the hydrogen atoms in the benzene ring. However, since paramagnetic shielding plays a prominent role in ^{13}C spectra, the chemical shifts of alkenes are close to those of substituted benzenes.

2. Similar to the order of $\delta_{C}>\delta_{CH}>\delta_{CH_2}>\delta_{CH_3}$ in an alkyl group, the order of $\delta_{-C=}>\delta_{-CH=}>\delta_{CH_2=}$ exists. The chemical shift of $=CH_2$ is less than that of =C– by 10–40 ppm.

3. The chemical shifts of β-, γ-, δ- and ε-carbon atoms in an alkene are slightly larger than those of the corresponding alkane by less than 1 ppm. The chemical shift of the α-carbon atom is larger than that of the corresponding alkane by about 4–5 ppm. The Greek letters above are used to denote the position of the carbon atom to be considered with respect to the substituent. From this explanation above, it can be shown that the chemical shifts of a substituted

alkene can be estimated using its corresponding alkane, except for the α-carbon atom.

4. Conjugation effect: when two double bonds are conjugated, the chemical shifts of the two middle carbon atoms will decrease because their bond orders decrease. For example, structure **15.**

The approximate calculation of substituted alkenes

The chemical shifts of a substituted alkene are approximately the same as those of its corresponding alkane except for the carbon atoms in the double bond and its α-carbon atom.

There are two equations and corresponding parameters for the calculation of the chemical shifts of alkene carbon atoms [1, 2].

3.2.5 Benzene and its Derivatives

The chemical shift of benzene is 128.5 ppm.

If a benzene molecule is substituted, the substituted C-1 carbon atom will have a downfield shift that can have a value as great as 35 ppm. The *ortho*- and *para*-carbon atoms have considerable shifts of less than 16.5 ppm. The *meta*-carbon atoms have almost no chemical shifts.

1. Factors affecting δ values
   i. The influence of substituents on the C-1 atom is uniform. The chemical value of the C-1 atom increases in increments with the electronegativity of the substituents.
   ii. The more branches the substituent has, the larger the δ value of the C-1 carbon atom.
   iii. The substitution by a heavy atom (I or Br) leads to an upfield shift, especially by I.
   iv. Mesomeric effect. The mesomeric effect is just the resonant effect. The classification of substituents for benzene in ^{1}H spectra can also be used in ^{13}C spectra. However, it should be noted that the influence of the third type of substituents is smaller in ^{13}C spectra than in ^{1}H spectra.
   v. The influence of electron field on the δ values. If the chemical shifts of a substituted benzene molecule are explained just by the resonant effect, it is difficult to elucidate the chemical shift of the *ortho*-carbon atom. Take nitrobenzene as an example. According to the resonant effect, the chemical shift of the *ortho*-carbon atom will increase: however, it decreases (to less

than that of benzene by 5.3 ppm). This is the influence of the electron field on the electrons of the adjacent C–H bond, which moves the electrons towards the carbon atom so that the δ_C decreases.

2. The approximate calculation of the chemical shifts of a substituted benzene molecule
The chemical shifts of a substituted benzene molecule can be calculated approximately by Eq. (3-2).

$$\delta_C(k) = 128.5 + \sum_i Z_{k_i}(R_i) \tag{3-2}$$

The related empirical parameters are listed in Tab. 3.2.

Tab. 3.2 Empirical parameters used in Eq. (3-2) for the calculation of the chemical shifts of a substituted benzene [3]

Substituent	Z_1	Z_2	Z_3	Z_4
–H	0.0	0.0	0.0	0.0
CH_3	9.3	0.6	0.0	–3.1
$–CH_2CH_3$	15.7	–0.6	–0.1	–2.8
$–CH(CH_3)_2$	20.1	–2.0	0.0	–2.5
$–CH_2CH_2CH_2CH_3$	14.2	–0.2	–0.2	–2.8
$–C(CH_3)_3$	22.1	–3.4	–0.4	–3.1
$–CH_2Cl$	9.1	0.0	0.2	–0.2
$–CH_2Br$	9.2	0.1	0.4	–0.3
$–CF_3$	2.6	–3.1	0.4	3.4
CH_2OH	13.0	–1.4	0.0	–1.2
$–C\equiv CH$	–6.1	3.8	0.4	–0.2
–PH	13.0	–1.1	0.5	–1.0
–F	35.1	–14.3	0.9	–4.4
–Cl	6.4	0.2	1.0	–2.0
–Br	–5.4	3.3	2.2	–1.0
–I	–32.3	9.9	2.6	–0.4
–OH	26.9	–12.7	1.4	–7.3
$–OCH_3$	30.2	–14.7	0.9	–8.1
$–NH_2$	19.2	–12.4	1.3	–9.5
$–NHCH_3$	21.7	–16.2	0.7	–11.8
$–N(CH_3)_2$	22.4	–15.7	0.8	–11.8
$–NO_2$	19.6	–5.3	0.8	6.0
–CHO	9.0	1.2	1.2	6.0
$–COCH_3$	9.3	0.2	0.2	4.2
–COOH	2.4	1.6	–0.1	4.8
$–COOCH_3$	2.1	1.2	0.0	4.4

3.2.6 Carbonyl Compounds

Carbonyl groups have the largest chemical shifts amongst the common functional groups. Therefore, they can be easily recognized.

The fact that carbonyl groups have the largest chemical shifts can be elucidated using the resonant effect (**16**). Because the carbonyl group lacks electrons, it resonates in the downfield region.

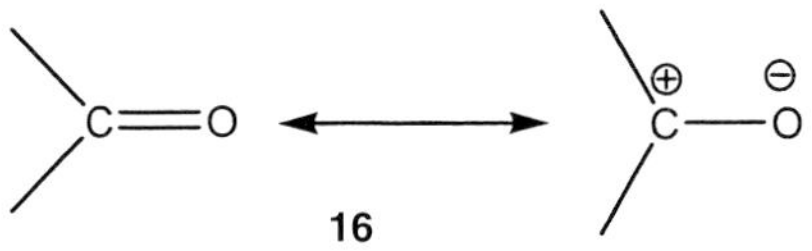

16

If a carbonyl group is connected to a heteroatom or an unsaturated functional group, the deficiency of the electrons in the carbonyl group is moderated. As a result, the chemical shift of the substituted carbonyl group will clearly decrease (**17**).

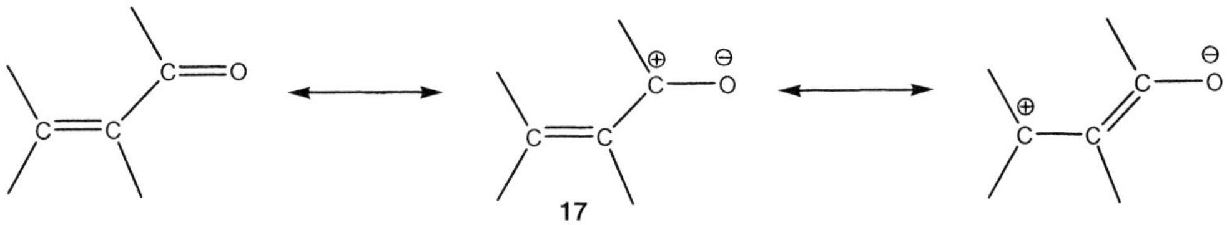

17

For the reason given above, aldehydes and ketones have the largest chemical shifts ($\delta > 195$ ppm) while acid halides, acid anhydrides, amides, esters, etc., have δ values of less than 185 ppm. α,β-Unsaturated aldehydes and ketones have a smaller δ value than the saturated ones. However, the function of the conjugation with a double bond is weaker than that of a heteroatom.

The distribution order of chemical shifts of common functional groups in ^{13}C spectra is similar to that in ^{1}H spectra. For example, alkanes have the smallest δ values both in ^{13}C spectra and in ^{1}H spectra while aldehydes have the largest δ values in both ^{13}C and ^{1}H spectra. The substitution by an electronegative functional group leads to an increase in the δ value of the substituted carbon and related hydrogen atom. The peaks of alkenes and substituted benzenes are situated between those of alkanes and carbonyl groups. These similarities can be used in the interpretation of both ^{13}C and ^{1}H spectra.

The distribution of the chemical shifts of functional groups in ^{13}C spectra is shown in Tab. 3.3.

Tab. 3.3 Distribution of the chemical shifts of functional groups in ^{13}C spectra

Functional group	δ_C *(ppm)*	*Functional group*	δ_C *(ppm)*
$-(CH_2)-CH_3^*$	10–15		110–150
$>C-CH_3^*$	25–30	N	125–155
$-C=C-CH_3^*$	15–28	$-C\equiv N$	110–130
CH_3^*	15–25	$R-C(=O)-OR'$	165–175
$-N-CH_3^*$	25–45	$R-C(=O)-Cl$	165–180
$-O-CH_3^*$	45–60	$R-C(=O)-OH$	172–185
$-CH_2^*-C<$	23–37	$R-C=C-C(=O)-H$	165–175
$-CH_2^*-N-$	41–60	$R-C(=O)-H$	200–205
$-CH_2-O-$	45–75	$R-C(=O)-C=C-R'$	195–205
$-C\equiv C-$	70–100	$R-C(=O)-R'$	205–220
$-HC=CH-$	110–150		

3.2.7
Influences of Hydrogen Bonds and the Medium

1. Hydrogen bonds

Hydrogen bonds include interhydrogen bonds and intrahydrogen bonds.

The formation of an intrahydrogen bond in a carbonyl compound moves the non-bonding electrons of the oxygen atom towards the hydrogen atom so that the carbonyl group is more deficient in electrons. As a result, the carbonyl group will have a larger δ value.

The function of interhydrogen bonds is similar to that of intrahydrogen bonds (structure **18**).

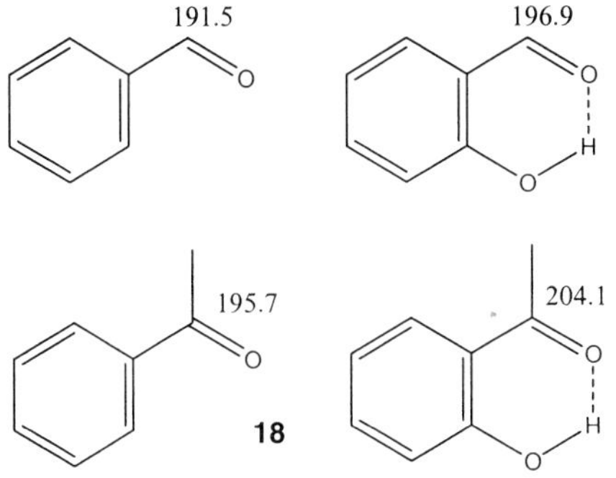

191.5 196.9

195.7 204.1

18

2. Medium

The dilution of a solution or changing the solvents will change the δ value of the solute. In general the change is less than several ppm, but in some instances the change could reach 10 ppm. This change is related to hydrogen bonds.

3.3
Coupling and Decoupling Methods in ^{13}C Spectra

3.3.1
Coupling in ^{13}C Spectra

Because the natural abundance of ^{13}C is only 1.1%, the ^{13}C–^{13}C coupling can be omitted. On the other hand, as the natural abundance of ^{1}H is 99.98%, the ^{13}C peaks will be split by ^{1}H if ^{1}H nuclei are not to be decoupled.

Because $\frac{\gamma_{^{13}C}}{\gamma_{^{1}H}} \approx \frac{1}{4}$, namely $\frac{\nu_{^{13}C}}{\nu_{^{1}H}} \approx \frac{1}{4}$, a CH_n group is a typical AX_n system.

The most important coupling between ^{13}C and ^{1}H is that from $^{1}J_{^{13}C-^{1}H}$, the value of which is determined by the percentage of s electrons in the C–H bond. The following expression holds approximately:

$$^{1}J_{^{13}C-^{1}H} = 5 \times (s\%)\ \text{Hz} \tag{3-3}$$

where *s*% is the percentage of s electrons in a C–H bond.

The following data are the elucidation of Eq. (3-3).

$CH_4(sp^3, s\% = 25\%)$	$^1J = 125$ Hz
$CH_2{=}CH^2(sp^2, s\% = 33\%)$	$^1J = 157$ Hz
$C_6H_6(sp_2, s\% = 33\%)$	$^1J = 157$ Hz
$CH{\equiv}CH(sp, s\% = 50\%)$	$^1J = 249$ Hz

Because the value of ^1J is large, the decoupling to ^{1}H is necessary to remove the significant overlapping of split ^{13}C peaks.

Besides the *s*%, the electronegativity of substituents also affects the ^1J value. The increase in the electronegativity will increase the ^1J value. For example, the ^1J value of a substituted methane can be enhanced by 41 Hz.

The enhancement of ^1J values by increasing *s*% and/or by increasing the electronegativity of substituents is in agreement with related theoretical calculations and predictions.

The variation of $^2J_{CH}$ ranges from –5 to 60 Hz. The value of $^3J_{CH}$ is less than 20 Hz, which is related to the substituents and their steric positions. The Karplus equation holds approximately for $^3J_{CH}$ (see Section 2.2).

In a substituted benzene, $|^3J_{CH}| > |^2J_{CH}|$.

3.3.2
Broadband Decoupling

Broadband decoupling is also called proton noise decoupling, which is used most frequently for the measurement of ^{13}C spectra.

During the measurement of the ^{13}C spectrum, a broad frequency band, which covers all frequencies of the hydrogen nuclei of the sample, is used to irradiate the sample. As a result, the coupling between ^{13}C and ^{1}H is removed completely. Under these conditions, each carbon atom has only one peak in the ^{13}C spectrum.

By using decoupling, split peaks are concentrated into one peak to improve its signal-to-noise ratio. At the same time, the irradiation leads to the nuclear Overhauser effect so that the ratio is further enhanced. According to Eq. (2-35)

$$\eta_I = \frac{\gamma_S}{2\gamma_I} \tag{2-35}$$

and the particular values of γ_S and γ_I: $\gamma_S = \gamma_H = 2.6752 \times 10^8$ kg^{-1} s A and $\gamma_I = \gamma_C = 0.6726 \times 10^8$ kg^{-1} s A, $\eta_I(\eta_C)$ can be calculated. Its value is 1.998, which means that the signal intensity of ^{13}C nuclei can be enhanced to three times that without the irradiation.

Broadband decoupling is most frequently used as a matter of routine.

It should be noted that the conditions for recording a ^{13}C spectrum are very important in obtaining a correct result. If a carbon atom has a rather long T_1 and the interval between the end of the acquisition and the next pulse is not long enough, the signal of the carbon atom will be greatly reduced because, before the next pulse, its magnetization vector is less than M_0. Therefore, the carbon atom with a very long T_1 will have no signal; the carbon atom with a long T_1 will have a weak signal; and the carbon atom with a short T_1 will have a strong signal.

Briefly, because of different T_1 values and the η of the carbon atoms, the peak areas (roughly the peak heights) in a ^{13}C spectrum are not proportional to the numbers of the carbon atoms in the sample. However, the number of carbon atoms in the sample can be estimated from the peak areas of the signals in a ^{13}C spectrum by an experienced person.

The ^{13}C spectrum of compound **C3-1** with broadband decoupling is illustrated in Fig. 3.3.

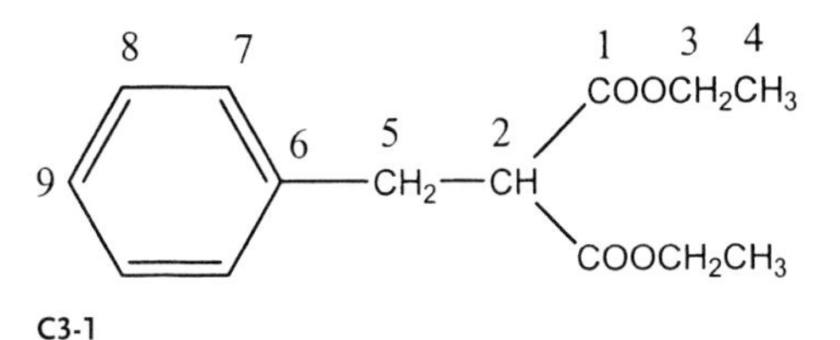

C3-1

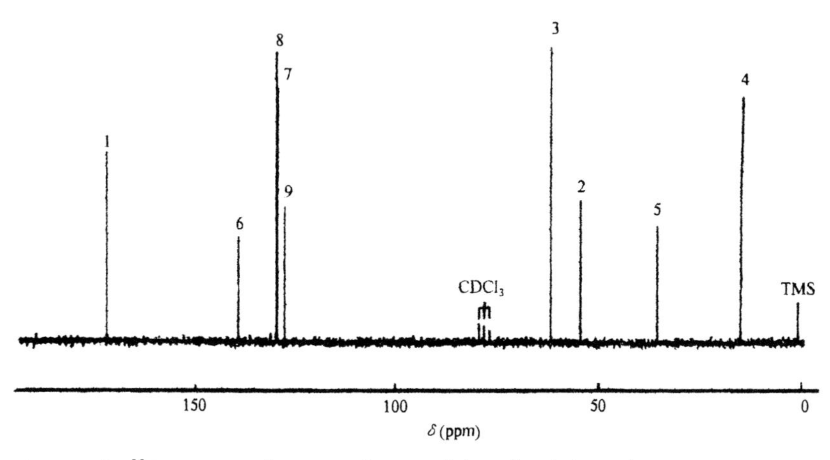

Fig. 3.3 The ^{13}C spectrum of compound **C3-1** with broadband decoupling.

3.3.3
Off-resonance Decoupling

Off-resonance decoupling is now an out-dated technique. It was applied to identify the order of carbon atoms with reduced coupling constants. Today the application of pulse sequences, such as DEPT, can determine the order of carbon atoms without split peaks. Therefore, off-resonance decoupling is no longer used.

3.3.4
Selective Decoupling

Selective decoupling is also an out-dated technique. It was used to find a single pair of correlated ^{13}C and ^{1}H atoms. It has been replaced by H, C-COSY which correlates all pairs of ^{13}C and ^{1}H in a spectrum.

3.3.5
Gated Decoupling

An emission gate of the NMR spectrometer is used to control the irradiation period of radiofrequency ν_1 while a receptor gate is used to control the time during which the receptor works. Gated decoupling is a method in which the emission gate and the receptor gate are controlled to obtain a decoupled spectrum.

Gated decoupling with suppressed NOE is used most frequently, providing a quantitative ^{13}C spectrum. Its principle is shown in Fig. 3.4.

The abscissa of Fig. 3.4 is time. Before the excitation of ^{13}C nuclei, a period of t_R is used to establish an equilibrium. Therefore, it is necessary that $t_R \geq 5T_1$ where T_1 is the longest T_1 among all ^{13}C nuclei of the sample. After the excitation pulse for ^{13}C nuclei, the receptor gate opens just for a short time to receive ^{13}C signals. In the same time period, the decoupling for ^{1}H works so that the ^{13}C signals obtained are decoupled, which means every carbon atom has only one peak. Because the irradiation time for ^{1}H is short compared with t_R, NOE can be neglected. As t_R is longer than five times the longest T_1 for all ^{13}C nuclei of the sample, all carbon atoms have their own M_0 under equilibrium condi-

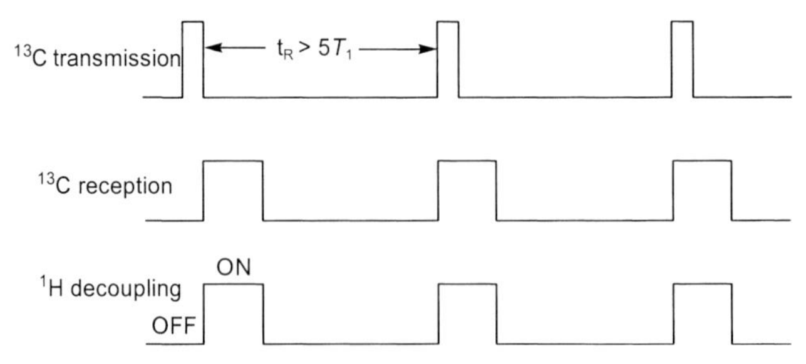

Fig. 3.4 The principle of gated decoupling with suppressed NOE.

tions before the excitation. Under these two conditions (a long t_R and a short irradiation), the recorded ^{13}C spectrum is a quantitative spectrum with a good approximation, in which peak areas are proportional to the number of carbon atoms. This type of ^{13}C spectra is very useful in many cases, such as the measurement of a sample that consists of two isomers in equilibrium.

3.4 Relaxation

3.4.1 Why does the Discussion of Relaxation of ^{13}C Nuclei Require a Whole Section?

It is for the following reasons.

1. ^{13}C nuclei have longer relaxation times than ^{1}H nuclei, which can be as long as 100 s. As a result, the relaxation times of ^{13}C nuclei can be measured with great precision.
2. FT-NMR spectrometers are used for the measurement of a ^{13}C spectrum. Because the intensity of the radiofrequency of FT instruments is stronger than that of CW instruments by several orders of magnitude, saturation occurs more easily for the measurement of ^{13}C spectra than for the measurement of ^{1}H spectra. As a result, the relaxation process is more critical for ^{13}C nuclei than for ^{1}H nuclei.

   In a routine ^{13}C spectrum, the fact that every carbon atom has only one peak favors the measurement of relaxation times.
3. Some structural information can be obtained from the values of relaxation times.

3.4.2 Basic Concepts of the Relaxation of ^{13}C Nuclei

The concept of relaxation was discussed in Section 1.5.

The sample to be measured by high resolution NMR is a solution. Under the influence of B_1, sample molecules are excited. If relaxation mechanisms do not exist, the saturation, in which the difference between the populations of a higher level and a lower level is zero, leads to the no NMR signals being produced.

The longitudinal relaxation of ^{13}C nuclei is achieved by the following mechanisms: dipole–dipole interaction, spin rotation, scalar coupling, chemical shift anisotropy, quadrupole moment relaxation and the effect of a paramagnetic substance.

The dipole–dipole interaction plays an important role in longitudinal relaxation. This mechanism is closely related to molecular motion. The concept of τ_C was introduced in Section 2.7.2. τ_C is a correlation time, a time constant, during which the sample molecule loses phase coherence in its solution. Sometimes τ_C

is defined as the average time required for a molecule to rotate through an angle of one radian. However, τ_C is a property of a molecule, which is related to the molecular weight, solution viscosity, etc.

3.4.3
Measurement of Relaxation Time

1. Measurement of T_1

The method used most frequently for the measurement of T_1 is inversion recovery. Results from this method have the best precision among all methods for the measurement of T_1.

The pulse sequence in the inversion recovery method is illustrated in the upper part of Fig. 3.5. It can be written as:

$$180^\circ - \tau\ (\text{varied}) - 90^\circ - T\ (>5T_1)$$

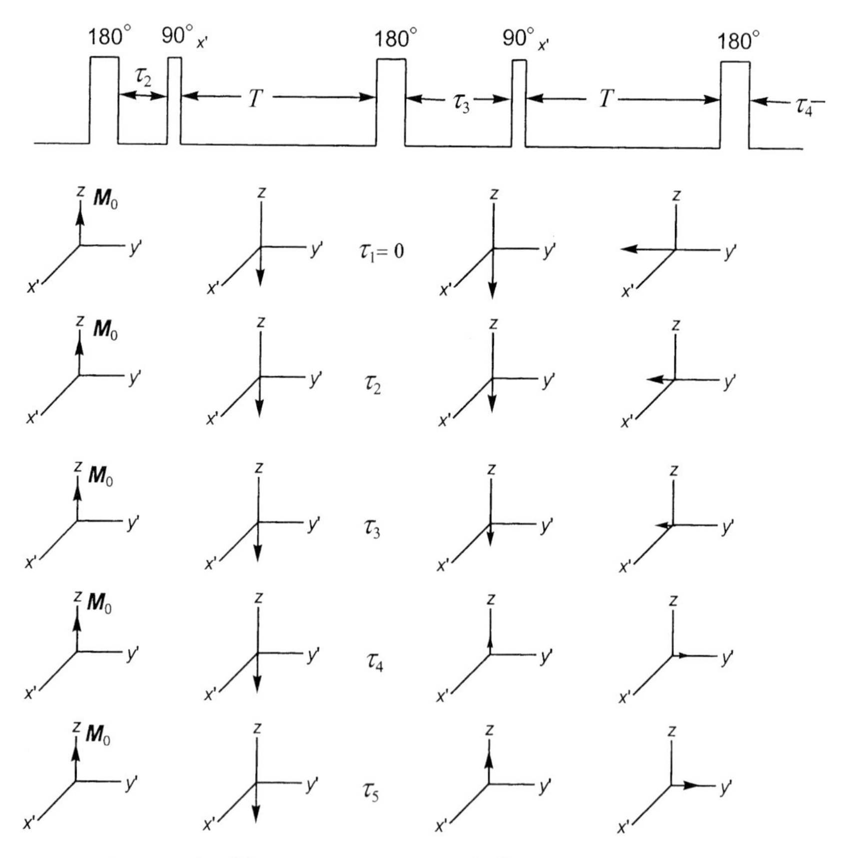

Fig. 3.5 The principle of the inversion recovery method.

where T_1 is the longest longitudinal relaxation time for all carbon atoms of the sample to be measured.

In the experiment, the value of τ is varied from a very short period to a long period.

The motion of the magnetization vector is shown in the lower part of Fig. 3.5 in the order of the values of τ. After a period of T of longer than $5T_1$, the magnetization vector relaxes to M_0. The 180° pulse rotates the magnetization vector from the z axis to the $-z$ axis. The magnetization vector M will relax towards M_0 during τ. If $\tau=0$, the 90° pulse rotates the magnetization vector from the $-z$ axis to the $-y'$ axis, resulting in a negative signal. If τ has a sufficiently long value, M will relax to M_0 and a normal positive signal will be recorded. This process can be described with the following equation:

$$\frac{dM_Z}{dt} = -\frac{M_Z - M_0}{T_1}$$

Integration of the equation above gives Eq. (3-4).

$$M_Z - M_0 = Ae^{-t/T_1} \tag{3-4}$$

The t in Eq. (3-4) corresponds to the τ in Fig. 3.5. It should be noted that t is counted from the end of the 180° pulse in Fig. 3.5.

The initial condition for Eq. (3-4) is that

$$M_Z(0) = -M_0$$

The substitution of the equation above into Eq. (3-4) leads to

$$A = -2M_0$$

Finally Eq. (3-4) becomes

$$M_Z = M_0(1 - 2e^{-t/T_1}) \tag{3-5}$$

It can be rewritten as

$$\ln\frac{M_0 - M_Z}{M_0} = -\frac{t}{T_1} + \ln 2 \tag{3-6}$$

By plotting $\ln\frac{M_0-M_Z}{M_0}$ versus t (namely τ), that is the slope of the straight line, $-\frac{1}{T}$, can be obtained. Therefore, T_1 can be calculated.

Fig. 3.6 is an example of the measurement of T_1 by the inversion recovery method. The measured sample is ethyl benzene. Because the sample is not deoxygenated, the measured T_1 values are fairly large.

Readers interested in other methods for the measurement of T_1 are referred to reference [7] in which five methods are presented.

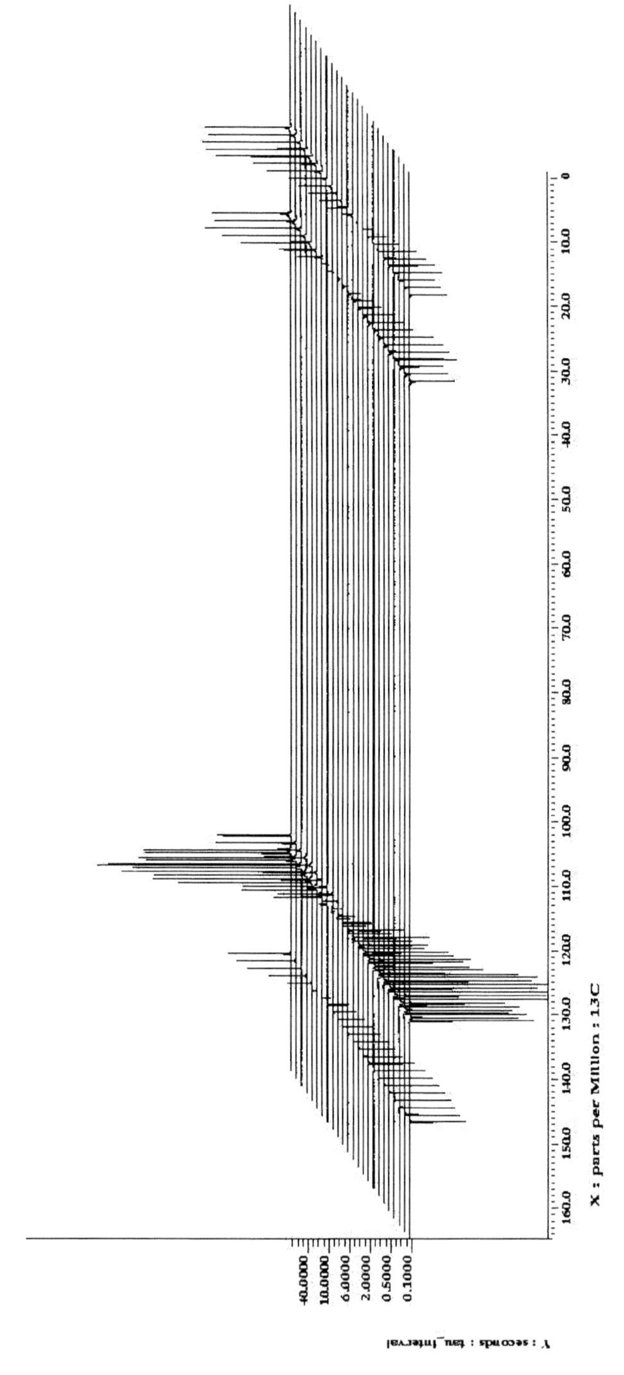

Fig. 3.6 The measurement of T_1 by the inversion recovery method.

2. The measurement of T_2

In Section 1.5 it was shown that T_2 can be calculated from the width of a peak at the half height. However, the measurement will be influenced greatly by the inhomogeneity of the magnetic field. The method of spin echo (see Section 4.1.3) overcomes this drawback and can measure T_2 with great precision.

Another method for the measurement of T_2 is spin locking (see Section 4.1.6) through which $T_{1\rho}$ can be measured, and where the subscript ρ represents the rotating frame. $T_{1\rho}$ is the longitudinal relaxation time in the rotating frame. It is true that $T_1 \approx T_{1\rho} \approx T_2$ for a non-viscous liquid and $T_1 > T_{1\rho} \approx T_2$ for a viscous liquid. In any case, $T_{1\rho}$ is approximately equal to T_2.

3.4.4 Application of T_1

1. Providing structural information

The values of T_1 can be used for the assignment of the ^{13}C spectrum of an unknown compound. Therefore, the reliability of the assignment can be improved.

A typical example is the assignment of the ^{13}C spectrum of reserpine (**19**). The numerals denoted beside its structural formula are the T_1 values (in seconds) of the carbon atoms, which are close to the numerals. From these numerals, the regularity of the T_1 values is obvious. Take the T_1 values of quaternary carbon atoms attached to methoxy group as examples:

12.8 s: Two neighboring carbon atoms are quaternary carbon atoms.

3.8 s: One adjacent carbon atom is a quaternary carbon atom and another is CH.

3.0 s: Two neighboring carbon atoms are CH.

The assignment based on T_1 values agrees with that based on the chemical shifts (structure **19**).

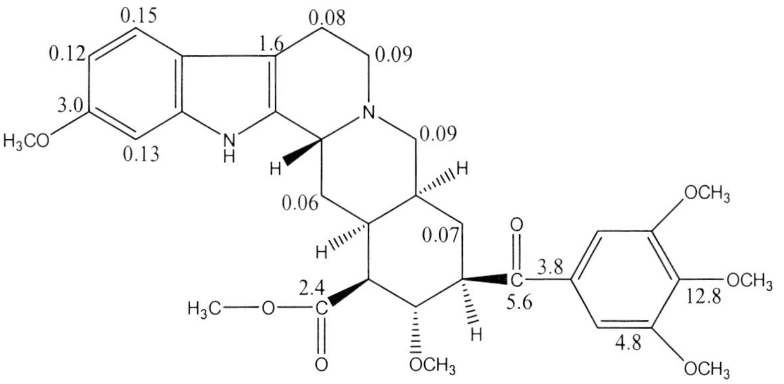

0.15 0.08 0.12 1.6 0.09 3.0 N H_3CO N H 0.13 0.09 H H 0.06 OCH_3 0.07 O H 3.8 2.4 C OCH_3 H_3C—O—C 5.6 12.8 O H OCH_3 4.8 OCH_3

19

2. Promoting studies of solutions and the motions of molecules
   The measurement of T_1 is an important method of studying a solution (such as segment motions of a polymer; interchange reactions).
   T_1 values offer the following information: size of molecules, anisotropy of molecular motion, intramolecular rotations, steric barriers, molecular flexibility, the association between molecules (or ions), and so forth [2].

3.5 Interpretation of ^{13}C NMR Spectra

^{13}C NMR spectra can be used in many domains. However, only the application to the structural identification of organic compounds is discussed in this section.

Before the interpretation of ^{13}C NMR spectra, sampling and plotting will be described.

3.5.1 Sampling and Plotting

1. Sampling
   It is necessary to prepare a non-viscous solution with an appropriate concentration for the measurement of a ^{13}C NMR spectrum. As the sensitivity of the measurement of a ^{13}C spectrum is much lower than that of a ^{1}H spectrum, the concentration of the solution to be measured should be high and the number of scan times sufficient. The lower the frequency of the spectrometer, the higher the requirements of these parameters. If the sample molecule has a large molecular weight, the data acquisition will be longer.
   Although the protons, which exist in the solvent, do not interfere with the measurement of the ^{13}C spectrum, deuterated solvents are still used in the measurement of the ^{13}C spectrum in order to measure the related ^{1}H spectrum and for locking of the magnetic field.
2. Plotting
   The measurement of a ^{13}C spectrum is related to the acquisition parameters (the tip angle by a pulse, the delay between two pulses, spectral width, etc.), accumulation, data treatment, and so forth [1,2].
   Routine measurements of ^{13}C spectra are those with complete decoupling. In general, DEPT (see Section 4.2) is necessary to determine the multiplicity of the carbon atoms of the sample. If quantitative information about the carbon atoms is required, the pulse sequence with gated decoupling with suppressed NOE is adopted.

3.5.2
Steps for the Interpretation of ^{13}C Spectra

The following steps are recommended.

1. Discerning real peaks of a ^{13}C spectrum
   i. Solvent peak
   The solvent peak in a ^{1}H spectrum originates from an imperfect deuteration of the solvent while the solvent peak in a ^{13}C spectrum originates from the solvent itself. Fortunately these solvents are small molecules, which have a small τ_C and a long T_1 so that the solvent peak in the ^{13}C spectrum is usually not strong.
   The solvent peak in a ^{13}C spectrum can be discerned from its chemical shift (Tab. 3.4).
   Note:
   - The spin quantum number of deuterium, I, is 1. Therefore, the splitting by deuterium should be calculated using $2nI+1$ and the intensities of split peaks cannot be estimated by the coefficients developed from a binomial.
   - According to the convention, s, d, t and q are used to show singlet, doublet, triplet and quartet signals. The peak set, which is split into more than four lines, is denoted by the Arabic numerals, for example, 7 implies a septet.

   ii. Impurity peaks
   Please refer to Section 2.9.2 (discerning impurity peaks in a ^{1}H spectrum).
   iii. Do not lose sample peaks
   When acquisition parameters are not set correctly, for example, the tip angle is large but the delay between two pulses is not long enough, and it is possible to lose peaks of the quaternary carbon atoms.
   When the spectral width is not wide enough, the peaks that are situated beyond this width will fold into the spectrum.
2. Calculate the unsaturation number of the unknown compound according to its elemental composition
   Please refer to Section 2.9.2.
3. Analyze molecular symmetry
   If the molecule to be measured has no symmetry, the peak number in its ^{13}C spectrum is equal to the number of carbon atoms in the molecule. If the peak

Tab. 3.4 Signals for common solvents in ^{13}C spectra

Deuterated solvent	*$CDCl_3$*	*CD_3COCD_3*		*CD_3OD*	*C_6D_6*	*C_5D_5N*			*CD_3SOCD_3*
δ_C	77.0	30.2	206.8	49.3	128.7	123.5	135.5	149.8	39.7
Peak shape	t	7	s	7	t	t	t	t	7

number is less than the number of the carbon atoms in the molecule, the molecule should have some symmetry.

4. Classified regions of δ_C

   δ_C can be classified into three regions.

   i. The region for carbonyl groups and accumulated alkenes where δ_C >150 ppm, and in general $\delta_C > 165$ ppm.

      If $\delta_C > 200$ ppm, it means an aliphatic aldehyde or ketone group is present. The connection of a carbonyl group to a heteroatom shifts δ_C to within the range of 160–180 ppm. The conjunction of a carbonyl group with a double or triple bond shifts δ_C in the upfield direction by less than 10 ppm.

      If a compound contains an accumulated alkene, C=C=C, the central carbon atom has a peak in this region. However, the two other carbon atoms have two peaks in the region of the double bonds. Therefore, accumulated double bonds can be differentiated from carbonyl groups.

   ii. The region for double bonds and phenyl rings where $\delta_C = 90–160$ ppm, and in general $\delta_C = 100–150$ ppm

      Alkenes, phenyl rings, nitrile groups (i.e., the carbon atom), etc., have their signals in this region.

      From the carbon atoms whose signals are situated in these two regions, their corresponding unsaturation number can be calculated. The difference between this number and that calculated from the molecular formula equals the number of the formed cycles of the molecule.

   iii. The region for saturated carbon atoms

      $\delta_C < 100$ ppm: if a carbon atom is not connected to a heteroatom, its chemical shift is usually less than 55 ppm.

      Alkyne groups have signals within the range 70–100 ppm, which is an exception for unsaturated carbon atoms.

5. Determination of the order of the carbon atoms

   By using DEPT (or another pulse sequence, see Section 4.2), the order can be determined. Therefore, the number of hydrogen atoms that are connected to carbon atoms can be determined. If this number is less than that in the molecular formula, the other hydrogen atoms should be connected to heteroatoms.

6. Postulate some structural units and combine them to obtain some possible structural formulae

7. The assignment of postulated possible structural formulae

   The most likely (most reasonable) structural formula can be selected from the assignment of postulated possible structural formulae.

The ^{1}H spectrum and the ^{13}C spectrum of an unknown compound are mutually complementary. If the ^{1}H spectrum of the unknown compound is available, it is necessary to interpret the ^{13}C spectrum together with the ^{1}H spectrum.

3.5.3
Examples of the Interpretation of ^{13}C Spectra

Example 1
An unknown compound has a molecular formula of $C_6H_{10}O_2$ and the ^{13}C NMR data shown in Tab. 3.5. Try to deduce its structure.

Tab. 3.5 ^{13}C NMR data for $C_6H_{10}O_2$

δ (ppm)	17.4	14.3	60.0	123.2	144.2	166.4
Multiplicity	q	q	t	d	d	s

Solution
From its molecular formula, the unsaturation number can be calculated as being 2, which agrees with the ^{13}C data (a double bond and a carbonyl group: 123.2, d; 144.2, d; and 166.4, s).

The chemical shift of 166.4 ppm indicates that the unknown compound should belong to a carbonyl group connected to a heteroatom (in this example, it should be an oxygen atom) and a double bond. Therefore, this unknown compound should contain the structural unit **20**.

$$—O—C(=O)—CH{=}CH—$$

20

Because the δ value of the CH_2 is 60.0 ppm, the CH_2 should be connected to an oxygen atom. As two remaining CH_3 moieties have different chemical shifts, they should be situated at two different positions.

Through combination of the postulations above, the structure of the unknown compound can be deduced to be **21**.

$$\underset{14.3}{H_3C}—\underset{60.0}{H_2C}—O—\underset{166.4}{C(=O)}—\underset{144.2}{CH}{=}\underset{123.2}{CH}—\underset{17.4}{CH_3}$$

21

Example 2
An unknown compound has a molecular formula of $C_{11}H_{14}O_2$ and the ^{13}C NMR data given in Tab. 3.6. Try to deduce its structure.

Tab. 3.6 ^{13}C NMR data for $C_{11}H_{14}O_2$

δ (ppm)	39.9	55.6	55.7	111.9	112.5	115.5	120.7	132.7	137.9	147.9	149.4
Multiplicity	t	q	q	d	d	t	d	s	d	s	s

Solution

From its molecular formula, the unsaturation number can be calculated as being 5.

Because 11 carbon atoms have 11 signals in the ^{13}C spectrum, the molecule has no symmetry. From the multiplicity it can be shown that all 14 hydrogen atoms are connected to carbon atoms. Therefore, there are no hydrogen atoms connected to heteroatoms.

Because eight carbon atoms have signals in the region of the double bond and the phenyl ring, it can be postulated that the unknown compound contains a phenyl ring and a double bond. It should be noted that the signal at 115.5 ppm has a multiplicity of “t”, which means the existence of a $CH_2(-C{=}CH_2)$.

On the basis of the two signals (55.6, q and 55.7, q) and the presence of two oxygen atoms in the molecular formula, it can be deduced that the compound contains two methoxy groups and these two groups have similar chemical environments. Therefore, the unknown compound should be the structural unit **22**.

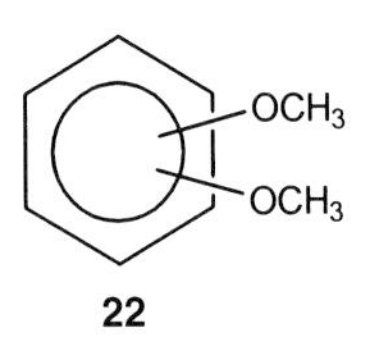

22

Compared with the molecular formula, only one carbon atom is left. As the phenyl ring has three substituents (three signals with multiplicity of “s” in the region of double bonds and phenyl rings), the three substituents should be two methoxy groups and the group that contains the remaining carbon atom and the double bonds is as follows:

$-CH_2-CH{=}CH_2$

Therefore, all that remains to in order to find the structure is the determination of the positions of the three substituents.

The analysis of the chemical shifts of the substituted carbon atoms of the phenyl ring is very important in finding the positions of substituents.

From Tab. 3.2, it can be seen that the substitution of a carbon atom into a phenyl group by a methoxy group will produce a large chemical shift:

$$128.5 + 30.2 = 158.7\ \text{(ppm)}$$

However, the two substituted carbon atoms have δ values of 147.9 and 149.4 ppm. Therefore, these two methoxy groups should be adjacent. Because two methoxy groups are adjacent, their chemical shifts can both decrease.

The third substituted carbon atom has a chemical shift of 132.7 ppm, which is close to that of the non-substituted benzene. Therefore, the third substituents is not adjacent to any methoxy group.

Through combination of the postulations above, the structure of the unknown compound can be deduced to be **23**.

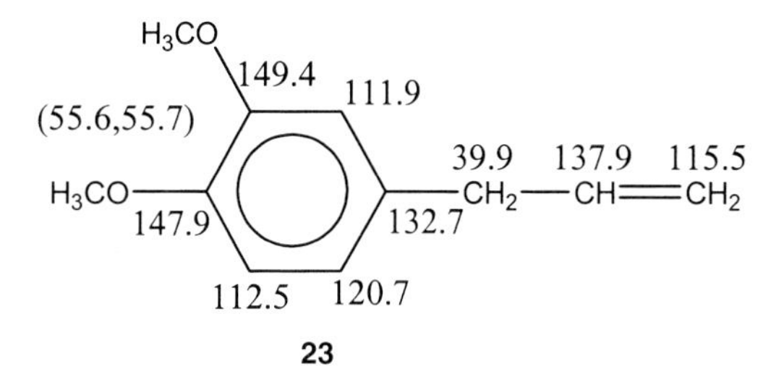

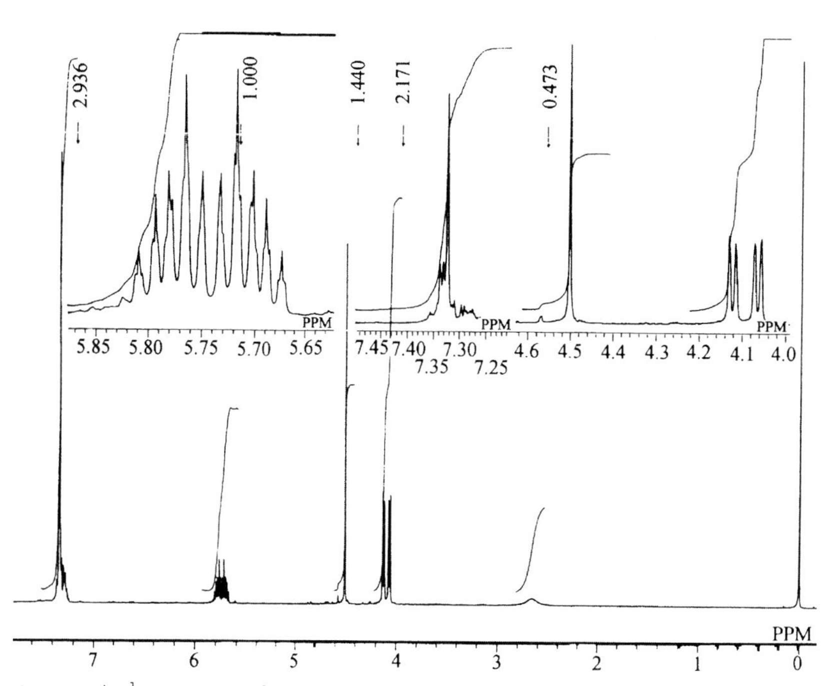

Fig. 3.7 The ^{1}H spectrum of an unknown compound.

The assignment of chemical shifts is denoted beside the related carbon atoms.

From these two examples, it has been shown that the structure of an unknown compound can be deduced from its molecular formula and the ^{13}C spectrum if the structure is simple.

Example 3

An unknown compound has a molecular formula of $C_{11}H_{14}O_2$. Its ^{1}H spectrum, ^{13}C spectrum and DEPT are shown in Figs. 3.7, 3.8 and 3.9. Try to deduce the structure.

Solution

Please read the discussion on DEPT in Section 4.2. A comparison of the ^{13}C spectrum and the DEPT (spectrum) shows that the unknown compound contains a quaternary carbon atom ($\delta = 137.8$ ppm, no signal in DEPT), three CH_2 (in the aliphatic region) and CH substituents (in the phenyl ring region).

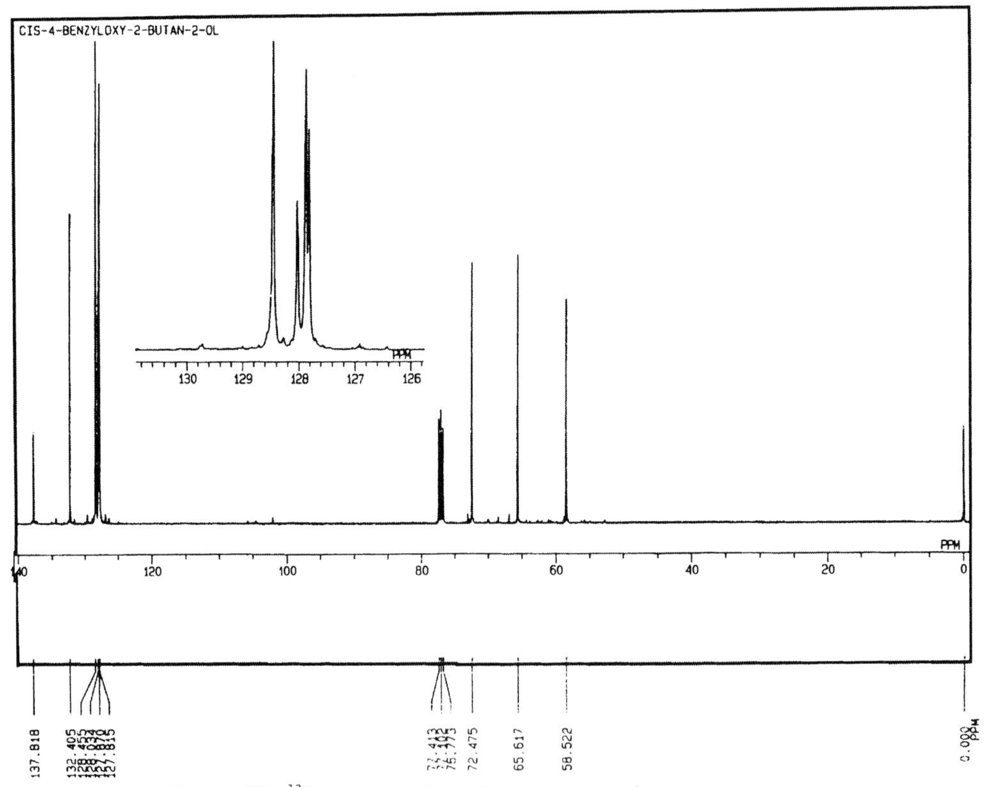

Fig. 3.8 The ^{13}C spectrum of an unknown compound.

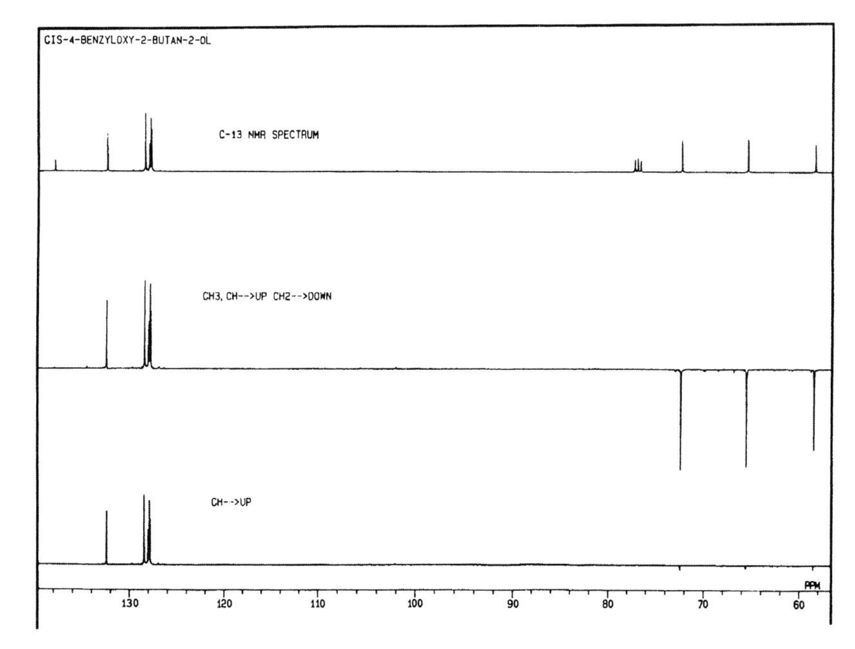

Fig. 3.9 The DEPT spectrum of an unknown compound.

Combination of a ^1H spectrum and its corresponding ^{13}C spectrum can help to deduce a rather complicated structure. It is recommended that the signals in the ^1H spectrum and those in the ^{13}C spectrum are correlated first.

From the molecular formula, the unsaturation number can be calculated as 5, which corresponds to a phenyl ring and a double bond. This postulation can be proved by the ^1H spectrum in which the pattern of the mono-substitution of a phenyl ring by a substituent of the first type is clearly shown at about 7.35 ppm, which means the phenyl ring is substituted directly by a carbon atom (see Section 2.5.1).

The peaks at about 5.74 ppm illustrate the existence of two alkene hydrogen atoms. From the enlarged peaks it is known that the peaks consist of an AB system ($J_{AB} \approx 11.9$ Hz) which is further split by two coupling constants ($J \approx 4.9$ Hz, *t*; $J \approx 1.2$ Hz, *t*). The value of 11.9 Hz illustrates a *cis*-configuration. The value of 4.9 Hz is the coupling constant of *vicinal* hydrogen atoms, while the value of 1.2 Hz is a ^5J coupling constant. The deductions above leads to the following structural unit:

$$-CH_2-CH=CH-CH_2-$$

The peak set at 4.04–4.14 ppm illustrates the existence of two CH_2 groups, whose peaks are split by a CH, respectively. This deduction agrees with the structural unit above. As both of these have a fairly large δ value, they should be connected to an oxygen atom, respectively.

The singlet at 4.50 ppm illustrates the existence of an isolated CH_2.

Through combining the postulations above, the structure of the unknown compound can be deduced to be the following structure.

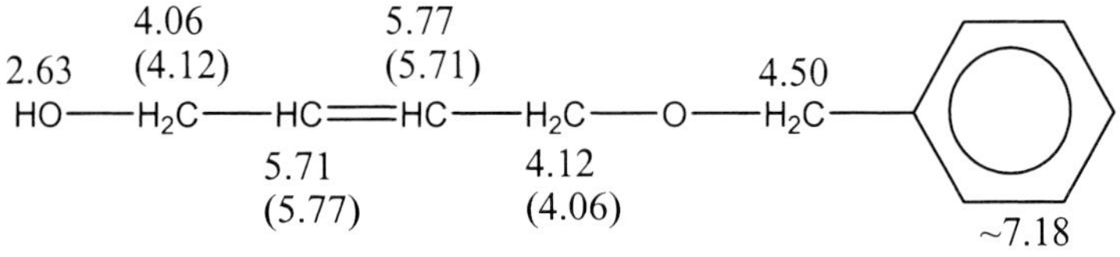

2.63 HO—H_2C 4.06 (4.12) —HC 5.71 (5.77) =HC 5.77 (5.71) —H_2C 4.12 (4.06) —O—H_2C 4.50 —C_6H_5 ~7.18

24

HO—H_2C 58.5 —HC 127.8 =HC 132.4 —H_2C 65.0 —O—H_2C 72.5 —C_6H_5 137.8, 127.9, 128.5, 128.0

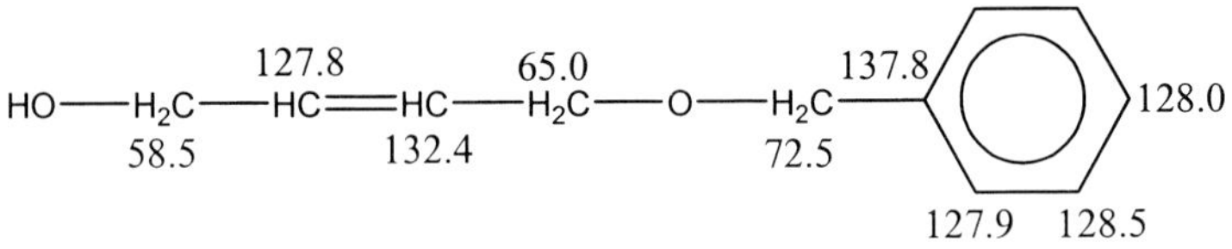

25

The assignment of chemical shifts is denoted beside the related carbon atoms or hydrogen atoms in the structural formula.

Example 4

An unknown compound has a relative molecular weight of 242. Its ^{13}C spectrum (including DEPT) and ^{1}H spectrum are shown in Figs. 3.10 and 3.11. Try to deduce its structure.

Solution

By combining the ^{13}C and ^{1}H spectra, the number of the carbon and hydrogen atoms in the unknown compound are obtained.

In the ^{1}H spectrum, the peaks from downfield to upfield correspond to the number of hydrogen atoms, that is 2, 10, 1, 1, 2 and 2, respectively. The peak at 2.87 ppm disappears by shaking the sample solution with several drops of heavy water, thus it is the peak of two active hydrogen atoms.

There are four lines in the ^{13}C spectrum in the region of the double bond and phenyl ring. Although this ^{13}C spectrum is not quantitative, it can be esti-

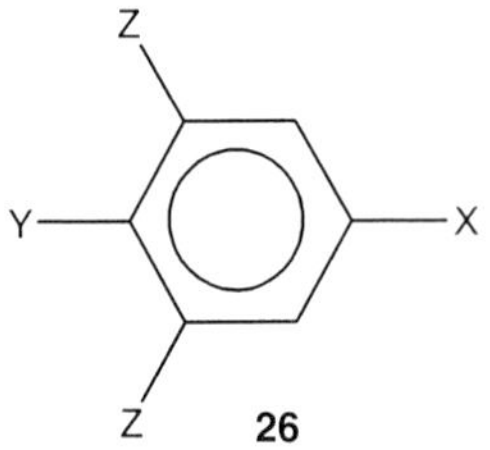

Z, Y, X, Z

26

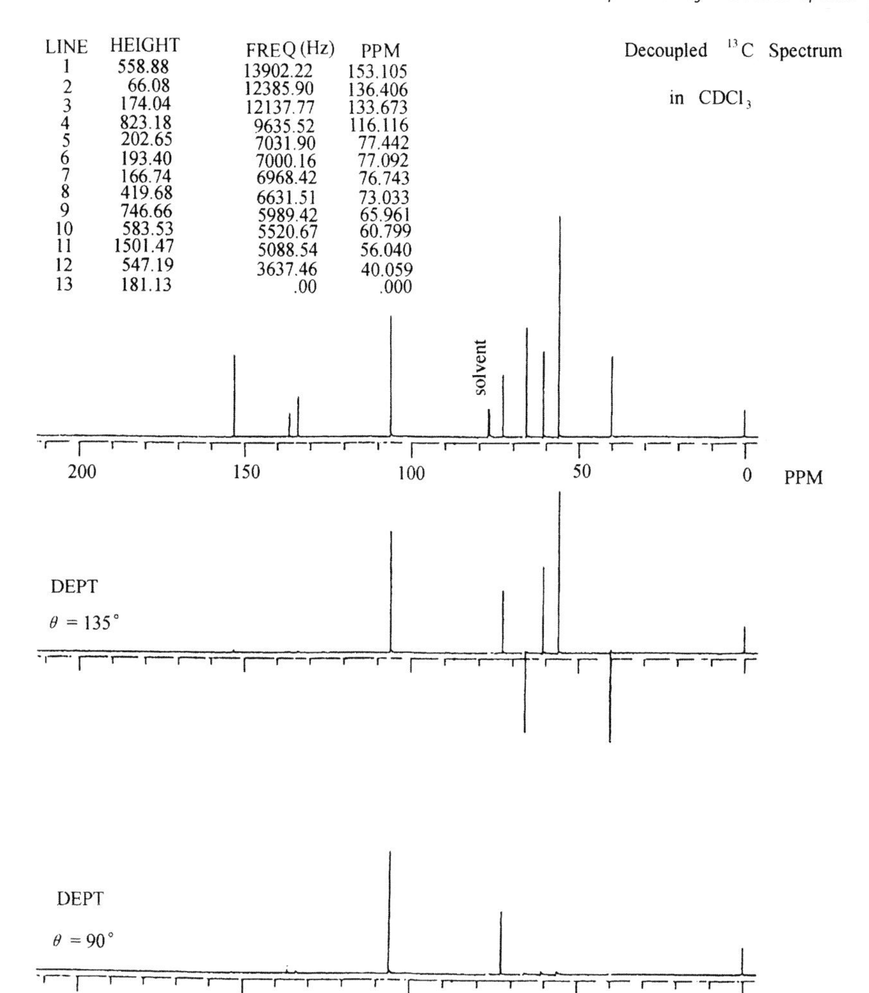

LINE	HEIGHT	FREQ (Hz)	PPM
1	558.88	13902.22	153.105
2	66.08	12385.90	136.406
3	174.04	12137.77	133.673
4	823.18	9635.52	116.116
5	202.65	7031.90	77.442
6	193.40	7000.16	77.092
7	166.74	6968.42	76.743
8	419.68	6631.51	73.033
9	746.66	5989.42	65.961
10	583.53	5520.67	60.799
11	1501.47	5088.54	56.040
12	547.19	3637.46	40.059
13	181.13	.00	.000

Fig. 3.10 The ^{13}C and DEPT spectra of an unknown compound.

mated that each of the two peaks at 106.1 and 153.1 ppm corresponds to two equivalent carbon atoms. Taking into account an isolated peak in the ^{1}H spectrum, structural unit **26** can be drawn.

There are five peaks in the upfield region of the ^{13}C spectrum. From DEPT $\theta=90^\circ$ (only CH groups have peaks) and DEPT $\theta=135^\circ$ (CH_2 groups have negative peaks, and CH and CH_3 have positive peaks), it can be shown that the compound has a CH and two CH_2 groups. Although the compound only has two peaks from the CH_3 groups, it can be shown that it has three CH_3 groups because the peak at 56.0 corresponds to two carbon atoms.

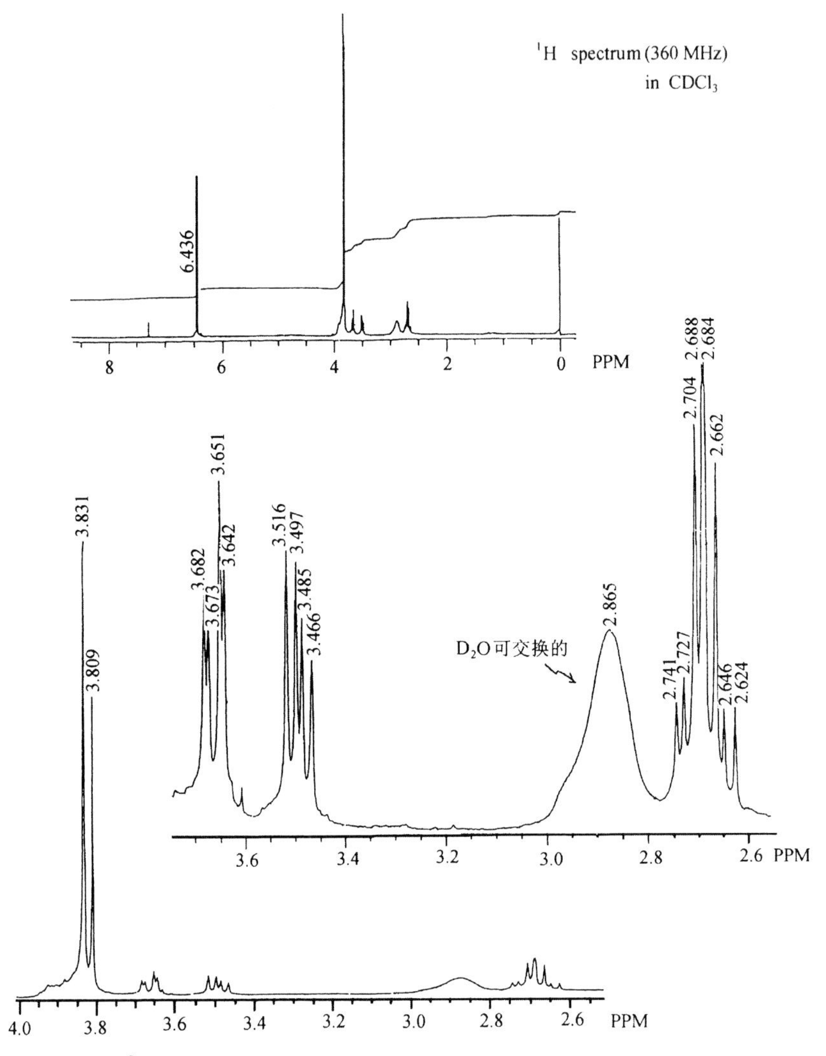

Fig. 3.11 The ¹H spectrum of an unknown compound.

The presence of two equivalent methyl groups is confirmed by the ¹H spectrum because the peak at 3.83 ppm corresponds to six hydrogen atoms. From its δ value and peak shape the two methyl groups should be two equivalent methoxy groups. The peak at 3.81 ppm should correspond to another methoxy group. These two sharp singlets are superimposed upon a broad band. The latter corresponds to the hydrogen atom that is connected to the carbon atom with a peak found at 73.0 ppm.

Let us consider the two peak sets at about 3.49 and 3.67 ppm in the ¹H spectrum. These two peak sets consist of eight peaks in which an equal interval appears four times: 3.682–3.651 = 3.673–3.642 = 3.516–3.485 = 3.497–3.446. Taking into account the frequency of the spectrometer of 360 MHz, the coupling constant can be calculated to be 11.2 Hz. From the pattern it can be shown that these eight lines belong to the AB part of an ABX system. These two hydrogen atoms belong to a CH_2. The coupling constant value of 11.2 Hz agrees with that of ^2J. From its δ_C value of 66 ppm it can be estimated that the CH_2 is connected to an oxygen atom.

The peak set with eight peaks at about 2.7 ppm can be analyzed in a similar manner to the discussion above. It also belongs to the AB part of an ABX system. These two hydrogen atoms are connected to the carbon atom whose peak is situated at 40.1 ppm.

As the compound contains only one CH, this CH should belong to the following structural unit:

$-CH_2-CH-CH_2-$

The CH has fairly large δ_C and δ_H (73.0 and 3.9 ppm, respectively), so this CH should be connected to an oxygen atom.

Combination of the postulations above, the structural units of the unknown compound can be deduced as structure units **27**.

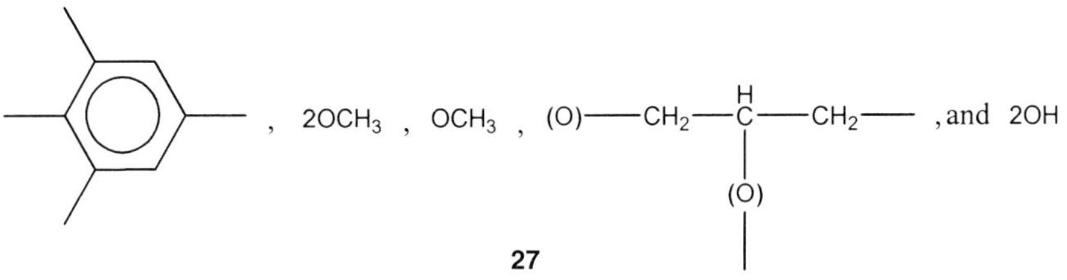

, $2OCH_3$, OCH_3 , (O)—CH_2—C(H)(O)—CH_2— , and 2OH

27

The total mass of these structural units is just 242 u, the same as the relative molecular weight, which means that there is no other heteroatom in this compound, an illustration of the accuracy of the postulations above.

What should be emphasized here is that in order to deduce the position of the substitution in the phenyl ring, it is necessary to analyze the chemical shifts of the substituted carbon atoms in the phenyl ring.

From 153.1 ppm (two equivalent carbon atoms), 136.4 ppm (one carbon atom) and 133.7 ppm (one carbon atom), the substituted positions of the phenyl ring can be determined to be as in structure **28**.

28

Combining the postulations above, the structure of the unknown compound can be deduced as structure **29**.

29

The assignment of chemical shifts is illustrated in structure **30**: the values of δ_C are denoted beside the structural formula and those of δ_H are denoted in parentheses.

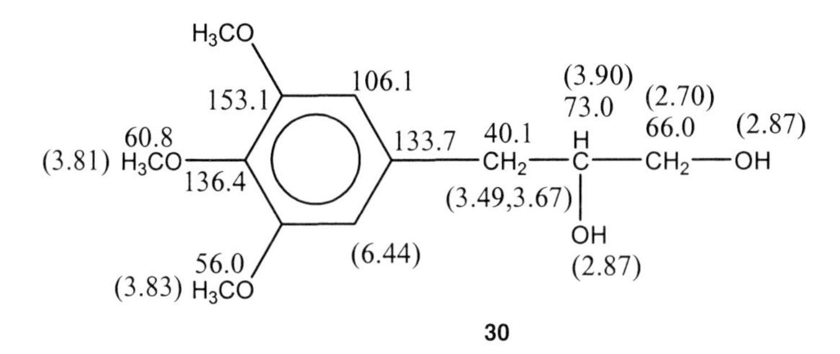

Example 5

An unknown compound has a molecular formula of $C_{13}H_{16}O_2N_2$. Its carbon NMR data are shown in Tab. 3.7. Its ^{1}H spectrum after isotropic exchange with D_2O is shown in Fig. 3.12. Before the exchange, there is one blunt peak at 5.8 ppm with one hydrogen atom and another blunt peak at 8.2 ppm with one hydrogen atom. The peak set at 3.58 ppm was a quartet before the exchange. Try to deduce the unknown structure.

Tab. 3.7 Carbon NMR data for $C_{13}H_{16}O_2N_2$

δ (ppm)	23.2	25.2	39.8	55.9	100.5	112.0	112.4	112.6	122.8	127.7	131.6	154.1	170.2
Multiplicity	q	t	t	q	d	d	d	s	d	s	s	s	s

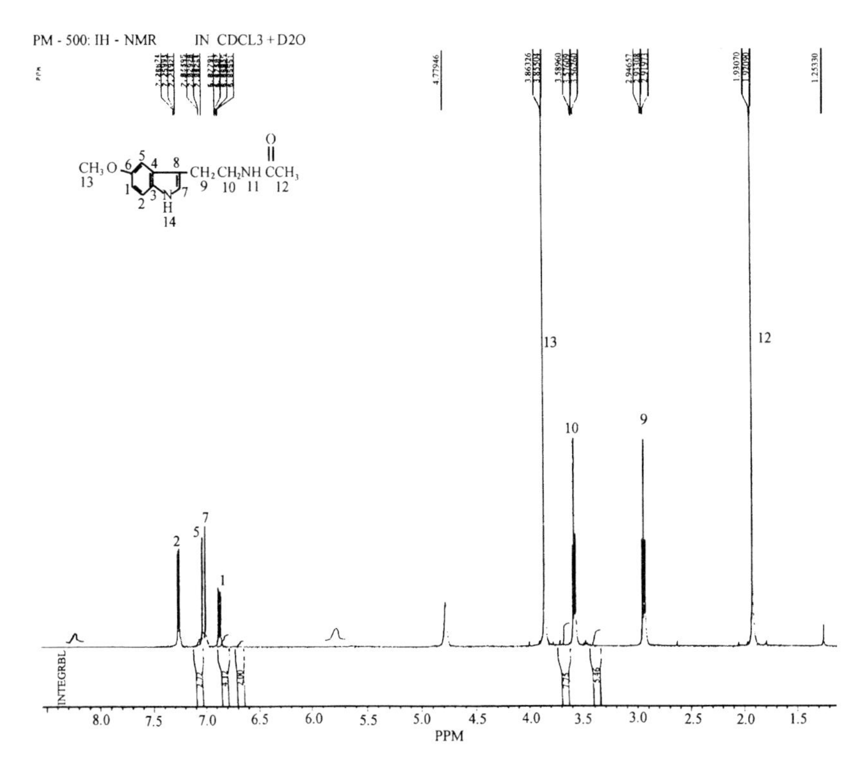

Fig. 3.12 The ¹H spectrum of an unknown compound.

Solution

From its molecular formula, the unsaturation number can be calculated as 7, which is a rather large number.

As 13 carbon atoms have 13 chemical shifts, this molecule has no symmetry. From the multiplicity it is known that 14 hydrogen atoms are connected to carbon atoms. Therefore, two hydrogen atoms should be connected to heteroatoms, which agrees with the two exchangeable peaks at 5.8 and 8.2 ppm, respectively. Obviously these two hydrogen atoms have slow exchange rates, which can be seen from their peak shapes (blunt).

The strong singlet at 3.86 ppm in the ¹H spectrum should be the peak of a methoxy group, whose peak in the ¹³C spectrum is situated at 55.9 ppm. Another singlet at 1.92 ppm in the ^{1}H spectrum should belong to a methyl group, which is connected to a double bond without ^{3}J coupling. The methyl group has a peak at 23.2 ppm in the ^{13}C spectrum.

From the remaining two triplets it can be seen that this compound has the structural unit:

$$–CH_2–CH_2–$$

Because the triplet at 3.58 ppm was a quartet before the isotropic exchange, this structural unit can be extended as:

$$–NH–CH_2–CH_2–$$

From the doublet at 7.26 ppm (J=8.75 Hz), the doublet at 7.03 ppm (J=2.38 Hz) and the peak set (d×d, J=8.76 and 2.40 Hz) at 6.87 ppm, it can be seen that this compound has a 1,2,4-substituted phenyl ring.

Three points must be considered here:

1. The calculated unsaturation number is 7 but that calculated from known structural units is only 6 (a phenyl ring, 4; a double bond, 1; and a carbonyl group, 1).
2. There is an NH peak at 8.2 ppm.
3. The presence of the structural unit of a 1,2,4-substituted phenyl ring.

On the basis of the three considerations above, structural unit **31** can be postulated.

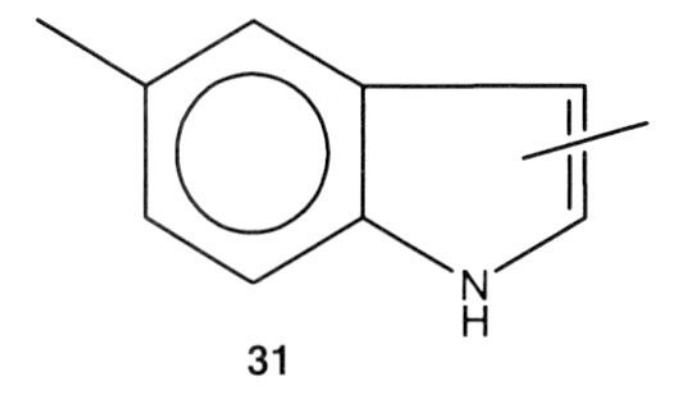

Because all three unsubstituted carbon atoms in the phenyl ring have fairly small δ_H and δ_C, they must be affected by substituents that shift their adjacent carbon atoms and the *para*-carbon atom upfield. From the δ_H and δ_C, the methoxy group should be connected to the phenyl ring. These considerations above can be extended to give structural unit **32**.

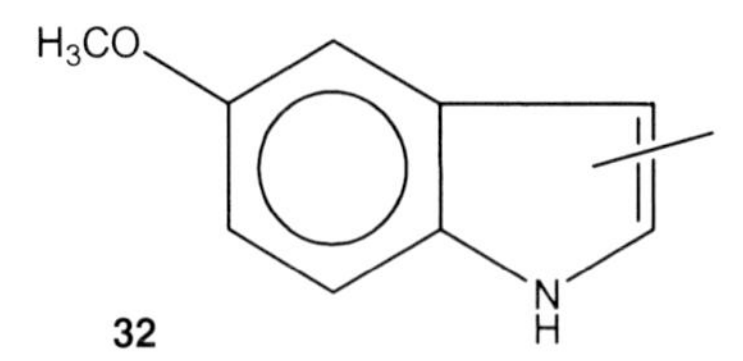

Combination of the aliphatic structural unit and the remaining signals in the ^{1}H and ^{13}C spectra can extend the structural unit into **33**.

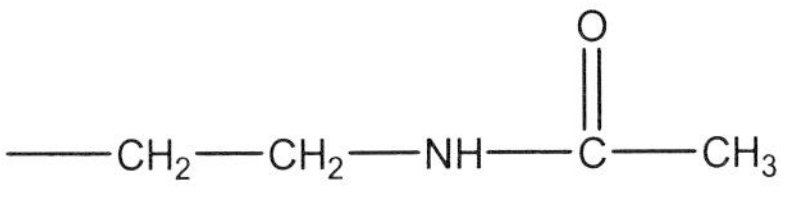

—CH₂—CH₂—NH—C(=O)—CH₃

33

A pyrrole (unsubstituted) has chemical shifts as follows: α-H, 6.68 ppm; β-H, 6.28 ppm. From the singlet at 7.00 ppm in the ^{1}H spectrum, it can be postulated that the substitution is at the β-position. Therefore, the structure of the unknown compound is completed as **34**.

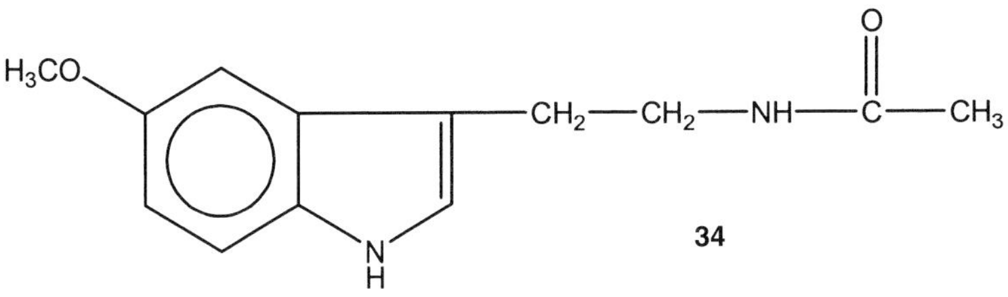

H3CO, CH2—CH2—NH—C(=O)—CH3, N H

34

The assignment of the ^{1}H spectrum is as in structure **35**.

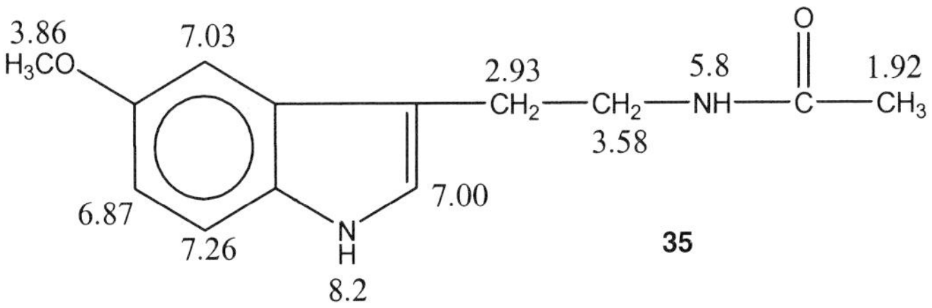

3.86 H3CO, 7.03, 2.93 CH2, 3.58 CH2, 5.8 NH, 1.92 CH3, 6.87, 7.26, 7.00, N H 8.2

35

For the assignment of the phenyl ring in the ^{13}C spectrum more data are required, although a tentative assignment of the ^{13}C spectrum is as in structure **36**.

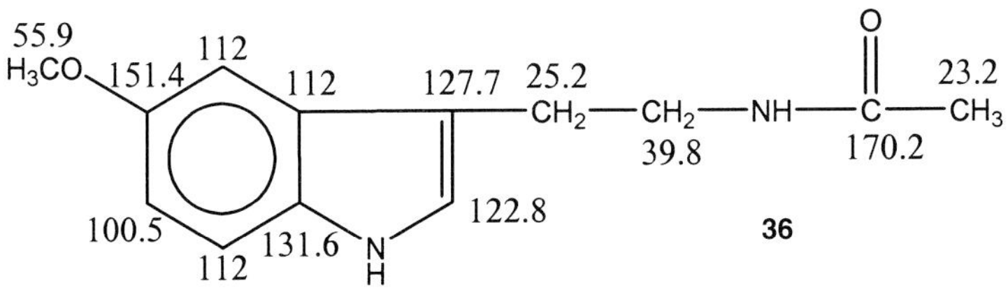

55.9 H3CO, 151.4, 112, 112, 127.7, 25.2 CH2, 39.8 CH2, NH, 170.2 C(=O), 23.2 CH3, 100.5, 112, 131.6, 122.8, N H

36

3.6 References

1 F.W. Wehrli, A.P. Marchand, S. Wehrli, *Interpretation of Carbon-13 NMR Spectra*, 2nd Edn. John Wiley & Sons, Chichester, 1988.
2 E. Breitmaier, W. Voelter, *^{13}C NMR Spectroscopy*, 2nd Edn. Verlag Chemie, Weinheim, 1978.
3 J.T. Clerc, *Structure Analysis of Organic Compounds*, Budapest Akademiai Kiado, 1981.
4 H. Günther, *NMR Spectroscopy, An Introduction*, Wiley, New York, 1980.
5 J.A. Pople, *Mol. Phys.* **1964**, *7*, 301–306.
6 M. Karplus, J.A. Pople, *J. Chem. Phys.* **1963**, *38*, 2803–2807.
7 M.L. Martin, G.J. Martin, J.J. Delpuech, *Practical NMR Spectroscopy*, Heyden, London, 1980.

4
Application of Pulse Sequences and Two-dimensional NMR Spectroscopy

The discovery of pulse-Fourier transform NMR spectroscopy made possible the detection of isotopes with low natural abundance and low sensitivity. Subsequently, the multiplicity of ^{13}C NMR peaks was successfully measured using particular pulse sequences. Then 2D NMR spectroscopy was developed as a perfect method, which offers an objective and accurate structural determination of organic compounds, a milestone in NMR spectroscopy. The Nobel Prize in chemistry in 1991 was awarded solely to Professor Richard R. Ernst for his significant contributions to the development of 2D NMR theory.

In this chapter, the discussion of a number of processes related to pulse sequences is based mainly on the model of macroscopic magnetization vectors, which can be easily understood by chemists because of their clear physical representations. The discussion will start with the basic units of common pulse sequences, thus improving the reader's understanding of the principles of pulse sequences.

Because some of the pulse sequences are hard to explain through the model of macroscopic magnetization vectors, a basic description of NMR experiments using the quantum mechanism is introduced briefly in this chapter. For convenience, the formalism of product operators is presented in Appendix 1.

References [1–8] are listed for the reader who wishes to understand the related topics further.

4.1
Fundamentals

4.1.1
Transverse Magnetization Vector

1. Formation of the transverse magnetization vector

In the rotating frame, a magnetization vector $\boldsymbol{M}$ is affected by an effective magnetic field $\boldsymbol{B}_{\text{eff}}$ (see Section 1.4.2). M tips from the z axis towards the y' axis, producing a transverse magnetization vector.

Structural Identification of Organic Compounds with Spectroscopic Techniques. Yong-Cheng Ning
Copyright © 2005 WILEY-VCH Verlag GmbH & Co. KGaA, Weinheim
ISBN: 3-527-31240-4

2. Transverse magnetization vector and NMR signal

The transverse magnetization vector actually rotates in the laboratory frame. It traverses the detector coil of an NMR instrument so that it produces an NMR signal in the coil. Therefore, it should be borne in mind that a transverse magnetization vector is always associated with an NMR signal. However, the signal shown in an NMR spectrum can take different forms. The four typical signals are the positive absorption signal, the positive dispersion signal, the negative absorption signal and the negative dispersion signal, as shown in Fig. 4.1. These four signals correspond to the fact that the transverse magnetization vector is along the y', x', $-y'$ and $-x'$ axes, respectively. An absorption signal implies that the signal is symmetrical about the axis perpendicular to the frequency axis through the resonant frequency ω_0. The term dispersion signal is that which is anti-symmetrical about the resonant frequency ω_0.

If a transverse magnetization vector is situated between two axes, for example, between the y' and the x' axes, the NMR signal is the addition of the two signals that correspond to these two axes, in which case the signal is the addition of a positive absorption and a positive dispersion. The direction of the transverse magnetization vector, which is determined by the angle between the $+y'$ axis and the transverse magnetization vector, is called the "phase" of a signal. The phase can be adjusted so that all signals are adjusted into positive absorption signals in an NMR spectrum when the spectrum is recorded.

3. Precession of transverse magnetization vectors in the x'–y' plane

It must be emphasized that each peak in an NMR spectrum has its corresponding magnetization vector and its transverse magnetization vector after a pulse.

Each transverse magnetization vector $M_\perp$ precesses in the x'–y' plane of the rotating frame except for the one that has the same frequency as the rotating frame. Theoretically, this frequency can be at any point on the frequency axis. If a peak is situated at the left of the point, its transverse magnetization vector will precess clockwise in the x'–y' plane and if at the right, anti-clockwise. Strictly speaking, this argument corresponds to the fact that the nuclei have positive γ values.

To be more specific, we can now discuss the peaks of A in an AX_2 system as an example. Through the coupling from X, A has three peaks: $\nu_{A1}=\nu_A+J$, $\nu_{A2}=\nu_A$ and $\nu_{A3}=\nu_A-J$. If the rotating frequency of the rotating frame, ω_0, is

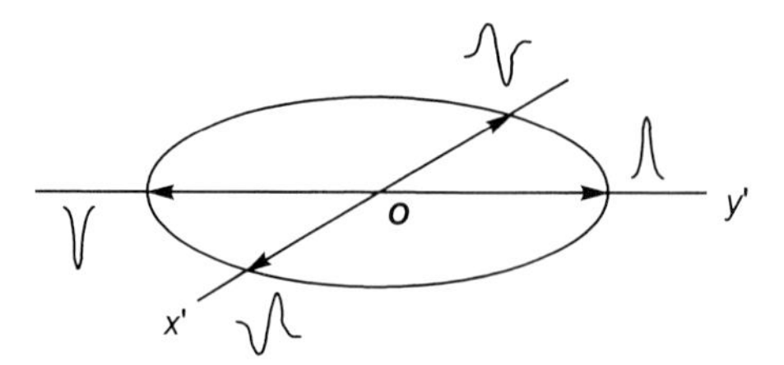

Fig. 4.1 Absorption and dispersion signals.

equal to ν_0, $M_{\perp 2}$ is relatively static in the x'–y' plane but $M_{\perp 1}$ precesses clockwise in the x'–y' plane with a relative frequency J, and $M_{\perp 3}$ anti-clockwise (with the same absolute value of J). If X is decoupled, $M_{\perp 1}$ and $M_{\perp 3}$ will precess with the same frequency ω_0. Their resultant transverse magnetization vector will give rise to one single peak.

4. Quadrature detection

The content from this paragraph to the end of Section 4.1.2 can be skipped in the first reading of the main part of this chapter. However, it is fundamental information for Sections 4.1.9, 4.6.2, 4.6.6, 4.9.1 and Appendix 1.

Under the action of a 90° pulse, a transverse magnetization vector is produced. It will precess in the rotating frame, which induces a voltage in the detector coil, an NMR signal.

If the transmitter frequency is set at an extreme edge of the spectral width, the detected frequency of the signal, which is the difference between the frequencies of the signal and the rotating frame, is fixed. However, if in experiments, the transmitter frequency is set at the center of the spectral width, the radio-frequency power can be reduced and the sensitivity (signal-to-noise ratio) can be improved. If only one detector set along the y' axis is used, it can not distinguish the two transverse magnetization vectors with the same absolute value of the rotating frequencies, which have opposite rotation directions (clockwise and anti-clockwise) because both give the same signal as in Eq. (4-1):

$$V = V_0 \cos 2\pi(\nu - \nu_0)t \tag{4-1}$$

where V is the detected signal induced by the transverse magnetization vectors; $V_0$is the detected maximum signal; ν is the rotating frequency of the transverse magnetization vectors; ν_0 is the frequency of the rotating frame; t is the time counted from the moment the transverse magnetization vectors are produced.

Because $\cos a = \cos(-a)$, the detector can not distinguish these two transverse magnetization vectors.

Two schemes have to be applied to distinguish between them. The first scheme is the use of two detectors that are set along the y' axis and the x' axis, respectively, and acquire data in the same time. The second detector (set along the x' axis) measures the component of the transverse magnetization vectors along the x' axis, Eq. (4-2).

$$V = V_0 \sin 2\pi(\nu - \nu_0)t \tag{4-2}$$

Because $\sin a = -\sin(-a)$, the directions of the rotations of the transverse magnetization vectors (clockwise or anti-clockwise), that is, the signs of their frequencies (positive or negative with respect to ν_0 value) can be determined. The characteristic of the first scheme is *simultaneous* acquisition (sampling).

The second scheme is in principle the use of a single detector [9].

Quadrature detection can be achieved with only one detector but at a rate twice that of the first scheme, and for every data sampling the phase of the detector has an increment of 90°, which means that the detection will be sequential along the axes of x', y', $-x'$, $-y'$..., that is, the detector rotates in the rotating frame anti-clockwise. In fact, this manipulation is performed by two detectors that are set along the x' and y' axes, respectively. The first pair of data samplings is acquired sequentially, first on the x' axis and then on the y' axis. The second pair of data samplings is acquired exactly as with the first pair of data samplings, but the acquired data are multiplied by –1. Therefore, the result of this second pair of data samplings is equivalent to that obtained by performing the sampling first on the $-x'$ axis and then on the $-y'$ axis. Because the increment of the phase is 90°, the acquired signal has an additional frequency of SW/2 where SW is the spectral width. Therefore, the frequency range is changed from ± SW/2 to 0 to SW. Under these conditions, the discrimination of the frequency is solved. From the discussion above, the second scheme is termed "pseudo-quadrature detection."

The characteristic of the second scheme is *sequential* acquisition.

4.1.2 Coherence and Related Topics

In this chapter, the mechanism of pulse sequences is discussed based mainly on the model of the magnetization vector. However, the discussion of the model is a simplified explanation. NMR experiments should be described thoroughly by the quantum mechanism, which means that the discussion should include the term "coherence".

Firstly, we will introduce the concept of coherence.

The formation of the transverse magnetization vector corresponds to a transition, during which the change of the (total) magnetic quantum number between two related energy levels is $\Delta m = \pm 1$.

The magnetization vector of an isolated nucleus is considered initially. Its transition occurs between the energy level of α ($m = +1/2$) and that of β ($m = -1/2$). The change of the magnetic quantum number between these two energy levels is $\Delta m = \pm 1$. This type of transition is called the single-quantum transition.

Then recall an AX system, the simplified energy level diagram of which is shown in Fig. 4.2.

The magnetization vector, which corresponds to the transition of X_1, is next taken as an example. The transition occurs between the energy level of $\alpha\alpha$ ($m = 1$) and that of $\alpha\beta$ ($m = 0$). It is also a single-quantum transition. Similarly, the transitions of the spectral lines of X_2, A_1 and A_2 also have $\Delta m = \pm 1$. These transitions are also single-quantum transitions.

A single pulse can excite a single-quantum transition, an allowed transition, which induces an NMR signal.

The transition between the energy levels of $\alpha\alpha$ and $\beta\beta$ results in a change in the magnetic quantum number of ± 2, that is $\Delta m = \pm 2$. This transition is called the double-quantum transition, or more generally, the multiple-quantum transition. The transition between the energy levels of $\alpha\beta$ and $\beta\alpha$ is called the zero-

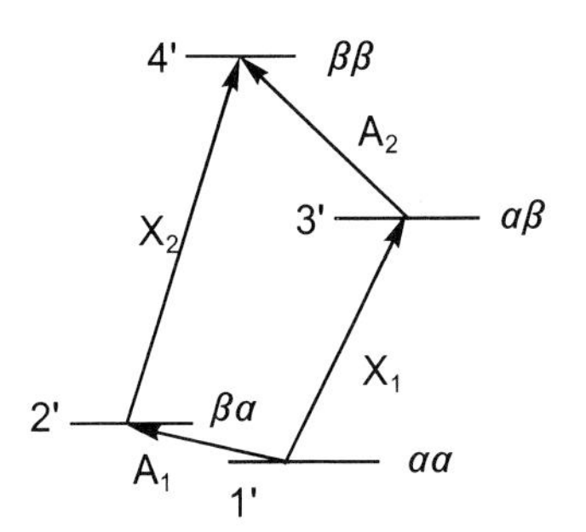

Fig. 4.2 The simplified energy level diagram of an AX system.

quantum transition since $\Delta m = 0$. Transitions with $\Delta m = \pm 2$ or $\Delta m = 0$ are non-allowed transitions. However, it is possible to excite a non-allowed transition by a pulse sequence.

The transverse magnetization vector can be used to describe the transition with $\Delta m = \pm 1$. In other words, only the transition with $\Delta m = \pm 1$ can be described by the transverse magnetization vector. On the other hand, transitions with $\Delta m = 0, \pm 2, \pm 3, \ldots$ can not be described using the model of the magnetization vector. Therefore, it is necessary to use a more generalized term, "coherence."

In quantum mechanics, the term "coherence" means a coherent superposition of two eigenstates. Coherence is used to describe any transition and has a coherence order, p. For a transition from $m = r$ to $m = s$, the coherence order, p, is $s - r$. From the discussion above, it follows that the coherence p is the change in the magnetic quantum number, Δm. When $\Delta m = \pm 1$, single-quantum coherence corresponds to the transverse magnetization vector.

Both zero-quantum coherence and multiple-quantum coherence can be termed multiple-quantum coherence.

Multiple-quantum coherence can be excited by a pulse sequence, but it cannot be detected directly. By using a pulse, multiple-quantum can be transferred into single-quantum coherence, which is then detected. An example of this is the "read pulse" in the pulse sequence of double-quantum filter COSY (see Section 4.6.6). A related discussion is presented in Appendix 1.

It is convenient to use a coherence transfer map (coherence transfer diagram) for the discussion on coherence. The coherence transfer map, which consists of horizontal lines with equal distances, is used to describe coherence transfer pathways. These horizontal lines represent coherence orders. In fact, the analysis of the mechanism of a pulse sequence is that of the coherence and coherence orders during the pulse sequence.

All NMR experiments begin at Boltzmann equilibrium, which corresponds to $p = 0$. The first pulse creates single-quantum coherence whose order can be +1 or –1, which means the hypothetical magnetization vectors precess clockwise or anti-clockwise in the rotating frame (with the same absolute value of velocity). During evolution (without the action of a pulse), coherence orders remain unchanged. The second pulse "branches" or "fans" out these two coherence orders ($p = +1$ and $p = -1$), which means several coherence transfer pathways will be created from one particular order of coherence. This result can be easily explained with Appendix 1. After the pulse, the coherence orders still stay unchanged during evolution. If the pulse sequence contains other pulses, the analysis is similar.

In an NMR experiment, only one desired coherence transfer pathway should be selected from several coherence transfer pathways produced after every pulse during a pulse sequence (in the phase-sensitive 2D NMR experiments, two pathways will be selected in a period) [10]. On the other hand, all other undesired coherence transfer pathways should be suppressed. For an observable response, the experiment must end with single-quantum coherence. It is customary to choose $p = -1$.

These selections (and suppressions) are performed by using phase cycling or pulsed-field gradients (see Section 4.1.9).

In a coherence transfer map, selected coherence transfer pathways are denoted by solid lines. Therefore, the produced coherence transfer pathways and selected coherence transfer pathways during a pulse sequence are described clearly in a coherence transfer map.

4.1.3 Spin Echo

Spin echo is frequently used in pulse sequences and thus it is an important unit for pulse sequences.

The sequence of the spin echo is

$$90^\circ_{x'} - \mathrm{DE} - 180^\circ_{x'} - \mathrm{DE}, \mathrm{AQT} \tag{4-3}$$

where 90° or 180° is the rotating angle of a magnetization vector; the subscript x' is the axis around which the magnetization vector rotates; DE is a delay; AQT is acquisition.

The pulse sequence in Eq. (4-3) is illustrated in Fig. 4.3 in which lower case letters indicate specific selected moments.

The principle of spin echo will be discussed for the following three cases.

1. Spin echo without couplings

This is the simplest case. Its principle is illustrated in Fig. 4.4.

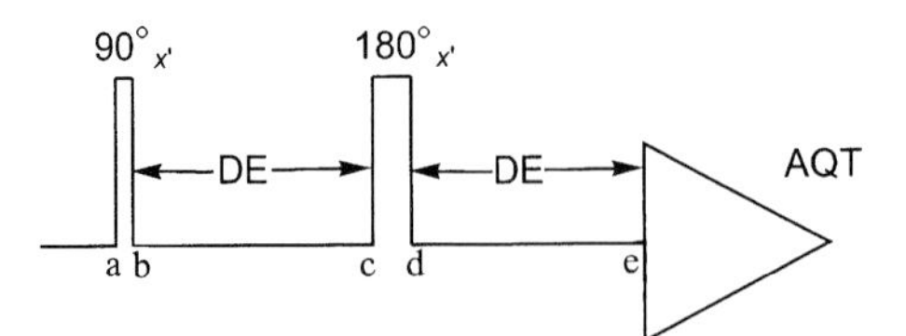

Fig. 4.3 Pulse sequence of the spin echo.

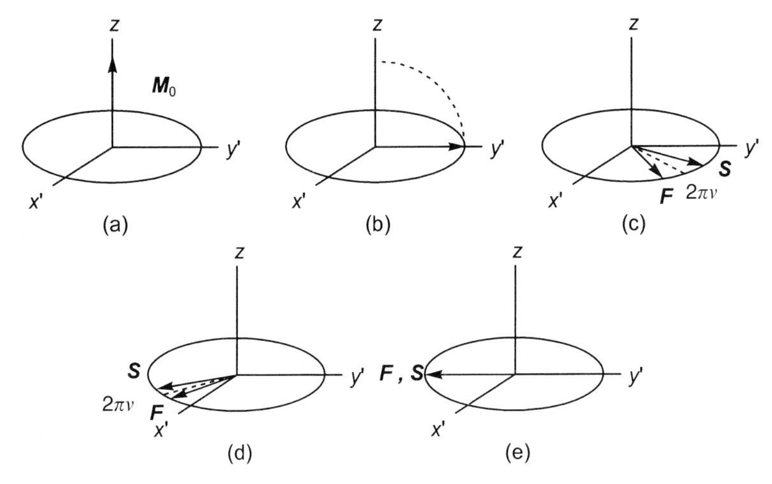

Fig. 4.4 Principle of the spin echo without couplings.

In Fig. 4.4 a, b, c, d and e correspond to a, b, c, d and e in Fig. 4.3, respectively. For simplicity, only one magnetization vector is considered.

i. Magnetization M_0 aligns with the z axis in an equilibrium state.
ii. Under the action of the $90^\circ_{x'}$ pulse, M_0 is rotated about the x' axis from the z axis to the y' axis, thus becoming $M_\perp$.

   Suppose $M_\perp$ precesses clockwise in the x'–y' plane. Because of magnetic field inhomogeneity in the sample solution to be measured, the precession frequencies in various volume elements are not exactly the same. Therefore, after a short time the transverse magnetization vectors form a fan shape. The concept of isochromate can be applied here, which means the transverse magnetization vector in an infinitely small volume unit precesses with a definite precession frequency. The highest rate is denoted as F and the lowest as S. F and S are symmetrical about $2\pi\nu$, where ν corresponds to the chemical shift of the nuclei. $2\pi\nu$ is the precession angle of the transverse magnetization vector. The angle between F and S increases with the delay.
iii. At the end of the first delay, F and S form a certain angle.
iv. Acted on by the $180^\circ_{x'}$, F and S are interchanged and F lags behind S. Because F precesses faster than S, the angle between F and S decreases with the second delay.
v. At the end of the second delay, F catches up with S and an echo appears. This process can also be called refocusing, which means the spin echo has overcome the inhomogeneity of the static magnetic field B_0.

If the transverse relaxation is neglected, the length of the transverse magnetization vector will not decrease after two delays. If the transverse relaxation is taken into account, the length of $M_\perp$ will decrease to $M_0 e^{-2\mathrm{DE}/T_2}$, where T_2 is the real transverse relaxation time. Therefore, the precise value of T_2 can be

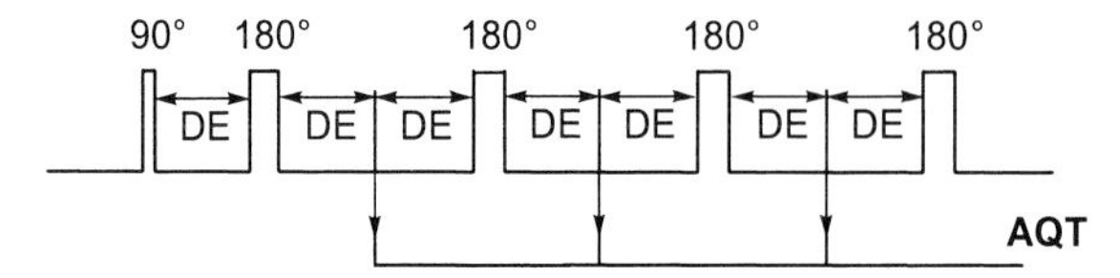

Fig. 4.5 The pulse sequence of the spin echo and related acquisitions for the measurement of T_2.

measured by the spin echo pulse sequence because the influence of the inhomogeneity of B_0 on T_2 has been removed. The pulse sequence of the spin echo and related acquisitions for the measurement of T_2 are shown in Fig. 4.5.

The pulse sequence (4-3) can be improved as (4-4).

$$90^\circ_{x'} - \mathrm{DE} - 180^\circ_{y'} - \mathrm{DE},\ \mathrm{AQT} \tag{4-4}$$

The advantage of pulse sequence (4-4) is that the error of the 180° pulse is not accumulated.

The second function of the spin echo is to refocus chemical shifts, which can be easily understood because a magnetization vector just reverses its movement during the second DE with respect to the first DE. Although $M_\perp$ has different δ values, they refocus at the end of the second delay.

The conclusion that the spin echo overcomes the inhomogeneity of the static magnetic field B_0 and refocuses chemical shifts is still valid for spin systems with couplings because the two types of spin systems are only different in peak number.

2. Spin echo for heteronuclear systems with couplings

For simplicity, we will discuss A in a heteronuclear AX system. Owing to the coupling between A and X, A has two peaks broadened by the inhomogeneity of B_0. We will concentrate on the function of the spin echo for split peaks. The discussion can be developed from two different aspects.

i. From the precession of transverse magnetization vectors in the x'–y' plane.
   A has two peaks at $\nu_{A1}=\nu_A+J/2$ and $\nu_{A2}=\nu_A-J/2$, respectively, which correspond to two magnetization vectors, M_{A1} and M_{A2}. We can set $M_\perp$ (A_1) to be F and $M_\perp$ (A_2) to be S. They will refocus at the end of the second delay.
ii. From the energy level diagram.
   The transitions of A_1 and A_2 are $\alpha\alpha\leftrightarrow\beta\alpha$ and $\alpha\beta\leftrightarrow\beta\beta$, respectively.
   The spin states of A, that is α and β, are exchanged by 180°_A, whereas the spin states of X are not, because X is another isotope with a different resonance frequency. Therefore, only the first Greek letter changes after the 180° pulse. Thus $\alpha\alpha\leftrightarrow\beta\alpha$ becomes $\beta\alpha\leftrightarrow\alpha\alpha$ and $\alpha\beta\leftrightarrow\beta\beta$ becomes $\beta\beta\leftrightarrow\alpha\beta$. The unchanged energy level differences mean an unchanged precession rate, which leads to the same conclusion that two transverse magnetization vectors $M_\perp$ (A_1) and $M_\perp$ (A_2) refocus at the end of the second delay.

3. Spin echo for homonuclear systems with couplings

For simplicity, we will discuss A in a homonuclear system AX.

Our attention is still concentrated on the function of the spin echo for split peaks: A_1 and A_2. The discussion goes as follows.

A has two peaks: A_1 at $\nu_{A1}=\nu_A+J/2$ and A_2 at $\nu_{A2}=\nu_A-J/2$. The two related transitions are $\alpha\alpha\leftrightarrow\beta\alpha$ and $\alpha\beta\leftrightarrow\beta\beta$, respectively.

We will discuss A_1 first. At the end of the first delay, the angle between the $M_\perp$ (A_1) and the y' axis is $2\pi(\nu_A+J/2)$DE, and after the $180^\circ_{x'}$ pulse, the angle is $\pi-2\pi(\nu_A+J/2)$ DE.

It has to be remembered that both nuclei in a homonuclear system are affected by the $180^\circ_{x'}$ pulse. Thus the transition $\alpha\alpha\leftrightarrow\beta\alpha$ becomes $\beta\beta\leftrightarrow\alpha\beta$, which means the precession rate of $M_\perp$ (A_1) changes from $\nu_A+J/2$ to $\nu_A-J/2$. At the end of the second delay, the angle between the $M_\perp$ (A_1) and the y' axis is

$$\pi - 2\pi(\nu_A + J/2)\mathrm{DE} + 2\pi(\nu_A - J/2)\mathrm{DE} = \pi - 2\pi J\mathrm{DE} \tag{4-5}$$

Similarly, at this time the angle between the $M_\perp$ (A_2) and the y' axis is

$$\pi - 2\pi(\nu_A - J/2)\mathrm{DE} + 2\pi(\nu_A + J/2)\mathrm{DE} = \pi + 2\pi J\mathrm{DE} \tag{4-6}$$

Therefore $M_\perp$ (A_1) and $M_\perp$ (A_2) do not refocus at the end of the second delay, but instead they form an angle of $4\pi J$DE symmetrical about the $-y'$ axis.

The NMR signal of A is the projection of $M_\perp$ (A) onto the y' axis. At the end of the second delay, the signal intensity of A is modulated by cos ($2\pi J$DE), or simply by J.

The process mentioned above is illustrated in Fig. 4.6.

J modulation can be applied in 2D NMR. So can a heteronuclear system achieve J modulation? The answer is positive if we use the following pulse sequence:

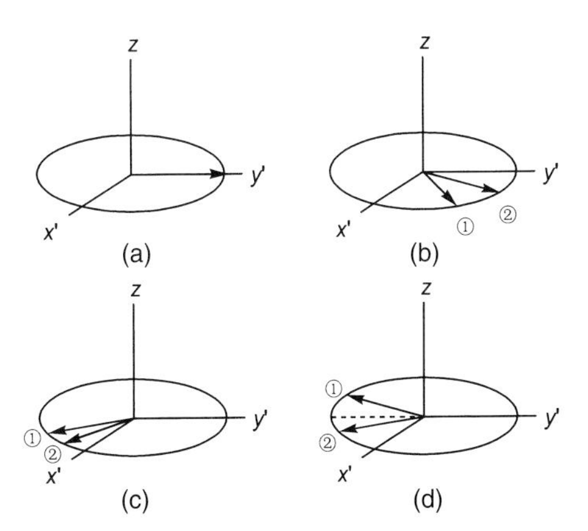

Fig. 4.6 NMR signal intensity is modulated by coupling constant J.

$$90^\circ_{x'} - \text{DE} - 180^\circ_{x'}(\text{A}) - \text{DE, AQT}_\text{A}$$
$$180^\circ_{x'}(\text{X}) \qquad (4\text{-}7)$$

The signal intensity of A will be modulated by the pulse sequence.

4.1.4
The Phase of an NMR Signal is Modulated by the Chemical Shift

For a heteronuclear system, the pulse sequence can be described as follows:

$$90^\circ_{x'}(\text{A}) - \text{DE} - 180^\circ_{x'}(\text{X}) - \text{DE, AQT}_\text{A} \qquad (4\text{-}8)$$

where $90^\circ_{x'}$ (A) and AQT_A are used for A and $180^\circ_{x'}$ (X) is used for X.

A heteronuclear AX system will be discussed as an example.

The pulse sequence and its function are illustrated in Figs. 4.7 and 4.8, respectively.

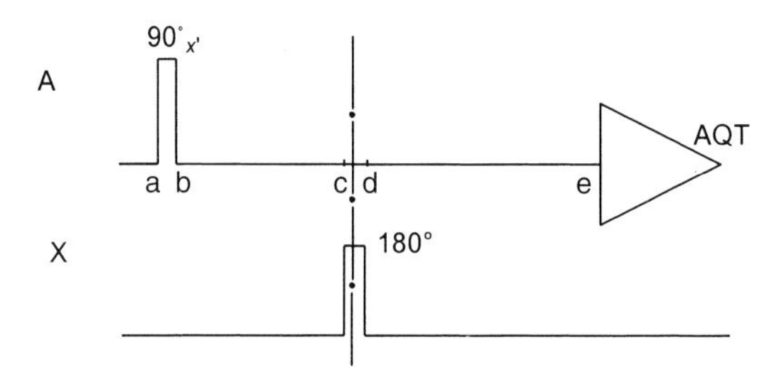

Fig. 4.7 Pulse sequence (4-8).

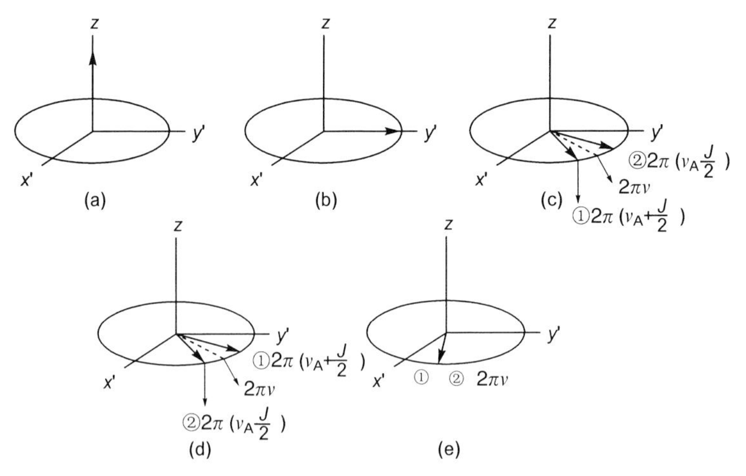

Fig. 4.8 The NMR signal is modulated by the chemical shift.

The a, b and c in Fig. 4.8 are the same as before. The $180^\circ_{x'}$ (X) pulse changes the spin state of X. Thus $\alpha\alpha \leftrightarrow \beta\alpha$ becomes $\alpha\beta \leftrightarrow \beta\beta$ and $\alpha\beta \leftrightarrow \beta\beta$ becomes $\alpha\alpha \leftrightarrow \beta\alpha$, which means the two transverse magnetization vectors exchange their precession rates. At the end of the second delay, the two transverse magnetization vectors refocus at $2\pi\nu_A$ DE, which means that an NMR signal is modulated by its chemical shift.

4.1.5
Bilinear Rotational Decoupling, BIRD

The pulse sequence of BIRD is shown in Fig. 4.9.

In fact the 180° pulse for the proton and the 180° pulse for ^{13}C are applied simultaneously. For the convenience of discussing the principle, we put the latter just after the former. This is reasonable because τ is a duration of milliseconds while pulses are only microseconds.

The BIRD pulse sequence has two functions, which are as follows:

1. It can distinguish the magnetization vector of hydrogen atoms connected to a ^{13}C directly from that of hydrogen atoms connected to a ^{12}C directly.
2. It can distinguish the magnetization vector of hydrogen atoms connected to a ^{13}C directly (with $^1J_{C\text{-}H}$) from that of hydrogen atoms connected to a ^{13}C indirectly (with $^2J_{C\text{-}H}$, $^3J_{C\text{-}H}$, etc.).

We will now discuss the first function; this principle is illustrated in Fig. 4.10.

In Fig. 4.10a, the two magnetization vectors of the two types of hydrogen atoms are along the z axis. Acted upon by the first $90^\circ_{x'}$, they have rotated from the z axis to the y' axis, as shown in Fig. 4.10b. Suppose that the rotating frequency of the rotating frame is equal to the chemical shift of the hydrogen atoms. The magnetization vector of hydrogen atoms connected to a ^{12}C will always be along the y' axis. As hydrogen atoms connected to a ^{13}C show two lines in a spectrum, they correspond to two magnetization vectors, one of them (α) rotating clockwise and the other (β) anti-clockwise. Both have an angular velocity of $\pm 2\pi \times {}^1J_{C\text{-}H}/2$, respectively. Because $\tau = 1/(2\,{}^1J_{C\text{-}H})$, both α and β have rotated $2\pi \times {}^1J_{C\text{-}H}/2 \times 1/2\,{}^1J_{C\text{-}H} = \pi/2$ at (c), which means that they are along the x' and $-x'$ axes, respectively. Now a pulse of 180°_x is applied to ^{1}H. The magnetization vector along the y' axis is rotated from the y' axis to the $-y'$ axis while the

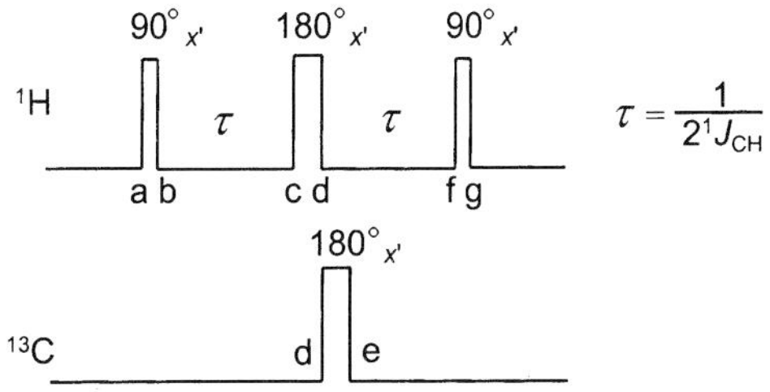

$$\tau = \frac{1}{2\,{}^1J_{CH}}$$

Fig. 4.9 Pulse sequence of BIRD.

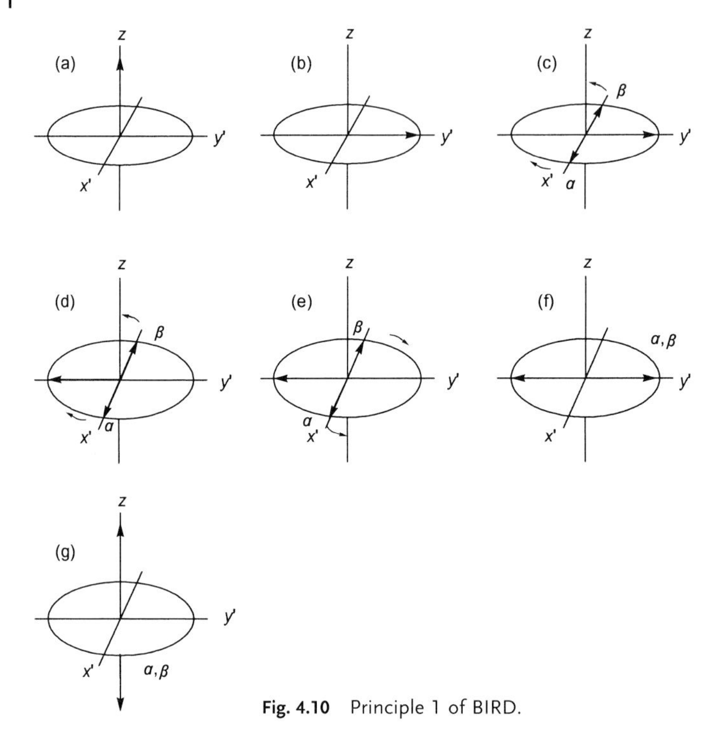

Fig. 4.10 Principle 1 of BIRD.

magnetization vectors along the $\pm x'$ axes is still along the same $\pm x'$ axes, as shown in Fig. 4.10d. Just after the $180^{\circ}_{x'}$ pulse is applied to ^{1}H, a $180^{\circ}_{x'}$ pulse is applied to ^{13}C (in fact these two pulses are applied simultaneously). The rotational directions of α and β are changed, as shown in Fig. 4.10 e. After a period of $\tau = 1/2\,^{1}J_{\text{C-H}}$, α and β meet on the y' axis while the magnetization vector of ^{1}H connected to ^{12}C is still along the $-y'$ axis, as shown in Fig. 4.10 f, which means that the magnetization vectors of hydrogen atoms associated with ^{13}C are different to that of hydrogen atoms associated with ^{12}C. After a $90^{\circ}_{-x'}$ pulse is applied to ^{1}H, the former is along the $-z'$ axis while the latter is along the z' axis, as shown in Fig. 4.10g.

The second function of a BIRD pulse sequence can be analyzed similarly. The principle is illustrated in Fig. 4.11.

4.1.6
Spin Locking

Spin locking was applied to the measurement of relaxation time $T_{1\rho}$ many years ago. Today spin locking is used mainly in some two-dimensional NMR spectra.

The pulse sequence of spin locking is illustrated in Fig. 4.12. The lowercase letters are used to denote specific moments.

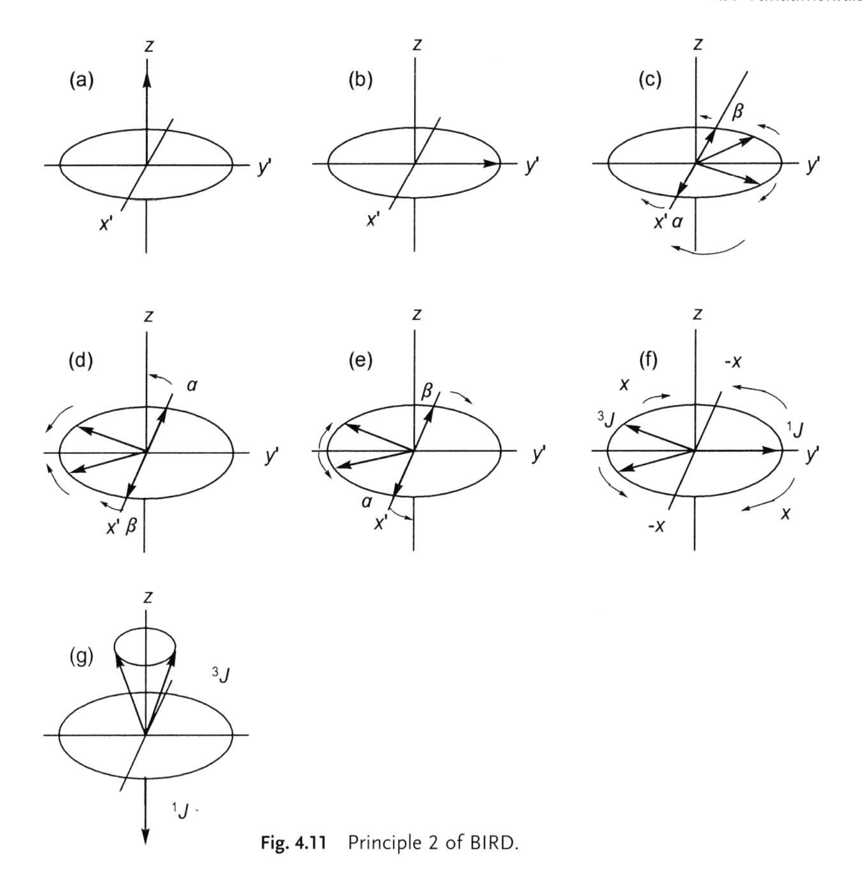

Fig. 4.11 Principle 2 of BIRD.

We shall discuss the pulse sequence for an isolated spin system first. The movements of the magnetization vector of the system are shown in Fig. 4.13.

In an equilibrium state, the magnetization vector is along the z axis, as shown in Fig. 4.13a. Influenced by a $90^{\circ}_{x'}$ pulse, the magnetization vector is rotated by 90° along the x' axis, arriving at the y' axis, as shown in Fig. 4.13b. Once the magnetization vector arrives at the y' axis, B_1 is changed from the x'

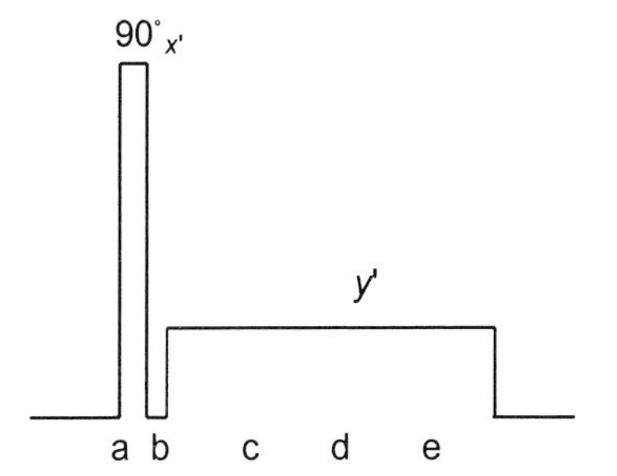

Fig. 4.12 The pulse sequence of spin locking.

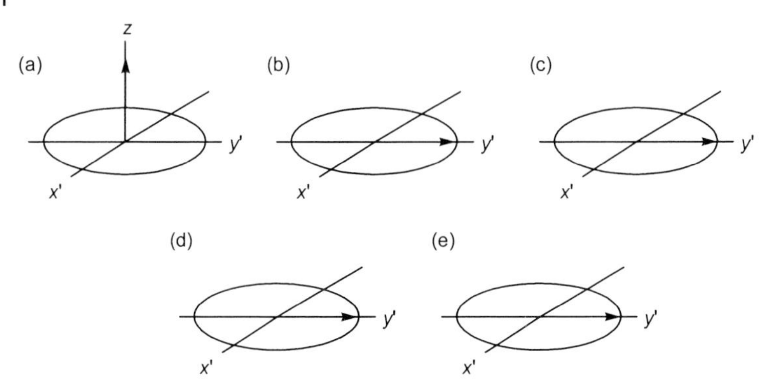

Fig. 4.13 Movements of the magnetization vector of an isolated spin system during spin locking.

axis onto the y' axis. B_1 will continue for a fairly long period of time compared with the relaxation time of the magnetization vector. Because the (transverse) magnetization vector M and B_1 have the same direction, they do not affect each other, that is, the moment acting on M is equal to zero. In this case M cannot change its direction, which means that M is locked along the y' axis as long as it exists. However, M is changing (reducing) its length because of relaxation. Therefore its length decreases continuously in the order c, d and e of Fig. 4.13, which is why this pulse sequence can be used to measure the relaxation time. In a sense, the relaxation along the y' axis in the rotating frame is similar to the longitudinal relaxation along the z' axis in the laboratory frame. The relaxation is called longitudinal relaxation in the rotating frame and denoted as $T_{1\rho}$ (ρ is used to denote the rotating frame).

Now we will discuss spin locking for a system consisting of two spins. In this context, the spin locking pulse sequence is another sequence that we will deal with later. At present it is sufficient to analyze magnetization vector changes for the two nuclei.

It is known from the above discussion that a transverse magnetization vector will be locked along the y' axis when B_1 is applied along the y' axis. Now there are two transverse magnetization vectors, both of which should be locked along the y' axis. As the two magnetization vectors are in the same direction, a dipole–dipole interaction will result, that is, their transfer and interchange, which is NOE except that the two magnetization vectors are not along the z axis but along the y' axis. We call it ROE (the rotating frame Overhauser effect). The two-dimensional NMR spectrum produced by using ROE is called ROESY, which will be discussed in Section 4.7.2.

4.1.7
Isotropic Mixing

The technique of isotropic mixing is the basis of the (2D NMR) HOHAHA (homonuclear Hartman-Hahn spectroscopy) and TOCSY (total correlation spectroscopy). While it could be classified as a "spin-locking" type of technique, the topic is discussed here because in experiments involving isotropic mixing, the parameter B_1 is usually much stronger than that used in spin-locking experiments, and the resultant spectra look different to those obtained by spin-locking techniques.

1. Hartmann-Hahn matching and cross-polarization

In solid-state NMR experiments, Hartmann and Hahn proposed that the pulse sequence shown in Fig. 4.14 and Eq. (4-9) be adopted to enhance the sensitivity of the ^{13}C nuclei.

In the pulse sequence shown in Fig. 4.14, a 90°_x pulse is applied to the ^{1}H channel and the ^{1}H magnetization vector moves from the z' axis to the y' axis. In the meantime B_1 is applied to the y' axis so that the transverse magnetization vector can be locked. At the time of locking, the ^{13}C channel is connected and the radio-frequency power is adjusted according to Eq. (4-9)

$$\gamma_H B_{1H} = \gamma_C B_{1C} \tag{4-9}$$

where γ_H and γ_C are the magnetogyric ratios of ^{1}H and ^{13}C, respectively, and B_{1H} and B_{1C} are the radio-frequency powers of the ^{1}H and ^{13}C channels, respectively.

Equation (4-9) is known as the Hartmann-Hahn correlation. Under such conditions, the ^{1}H and ^{13}C exchange energy and they are said to "make contact"

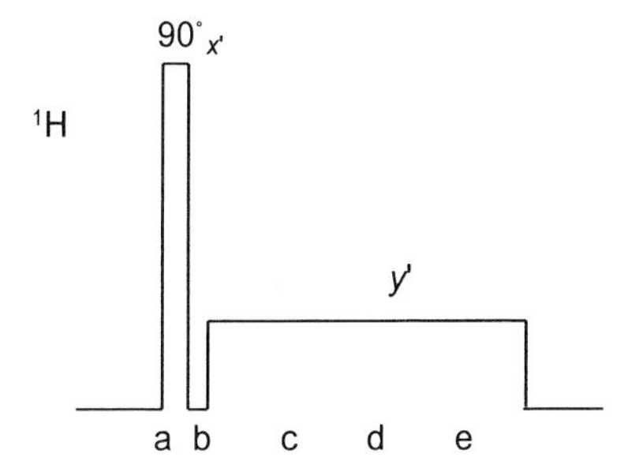

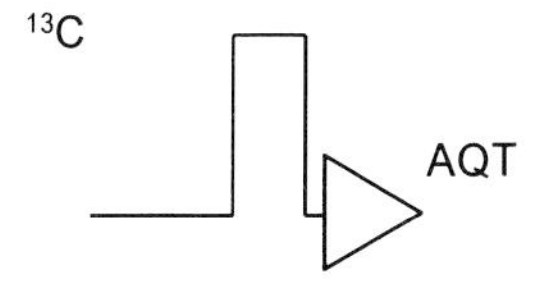

Fig. 4.14 Pulse sequence of the Hartmann-Hahn matching experiment.

with each other. Consequently the ^{13}C signals are enhanced due to the great abundance and the high magnetogyric ratio of ^{1}H (the ^{1}H signals will become weaker after "making contact"). This process of energy transfer between the two nuclei is called cross-polarization.

2. Cross-polarization in the homonuclear system

In a heteronuclear ^{1}H and ^{13}C system, due to the significant difference between the values of γ_H and γ_C (the former being four times greater than the latter), the resonance frequencies of the two nuclei differ greatly. Thus B_{1C} has little influence on ^{1}H and vice versa.

Davis and Bax reported that the cross-polarization by Hartmann-Hahn matching can also occur in a homonuclear system [11, 12] in which only B_{1H} is involved. In a homonuclear system, protons belonging to various functional groups have different chemical shifts. In other words, the transverse magnetization vectors have different precession frequencies $\nu_i - \nu_0$, where ν_i is the resonance frequency of ^{1}H in different functional groups and ν_0 is the precession frequency of the rotating frame relative to the laboratory frame. The term $\nu_i - \nu_0$ is known as "offset", with nuclei of different functional groups having different offset values so that the conditions of Eq. (4-9) are difficult to meet. However, cross-polarization can occur when the condition approaches that required for the Hartmann-Hahn matching.

Let us consider two ^{1}H nuclei, A and X, assuming that their offset frequencies are Δ_A and Δ_X, where $\Delta_A = \nu_A - \nu_0$ and $\Delta_X = \nu_X - \nu_0$. When $\nu_0 \gg \Delta_A$, the effective field strength of radio-frequency of ν_A can be expressed as [11]

$$\nu_A = \nu_0 + \Delta_A^2 / 2\nu_0 \tag{4-10}$$

For the same reason,

$$\nu_X = \nu_0 + \Delta_x^2 / 2\nu_0 \tag{4-11}$$

Subtracting ν_X from ν_A will give

$$\nu_A - \nu_X = (\Delta_A^2 - \Delta_X^2)/2\nu \tag{4-12}$$

If the coupling constant between A and X is J_{AX}, and Eq. (4-13) is fulfilled,

$$\frac{|\Delta_A^2 - \Delta_X^2|}{2\nu_0} < |J_{AX}| \tag{4-13}$$

A and X can transfer their magnetization vectors effectively. The physical implication of this relationship is that when a strong spin locking field is applied, the influence caused by the difference in chemical shift becomes much smaller than that caused by the coupling effect, leading to a situation where an AX system approaches an AA′ system so as to form a strong coupling system. For a big coupling system, the magnetization vectors of this system can fully interact with one another through coupling constants.

Since the effect of chemical shifts is temporarily "removed," the mixing under such a condition is called "isotropic mixing." The intensity of the spin locking field must be higher than that required for ROE in order to achieve isotropic mixing.

3. Isotropic mixing

The concept of "isotropic mixing" was first proposed by Braunschweiler and Ernst who put forward the term TOCSY (total correlation spectroscopy), and described its pulse sequence, principles and examples. During the evolution period of a 2D experiment, the effect of coupling is weak and both the chemical shift and the coupling constant play a significant role. This condition is referred to as the single spin mode. During the mixing period, owing to the application of a pulse sequence, the system becomes a strong coupling environment. The effect of chemical shift is diminished at this time. From the aspect of quantum mechanics, only the H_J term remains in the Hamilton equation, and the single spin mode is no longer present; it becomes a collective spin mode. If the mixing period is long enough, the effect can be extended to the entire coupling system, and isotropic mixing occurs during this period. The detection period that follows has a weak coupling effect again, and the chemical shift and coupling constants exert their effects.

The cross-polarization in homonuclear Hartmann-Hahn matching and isotropic mixing will be further discussed in subsequent sections on HOHAHA and TOCSY (Section 4.8.3).

4.1.8 Selective Population Inversion

Selective population inversion (SPI) refers to the inversion of the particular magnetization vector of a spectral line.

Let us consider the $^{13}C^{1}H$ system, a typical AX system, as an example to illustrate the process of SPI.

The energy diagram of an AX system is shown in Fig. 4.15.

In the diagram, the population of each energy level is clearly marked.

The population n_i can be described by the Boltzmann distribution, that is:

$$\frac{n_i}{n_j} \propto e^{-\frac{\Delta E}{kT}} \qquad (4\text{-}14)$$

where k is the Boltzmann constant; T is the absolute temperature; E is the energy difference between energy levels, and in NMR experiments, $\Delta E = h\nu$

Since $\Delta E \ll kT$, only the first two terms need to be considered.

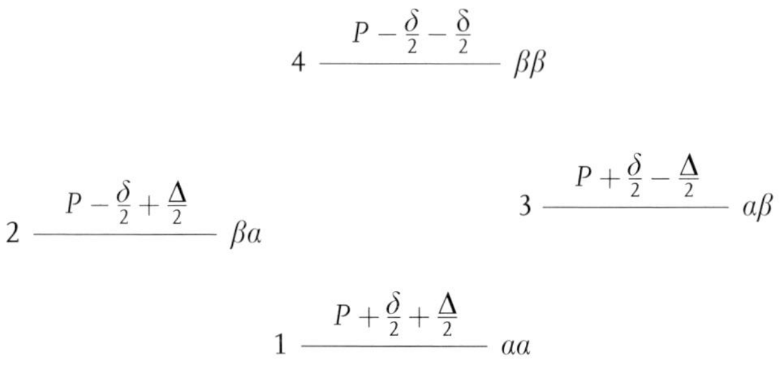

$$4 \quad \underline{P - \frac{\delta}{2} - \frac{\delta}{2}} \quad \beta\beta$$

$$2 \quad \underline{P - \frac{\delta}{2} + \frac{\Delta}{2}} \quad \beta a \qquad\qquad 3 \quad \underline{P + \frac{\delta}{2} - \frac{\Delta}{2}} \quad a\beta$$

$$1 \quad \underline{P + \frac{\delta}{2} + \frac{\Delta}{2}} \quad aa$$

Fig. 4.15 Four energy levels of the $^{13}C^{1}H$ system and its population.

$$\begin{aligned} e^{-\frac{\delta E}{kT}} &= 1 - \frac{\Delta E}{kT} + \left(\frac{\Delta E}{kt}\right)^2 - \cdots \\ &\cong 1 - \frac{\Delta E}{kT} \\ &\cong 1 - \frac{hv}{kT} \end{aligned} \tag{4-15}$$

From Eq. (4-15), it is seen that the difference in population between two transition energy levels of one type of nuclei (which relates to the intensities of spectral lines) is expressed as hv/kT. When hv_H/kT is expressed as Δ, and hv_C/kT as δ, and assuming that the average population in the four energy levels in Fig. 4.15 is p, then the populations of the four energy levels under the equilibrium condition are

$$\begin{aligned} p_1 &= p + \delta/2 + \Delta/2 \\ p_2 &= p - \delta/2 + \Delta/2 \\ p_3 &= p + \delta/2 - \Delta/2 \\ p_4 &= p - \delta/2 - \Delta/2 \end{aligned}$$

These population values are shown in Fig. 4.15.

We can further simplify the above expressions as follows.

Because $\gamma_H = 4\gamma_C$, so $v_H = 4v_C$ and $\Delta = 4\delta$. When we set $\delta = 2$, the populations of the system is shown in Fig. 4.16.

From the diagram it can be seen that the intensities of the two spectral lines of ^{13}C ($aa \leftrightarrow \beta a$, $a\beta \leftrightarrow \beta\beta$) have a ratio of 1:1, and those of ^{1}H ($aa \leftrightarrow a\beta$, $\beta a \leftrightarrow \beta\beta$) have a ratio of 4:4. (We have set the transition intensity for ^{13}C as 1.)

We shall now discuss the process of selective population inversion. The term "selective" refers to the use of a selective pulse focusing on a specific transition. Let our discussion focus on the ^{1}H transition between $aa \leftrightarrow a\beta$ and assume that its magnetization vector is rotated by 180°. Contrary to what we have discussed about non-selective pulses, in which all nuclei of the same isotope (e.g., ^{1}H or ^{13}C) are affected, the selective pulse only affects a selective transition of selected

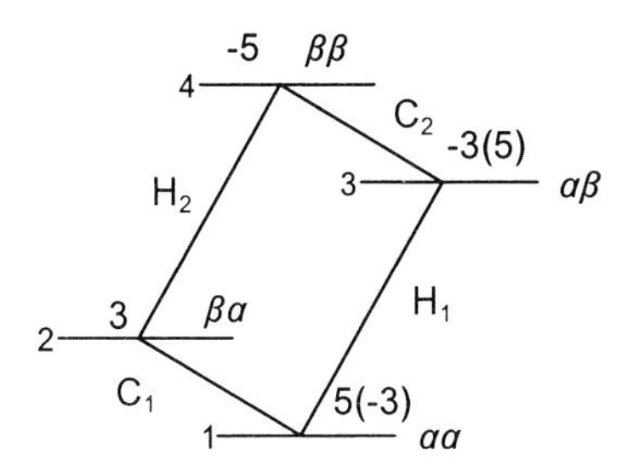

Fig. 4.16 Four energy levels of a $^{13}C^{1}H$ system and its population and transitions.

nuclei and has no effect on other transitions. Experimentally this situation is achievable. Based on the Fourier transform principle, the bandwidth excited by a pulse is proportional to $1/t_p$ (where t_p is the pulse width). When the pulse width is long, the pulse-excited bandwidth will be narrow. The shaped pulse, which is more suitable for selective induction, will be discussed in a later section (Section 4.1.10). A later discussion (Section 4.2.2) also shows that a selective pulse can be replaced by a pulse sequence composed of a series of non-selective pulses.

Our discussion now focuses on a selective 180° pulse applied to the transition between the energy levels 1 and 3 leading to the change of the magnetization vector from the z' axis to the $-z'$ axis. It has to be emphasized again that a selective 180° pulse will invert only a particular magnetization vector. This phenomenon can be illustrated in the energy level diagram, in which the changed populations of these two energy levels are denoted in parentheses while the populations before the selected excitation are denoted before the parentheses. The two other energy levels are not affected by such a selective pulse, and therefore their populations remain unchanged.

A comparison of the population under equilibrium conditions with that after applying a selective 180° pulse shows the following differences.

Under equilibrium conditions:

$$p_1 - p_2 = (p + 5\delta/2) - (p + 3\delta/2) = \delta = 2$$
$$p_3 - p_4 = (p - 3\delta/2) - (p - 5\delta/2) = \delta = 2$$

After applying a selective 180° pulse:

$$p_1 - p_2 = (p - 3\delta/2) - (p + 3\delta/2) = -3\delta = -6$$
$$p_3 - p_4 = (p + 5\delta/2) - (p - 5\delta/2) = 5\delta = 10$$

It is clear that the transition strength of ^{13}C is increased from 1 and 1 to –3 and 5, respectively, a phenomenon termed polarization transfer. Since an isotope with a small magnetogyric ratio γ (e.g., ^{13}C) will have a small nuclear magnetic energy difference, the difference in population defined by the Boltzmann distribution is thus small under equilibrium conditions, and the intensities of the spectral lines will be low. When a selective 180° pulse is applied to the

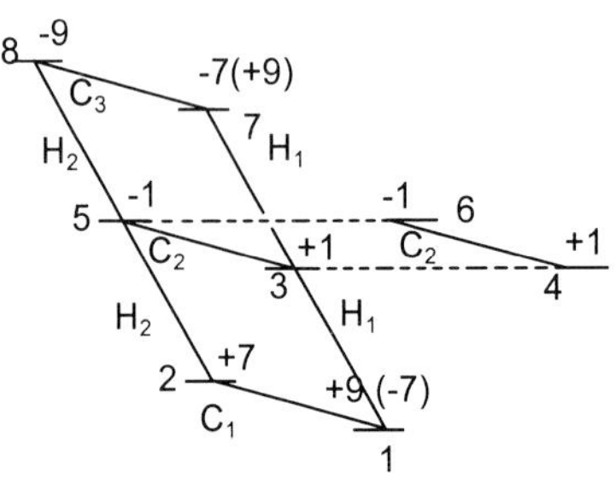

Fig. 4.17 Population of energy levels of the $^{13}CH_2$ system (in numbers) and their difference.

coupled nuclei with greater γ values (e.g., ^1H), the latter nuclei can transfer the greater population difference to the former nuclei. As a result, the signal intensities of the former are enhanced.

In conclusion, for a $^{13}C^1H$ system, the intensities of the ^{13}C spectral lines are increased from 1:1 to $[1-\gamma(^1H)/\gamma(^{13}C)]:[1+\gamma(^1H)/\gamma(^{13}C)]$. After subtracting the initial polarization (i.e., the population difference under the equilibrium condition), the spectral intensity ratio becomes $-[\gamma(^1H)/\gamma(^{13}C)]:[\gamma(^1H)/\gamma(^{13}C)]$. Thus, the smaller the γ value of an isotope, the greater the signal enhancement of the isotope after polarization transfer.

For an AX_2 system (a $^{13}CH_2$ group is used as an example), the populations of energy levels of the system are shown in Fig. 4.17.

A similar discussion leads to that the intensities of the three ^{13}C spectral lines as being –7:2:9. After subtracting the initial polarization (1:2:1), the result becomes –8:0:+8.

For an AX_3 system, we will use $^{13}CH_3$ as an example (Fig. 4.18).

A similar discussion leads to that the intensities of the four ^{13}C spectral lines as thus being –11, –9, +15 and +13, respectively. After subtracting the initial polarization (1, 3, 3, 1), the result will become –12:–12:+12:+12.

The above SPI results of the ^{13}CH, $^{13}CH_2$ and $^{13}CH_3$ systems will be applied in the analysis of INEPT (Section 4.2.2).

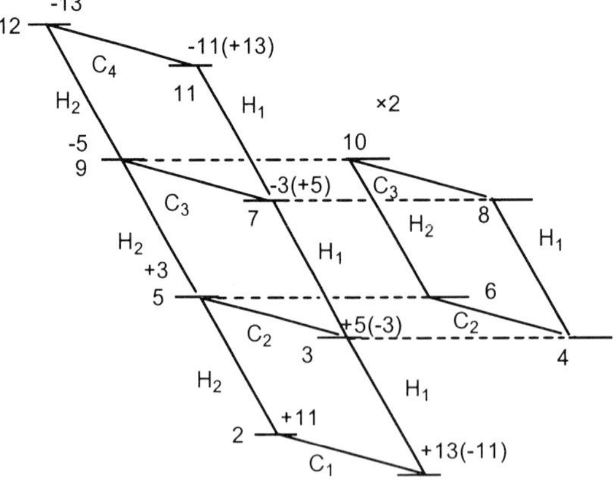

Fig. 4.18 Populations of energy levels of the $^{13}CH_3$ system (in numbers) and their difference.

4.1.9 Pulsed-field Gradient

In recent years, the technique of pulsed-field gradient (PFG), which refers to the application of a gradient along an axis during the time interval τ, has been widely applied in two-dimensional and multi-dimensional NMR experiments. We will use the z axis as an example (Fig. 4.19).

When a sample is placed inside a magnetic field along the z axis and assuming that the inhomogeneity of the magnetic field is negligible, nuclei located in different positions on the z axis (i.e., at different heights in the sample tube) will experience the same magnetic field B_0, as shown in Fig. 4.19a. When a gradient field $B'(z)$ is applied along the z axis

$$B'(z) = g(z)z \tag{4-16}$$

where $g(z)$ is the intensity of the magnetic field gradient along the z axis; z is the height (the origin being O, the upper part represented by a positive sign and the lower part by a negative sign).

The total magnetic field experienced by the sample at height z is

$$B_0 + B'(z) = B_0 + g(z)z \tag{4-17}$$

as shown in Fig. 4.19b.

We will now discuss a single magnetization vector. Before the application of the magnetic field gradient, the magnetization vector will have the same precession frequency at different heights:

$$\omega = \gamma B_0/2\pi \tag{4-18}$$

After applying the magnetic field gradient, due to the different values of $B'(z)$ at different heights, the magnetization vector will precess at the same frequency only at zero height. At positions above the O level, the magnetization vector will precess faster, while at positions below the O level, the magnetization vector will

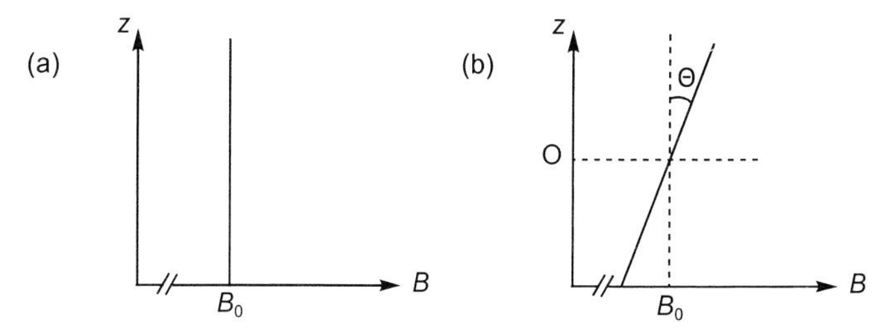

Fig. 4.19 (a) Original uniform magnetic field B_0 along the z axis. (b) Application of a gradient field along the z axis.

precess more slowly, because the precession frequency is varied according to the following equation:

$$\omega = \frac{\gamma}{2\pi} \times [B_0 + g(z)z] \tag{4-19}$$

From this equation it can be seen that the magnetization vectors at different heights can be "marked" by the application of a magnetic field gradient.

On the basis of the previous discussion, it is easier to understand the process of dephasing the magnetization vector by the application of the field gradient.

A magnetization vector will precess along the z direction under equilibrium conditions, as shown in Fig. 4.20 a. When a 90°_x pulse is applied, the magnetization vector is rotated from the z axis to the y' axis, as shown in Fig. 4.20 b. If a pulsed-field gradient is applied immediately after time b, the magnetization vectors at point c at different heights will stay in the y' direction because points c and b are very close together.

With the thin disk at point O as the center, five thin disks an equal distance apart are cut at unit height (A, B, O, C and D from top to bottom). Suppose the angular frequency of the magnetization vector equals that of the rotating frame at point O, the magnetization vector will precess clockwise at points A and B and anti-clockwise at points C and D, as shown in Fig. 4.20 d. At the end of the

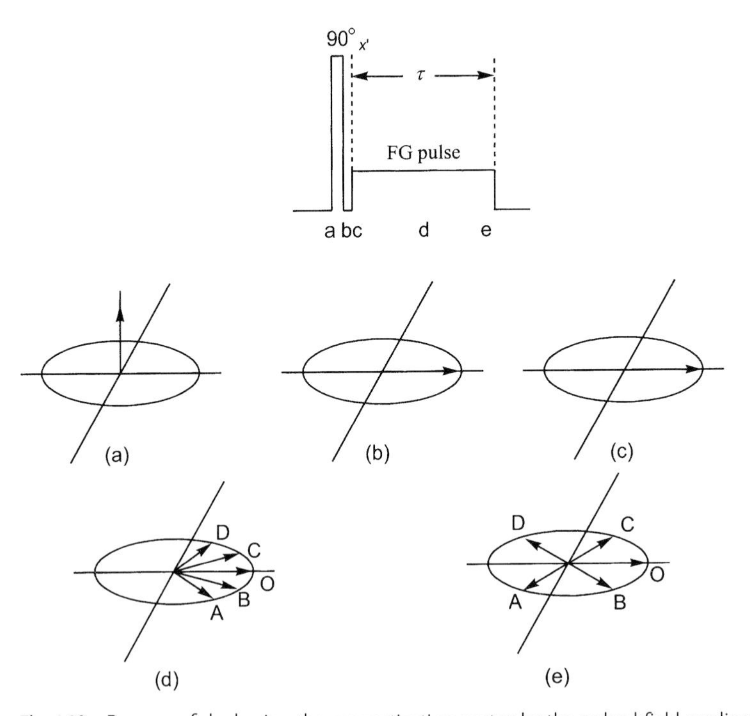

Fig. 4.20 Process of dephasing the magnetization vector by the pulsed-field gradient.

pulsed-field gradient, as shown in Fig. 4.20e, the magnetization vectors in A, B, O, C and D will disperse widely, the radians being $2g\tau$, $g\tau$, 0, $-g\tau$, and $-2g\tau$, respectively, relative to the y' axis. If τ is longer than a certain time, the sum of all magnetization vectors becomes zero, that is, they are dephased completely.

If, immediately after point e, a pulsed-field gradient is applied in the reverse direction with the same absolute intensity and duration, the magnetization vectors at different heights will then precess in the opposite direction. At the end of the second pulsed-field gradient, the magnetization vectors at different heights will reunite on the y' axis, a phenomenon referred to as rephasing or refocusing.

The unit of the field gradient is T m^{-1} (telsas per meter) or G cm^{-1} (1 telsa = 10000 gauss). The commonly used intensity lies in the range of 1–60 G cm^{-1} while τ (the time period during which a field gradient is applied) varies from 100 μs to 20 ms. Because of the difficulty of the instrumental hardware producing a "direct upward" or "direct downward" field gradient, a half sine wave curve is frequently applied. Irrespective of the type of curves used, the total intensity of the field gradient G is

$$G = \int_0^{\tau} g(t)\mathrm{d}t \tag{4-20}$$

A field gradient can be applied not only along the z axis, but also along the x or y axes. Since the distance of the sample tube along the z axis is much greater than the diameter of the tube, the magnitude of the applied field gradient should be much larger along the x or y axis. Whenever a field gradient is applied along the x or y axis, the sample tube should remain still with no spinning. Similar to the case when a field gradient is applied along the z axis, the magnetic field experienced at the center remains unchanged, that is, B_0, while an additional B' is added at other locations:

$$B'(x) = g(x)x \tag{4-21}$$

and

$$B'(y) = g(y)y \tag{4-22}$$

where $g(x)$ and $g(y)$ are the intensities of the field gradient along the x and y axis, respectively.

Hence, the total magnetic field is varied when a field gradient is applied along the x or y axis, giving a "regional marking" to the transverse magnetization vectors.

We will now discuss the applications of pulsed-field gradients.

The first application, which is easy to understand, is suppressing the solvent signals. Given that the diffusion of large molecules is relatively slow, the sample molecules remain in the same unit volume for a short time period, for example, at the end of the second pulsed-field gradient (or at the end of the last pulsed-field gradient in a complicated pulse sequence). At this time, the dephasing caused by the first field gradient refocuses and an NMR signal results. On the other hand, the diffusion of water molecules is fast and the magnitude of their motion is large. As a result, their magnetization vectors cannot be refocused and the signals will be suppressed. It has to be stressed here that refocusing can only occur within the same unit volume. Although our discussion on dephasing and refocusing involves two consecutive pulsed-field gradients (in opposite directions), other pulses (e.g., 90° and 180°) can appear in between.

Secondly, the pulsed-field gradient can be applied during the process of selecting a desired signal from useful and useless signals produced by a pulse sequence. This function, which is the main reason for its applications in 2D and multi-dimensional NMR, was achieved by the phase cycling technique before the adoption of the pulse-field gradient technique.

The concept of phase cycling will now be introduced. Phase cycling is a procedure often used in NMR experiments. The phases of the pulse to excite spins and the phases of the detector are changed systematically so that desired signals are accumulated and unwanted signals are deleted.

One-dimensional NMR experiments are initially considered. As was described in Section 4.1.1, quadrature detection is necessary. Quadrature detection with simultaneous acquisition by two detectors will now be discussed. If the phase difference of these two detectors is not exactly 90° and/or the gains of these two detectors are slightly different, a quadrature imaginary image, or a quad imaginary image, which is a false signal, will appear together with a real signal. Obviously it will interfere with the experimental result. If the procedure of spectral difference is used, this problem can be resolved. Therefore, the data are accumulated, during which time the phase of the pulse and those of the detectors are changed systematically. The systematic change of the phase of the pulse and those of the detectors is phase cycling, which performs the operation of spectral differences. Quadrature imaginary images are deleted in this way.

In a two-dimensional NMR experiment, a pulse sequence is used. Under the action of a pulse sequence, there are several coherence transfer pathways of which only one is desired in general. The phase cycling has the task of choosing the desired coherence transfer pathway that possesses the correct coherence order (including the sign).

Suppose that the phase change of a pulse, that is, its phase increment, is φ. If the pulse causes a change Δp in coherence order, the resulting coherence suffers a phase shift $\varphi\Delta p$. Therefore, if the phase of the detector changes according to the variation, the signal, which corresponds to the chosen coherence, will be recorded while other signals will not be collected. This is the principle of the selection of the desired signal by the phase cycling.

Now we can take a multi-quantum 2D NMR spectrum as an example to illustrate the selection of the specific signal. Multi-quantum coherence appears after the application of a pulse sequence. The n-quantum coherence is n times more sensitive to the phase change than the single quantum coherence. In other words, when the phase of a pulse is changed (e.g., from x' to $-y'$, $-x'$, or y' axes), the spinning rate of the n-quantum coherence is n times that of the single quantum coherence. Therefore, varying the phase of the receiver at the corresponding rate can lead to the removal of the single quantum coherence.

In applying phase cycling to a 2D NMR experiment, multiple sampling (normally a multiple of four) is necessary for every t_1. Consequently, the signal-to-noise ratio is improved. However, the operation of the phase cycling is time-consuming.

Next we will discuss how to select the desired NMR signals by using the pulsed-field gradient.

Recall the suppression of the solvent peak in an NMR experiment, which is based on the fact that the sum of the precession angles of the magnetization vector (in a volume element) acted on by the pair of pulsed-field gradients is zero. This means that under the action of the first pulsed-field gradient the magnetization vector precesses by an angle (θ) but under the action of the second pulsed-field gradient the magnetization vector precesses in the opposite sense by the same angle ($-\theta$). This situation can be extended to a general consideration. The condition for refocusing a magnetization vector by pulsed-field gradients is

$$\sum_i \theta_i = 0 \tag{4-23}$$

From Eq. (1-12), $\omega = \gamma B_0$, θ_i can be written as

$$\theta_i = \gamma G_i \tau_i \tag{4-24}$$

where θ_i is the precession angle of the magnetization vector under the action of the i-th pulsed-field gradient; γ is the magnetogyric ratio; τ_i is the time duration in which the i-th pulsed-field gradient is applied. In PFG experiments, τ_i is usually kept constant.

G_i is the amplitude of the i-th pulsed-field gradient. It is assumed that all pulsed-field gradients have the same shape, which is true in PFG experiments. Since τ_i is normally kept constant in a PFG experiment, G_i will change.

Equation (4-24) is just a preliminary result. For general considerations, magnetization will be replaced by coherence, in which case coherence order, p, will be introduced into the right side of Eq. (4-24), that is:

$$\theta_i = p\gamma G_i \tau \tag{4-25}$$

Since τ_i is usually kept constant, it is changed to τ.

Equation (4-25) describes the function of a pulsed-field gradient. For the function of the pulsed-field gradients, which are used in a pulse sequence, the condition for refocusing a selected coherence is

$$\sum_i \theta = \sum_i p_i \gamma G_i \tau = 0 \tag{4-26}$$

Because the coherence order changes during the pulse sequence, p has the appropriate subscript.

The evolution of the coherence order is determined by the applied pulse sequence. Pulsed-field gradients cannot change the coherence order. However, a desired coherence order can be selected by pulsed-field gradients according to Eq. (4-26). For example, in the DQF-COSY (see Section 4.6.6), the desired coherence order before the read pulse is 2 while that after the read pulse is –1. If a pair of pulsed-field gradients are used before and after the read pulse, respectively, their amplitudes can be calculated according to Eq. (4-26). Suppose that G_1 and G_2 are the amplitudes of the first and the second pulsed-field gradients, respectively (before and after the read pulse). The precession angle of the coherence during the first pulsed-field gradient is

$$\theta_1 = 2\gamma G_1 \tau$$

and similarly

$$\theta_2 = -1\gamma G_2 \tau$$

From Eq. (4-26),

$$\theta_1 + \theta_2 = 0$$

which leads to

$$G_2 = 2G_1$$

Therefore, under these conditions, the desired coherence can be selected.

If the pulse sequence is related to a heteronuclear system, the term γ_i has to be considered.

If the signal-to-noise ratio permits it, acquisition is achieved once for each t_1 with pulsed-field gradients, thereby significantly shortening the running time. As only a particular signal is selected by a pulsed-field gradient, the t_1 noise (the noise parallel to the F_1 axis) will also diminish.

The third application of the pulsed-field gradient is to match the selective excitation (see Section 4.1.10 for a discussion on this).

4.1.10
Shaped Pulse

A shaped pulse refers to a pulse that is different from the square pulse, such as those of the Gaussian or half-Gaussian shapes.

The application of high-power square waves in 1D or in multi-dimensional NMR experiments leads to non-selective excitation. However, under particular circumstances it may be desirable to excite a specific spectral line by applying selective excitation, which will be discussed below.

Based on the Fourier transform principle, the spectral widths to be excited by a shaped pulse or by a square pulse are fairly different. The spectral width to be excited can be determined by using a shaped pulse of a defined shape and pulse width, and the rotation angle of a magnetization vector is selected by choosing a particular power for the emission coil (through the control of its attenuation value). The selection of a spectral line can be achieved by using an appropriate offset value. Combination of the above conditions will lead to a fixed rotation of the magnetization vector of a particular spectral line (or a narrow spectral region).

Real situations can be much more complicated. For example, in a Gaussian shaped pulse, the excited spectral width is likely to be influenced by the rotation angle of the magnetization vector and by varying power. In the case of the E-BURP pulse, which is non-linear, only a 90° rotation can be achieved.

The application of selective excitation, which is mainly used to reduce the number of dimensions of multi-dimensional NMR experiments, will now be discussed. For instance, a 2D process can be reduced to 1D, a 3D to 2D, and so forth. The conversion of the 2D spectrum into 1D will be taken as an example.

It should be noted that the 1D spectrum referred to here is not the usual 1D technique of NMR spectroscopy, but rather a 1D expression derived from a 2D spectrum. A 1D spectrum can be obtained from a 2D experiment by plotting the cross section at a particular frequency. We have to bear in mind that such a 1D spectrum has a low digital resolution because of the small number of data points in any one dimension of a 2D NMR experiment, much smaller than that in a conventional 1D experiment.

Let us take a COSY experiment as an example (see Section 4.6.1). The COSY pulse sequence consists of two 90° (non-selective) pulses, with the result that the ^3J coupling relationship among all ^{1}H moieties in the sample molecule is obtained. A simplified 1D diagram can be constructed by changing the first 90° pulse into a selective 90° pulse and using an appropriate offset to fix the position of the selected spectral line. As a particular set of spectral lines is selected, only spectral lines showing ^3J coupling with the specific set will appear. Because of the better digital resolution, the coupling constants can be determined accurately. This method can be used for TOCSY (see Section 4.8.3) so that the couplings between all of the ^{1}H belonging to the same coupling system and the particular ^{1}H are shown with good resolution.

We will now take GOESY (gradient enhanced NOE spectroscopy) [14] as another example. Although NOESY (see Section 4.7.1) is an important 2D technique that measures NOE, it is often replaced by the simple difference NOE (1D) spectrum for the following reasons: (1) There is a mixing time τ_m in the pulse sequence of NOESY, which has to be precisely selected in order to obtain good results. (2) The experimentation time needed for the 2D experiment is also longer than that needed for a 1D measurement. (3) A spectroscopist is often interested in the NOE of some, but not all, of the ^{1}H in the molecule. (4) The 1D difference NOE spectra have a higher sensitivity.

On the other hand, the 1D NOE method has certain drawbacks. For example, positive or negative false peaks could appear along the baseline. A modification of the pulse sequence of NOESY by using a selective pulse together with a pulsed-field gradient will lead to the GOESY spectrum [14]. In the process, selective excitation of the magnetic vector will refocus (by forming the echo) under the influence of the pulsed-field gradient, thus creating a signal. Other signals are suppressed due to the dephasing the magnetization vectors. A GOESY spectrum is equivalent to a difference NOE spectrum, with the advantage of being able to clearly reveal weak NOE (about 0.03%) signals.

Similarly, when selective excitation is applied to 3D NMR experiments, the 2D information of selected spectral lines can be obtained.

In addition to their major application to reduce the "dimension" of multi-dimensional NMR experiments, shaped pulses can be used for other purposes such as suppression of the water signal.

4.2 Spectrum Editing

This section deals with the determination of the multiplicity of ^{13}C, that is, the order of ^{13}C, which means finding how many hydrogen atoms are connected to related carbon atoms. It is necessary to deduce an unknown structure.

4.2.1 *J* Modulation or APT

In fact *J* modulation and APT are very similar and they were developed almost simultaneously. *J* modulation was developed by le Cocq and Lallemand [15] and APT (proton attached test) by Patt and Shoolery [16].

This method, although having a simple pulse sequence, is effective when used.

The simplest pulse sequence for APT is shown in Fig. 4.21.

The movements of magnetization vectors of carbon atoms with different orders under the influence of the pulse sequence are illustrated in Fig. 4.22.

For simplicity, assume that the rotating angular velocity of the rotating frame is equal to $2\tau\nu_C$ (where ν_C is the chemical shift of the carbon atom being considered). The principle will be discussed in terms of carbon atoms with different orders.

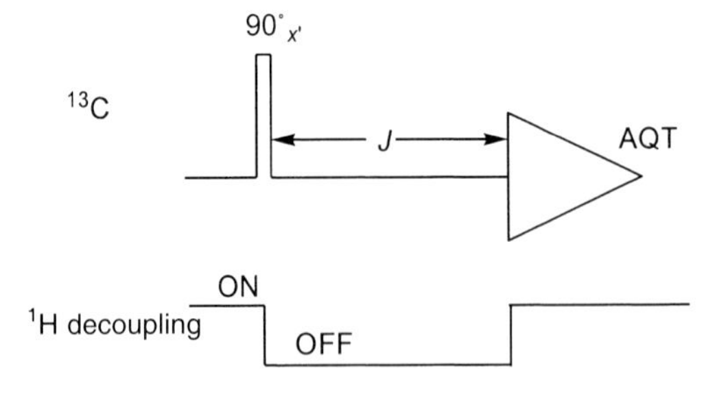

Fig. 4.21 The simplest pulse sequence for APT.

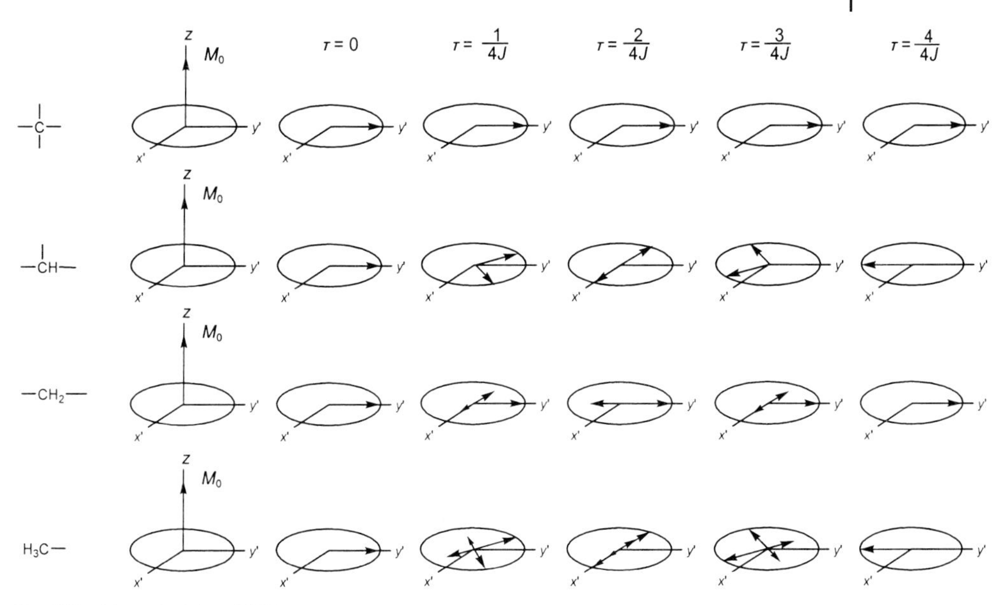

Fig. 4.22 The principle of APT.

For the quaternary carbon atom, the 90° pulse rotates its magnetization vector from the z axis to the y' axis, producing a transverse magnetization vector. As the quaternary atom is not connected to hydrogen atoms, there are no split transverse magnetization vectors, and the same is true of the spectral line of the quaternary atom. Under these conditions, the transverse magnetization vector will always be along the y' axis.

For the tertiary carbon atom CH, the 90° pulse rotates its magnetization vector from the z axis to the y' axis, producing a transverse magnetization vector. Because the decoupling of hydrogen atoms is cut off after the 90° pulse, the transverse magnetization vector is split into two components once it arrives at the y' axis. Their angular velocities in the x'–y' plane are $\pm 2\pi \times J/2 = \pm \pi J$, respectively. They form an angle of $\pm\pi/4$, $\pm\pi/2$, $\pm 3\pi/4$ or $\pm\pi$ with respect to the y' axis when $\tau = 1/(4J)$, $2/(4J)$, $3/(4J)$ or $1/J$, respectively. During acquisition decoupling is applied. The two components will be fixed relatively. Their resultant vector produces a decoupled NMR signal. A signal for a maximum absolute value is obtained when $\tau = 1/J$ because the angle formed by the two components is zero.

The secondary carbon atom for $-CH_2$, and the primary carbon atom for CH_3 can be discussed similarly. The signals for the four types of carbon atoms can be written as follows:

- C: 1
- CH: $\cos(\pi J \tau)$
- CH: $0.5 + 0.5\cos(2\pi J \tau)$ (the former corresponding to the central component, and the latter to the two other components)

- CH: $0.25 \cos(3\pi J\tau) + 0.75 \cos(\pi J\tau)$ (the former corresponding to the components ascribed to the outer two lines, and the latter to the components ascribed to the inner two lines).

From the above discussion, we know that carbon atoms can be classified into two types: C and CH_2; CH and CH_3 by using the simple pulse sequence. This is because the former produces positive signals while the latter produces negative signals when $\tau = 1/J$. Quaternary carbon atoms can be differentiated from secondary carbon atoms by the following two methods: (1) Quaternary carbon atoms produce weak signals and secondary carbon atoms produce strong signals in a ^{13}C spectrum with broadband decoupling. (2) If acquisition (with decoupling) starts at $\tau = 1/(2J)$, quaternary carbon atoms produce positive signals while secondary carbon atoms produce no signals. In general, primary carbon atoms can be distinguished from tertiary carbon atoms because the former has fairly small chemical shifts. It is difficult to differentiate CH_3 from CH when CH_3 has a δ value greater than 24.8 ppm, which is the minimum possible δ value of CH.

By using APT, all signals except that of quaternary carbon atoms are enhanced by NOE.

The prerequisite for applying APT is that carbon atoms have approximate $^1J_{C-H}$ values. Fig. 4.22 is drawn under these conditions.

Two other pulse sequences for APT are shown in Fig. 4.23.

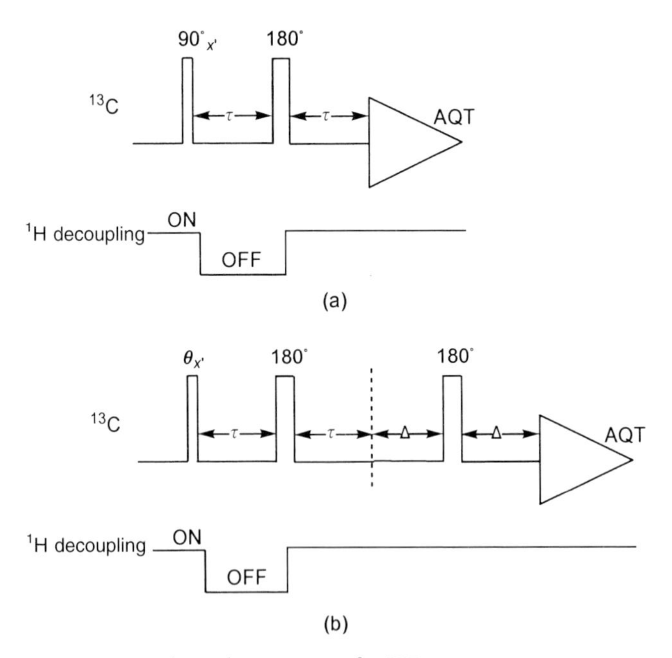

Fig. 4.23 Two other pulse sequences for APT.

4.2.2
INEPT (Insensitive Nuclei Enhancement by Polarization Transfer)

The pulse sequence for INEPT is shown in Fig. 4.24.

The τ value in Fig. 4.24 is not a constant. Three spectra are obtained with τ values of $1/4J$, $2/4J$ and $3/4J$.

We will now take a system consisting of $^{13}C^{1}H$ as an example. The principle of INEPT is illustrated in Figs. 4.25 and 4.26.

The movements of a ^{1}H magnetization vector are shown at the top of Fig. 4.25 while those of a ^{13}C magnetization vector are at the bottom. For simplicity we assume that the rotating frequency of the rotating frame is equal to the chemical shift of the ^{1}H.

At point (a) of the time axis, the two components of the ^{1}H magnetization vector, M_{01} and M_{02}, are along the z' axis. They are rotated to the y' axis under the influence of the $90^{\circ}_{x'}$ pulse, as shown in Fig. 4.25 b. After a duration of $1/(4J)$, they have rotated $\pm\pi/4$, respectively, in the x'–y' plane, as shown in Fig. 4.25 c. At this time we simultaneously apply a $180^{\circ}_{x'}$ pulse to ^{1}H and a $180^{\circ}_{x'}$ pulse to ^{13}C. At point (d) M_{01} and M_{02} have rotated 180° along the x' axis. They also exchange their rotating directions because of the 180° pulse for ^{13}C (see Section 4.1.3). Again after $1/(4J)$, they form an angle of 180°, as shown in Fig. 4.25 e. At this moment we apply two $90^{\circ}_{y'}$ pulses: one to ^{1}H and one to ^{13}C. For simplicity, it is assumed that the $90^{\circ}_{y'}$ pulse is applied to ^{1}H first. M_{01} and M_{02} have rotated from the x' axis to the z axis, as shown in Fig. 4.25 f. Compare the situation at (f) with that at (a). At (a), both M_{01} and M_{02} are along the z' axis. However, at (f) one of them has been inverted, which means that the two populations for the two energy levels have been exchanged. This is just the SPI (selective population inversion, see Section 4.1.8), which leads to transferring polarization from ^{1}H into ^{13}C. Therefore the two components of the ^{13}C magnetization vector are along the $\pm z'$ axis, respectively [at the

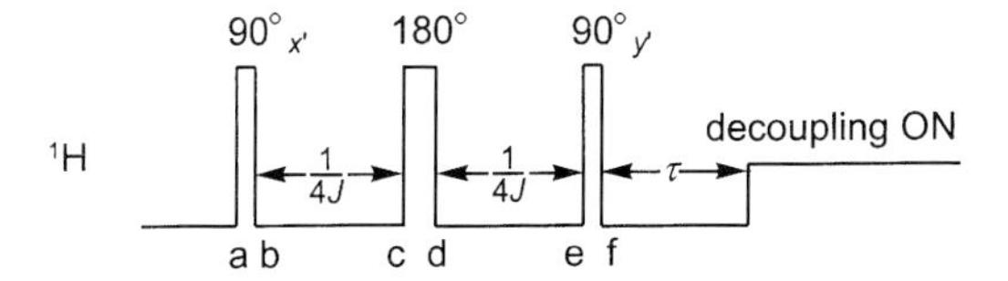

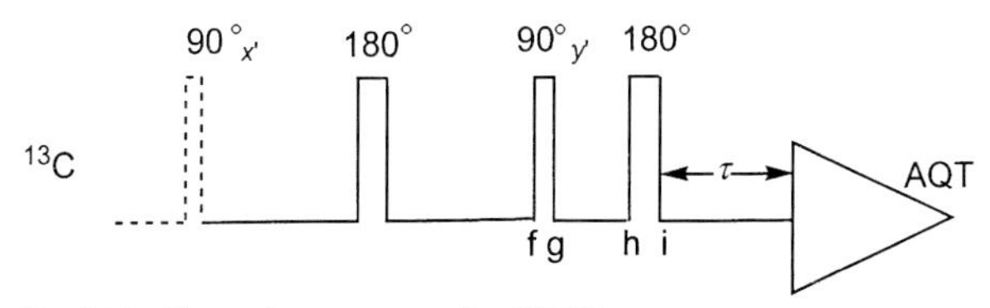

Fig. 4.24 The pulse sequence for INEPT.

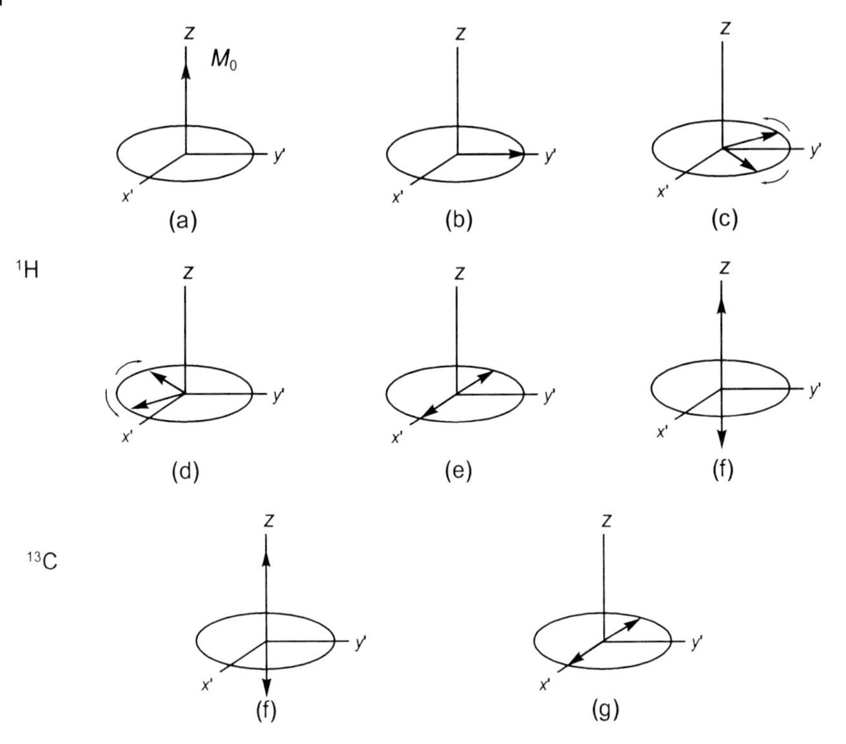

Fig. 4.25 Principle (1) of INEPT.

bottom of (f)]. Their intensities are enhanced from +1, +1 to −3, +5. Then we consider the $90^\circ_{y'}$ pulse for ^{13}C which rotates the two components from the z axis to the x' axis, as shown in Fig. 4.25 g.

It is better to obtain a pair of signals with intensities of +4 and −4, which is why $90^\circ_{y'}$ and $90^\circ_{-y'}$ are applied alternately as the second 90° pulse to ^{1}H. Another method is to apply a 90° pulse, a pre-saturated pulse, before (a), as shown by the dotted line in Fig. 4.25.

If acquisition starts at (g), there is no ^{13}C signal, because the resultant of the two components is equal to zero. It is necessary to wait until the two components are overlapped or at least approach each other.

So far we have discussed the CH system. In fact, CH_2 and CH_3 are similar to CH in SPI (see Section 4.1.8).

Now we shall deal with τ, that is from (g) to (h). Fig. 4.22 can be used as a reference. However, the following facts should be noted.

1. Magnetization vector components after SPI are different in number and/or in intensities from those without SPI. For a CH_2, there are two components in the former case but in the latter there are three components. A CH_3 has a quartet signal with intensities of 1, 3, 3 and 1, while after SPI it has a quartet signal with intensities of +1, +1, −1 and −1.

2. In Fig. 4.22 components are positioned in one direction (the y' axis) just after the 90° pulse. However, in Fig. 4.25 g two components are opposite.

The movements of magnetization vector components of CH, CH_2 and CH_3 in the x'–y' plane are shown in Fig. 4.26.

1. When $\tau = 1/(4J)$
   For CH, its two components have rotated 45°. If acquisition with decoupling starts, their resultant along the y' axis gives a positive signal with a factor of cos 45°.
   For CH_2, its two components have rotated 90°, thus overlapping. With decoupling, they give a positive maximum signal.
   For CH_3, it is similar to CH.
   Therefore, when $\tau = 1/(4J)$, INEPT gives positive signals for CH, CH_2 and CH_3, of which the signal of CH_2 is maximum.
2. When $\tau = 2/(4J)$
   Similar to the discussion above, the conclusion is: when $\tau = 2/(4J)$, INEPT only results in positive maximum signals for CH.
3. When $\tau = 3/(4J)$
   The reader can follow the above-mentioned discussion to reach a solution. INEPT gives negative maximum signals to CH_2, and positive signals with a factor of cos 45° to CH and CH_3.
   The 180° pulse between (h) and (i) and two time intervals of τ in Fig. 4.24 are used to refocus chemical shifts.

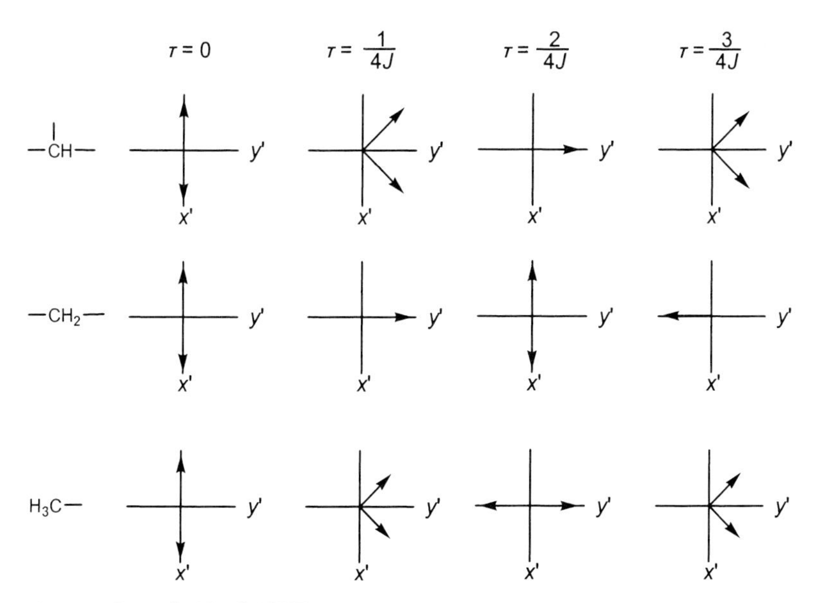

Fig. 4.26 Principle (2) of INEPT.

Quaternary carbon atoms have no signals for any τ values because the signals are averaged to zero when $90^\circ_{y'}$ and $90^\circ_{-y'}$ are applied alternately. They have no signals either when the pre-saturated pulse is applied.

With $\tau=2/(4J)$ only the signals for CH are obtained from INEPT. In an INEPT spectrum with $\tau=3/(4J)$, CH and CH_3 are determined from their positive signals while CH_2 is determined from its negative signal. CH_3 can be differentiated from CH by comparing the spectrum of INEPT with $\tau=3/(4J)$ and that with $\tau=2/(4J)$. By comparing the broadband decoupling spectrum and the INEPT spectrum with $\tau=3/(4J)$, quaternary carbon atoms can be determined.

INEPT is also applicable to 2D NMR.

As INEPT utilizes SPI, the smaller γ_X is, the stronger enhancement the nuclei have.

An INEPT spectrum is similar to that of DEPT with an example of the latter shown in Fig. 4.29.

4.2.3
DEPT (Distortionless Enhancement by Polarization Transfer)

Here "distortionless" means that phases of signals are distortionless, that is, the requirements for the non-variation of J values for different functions and for the accuracy of pulse widths are not as strict when adopting DEPT compared with those when adopting INEPT and APT.

The pulse sequence for DEPT is shown in Fig. 4.27.

The principles of APT and INEPT can be explained well by use of the magnetization vector model. However, it is difficult to explain the principle of DEPT by use of the magnetization vector model because a pulse of 45° or 135° is applied, which is hard to explain by use of this model.

In the discussion of DEPT the product operator formalism (Appendix 1) should be used.

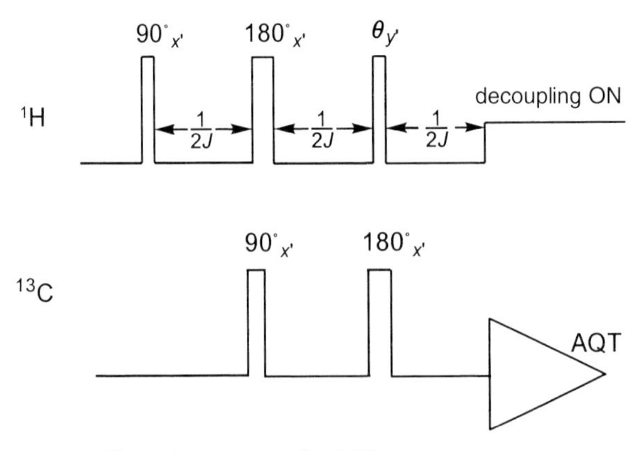

Fig. 4.27 The pulse sequence for DEPT.

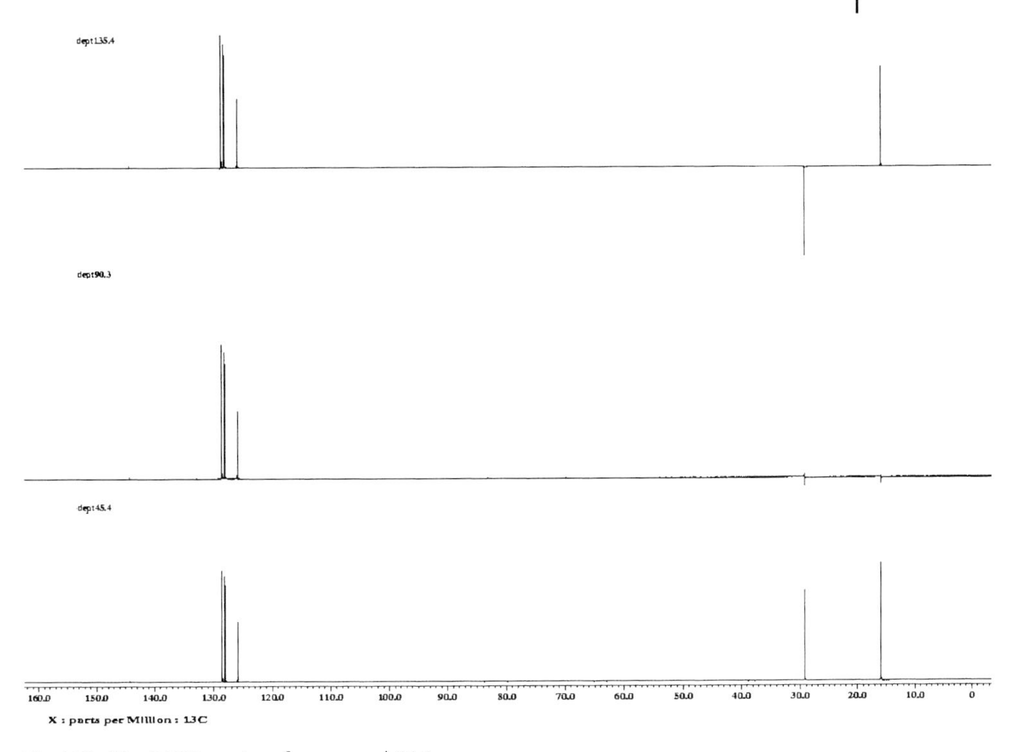

Fig. 4.28 The DEPT spectra of compound **C4-1**.

The result for DEPT is similar to that for INEPT.

When $\theta = 45°$, the spectrum for DEPT is similar to the spectrum for INEPT with $\tau = 1/(4J)$.

When $\theta = 90°$, the spectrum for DEPT is similar to the spectrum for INEPT with $\tau = 2/(4J)$.

When $\theta = 135°$, the spectrum for DEPT is similar to the spectrum for INEPT with $\tau = 3/(4J)$.

The sub-spectrum which consists only of CH signals is obtained from DEPT with $\theta = 90°$.

As for INEPT, there are no signals for quaternary carbon atoms in DEPT spectra.

Fig. 4.28 shows the DEPT spectra of compound **C4-1**.

$-CH_2CH_3$

C4-1

DEPT is used most frequently.

The above-mentioned three methods were developed at the beginning of the 1980s. Subsequently the study of pulse sequences was concentrated mainly on 2D NMR. However, some new pulse sequences have been developed for the determination of carbon atom orders. PENDANT (polarization enhancement during attached nucleus testing) [17, 18] is a successful example, which gives the signals of quaternary carbon atoms and does not produce distortion of the signal phases.

4.3 Introduction to 2D NMR

4.3.1 What are 2D NMR Spectra?

Through Fourier transform, free induction decay (FID), a function of time, is transformed into a spectrum, a function of frequency. This spectrum is one-dimensional because there is only one variable: frequency.

If we obtain a series of spectra by changing particular parameters, for example, temperature, concentration, pH values, and so forth, to form a "cluster" of spectra, the assembly of spectra is still a one-dimensional spectra, because there is only one frequency variable. Therefore the spectrum cluster obtained from the inversion recovery method for the measurement of T_1 is also a one-dimensional spectra.

The two-dimensional spectrum is one which shows the correlation of its two frequency variables. A 2D NMR spectrum is obtained by performing the Fourier transform twice. Before the Fourier transform, the two variables are two time variables: t_1 and t_2, where t_2 is the acquisition time while t_1 is a time space in a pulse sequence; for example, t_1 is τ in the spin echo. The two time variables, t_1 and t_2, are independent.

The formation of a 2D NMR spectrum is illustrated in Fig. 4.29.

In Fig. 4.29 a, the abscissa is t_2, which increases from left to right. Each curve is recorded with a t_1 value. The curve cluster is recorded from bottom to top as t_1 increases. In brief, Fig. 4.29 a shows $S(t_1,t_2)$ which is a function of t_1 and t_2.

The result of transforming t_2 into ω_2 through the first Fourier transform, with t_1 as a constant is shown in Fig. 4.29 b. If we take a cross-section through the right end of the ω_2 axis, we can see a cyclic curve (cosine or sine). Therefore this cyclic function can again be treated by a Fourier transform. The variable t_1 is then transformed into ω_1. Finally we obtain Fig. 4.29 c, which illustrates a function with two frequency variables, $S(\omega_1,\omega_2)$. Therefore Fig. 4.29 c is a two-dimensional spectrum.

The above-mentioned process can be expressed by the following mathematical operations:

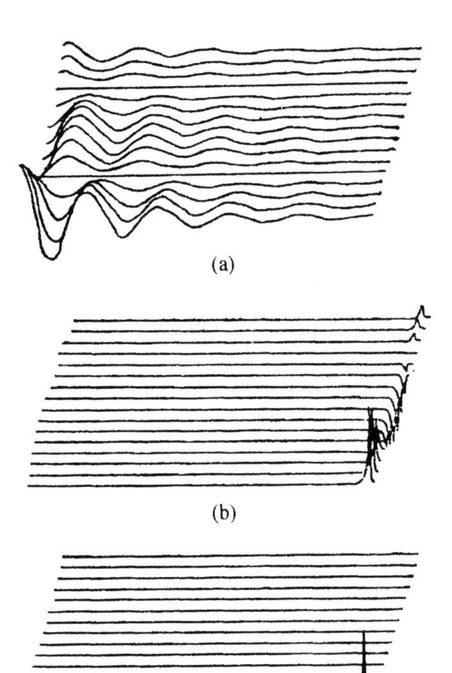

Fig. 4.29 $S(\omega_1,\omega_2)$ is transformed from $S(t_1,t_2)$ by performing a Fourier transform twice.

$$\int_{-\infty}^{\infty} dt_1 e^{-i\overline{\omega}_1 t_1} \int_{-\infty}^{\infty} dt_2 e^{-i\omega_2 t_2} S(t_1, t_2)$$
$$= \int_{-\infty}^{\infty} dt_1 e^{-i\omega_1 t_1} S(t_1, \omega_2)$$
$$= S(\omega_1, \omega_2) \tag{4-27}$$

Fig. 4.29 c shows a typical example of a 2D NMR spectrum. On the other hand, we have the relationship for the measurement of T_1 by the inversion recovery method:

$$S(t_1, t_2) = e^{-\frac{t_1}{T_1}} \cos \omega_2 t_2 \tag{4-28}$$

Although the function S has two time variables, it is not a cyclic function of t_1 hence it is not a function of 2D NMR.

4.3.2
Time Axis of 2D NMR

There are several types of 2D NMR spectra. However all 2D NMR experiments can be divided into four periods along the time axis:

Preparation period → evolution period → mixing period → detection period

- Preparation period: a fairly long period without pulses. After this period the spin system returns to thermal equilibrium and has the reproducible starting conditions.
- Evolution period (t_1): at the beginning of this period the spin system is excited by one or more pulses. The components of the magnetization vectors then evolve. The evolution period, t_1, increases gradually.
- Mixing period (t_m): in this period the detection conditions are built up. Some 2D NMR experiments have no mixing period.
- Detection period (t_2): FID signals are detected in this period.

After the first Fourier transform, t_2 is transformed into ω_2, which is the usual resonant frequency. After the second Fourier transform, t_1 is transformed into ω_1, which has a different meaning, depending on the pulse sequence to be used.

4.3.3 Classification of 2D NMR Spectra

2D NMR spectra can be classified into the following three types:

1. *J* resolved spectra
   The *J* resolved spectrum is also called the J–δ spectrum, which distinguishes between chemical shifts and coupling splitting. *J* resolved spectra include homonuclear and heteronuclear *J* resolved spectra.
2. Chemical shift correlation spectra
   The chemical shift correlation spectrum is also termed the δ–δ spectrum, which illustrates the correlation between resonant signals. It is the most important of all 2D NMR spectra. Chemical shift correlation spectra are divided into three categories: homonuclear chemical shift correlation spectra, heteronuclear chemical shift correlation spectra and 2D NMR spectra of NOE or chemical exchanges.
3. Multiple quantum spectra
   Usually NMR signals are measured under the conditions of $\Delta m = \pm 1$. It is called a single quantum transition. By using particular pulse sequences, multiple quantum transitions (Δm is an integer greater than 1) can be detected to produce the multiple quantum 2D NMR spectra.

4.3.4 Illustration of 2D NMR Spectra

1. Stacked trace plot
   The stacked trace plot consists of tightly arranged normal ("one dimensional") spectra. It is similar to the spectrum cluster of T_1 measurement by the inversion recovery method. Stacked trace plots are seldom used today.
2. Contour plot
   The contour plot is similar to the contour map. The smallest central circle denotes the frequency of a peak. The number of circles denotes the intensity of the peak.

The advantages of the contour plot are precise resonant frequency display and savings in time. In general, for 2D NMR spectra contour plots are used today.

Sections and projections can be used for local illustration in addition to a general representation of a 2D NMR spectrum.

As there are many types of 2D NMR spectra, only the 2D NMR spectra that are frequently applied are discussed.

4.4 *J* Resolved Spectra

4.4.1 Homonuclear *J* Resolved Spectra

The pulse sequence for the homonuclear *J* resolved spectrum is shown in Fig. 4.30.

From Fig. 4.30 it can be seen that the pulse sequence is the sequence of the spin echo, except that now DE is changed into $t_1/2$ and acquisition time is denoted as t_2. Therefore, the conclusions on the spin echo for a system with homonuclear coupling can be utilized directly.

Related calculations can lead to a result for the pulse sequence. Performing a Fourier transform twice on the result will lead to Fig. 4.31.

With a computer, we can tilt the two lines that connect the two related points so that the two lines are vertical to the base side, which leads to Fig. 4.32.

From Fig. 4.32 it follows that the functions of δ and J have been separated: the chemical shift is illustrated by the ω_2 axis while the coupling splitting and coupling constants are shown by the ω_1 axis. The projection of a homonuclear J spectrum onto the ω_1 axis gives a decoupled hydrogen spectrum.

The systems of AX_2 and AX_3 can be dealt with similarly with the same result.

In a normal (one dimensional) ^1H spectrum, split signals can overlap each other partially or completely when they have approximate δ values. It is difficult to discern the split pattern and coupling relationship and hence to read out the coupling constants. The homonuclear J spectrum resolves this problem perfectly, because split signals can be readily distinguished if they have a δ value slightly different to the others.

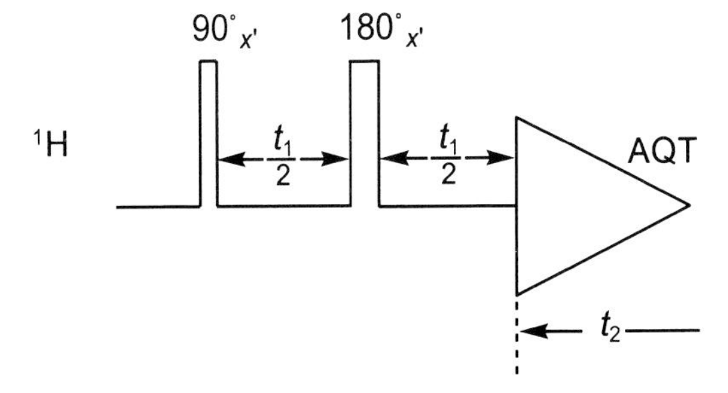

Fig. 4.30 The pulse sequence for the homonuclear *J* resolved spectrum.

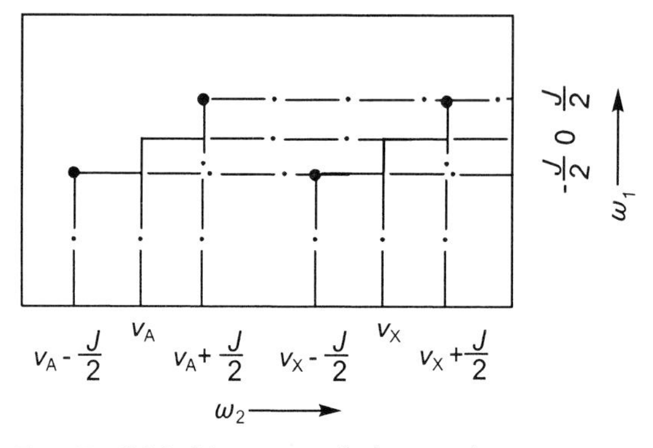

Fig. 4.31 Original *J* spectrum of a homonuclear AX system.

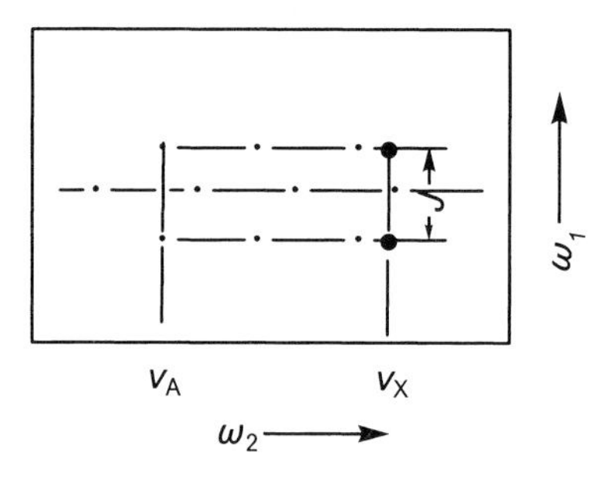

Fig. 4.32 The *J* resolved spectrum of a homonuclear AX system.

For a strongly coupled system, the *J* resolved spectrum will appear complex.

Now we explain, as an example, the application of *J* resolved spectra with tobramycin, which is an antibiotic drug and whose structural formula is shown as **C4-2.**

CH_2OH
CH_2NH
O
OH
O
OH
O
NH_2
OH
NH_2
NH_2
OH
NH_2
NH_2

C4-2

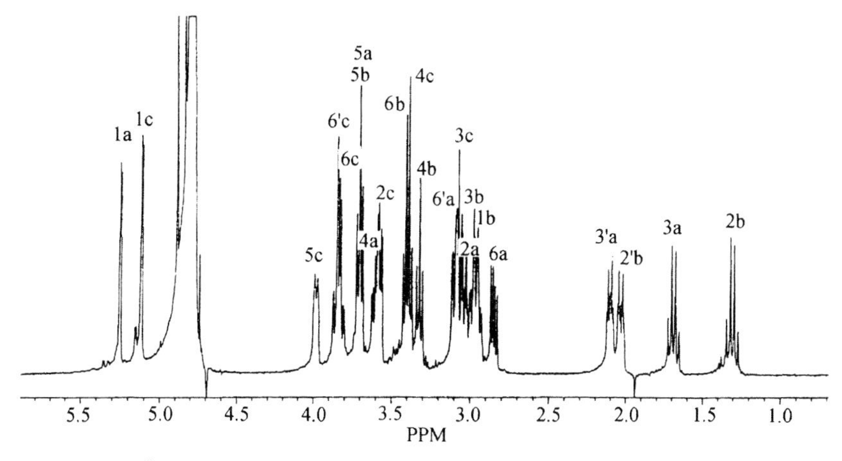

Fig. 4.33 The ^{1}H spectrum of tobramycin.

Although tobramycin has a known structure, its conformation is unknown. That is to say, it is not known which substituents are in the axial or the equatorial positions, which can be determined if all remaining hydrogen atoms in the rings are assigned. As $J_{aa} > J_{ae} \geq J_{ee}$ in a six-membered ring, the substitution direction of the hydrogen atoms can be determined according to their J values. Obviously if the position of every hydrogen atom in the rings is known, the substitution directions of substituents are known.

Although the ^{1}H spectrum of tobramycin shown in Fig. 4.33 is recorded with a 500 MHz instrument, it is still difficult to resolve some overlapped peaks.

The assignment of the ^{1}H spectrum of tobramycin is completed by a combination of several 2D NMR spectra. The J values of hydrogen atoms are acquired from Fig. 4.34.

In Fig. 4.34, most peaks have clear splitting patterns. For example, the patterns of q, q, d×t, and d×t are illustrated clearly from the far right end to the left of the ω_2 axis.

Because the full range of the ω_1 axis is only ±20 Hz, all J values can be read out precisely.

A complex pattern is displayed where $\omega_2 \approx 3.8$ ppm because three adjacent hydrogen atoms with approximate δ values form a strongly coupled system (a section of which is shown on the right). Therefore, homonuclear J resolved spectra are not quite appropriate for strongly coupled systems.

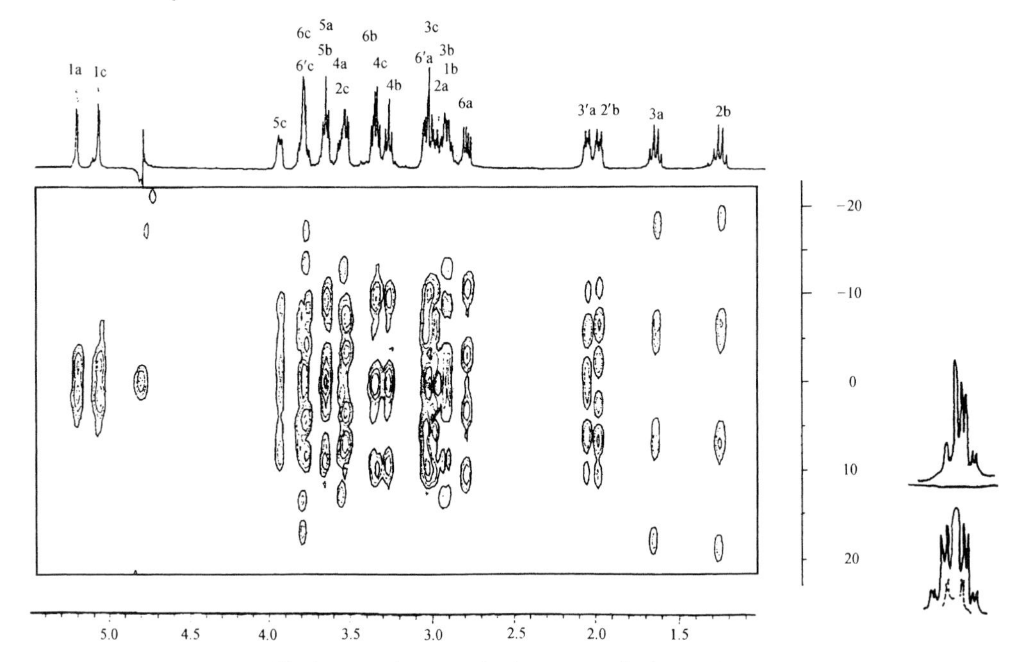

Fig. 4.34 The homonuclear J resolved spectrum of tobramycin.

4.4.2 Heteronuclear J Resolved Spectra

The pulse sequence for the heteronuclear J resolved spectrum is shown in Fig. 4.35.

The conclusion from Section 4.1.3 can be used to discuss the subject in a way similar to the homonuclear J resolved spectrum.

Related calculations can lead to the result of the pulse sequence. Performing a Fourier transform twice on the result will lead to Fig. 4.36, which is similar to the homonuclear J resolved spectrum (after the tilt).

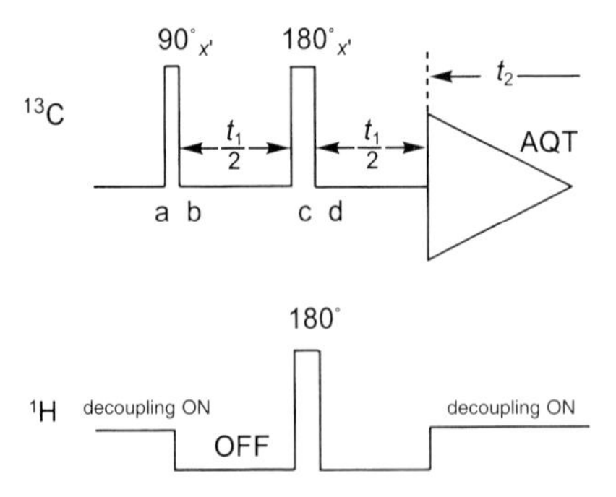

Fig. 4.35 The pulse sequence of the heteronuclear J resolved spectrum.

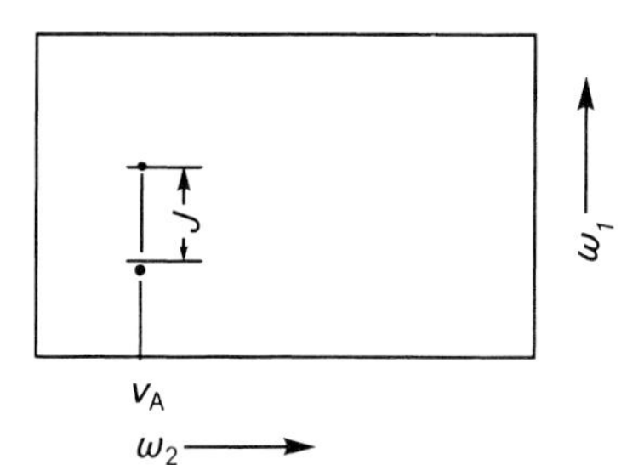

Fig. 4.36 The J resolved spectrum of a heteronuclear AX system.

Since DEPT is simpler and saves more time than the heteronuclear resolved J spectrum, the latter is used infrequently.

We will now discuss correlation spectroscopy, which is the core of 2D NMR experiments. Correlation spectroscopy, especially homonuclear correlation spectroscopy, or COSY, occupies a most important position in 2D NMR spectra.

4.5 Heteronuclear Shift Correlation Spectroscopy

4.5.1 H,C-COSY

COSY is the abbreviation for correlation spectroscopy. H,C-COSY is correlation spectroscopy of ^{1}H and ^{13}C, which shows the connections between ^{1}H and ^{13}C through one chemical bond, so that it is necessary to postulate an unknown structure. An H,C-COSY spectrum is equivalent to a whole series of selective decoupling spectra.

The pulse sequence for H,C-COSY is shown in Fig. 4.37. Note that the former part [from (a) to (e)] is identical to the pulse sequence (4-12) (see Section 4.1.4).

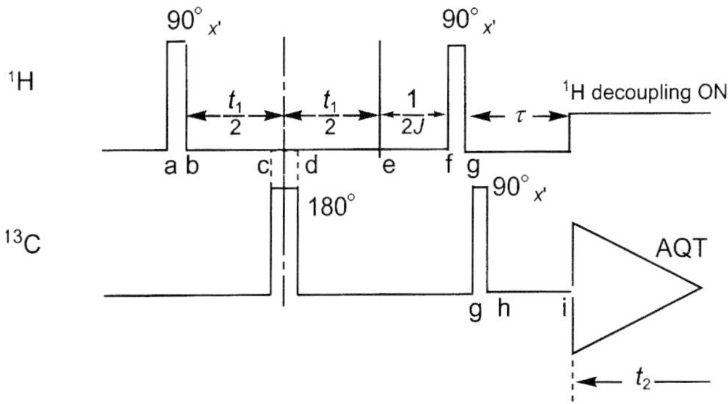

Fig. 4.37 The pulse sequence for the H,C-COSY.

We will take a CH system as an example. The movement of components of the ^{1}H magnetization vector is illustrated in the upper part of Fig. 4.38, and that of ^{13}C in the lower part.

Fig. 4.38 will be explained from the top. The explanation of (a) to (e) of Fig. 4.38 is the same as that of the pulse sequence in Eq. (4-8). Two components refocus at (e). However, their phase is modulated by their δ value. The refocused components form an angle of $2\pi\nu_H t_1$ with the y' axis. After $1/(2J)$, the two components make the angles of $2\pi\nu_H t_1 + 2\pi\nu_H \times 1/(2J) \pm \pi/2$, respectively, as shown in Fig. 4.38 f. Under the influence of $90^\circ_{x'}$, the two components have rotated from the x'–y' plane to the x'–z plane, as shown in Fig. 4.38 g. The angle made by the two components with the $\pm z$ axis is equal to that made by them with the $\pm y'$ axis. As there are two opposite components of ^{1}H magnetization along the $\pm z$ axis, SPI will produce two enhanced components of the ^{13}C magnetization vector, as shown in (g) at the bottom of Fig. 4.38. However, the enhancement factor is $\cos(2\pi\nu_H t_1)\gamma_H/\gamma_C$ at present because the two components of ^{1}H magnetization vector make the angles of $2\pi\nu_H t_1 + 2\pi\nu_H \times 1/(2J) \pm \pi/2$ with the $\pm z$ axis (the last two constant terms are deleted through the Fourier transform).

Now we will discuss the movement of the two components of the ^{13}C magnetization vector. They are along the z axis, as shown in Fig. 4.38 a. They have been rotated from the z axis to the $-z$ axis by the 180° pulse for ^{13}C, as shown in Fig. 4.38 d. The two components are enhanced by SPI, as shown in Fig. 4.38 g. They have been rotated from the $\pm z$ axis to the $\pm y'$ axis by the $90^\circ_{x'}$ pulse for ^{13}C, as shown in Fig. 4.38 (h). As the two components are opposite,

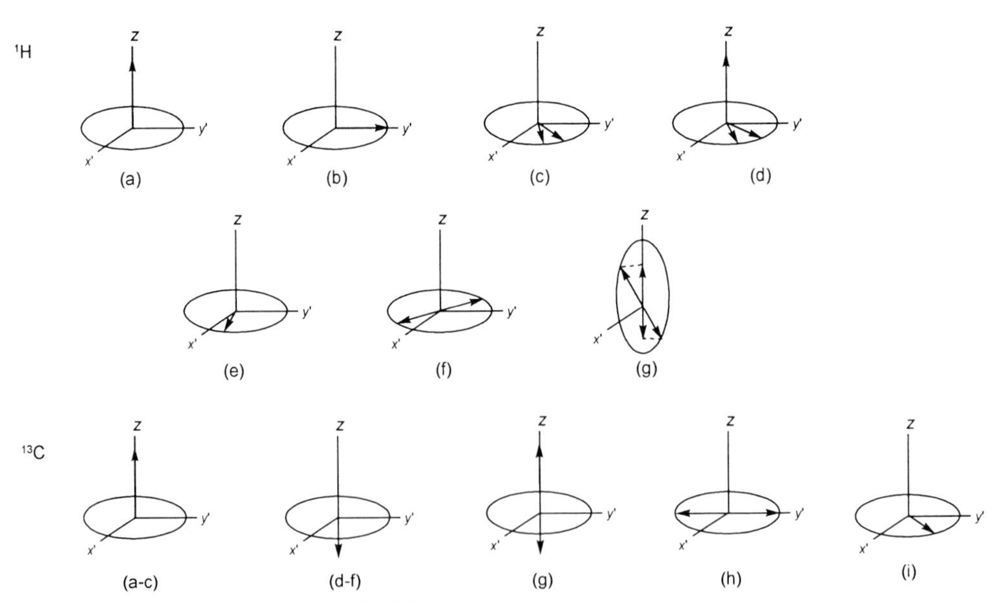

Fig. 4.38 The principle of the H,C-COSY.

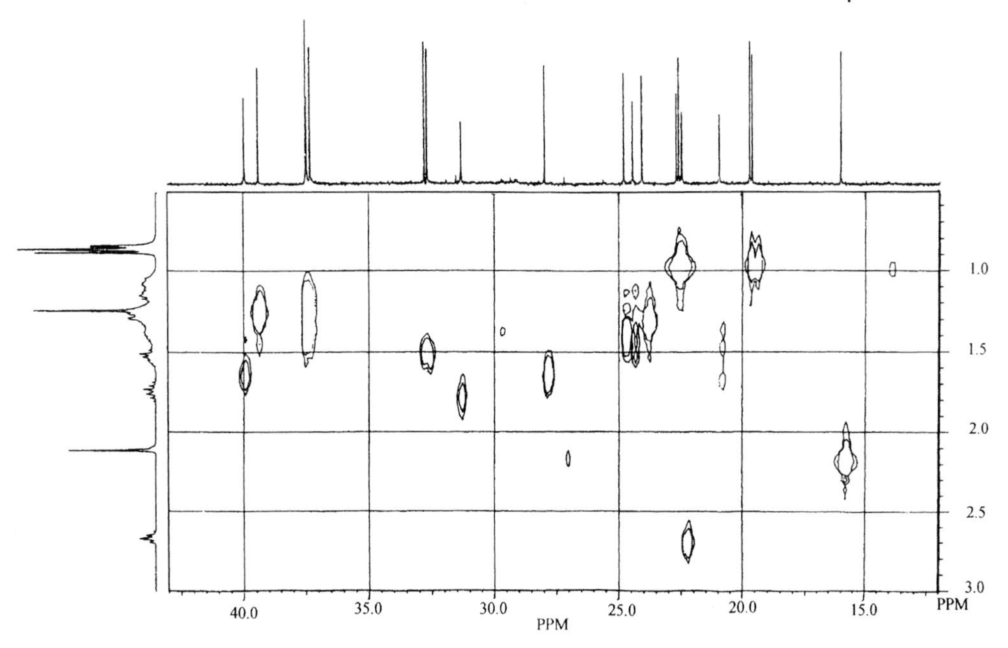

Fig. 4.39 H,C-COSY (high field part) of δ-vitamin E.

we have to wait a period of time so that the two components meet (or at least get close to each other) to produce a ^{13}C signal, as shown in Fig. 4.38 (i).

From Fig. 4.26 it follows that the two components meet if $\tau = 2/(4J)$. However, we take $\tau = 1/(3J)$ to give a compromise between CH and CH_2 or CH_3.

Acquisition with the decoupling for ^1H starts at (i). The movement of the ^{13}C magnetization vector is determined only by chemical shift values of ^{13}C, ν_C, during t_2. After the Fourier transform, the chemical shift value ν_C is illustrated by the ω_2 axis. On the other hand, the peak intensity is modulated by $\cos\,[2\pi\nu_H t_1 + 2\pi\nu_H \times 1/(2J) + \pi/2]$. After the second Fourier transform for t_1, only ν_H remains. Therefore, a peak in a ^1H–^{13}C correlation spectrum is obtained. ν_C is shown by the ω_2 axis and ν_H is shown by the ω_1 axis, where ν_H is the chemical shift value of the ^1H connected to the ^{13}C, which means the relationship between a pair of coupled ^{13}C and ^1H is illustrated.

An example of H,C-COSY is shown in Fig. 4.39.

The structural formula of δ-vitamin E is as shown in **C4-3**.

HO
CH_3
O
CH_2—CH_2—CH_2—CH(CH_3)—CH_2—CH_2—CH_2—CH(CH_3)—CH_2—CH_2—CH_2—CH(CH_3)—CH_3
CH_3

C4-3

H,C-COSY has a rectangular form. The ^{13}C spectrum with broadband decoupling is placed at the top while the ^{1}H spectrum is beside it. Peaks in the rectangle are called cross peaks or correlated peaks, each of which shows a correlation of a pair of ^{13}C and ^{1}H which are connected directly. Every CH_3 or CH has a cross peak. A CH_2 group shows one cross peak when the two hydrogen atoms are chemically equivalent, or two cross peaks when they are not.

4.5.2
COLOC

COLOC is the abbreviation for (heteronuclear shift) correlation spectroscopy via long range couplings. Kessler et al. first proposed the term and the pulse sequence in 1984 [19]. Later pulse sequences belonging to COLOC were proposed [2].

If the J value in pulse sequence (Fig. 4.37) is replaced by a long range coupling constant between ^{13}C and ^{1}H, the pulse sequence for the ^{1}H–^{13}C long range correlation spectroscopy can be obtained. However, COLOC can give a better result.

Through the combination of H,C-COSY and COSY, it is possible to find the connections between H and C. However, the method is no longer valid for quaternary carbon atoms because no hydrogen atoms are connected to them. Therefore, it is very important to find the connections around the quaternary carbon atoms. Fortunately COLOC can do this. COLOC reveals couplings across two and/or three chemical bonds in which a heteroatom or a quaternary carbon atom can exist.

The pulse sequence for COLOC is shown in Fig. 4.40.

Emphasis is given to the comparison of COLOC with ^{1}H–^{13}C long range correlation spectroscopy.

1. COLOC is shorter than ^{1}H–^{13}C long range correlation spectroscopy on the time axis so that COLOC has a stronger signal than the latter because the effect of relaxation is decreased.

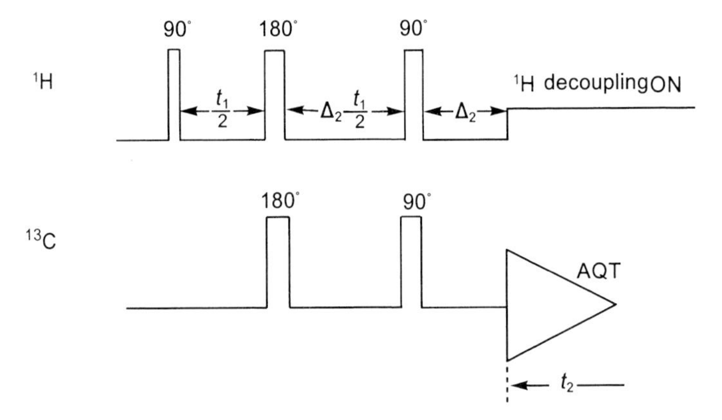

Fig. 4.40 The pulse sequence for COLOC.

2. There is no coupling splitting in the ω_1 axis so that the signal-to-noise ratio as well as resolution increases.

A delay of Δ_2 is set after the second 90° pulse. The application of Δ_2 has the following effect. The magnetization vectors of CH_2 and CH_3 are reduced because of their short T_2 for a fairly long period of Δ_2. However, the magnetization vectors of quaternary carbon atoms have a fairly long T_2 so that their signals are relatively enhanced after Δ_2.

Figures 8.21 and 8.25 are examples of COLOC.

4.5.3
H,X-COSY

We will deal with the correlation spectra between hydrogen atoms and heteroatoms (^{15}N, ^{31}P, etc.) in this section. Obviously the pulse sequence for H,C-COSY can be used except that the $^1J_{CH}$ value should be replaced by $^1J_{NH}$ or $^1J_{PH}$ value.

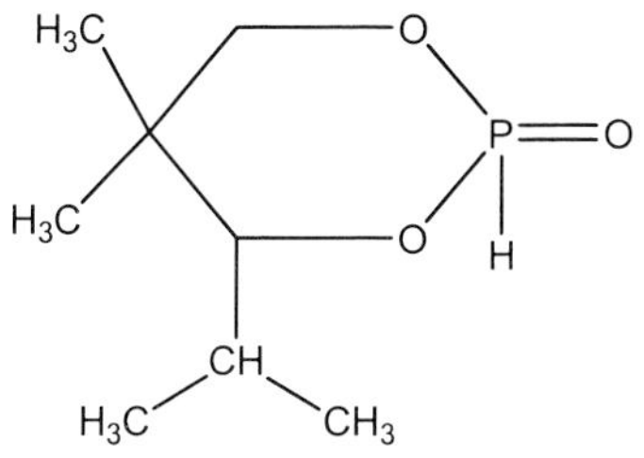

C4-4

The H,P-COSY of **C4-4** is shown in Fig. 4.41.

The projection of Fig. 4.41 onto the ω_2 axis shows the ^{31}P spectrum. We can see that there are two peaks at $\delta = 38.6$ and 35.7 ppm, respectively, which means the existence of two stereoisomers. Their peak area ratio is 8 : 1.

The projection of Fig. 4.41 onto the ω_1 axis shows the ^1H spectrum, which is complex because of the following factors.

1. Because the molecule has no symmetry plane, two methyl groups in an isopropyl are not chemically equivalent, nor are two methyl groups at the 5- or two hydrogen atoms at the 6-position.
2. ^{31}P is coupled with ^1H.
3. Two isomers exist.

In Fig. 4.41, there are two cross peaks which correspond to two ^{31}P signals. The strong signal is situated at $\omega_2 = 38.6$ ppm and $\omega_1 = 6.83$ ppm, which is an average value of 8.50 and 5.17 ppm where there are two peaks in the ^1H spectrum. Therefore, we can assign them to the hydrogen atom attached to the phos-

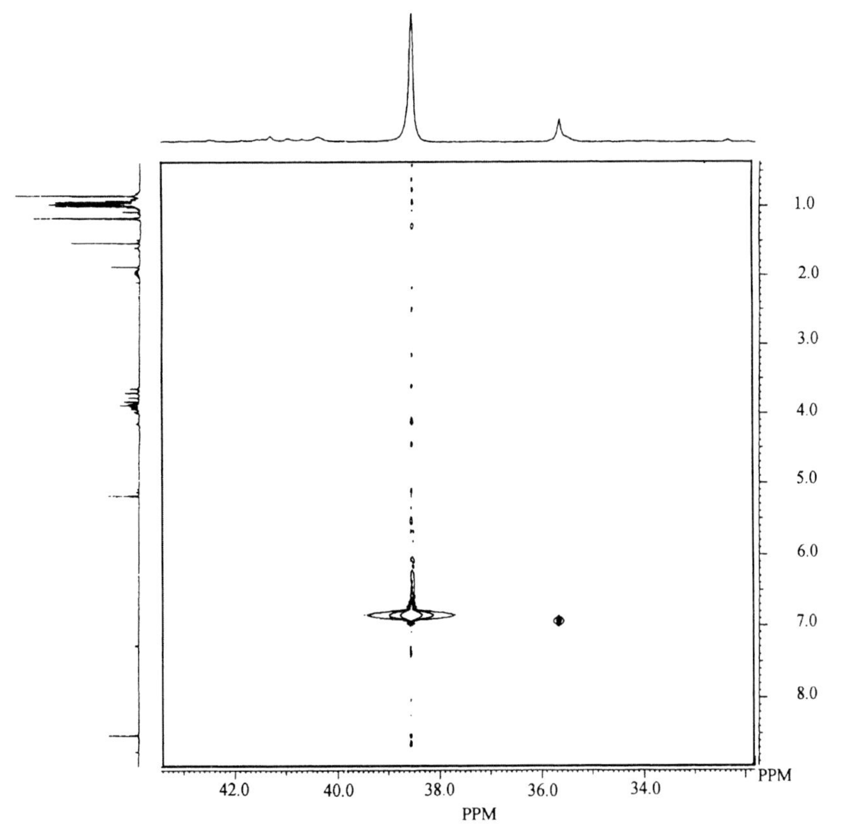

Fig. 4.41 H,P-COSY of **C4-4**.

phorus atom. In addition, we can calculate the heteronuclear coupling constant $^1J_{PH}$ (667 Hz) from the instrumental frequency of 200 MHz. Similarly, $^1J_{PH}$ of the minor isomer can be calculated (720 Hz).

In summary, H,X-COSY has two functions: one is to assign the peaks of hydrogen atoms attached to the heteroatom and the other is to calculate heteronuclear coupling constants.

4.6 Homonuclear Shift Correlation Spectroscopy

The homonuclear correlation spectrum is the core of 2D NMR and is always applied in routine measurement. The homonuclear correlation spectrum is abbreviated to H,H-COSY, or more simply, COSY.

4.6.1
COSY

The pulse sequence of COSY is shown in Fig. 4.42.

It is appropriate to explain the principle of COSY by the product operator formalism (see Appendix 1).

The function of COSY and the method of interpreting COSY are illustrated by the following example.

C4-5

The ^1H spectrum of compound **C4-5** is shown in Fig. 4.43.

It would appear that Fig. 4.43 is not in accord with the structure. The complexity of the ^1H spectrum is due to the existence of stereoisomers. In addition, the molecule has two chiral carbon atoms, one of which leads to the chemical non-equivalence of two pairs of hydrogen atoms in two CH_2 groups, which are next to the chiral carbon atoms.

From the peak area and its blunt shape the peak at 1.63 ppm can be assigned to the hydroxyl group. The peaks at 1.38 and 7.27 ppm can be assigned to the methyl group and the substituted benzene ring, respectively. The peaks of the CH_2 between the benzene ring and the quaternary carbon atom can be assigned to those at 2.93 and 3.02 ppm for the two isomers, respectively.

It is difficult to assign the numerous peaks between 3.1 and 4.3 ppm, which are ascribed to the other five hydrogen atoms. We use a, b, c, d and e to denote the five hydrogen atoms in an isomer and a′, b′, c′, d′ and e′ in the other isomer as shown in the structural formula.

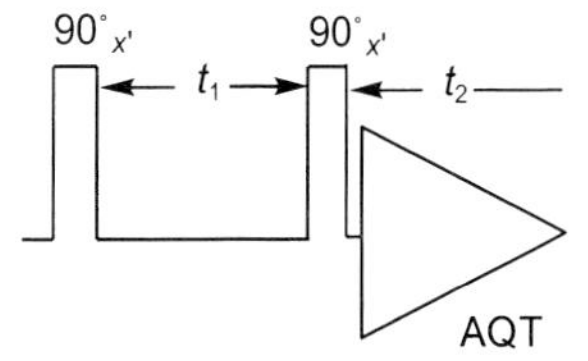

Fig. 4.42 The pulse sequence of COSY.

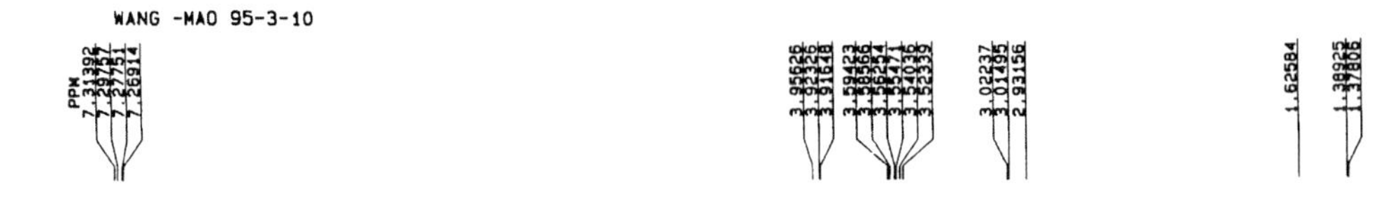

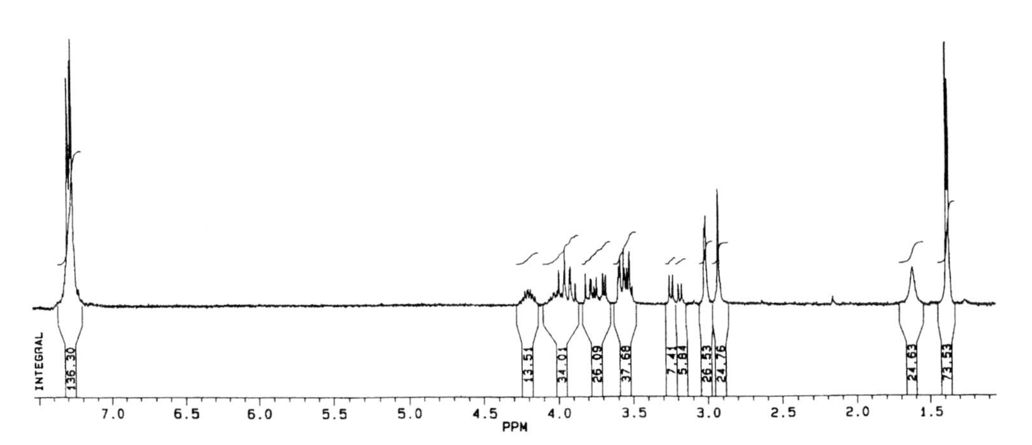

Fig. 4.43 The ^{1}H spectrum of compound **C4-5**.

From the ^{1}H spectrum it follows that the quantitative ratio of the two stereoisomers is approximately 1:1. Every hydrogen atom of an isomer has an integration value of about 12.5 in the ^{1}H spectrum.

It is impossible to analyze the coupling relationships in such a ^{1}H spectrum by establishing that the same J value exists in two split peak sets, because the peak shapes are too complicated to analyze and some peaks are indiscernible. The complexity of the ^{1}H spectrum results from the following factors:

1. The compound contains two chiral carbon atoms.
2. A complicated coupling relationship results from the fact that there are both ^2J and ^3J, which have several values, such as $J_{ac} \neq J_{bc}$ and $J_{dc} \neq J_{ec}$.
3. Some hydrogen atoms form strong coupling systems because of the small difference between their δ values.

However, we can assign a → e and a′ → e′ by using the COSY. The high field part of the COSY is shown in Fig. 4.44 in which the coupling relationships are clearly illustrated.

We will first describe how to acquire information from a COSY.

The projection of a COSY on the ω_2 (F_2, horizontal) axis or on the ω_1 (F_1, vertical) axis is the ^{1}H spectrum of the sample, which is set at the top and the

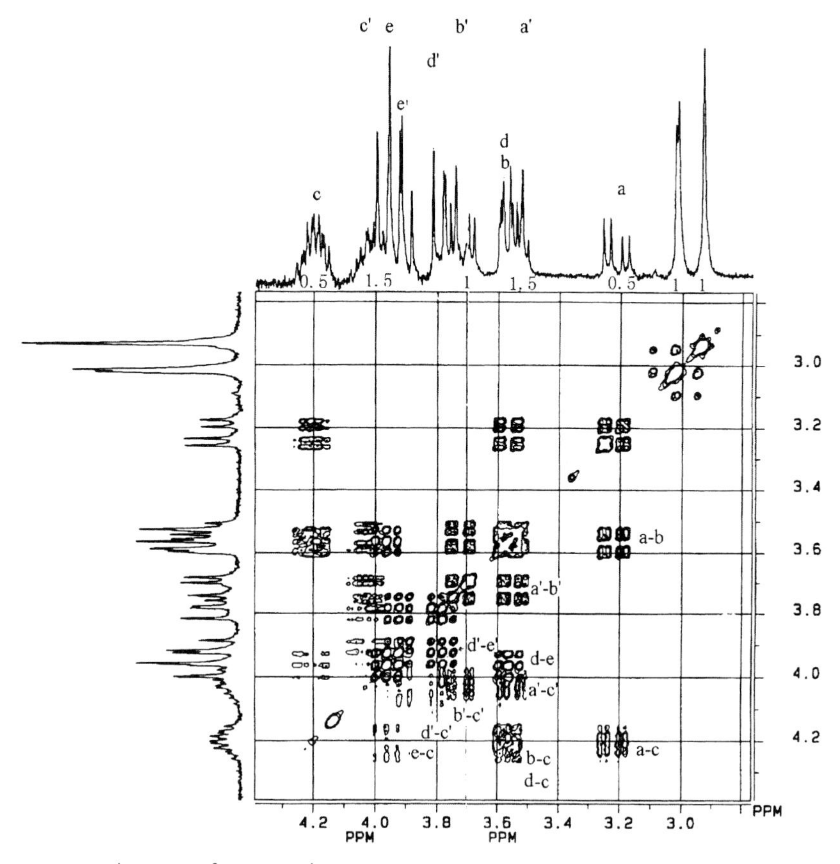

Fig. 4.44 The COSY of compound **C4-5**.

right (or left) of the COSY. The lateral ^1H spectrum can be omitted. A COSY has the form of a square or a rectangle (when F_2 and F_1 are not on the same scale). There is a diagonal, which is the symmetrical axis of the COSY. Frequently the diagonal rises from left to right. The peaks on the diagonal are called diagonal peaks or autocorrelated peaks, while the peaks that are situated off the diagonal, are called cross peaks or correlated peaks. The cross peaks are symmetrically distributed about the diagonal. Therefore, all information of the COSY is obtained from the cross peaks under (or above) the diagonal. The COSY mainly illustrates coupling relationships from ^3J. If we draw a vertical line through the center of a cross peak to be discussed, the vertical line will pass through the center of a peak set in the ^1H spectrum at the top of the COSY. This peak set is one of the two coupled peak sets that lead to the cross peak. Then we draw a horizontal line through the center of the cross peak, which will pass through a diagonal peak. A vertical line passing through the cross peak will cross the center of another peak set in the ^1H spectrum. This is another peak set contributing to the cross peak. Therefore, any pair of coupled peak sets can be found from its cross peak of a COSY regardless of their peak shapes.

Although the COSY reveals the coupling relationships through ^3J, some couplings through ^4J with a rather large ^4J value can be found in a COSY. On the other hand, some couplings from ^3J with a rather small value may not be found in COSY, for example, when the related dihedral angle is near 90°.

For some reasons such as the instability of the spectrometer, it is possible for there to be some vertical lines consisting of spots, which are called t_1 noises. In this case, the COSY should be treated with symmetrization in order that all spots that are not related to the symmetrical signals about the diagonal can be deleted. The prerequisite for symmetrization is that the numerical resolution on the F_2 axis is equal to that on the F_1 axis, which should be taken into account when one sets the parameters of the COSY.

Now we will analyze Fig. 4.44. Because of the symmetrical distribution of signals about the diagonal in the COSY, all we need to do is to analyze the signals under the diagonal. We start from the cross peak at $\omega_2=3.22$ ppm and $\omega_1=3.57$ ppm, which illustrates the coupling between the peak set at 3.22 ppm and that at 3.57 ppm. Under the cross peak, there is another cross peak at $\omega_2=3.22$ ppm and $\omega_1=4.20$ ppm. According to the structure, the peak set at 3.22 ppm can be assigned as "a", for the following three reasons: (1) The CH_2 in a ring has a greater chemical non-equivalence than that in a chain CH_2 of "d" and "e". (2) "d" and "e" have greater δ values than "a" and "b" because the former are more deshielded by the hydroxyl group than the latter by the ether bond. (3) "a" couples "b" and "c" with ^2J and ^3J, respectively. It coincides with the two cross peaks at $\omega_2=3.22$ ppm. Consequently, the peak set at 3.57 ppm is assigned as "b" and that at 4.20 ppm as "c".

There are three cross peaks at $\omega_1=4.20$ ppm, whose ω_2 values are 3.96, 3.57 and 3.22 ppm, respectively. Note that the peak set at 3.57 ppm corresponds to two hydrogen atoms. Therefore, the peak set at 3.57 ppm contains the signal of "d" in addition to "b". The peak set at 3.96 ppm can be assigned as "e" because the five hydrogen atoms (a, b, c, d and e) form a coupling system. In addition, there is the cross peak of "d" and "e" at ω_2 = 3.57 ppm and ω_1 = 3.96 ppm.

Similarly, "a′", "b′", "c′", "d′" and "e′" can be assigned. Their assignments are denoted above the ^1H spectrum.

From this example, we can see the function of COSY in establishing coupling systems.

4.6.2
Phase-sensitive Homonuclear Shift Correlation Spectroscopy

Phase-sensitive COSY is that which has cross peaks with pure absorption line-shapes in both F_1 and F_2 dimensions.

The conventional COSY, also known as magnitude COSY, has a few disadvantages. Because it is obtained by using a simple phase cycling, the cross section of the cross peak parallel to the F_1 or F_2 dimension is an absorption signal only if the cross section passes through the apex of the cross peak. All other cross sections, which do not pass through the apex, are the signals that contain a dispersion component, so that the cross peaks are twisted. In order to overcome this drawback, data are treated by Eq. (4-29).

$$M = \sqrt{R^2 + I^2} \tag{4-29}$$

where M is the intensity of a cross peak; R is the real portion of data obtained by the Fourier transform; and I is the imaginary portion of the data obtained by the Fourier transform.

When data are treated by using Eq. (4-29), the conventional COSY is called magnitude COSY.

The plot drawn after the process using Eq. (4-29) has the cross peaks with rather long tails because of the contribution of the dispersion component. The cross peaks have a fairly low resolution. Therefore, before the Fourier transform, data are multiplied by a particular window function. Consequently the resolution of the cross peaks will be improved but at the expense of the decreased signal-to-noise ratio.

The last step in the one-dimensional NMR experiment is the adjustment of the phase of the signals. Because the data are complete after the Fourier transform, any peak with a dispersion component can be adjusted as a signal with a pure absorption line-shape. In the experiment with conventional COSY, data are treated by using Eq. (4-29) so that the phases of the cross peaks can not be adjusted.

It is necessary to adopt quadrature detection in the one-dimensional NMR experiment (see Section 4.1.1). For the same reason, quadrature detection should be applied in 2D NMR experiments. Therefore, the aim of an ideal COSY experiment is to realize the quadrature detection and to ensure the cross peaks have pure absorption line-shapes in both F_1 and F_2 dimensions.

The quadrature detection in the F_2 dimension can be carried out as in the one-dimensional NMR experiment. Namely, two methods can be used: data are acquired at the x' axis and the y' axis simultaneously or sequentially. Note that the F_1 dimension is generated artificially. Two methods for the phase-sensitive COSY, which are related to the two methods mentioned above, were proposed. The first method corresponds to the simultaneous acquisition. For every t_1 increment, two sets of data are acquired and stored separately. One set of data is acquired with the first 90° pulse in the COSY pulse sequence along the x' axis and the other set of data is acquired with the first 90° pulse along the y' axis [20]. The acquired data are treated by the Fourier transform of complex numbers. The second method corresponds to the sequential acquisition. The acquisition is achieved by incrementing the phase of the first 90° pulse in the COSY pulse sequence by 90° for every t_1 increment and sampling the data twice as fast as with the first method. Because the first 90° pulse is 0°, 90°, 180°, 270°, and so forth, this method is called TPPI (time proportional phases increment)

[21]. The acquired data are treated by the Fourier transform of real numbers. In fact, these two methods are equivalent [22, 23]. Using either method, quadrature detection in the F_1 dimension can be realized. Because the necessary data are stored, the phases of cross peaks can be adjusted. Their line-shapes can be adjusted as the pure absorption mode. Therefore, the phase-sensitive COSY is obtained.

For further information, the reader is referred to references [4] and [24].

The phase-sensitive COSY has the following advantages: a rather high resolution, the readability of the coupling constants from its fine structures and the improvement of the signal-to-noise ratio.

The phase-sensitive COSY of AX and AMX systems is shown in Fig. 4.45.

The explanations of Fig. 4.45 are as follows [25]:

1. Only cross peaks are shown while diagonal peaks are not. In the phase-sensitive COSY, the cross peaks are adjusted as the pure absorption signals while the diagonal peaks are the dispersion signals.
2. The solid circles and the hollow circles represent absorption signals with positive and negative signs, respectively.
3. The areas of the circles show the intensities of related peaks.
4. The lengths of thick lines denote the active coupling constants, and those of the thin lines the passive coupling constants.

Coupling constants can be clearly read from the cross peaks of the phase-sensitive COSY, the fine structures of which are shown more distinctly than those of the ordinary COSY.

The "phase-sensitive" mode can be used in other types of 2D NMR.

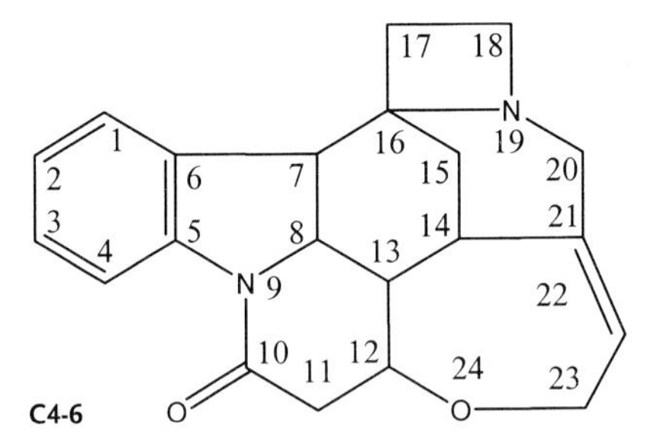

The high field part of the phase-sensitive COSY of strychnine (**C4-6**) is shown in Fig. 4.46.

In Fig. 4.46 we see that the cross peaks appear to be rectangles, in which fine structures are shown distinctly. Coupling constants can be easily read from the fine structures. The assignment of strychnine is denoted above the ^{1}H spectrum, which is at the top of the phase-sensitive COSY. The peak set of the hydrogen atom, which has a greater δ value than that of the other hydrogen atom in the same CH_2 group, is denoted with "′". Compared with the structural for-

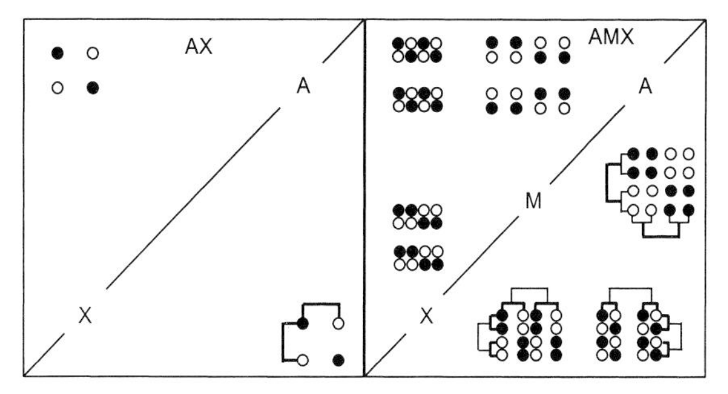

Fig. 4.45 The phase-sensitive COSYs of AX and AMX systems.

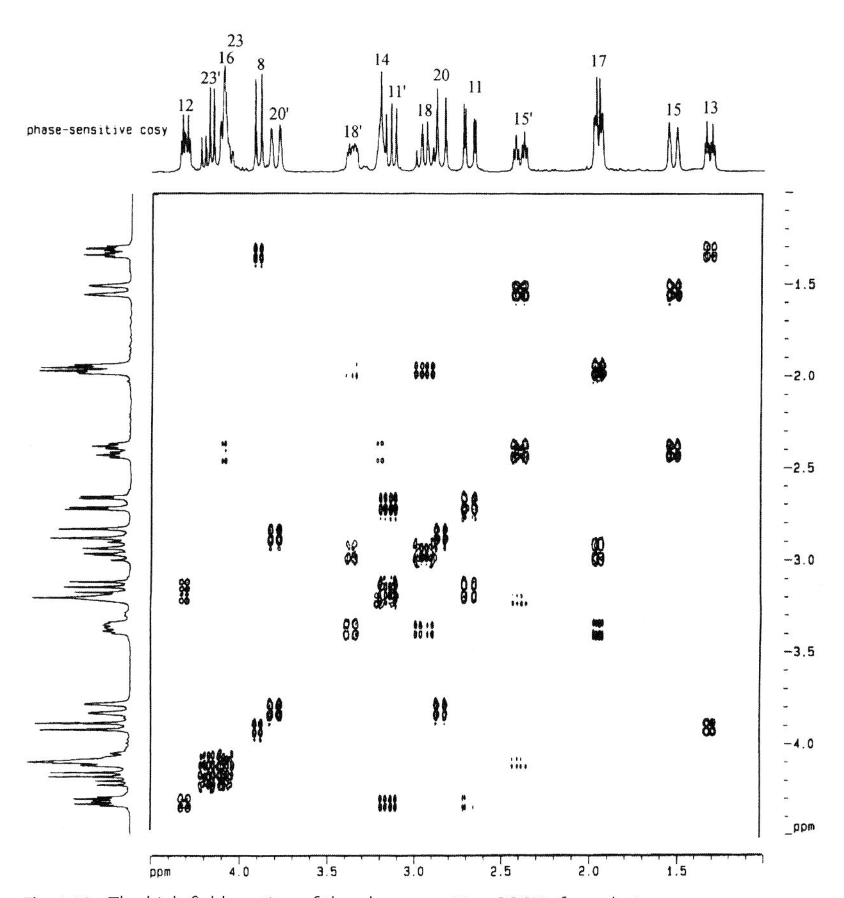

Fig. 4.46 The high field portion of the phase-sensitive COSY of strychnine.

mula, it can be found that some cross peaks are missing because the lowest cross sections for drawing the cross peaks are taken in slightly too high a plane.

The intensities of cross peaks in COSY are closely related to their J values. The larger J is, the stronger its related cross peak will be. Take H-13 for example. It couples with H-8, H-12 and H-14, respectively. It would appear that there should be three cross peaks below the peak set of H-13 of the ^{1}H spectrum. However, there is only one cross peak in this region because the coupling constant J_{13-8} is just enough to show the cross peak under the recording conditions. On the other hand, both J_{13-12} and J_{13-14} are so small (2.6 Hz) that the two related cross peaks are missing. All cross peaks produced from ^2J exist (H-15, H-15′; H-18, H-18′; H-20, H-20′; and H-23, H-23′) because all ^2J values are rather great.

4.6.3 COSY-45 (β-COSY)

The pulse sequence of COSY consists of two 90° pulses. If the width of the second pulse is decreased, the fine structures of the cross peaks and diagonal peaks in the two-dimensional spectrum will be simplified.

The second pulse can be taken as 30–60°, but usually as 45°, hence the name COSY-45, or generally β-COSY.

The function of COSY-45 can be illustrated by Fig. 4.47.

The high field portion of the COSY-45 of strychnine is shown in Fig. 4.47. Two functions of COSY-45 are listed as follows.

1. The widths of the diagonal peaks along the diagonal are decreased. Consequently, strong couplings, whose cross peaks are near the diagonal, can be found in the COSY-45 more easily than in an ordinary COSY, which can clearly be shown by comparing Fig. 4.47 with Fig. 4.46.

   It appears that the areas of the diagonal peaks in Fig. 4.47 are not decreased. In fact, the lowest cross section to illustrate the figure is taken at a fairly low position, which leads to the increase in the areas of the diagonal peaks, a fact that can be found from the shapes of the diagonal peaks.
2. The signs of coupling constants can be distinguished from a COSY-45. Some cross peaks in Fig. 4.47 appear in an overlap of two rectangles. Draw a line through the center of the lower rectangle and that of the higher rectangle. If the top of the line declines towards the left, the cross peak is ascribed to a coupling from ^3J. The cross peaks from H-17, H-18; H-17, H-18′; H-15′, H-14; H-15′, H-16, and so forth belong to this group. If the top of the line declines towards the right, the cross peak is ascribed to a coupling from ^2J, for example, the cross peak from H-18 and H-18′.

The observations discussed above agree with our knowledge that ^3J has a sign opposite to that of ^2J.

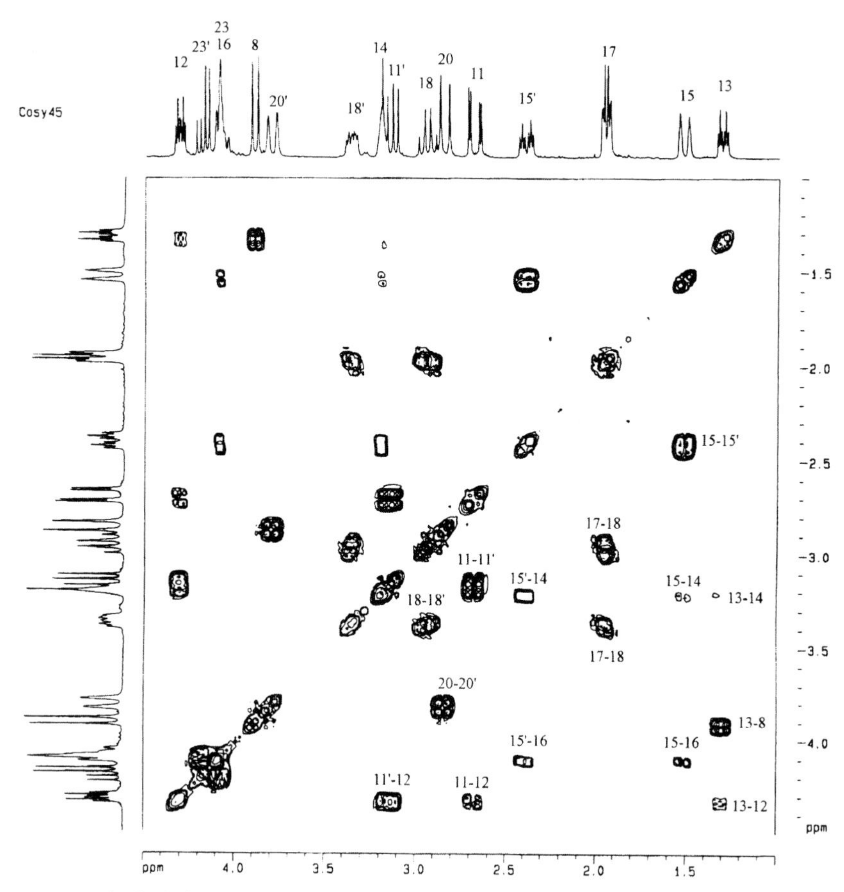

Fig. 4.47 The high field portion of the COSY-45 of strychnine.

4.6.4
COSY with Decoupling on the ω_1 Axis

If an ^{1}H spectrum consists of numerous and overlapped peak sets, the cross peaks in its related COSY will be considerably overlapped. This situation can be improved by the COSY with decoupling on the ω_1 axis.

The pulse sequence of the COSY with decoupling on the ω_1 axis is shown in Fig. 4.48.

The pulse sequence shown in Fig. 4.48 is different from that shown in Fig. 4.42 in the following two respects.

1. A 180° pulse is added between the two 90° pulses to adjust the t_1 value.
2. The time duration between the two 90° pulses is a constant, Δ, for which reason the decoupling on the ω_1 axis is achieved and it is known as COSY with a constant time.

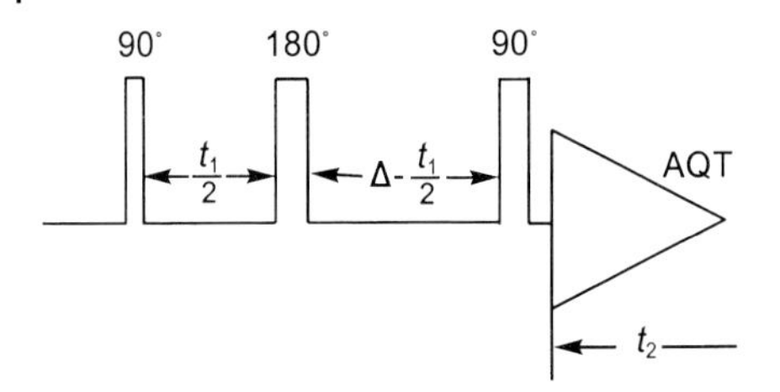

Fig. 4.48 The pulse sequence of the COSY with decoupling on the ω_1 axis.

The principle of the pulse sequence shown in Fig. 4.48 should be analyzed by the product operator formalism (see Appendix 1).

The projection of the COSY with decoupling on the ω_1 axis on the ω_1 axis is a decoupled ^{1}H spectrum, which consists only of singlets. This means that all cross peaks are "pressed" into related "lines" parallel to the ω_2 axis. Therefore, the COSY resolution on the ω_1 axis is greatly improved.

4.6.5 COSYLR

In an ordinary ^{1}H spectrum, long-range couplings producing a slight increase in width at the half-height of peaks are difficult to detect. Consequently, it is hard to determine the existence of a long-range coupling between two peak sets.

Two-dimensional NMR spectra create a new way to reveal long-range couplings, in which long-range couplings are shown in cross peaks, which are much more reliable than the slight increments in width at the half-height of the peaks. For this reason, a special COSY, the COSY optimized for long-range couplings or simply COSYLR (LRCOSY), has been developed.

The pulse sequence of COSYLR is shown in Fig. 4.49.

A comparison of Fig. 4.49 with Fig. 4.42 shows that the pulse sequence of COSYLR is that of COSY plus the time duration, Δ, which is situated before and after the second 90° pulse. For this reason, the COSYLR is also called delayed COSY or COSY with delay.

The principle of COSYLR can be explained by the product operator formalism. The intensity of a cross peak obtained by this theory is proportional to

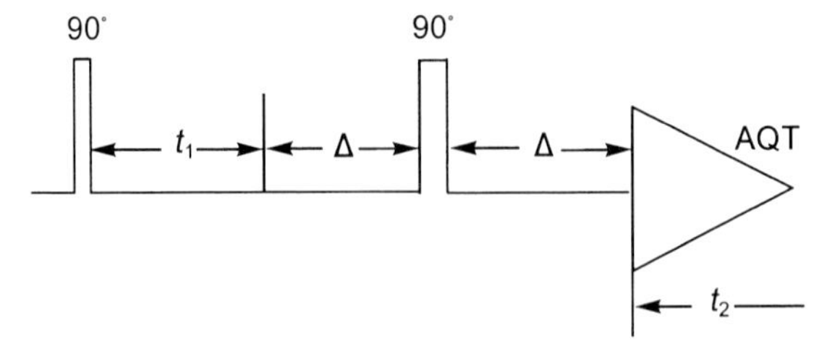

Fig. 4.49 The pulse sequence of COSYLR.

sin πJt, where t is t_1 or t_2. The coupling constants of long-range couplings are small. To record the cross peaks of long-range couplings, the numbers of both t_1 and t_2 should be greatly increased, which will lead to an impracticable internal memory of the computer performing the 2D NMR experiment and too long a calculation duration. The addition of Δ values before and after the second 90° pulse prolongs both t_1 and t_2 without increased calculations. The enhancement of intensities of cross peaks of COSYLR can be understood from the point of view that the transfers of long-range couplings are enhanced by prolonging the transfer time. On the other hand, the intensities of the cross peaks produced from ^3J, which are strong enough to give cross peaks in normal t_1 and t_2, will decay significantly with their transverse relaxation time T_2 through the two additional Δ values. If the Δ is increased to a certain value, both cross peaks and diagonal peaks produced from long-range couplings can dominate the spectrum. The value of Δ can be found by experiment, which is approximately 200 ms.

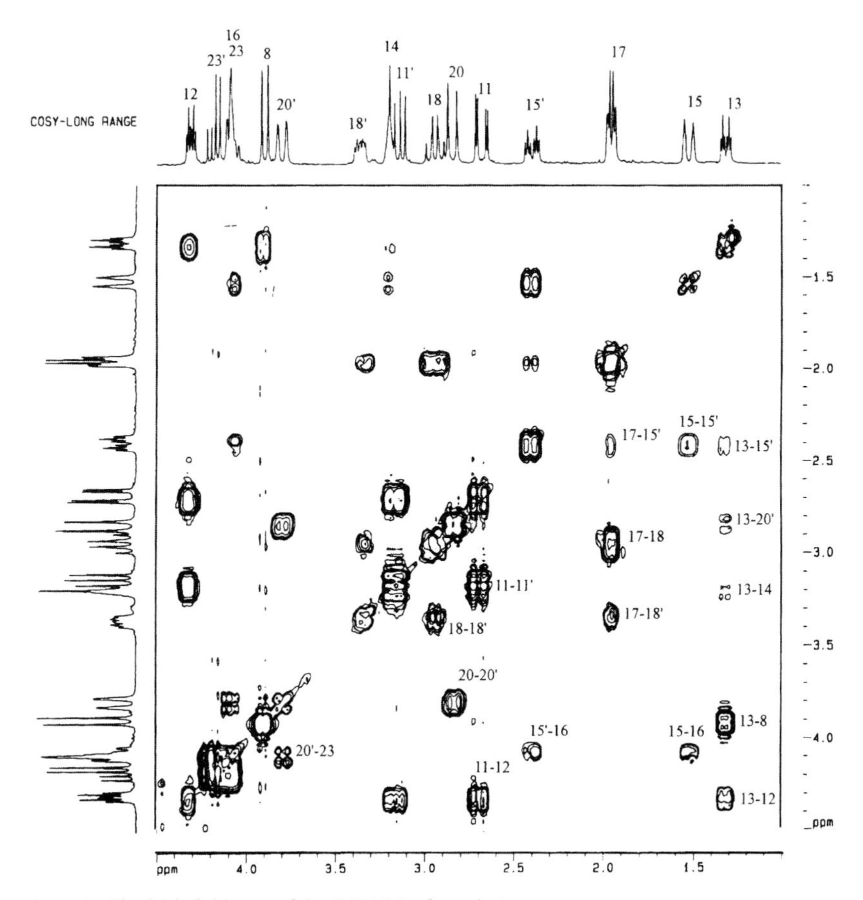

Fig. 4.50 The high field part of the COSYLR of strychnine.

The high field portion of the COSYLR of strychnine is shown in Fig. 4.50.

From a comparison of Figs. 4.46 and 4.47 with 4.50, it follows that cross peaks resulting from long-range couplings (H-13, H-15′; H-17, H-15′, etc.) appear as in Fig. 4.50. As the COSYLR is apt to reveal couplings with small coupling constants, the cross peaks, which are missing in Fig. 4.46, such as those produced from H-13, H-14; H-13, H-12; etc., appear as in Fig. 4.50. Under the experimental conditions, cross peaks produced from ^3J still exist because it is impossible to suppress all of them by prolonging t_1 and t_2. However, their intensities are decreased.

4.6.6
DQF-COSY

DQF-COSY, the abbreviation for double-quantum filtered COSY, belongs to double-quantum two-dimensional NMR spectra from the point of view of its principle. However, it is advisable to include it in this section since DQF-COSY is similar to an ordinary COSY in appearance.

If an organic compound contains a *tert*-butyl group, or a methoxy group, etc., which produces a strong singlet in its ^{1}H spectrum, then in its COSY the cross peaks produced from weak peaks of the ^{1}H spectrum have weak intensities. In this instance, DQF-COSY can give a better result than an ordinary COSY.

The pulse sequence of DQF-COSY is shown in Fig. 4.51.

A comparison of Fig. 4.51 with Fig. 4.42 shows that the pulse sequence of DQF-COSY consists of that of COSY plus a 90° pulse soon after the second 90° pulse of the pulse sequence of COSY. The time interval between the two 90° pulses is only several microseconds, which is determined by the time to reset the subsequent 90° pulse for the spectrometer. The product operator formalism should be used to analyze the principle of DQF-COSY. In Appendix 1, the result of the pulse sequence of COSY is given by using the formalism. Some detectable signals produced by the use of the second 90° pulse become undetectable signals after the third 90° pulse. On the contrary, some "double-quantum coherence" terms that are undetectable, are transformed into detectable signals to produce cross peaks by the use of the third 90° pulse.

Note that the first two 90° pulses have the same phase of φ, which is denoted at the bottom right-hand of the pulses, and that the phase of the third 90° pulse is denoted as x, which means a phase cycling. With the phase cycling the trans-

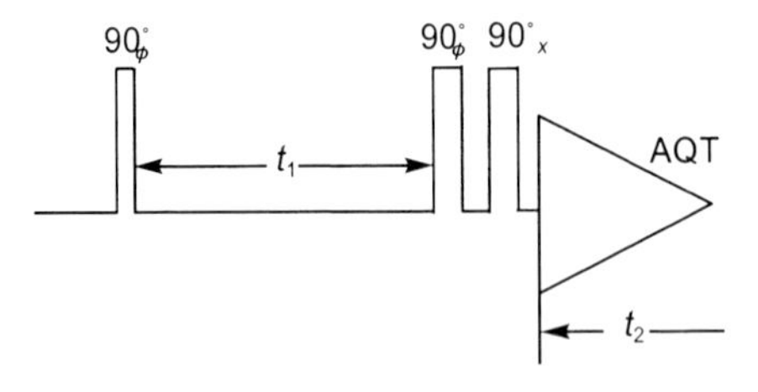

Fig. 4.51 The pulse sequence of DQF-COSY.

formed detectable signals can be measured. If the pulsed-field gradient is applied, the phase cycling can be omitted.

DQF-COSY belongs to MQF (multiple-quantum filtered)-COSY, or PQF-COSY, where p is the order of the multiple-quantum coherence. The pulse sequence of MQF-COSY is also that shown in Fig. 4.51, but with different phase cycling. The greater the value of p, the more cross peaks, which have coherence less than p–1, are deleted. For example, in the triple-quantum filtered COSY, cross peaks produced from single and double-quantum coherence are deleted. However, the greater the value of p, the smaller intensities the remaining cross peaks have. Therefore, DQF-COSY is most commonly used.

Although DQF-COSY is similar to an ordinary COSY in appearance, the former has the following two advantages:

1. The cross peaks produced from strong peaks in its ^{1}H spectrum are suppressed. The solvent peak, which is a strong peak in the ^{1}H spectrum in general, does not produce interfering peaks in the DQF-COSY.
2. The peak shapes are better in DQF-COSY than in phase-sensitive COSY. In the phase-sensitive COSY, if the cross peaks are adjusted into absorption lineshapes, the diagonal peaks of the phase-sensitive COSY have dispersion lineshapes with fairly long tails so that some cross peaks close to the diagonal may be subject to interference. However, in DQF-COSY both cross peaks and diagonal peaks can be adjusted into absorption signals, and at least the lineshapes of the diagonal peaks have been improved.

There are other homonuclear shift correlated spectra. The reader who is interested in them can consult reference [3].

4.7 NOESY and its Variations

NOESY is the abbreviation for nuclear Overhauser effect (two-dimensional) spectroscopy.

It is recommended that the reader understands NOE (discussed in Section 2.7.2) before reading this section.

It is convenient to display NOE information through a two-dimensional spectrum even though it has a lower sensitivity than that recorded in the one-dimensional experiment.

NOESY plays an important role in determining structures, configurations and conformations of organic compounds, and offers structural information on biological molecules, so that it occupies a significant position in 2D NMR spectra.

4.7.1 NOESY

The pulse sequence of NOESY consists of three non-selective 90° pulses, as shown in Eq. (4-30).

$$90^\circ - t_1 - 90^\circ - \tau_m - 90^\circ - t_2 \qquad (4\text{-}30)$$

From the pulse sequence or its improved version, an NOESY or a two-dimensional chemical exchange spectrum, which illustrates chemical exchanges or conformational interchanges of organic compounds, is obtained.

The mechanism of the pulse sequence producing a NOESY (spectrum) can be analyzed as follows.

When acted on by the first 90° pulse, the magnetization vectors are rotated from the z axis onto the y' axis to produce related transverse magnetization vectors, which precess in the x'–y' plane with their characteristic frequencies (the difference between their resonant frequency and that of the rotating frame) during the evolution time t_1, thus functioning as the labels of spins or frequencies. Acted upon by the second 90° pulse, these transverse magnetization vectors are rotated from the x'–y' plane onto the x'–z' plane to produce their longitudinal magnetization vectors. If some components of the transverse magnetization vectors remain, they must be removed. By omitting the transverse relaxation during t_1, at the end of the second 90° pulse the longitudinal magnetization vector of nuclei A, M_{ZA} is represented by Eq. (4-31):

$$M_{ZA}(t_1) = -M_{0A}\cos\left[(\omega_A - \omega)t_1\right] \qquad (4\text{-}31)$$

where M_{ZA} (t_1) is the longitudinal magnetization vector of nucleus A at the end of t_1 (when the second 90° pulse is just over); M_{0A} is the magnetization vector of A under equilibrium conditions; ω_A and ω are the resonant frequency of A and the rotating frequency of the rotating frame, respectively.

The period of the mixing time, τ_m, begins after the second 90° pulse. If the nucleus X is close to A, there is a dipole–dipole interaction between them, which is the cross relaxation (see Section 2.7.2) to produce NOE. Acted on by this cross relaxation, M_{ZA} decreases gradually with an increase in M_{ZX}. At the end of τ_m, the longitudinal magnetization vector of X has the value $CM_{ZA}(t_1)$, where C is determined by the values of τ_m and the velocity of the cross relaxation.

Acted on by the third 90° pulse, M_{ZX} is rotated from the x'–z plane onto the x'–y' plane to produce the transverse magnetization vector, which precesses in the x'–y' plane with the frequency of ω_X–ω to produce a detectable signal during t_2 starting from the third 90° pulse.

From the discussion above, the signal of X in the time domain can be expressed by Eq. (4-32).

$$S_X(t_1, t_2) = CM_{0A}\cos\left[(\omega_A - \omega)t_1\right]e^{i(\omega_X - \omega)t_2} \qquad (4\text{-}32)$$

where $S_X(t_1, t_2)$ is the signal of X in the time domain, which is a function of t_1 and t_2.

From Eq. (4-32), it follows that the signal of X is modulated by the resonant frequency of A. Therefore, cross peaks in such a 2D NMR spectrum correlate pairs of nuclei, between which there is NOE.

If there are any remaining transverse magnetization vectors after the second 90° pulse, they can be removed by using a pulsed-field gradient or by changing the phases of the second and the third 90° pulses.

In the pulse sequence [Eq. (4-30)], both t_1 and t_2 are independent time variables, while the mixing time, τ_m, is a selected time duration found by experiment. If τ_m is too short, cross relaxation will not take place completely. However, if τ_m is too long, the signal of M_{ZX} will decay too much because of its relaxation. Therefore, NOESY is more difficult to carry out than other 2D NMR experiments because the result of NOESY depends on whether τ_m is selected correctly.

From the discussion in Section 2.7.2, it follows that for a homonuclear system, $\omega\tau_C$ is important to the absolute value and the sign of NOE, where ω is the frequency of the spectrometer to be used and τ_C is the rotational correlation time of the sample molecules to be measured, which describes the velocity of molecular tumbling. Sample molecules with a rather large molecular weight tumble slowly. For example, protein molecules have tumbling frequencies of the order of 10^7 Hz. On the other hand, molecules with a rather lower molecular weight tumble rapidly so that in a non-sticky solvent their tumbling frequencies can reach 10^{11} Hz [26]. In general, τ_m is related to the molecular weight of the sample to be measured, the viscosity of the solution to be measured and the frequency of the spectrometer to be used. Therefore, τ_m is selected by experiment.

By using the pulse sequence in Eq. (4-30), NOESY is obtained if cross relaxation takes place in τ_m, while a two-dimensional chemical exchange spectrum is obtained if chemical exchanges occur in τ_m.

It is important to remove as many cross peaks produced by couplings in the experiment as possible, which can be achieved by using an improved pulse sequence and/or special phase cycling [2].

NOESY can be illustrated in a phase-sensitive mode in which the signs of the cross peaks show the signs of NOE.

NOESY is similar to COSY in appearance in both the phase-sensitive mode and the non-phase-sensitive mode (magnitude mode). Their difference is that the cross peaks of NOESY show the NOE while those of COSY show the couplings.

4.7.2
ROESY

ROESY is the abbreviation for rotating frame Overhauser effect spectroscopy.

From the discussion in Section 2.7.2 it follows that the enhancement of NOE will be close to zero when $\omega\tau_C \approx 1.12$. In the case of the measurement of NOE of a sample with a moderate molecular weight, the cross peaks of NOESY could be missing. However, compounds with a moderate molecular weight are frequently objects of interest in organic chemistry. Therefore, ROESY is an ideal solution in this situation.

It is necessary to remember the principle of spin locking, this is the basis from which ROESY can be understood.

Acted on by the $90^\circ_{x'}$ pulse, the magnetization vectors of sample molecules are rotated from the z axis onto the y' axis to produce related transverse magnetization vectors, which precess in the x'–y' plane with their angular frequencies of $\nu_i - \nu_0$. At this time, if $\boldsymbol{B}_1$ is applied along the y' axis, the effective field for the i-th magnetization vector is given by Eq. (4-33) (see Section 1.4.2).

$$\boldsymbol{B}_{\text{eff}}(i) = [4\pi^2(\nu_i - \nu_0)^2\boldsymbol{k}/\gamma^2 + \boldsymbol{B}_1^2]^{1/2} \quad (4\text{-}33)$$

where $\boldsymbol{B}_{\text{eff}}(i)$ is the effective field of the i-th nuclei and γ is the magnetogyric ratio of the nuclei.

When $\gamma B_1/(2\pi) >> (\nu_i - \nu_0)$, all magnetization vectors are locked along the y' axis. However, the condition of $\gamma B_1/(2\pi) >> (\nu_i - \nu_0)$ is not satisfied in ROESY experiments because it is difficult for B_1 to reach such a great value. Therefore, every $\boldsymbol{B}_{\text{eff}}(i)$ and the y' axis form a small angle, which is determined by $\nu_i - \nu_0$, indicating that the real situation is slightly different from that described in Section 4.1.6. The present circumstance can be explained as follows: The "main components" of magnetization vectors are locked along the y' axis. Consequently, cross relaxation, that is NOE, takes place between these magnetization vectors. On the other hand, the small components, which are vertical to the main components, dephase so quickly that they can be omitted.

The pulse sequence of ROESY is shown in Fig. 4.52.

The principle of the pulse sequence of ROESY can be easily understood. The 90° pulse produces transverse magnetization vectors, at which time t_1 starts. The transverse magnetization vectors are labeled with their resonant frequencies during t_1. The cross relaxation on the y' axis, that is, NOE in the rotating frame, occurs during the period of spin locking. The acquisitions are taken during t_2.

The following equations, which should be compared with Eqs. (2-39) to (2-42), hold in ROESY [3].

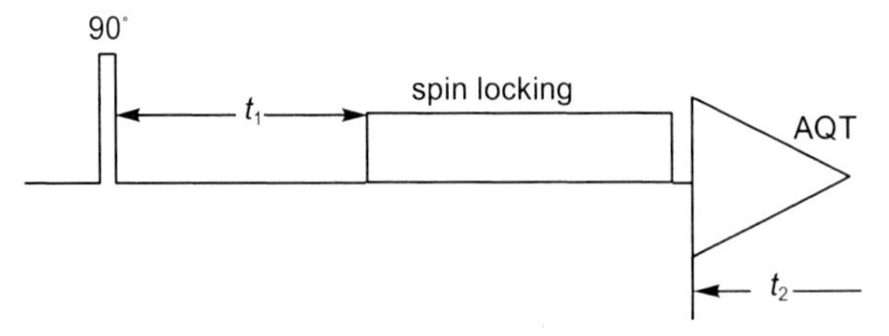

Fig. 4.52 The pulse sequence of ROESY.

$$W_1 = \frac{3}{40}\frac{\gamma_I^2\gamma_S^2\hbar^2\tau_C}{r^6}\left(\frac{1}{1+\omega^2\tau_C^2}+\frac{1}{1+4\omega^2\tau_C^2}\right) \tag{4-34}$$

$$W_0 = \frac{3}{40}\frac{\gamma_I^2\gamma_S^2\hbar^2\tau_C}{r^6}\left(\frac{1}{3}+\frac{1}{1+4\omega^2\tau_C}\right) \tag{4-35}$$

$$W_2 = \frac{3}{4}\frac{\gamma_I^2\gamma_S^2\hbar^2\tau_C}{r^6}\left(3+\frac{4}{1+\omega^2\tau_C^2}+\frac{1}{1+4\omega^2\tau_C}\right) \tag{4-36}$$

$$W_2 - W_0 = \frac{\gamma_I^2\gamma_S^2\hbar^2\tau_C}{10r^6}\left(2+\frac{3}{1+\omega^2\tau_C^2}\right) \tag{4-37}$$

It should be noted that the right side of Eq. (4-37) can not be equal to zero, which means the intensities of the cross peaks of ROESY are never equal to zero. However, the intensities of cross peaks of NOESY are greater than those of ROESY when $\omega\tau_C \to 0$ or $\omega\tau_C \to \infty$.

ROESY used to be known as CAMELSPIN, which is the abbreviation for cross-relaxation appropriate for minimolecules emulated by locked spins.

4.7.3 HOESY

HOESY is the abbreviation for heteronuclear NOE spectroscopy.

HOESY is similar to NOESY in that the former shows two nuclei close in space, which belong to different types of nuclei while the latter shows two nuclei close in space, which are of the same type.

The pulse sequence of HOESY is shown in Fig. 4.53.

The principle of the pulse sequence of HOESY can be explained as follows: The first 90° pulse applied to ^1H produces the transverse magnetization vectors of ^1H and t_1 starts from that time. Acted on by the 180° pulse applied at the middle point of t_1, the transverse magnetization vectors are modulated by their

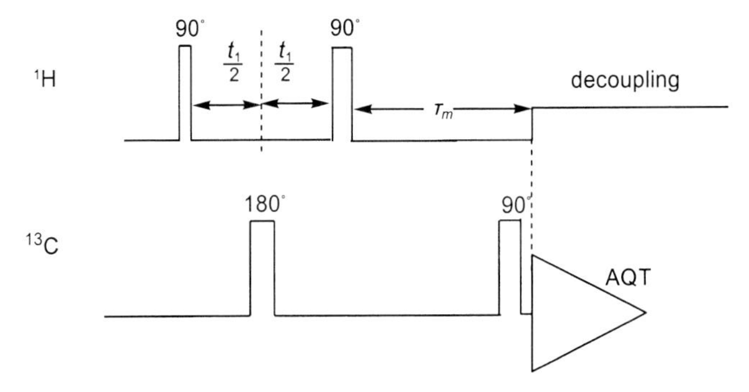

Fig. 4.53 The pulse sequence of HOESY.

δ values at the end of t_1 (see Fig. 4.37). The splittings from J are refocused at this point. The second 90° pulse applied to ^1H produces the longitudinal magnetization vectors from their transverse magnetization vectors. The cross relaxation between ^1H and ^{13}C, that is NOE between them, takes place during τ_m. The 90° pulse applied to ^{13}C changes the longitudinal magnetization vectors of ^{13}C into their detectable transverse magnetization vectors, which are related to ^1H nuclei close to the ^{13}C nuclei.

HOESY is similar to H,C-COSY in appearance. However, the cross peaks of the former show the NOE relationships between ^1H and ^{13}C.

4.8 Relayed Correlation Spectra and Total Correlation Spectra

Relayed correlation spectra are an extension of correlation spectra. Total correlation spectra are an extension of relayed correlation to the entire spin system.

4.8.1 RCOSY

RCOSY is the abbreviation for relayed COSY.

RCOSY is an extension of COSY. Take an AMX system as an example. In an ordinary COSY there are the cross peaks produced from J_{AM} and J_{MX}, while there are no cross peaks produced from J_{AX} as J_{AX} is equal to zero.

Suppose that there is another AMX system labeled as A′M′X′ to differentiate it from the former one. If the δ value of M′ is very close to that of M, the connection, A → M → X, can be confused with A → M′ → X′, which is shown in Fig. 4.54.

The drawback of the COSY can be overcome by the RCOSY in which the additional cross peak between A and X through M confirms the connection of A–M–X.

The pulse sequence of RCOSY is shown in Fig. 4.55.

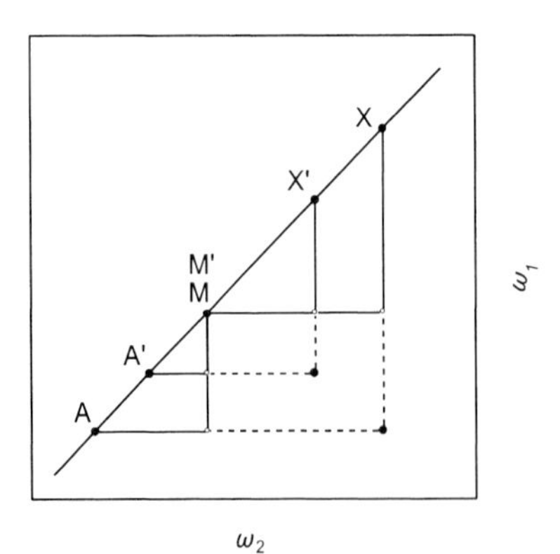

Fig. 4.54 The function of RCOSY. ■, Diagonal peaks; ○ cross peaks of COSY; ●, additional cross peaks of RCOSY.

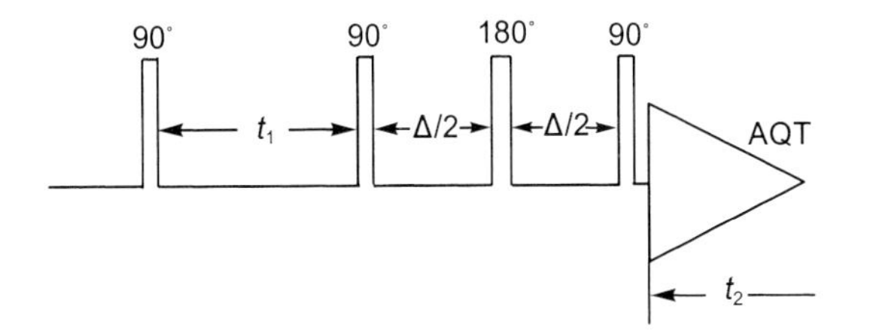

Fig. 4.55 The pulse sequence of RCOSY.

From a comparison of Fig. 4.55 with Fig. 4.42, it follows that the second 90° pulse in the pulse sequence of COSY is replaced by the sequence of 90°, $\Delta/2$, 180°, $\Delta/2$ and 90°.

The principle of RCOSY can be explained by the product operator formalism. Appendix 1 in this book deals only with the two-spin system, while RCOSY concerns at least the three-spin system, so the principle of RCOSY will be explained qualitatively here. The first 90° pulse produces transverse magnetization vectors. Coherence transfer from A to M takes place during t_1. The preceding part of the RCOSY pulse sequence is the same as that of COSY. Acted on by the 180° pulse, chemical shifts are refocused at the end of $\Delta/2$, 180° and $\Delta/2$, during which coherence transfer is extended into X. The third 90° pulse produces detectable components.

The intensities of cross peaks of RCOSY are proportional to $\sin(J_{AM}\Delta)\sin(J_{MX}\Delta)\exp(-\Delta/T_2)$. If Δ is too short, it is unfavorable for the coherence transfer. However, if Δ is too long, signal intensities will decrease through the transverse relaxation. Therefore, Δ should be an appropriate value, which is taken as 3.2–4.0 times that of $1/J_{max}$.

The series of $\Delta/2$, 180° and $\Delta/2$ can be used repeatedly. In this case, it forms the pulse sequence of multi-step RCOSY during which the coherence transfer is extended further. However, sensitivity will drop rapidly with the length of the coherence transfer. By contrast, TOCSY has the same function but a high sensitivity. Therefore, TOCSY with different isotropic mixing times, which have the functions of RCOSY and multi-step RCOSY, have replaced the latter.

4.8.2 Heteronuclear Relayed COSY

The key to the identification of an organic compound is the determination of the connective sequence of the carbon atoms. Therefore, 2D INADEQUATE (see Section 4.9.1) is the most attractive method because it can determine the connections between the carbon atoms directly. However, as 2D INADEQUATE measures the couplings between ^{13}C–^{13}C, whose probability of appearing in a molecule is 1/10000, its measurement requires a lengthy time acquisition and a large amount of sample to reach an appropriate signal-to-noise ratio. On the other hand, although the connections between carbon atoms can be determined by using the combination of COSY and H,C-COSY, these two independent experiments have to be per-

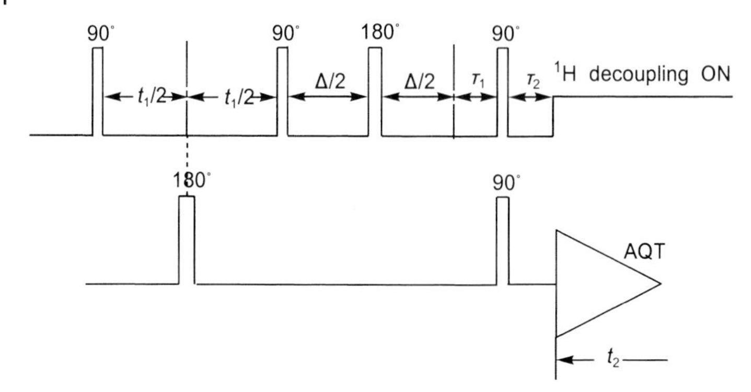

Fig. 4.56 The pulse sequence of the heteronuclear relayed COSY.

formed. The heteronuclear relayed COSY, another method for determining the connections, has a higher sensitivity than 2D INADEQUATE and is different from the combination of COSY and H,C-COSY, and requires just one experiment.

The pulse sequence of the heteronuclear relayed COSY is shown in Fig. 4.56.

The principle of pulse sequence as shown in Fig. 4.56 can be easily understood. This pulse sequence is constructed on the basis of the pulse sequence of H,C-COSY (shown in Fig. 4.37), with the series of 90°, $\Delta/2$, 180° and $\Delta/2$ inserted, which is necessary for relays.

In the structural unit of $-C_A(H_A)-C_B(H_B)-$, the function of the heteronuclear relayed COSY is to produce the transfer of $H_A \rightarrow H_B \rightarrow C_B$ and $H_B \rightarrow H_A \rightarrow C_A$, which is shown in Fig. 4.57.

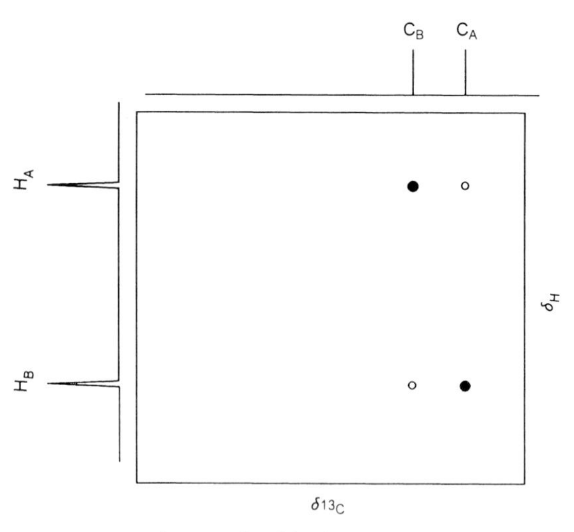

Fig. 4.57 A simple example of the heteronuclear relayed COSY. ○, Cross peaks of the heteronuclear COSY without relays; ● cross peaks of the heteronuclear COSY with relays.

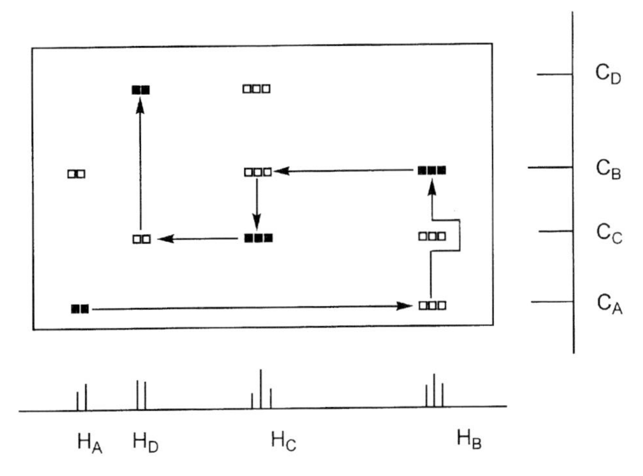

Fig. 4.58 The connections between carbon atoms are deduced by using the heteronuclear relayed COSY. □, Cross peaks of the heteronuclear COSY without relays; ■, cross peaks of the heteronuclear COSY with relays.

In Fig. 4.57, the empty circles, which are situated at δ_C on the ω_2 axis and at δ_H on the ω_1 axis, show the correlation between directly connected ^{13}C and ^{1}H, and the solid circles the correlation through relays between indirectly connected ^{13}C and ^{1}H. These two empty circles and two solid circles form a rectangle, which shows the existence of the structural unit.

Using this type of rectangles, a structural unit can be extended to a greater structural unit, even to the entire structure of the unknown compound. This process is shown in Fig. 4.58.

In fact, Fig. 4.58 is an extension or the combination of Fig. 4.57. The connection of two carbon atoms is identified by such a rectangle. Repeat this process, and the connection from C_A to C_D can be determined.

4.8.3
Total Correlation Spectroscopy (TOCSY)

Obviously, it is very useful to find all couplings in a spin system from one peak set of a hydrogen atom of the spin system, in which coupling constants between two hydrogen atoms separated by more than three bonds can be zero. Braunschweiler and Ernst proposed the term TOCSY according to the function of the pulse sequence in their paper [13]. Later Bax and Davis proposed the term HOHAHA (homonuclear Hartmann–Hahn spectroscopy) through homonuclear Hartmann–Hahn cross polarization [12]. They pointed out that HOHAHA is closely related to TOCSY. In fact, these two terms are used interchangeably in the literature. Because these two experiments are discussed from different aspects, they are presented in two parts of this section. It is recommended that the reader understands isotropic mixing (Section 4.1.7) before reading this section.

1. TOCSY

The pulse sequence of TOCSY is shown in Fig. 4.59.

The principle of Fig. 4.59 can be explained as follows: The evolution time t_1 starts from the end of the 90° pulse. Transverse magnetization vectors precess with their offset frequencies ($\nu_i - \nu_0$) in the x'–y' plane, respectively, thus functioning as a spin labeling. Here ν_i is the resonant frequency of i-th nuclei and ν_0 is the rotating frequency of the rotating frame. During the evolution time, there are only weak interactions, which are in a "single spin mode."

The isotropic mixing time, τ_m, is a characteristic of TOCSY. The specific pulse sequence in τ_m of TOCSY is not presented in Fig. 4.60 because four pulse sequences were proposed in the original reference. Of the four, the two rather simple sequences are four repetitive $180^\circ_{x'}$ pulses and a series consisting of $180^\circ_{x'}$, $180^\circ_{x'}$, $180^\circ_{-x'}$ and $180^\circ_{-x'}$. During the isotropic mixing time, the term ($\nu_i - \nu_0$) in the Hamiltonian is temporarily removed with only the isotropic coupling term left. The single spin mode is changed to a collective spin mode. The couplings will transfer, during the isotropic mixing time, to the directly coupled nuclei for a short isotropic mixing time, such as 20 ms, and to the entire spin system for a long isotropic mixing time, such as 50–100 ms.

During the detection time, the strong isotropic couplings return to weak couplings. Both chemical shifts and couplings play a role.

By using a TOCSY with a long isotropic mixing time, cross peaks, which belong to the other hydrogen atoms in the same spin system, can be found from one peak of a certain hydrogen atom.

2. HOHAHA

The original pulse sequence of HOHAHA is shown in Fig. 4.60.

The principle of Fig. 4.60 can be explained as follows: During the evolution time t_1 starting from the end of the 90° pulse, transverse magnetization vectors are labeled. During the spin locking time, cross polarization under Hartmann–Hahn matching (see Section 4.1.7), that is isotropic mixing, takes place, followed by the detection time, t_2.

It appears that the pulse sequence of ROESY (Fig. 4.53) is identical to that of HOHAHA. In fact, the field intensity of HOHAHA is much stronger than that of ROESY.

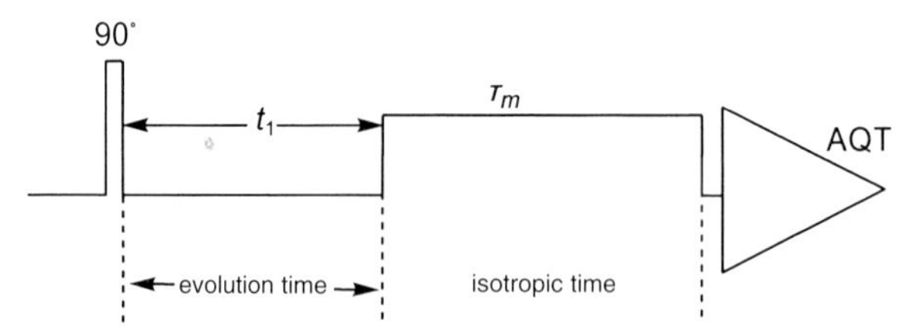

Fig. 4.59 The pulse sequence of TOCSY.

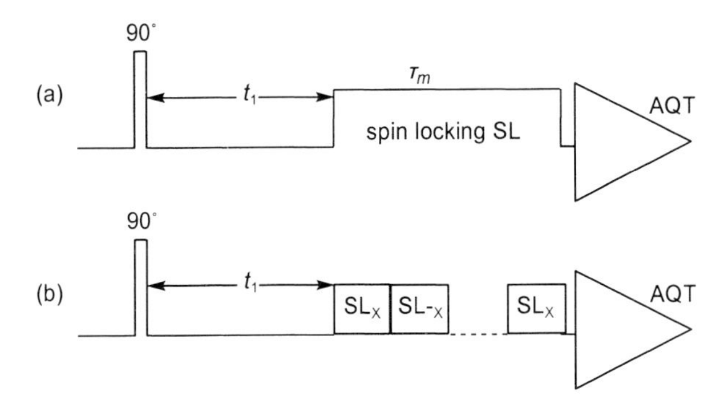

Fig. 4.60 The original pulse sequence of HOHAHA. (a) Spin locking with constant phase. (b) Spin locking with alternative phases.

By using the pulse sequence of Fig. 4.60 b, the effective frequency width can be extended. However, it is still not satisfactory, and therefore the pulse sequence is replaced by that of MLEV (Malcon Levitt)-17, as shown in Fig. 4.61.

Because the difference between Figs. 4.60 and 4.61 lies only in the replacement of the pulse sequence of spin locking by MLEV-17, all one needs to do is to explain the function of MLEV-17 in order to understand the principle. MLEV-17 consists of the first "trim" pulse, MLEV-16, an uncompensated 180° pulse, and the second "trim" pulse. The two "trim" pulses are denoted as "SL" and the uncompensated 180° pulse is indicated by "↓" in Fig. 4.61.

The discussion starts from the sequence of MLEV-16, or the cycle of MLEV-16, which consists of an integral multiple of cycles of following composite pulses:

ABBA BBAA BAAB AABB

where $A = 90^\circ_{-y} - 180^\circ_{x'} - 90^\circ_{-y}$ and $B = 90^\circ_{y} - 180^\circ_{-x'} - 90^\circ_{y'}$.

Both A and B are composite pulses, which consist of sequential pulses instead of one pulse, in order to overcome the imperfection of the pulse.

MLEV-17 is an improvement over MLEV-16. The two "trim" pulses, which are situated at the beginning and the end of the mixing time, can defocus any magnetization vector, which is not parallel to the x' axis so as to easily

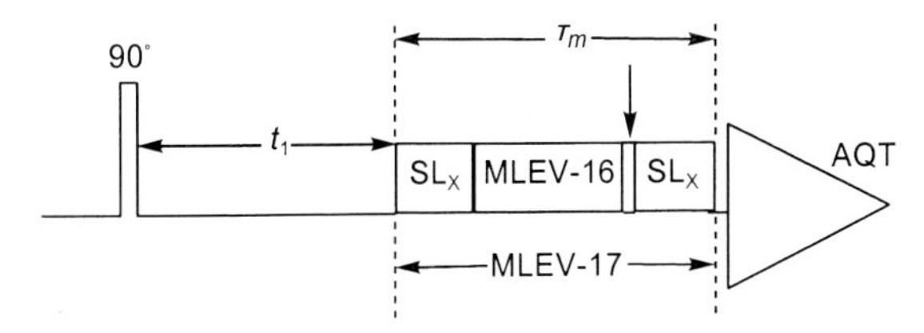

Fig. 4.61 The improved pulse sequence of HOHAHA.

produce the 2D NMR spectrum with absorption line shapes. The addition of the uncompensated 180° pulse can prevent the accumulation of phase angle errors during the cycling of composite pulses.

As with TOCSY, the coupling action will transfer longer distances with the increasing τ_m.

The appearance of HOHAHA is identical to that of TOCSY.

4.9 Multiple Quantum 2D NMR Spectra

A multiple quantum transition can be detected by using particular pulse sequences so that a multiple quantum 2D NMR spectrum, which provides important structural information, can be obtained.

4.9.1 2D INADEQUATE

INADEQUATE is the abbreviation for incredible natural abundance double quantum transfer experiment, through which the connections between the carbon atoms can be determined [27].

By using INADEQUATE the connection of two carbon atoms through the coupling between ^{13}C–^{13}C is determined, the probability of which is only 1/10000 since the natural abundance of ^{13}C is 1.1%. Fortunately, INADEQUATE can be improved, through the use of a spectrometer with a high frequency (hence with a high sensitivity) and/or an improved probe, and techniques such as magic angle spinning (see Section 8.3.3).

From the discussion above, it follows that the key to the accomplishment of INADEQUATE is the suppression of the signals produced from the individual carbon atoms without the coupling between ^{13}C–^{13}C. These signals are 100 times stronger than those from INADEQUATE!

As the connection of three ^{13}C atoms can be omitted, only the system of an AX or AB system is considered.

The pulse sequence of INADEQUATE is shown in Fig. 4.62.

In Fig. 4.62, τ and Δ are time durations while $90°(\varphi)$ is a "read" pulse. AQT (ψ) represents the acquisition according to a particular phase ψ, which is related to φ. The cycling between φ,ψ, and the phases of three types of signals are shown in Tab. 4.1.

The principle of INADEQUATE should be discussed in terms of the product operator formalism (see Appendix 1). However, some concepts, in particular that of phase cycling, are presented here.

Firstly, it should be noted that the INADEQUATE experiment is performed with ^{1}H decoupled, so that the signals of ^{13}C are not split by ^{1}H.

The series of $90^\circ_{x'}$–τ–$180^\circ_{y'}$–τ–$90^\circ_{x'}$ produces two types of signals: S_2 and S_0. The latter results from the uncoupled ^{13}C. The first $90^\circ_{x'}$ pulse rotates the magnetiza-

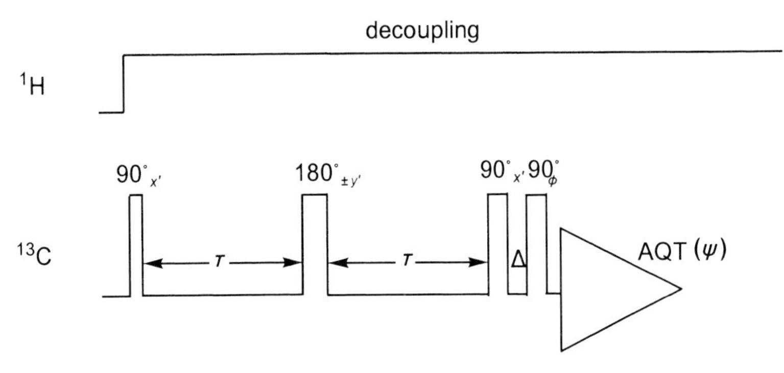

Fig. 4.62 The pulse sequence of INADEQUATE.

tion vector of ^{13}C from the z axis onto the y' axis. Suppose that the rotating frequency of the rotating frame is equal to the chemical shift value of the ^{13}C nuclei, the transverse magnetization vector will be along the y' axis after the $90^\circ_{x'}$ pulse. The series of τ–$180^\circ_{y'}$–τ refocuses the chemical shifts and overcomes the inhomogeneity of B_0 (see Section 4.1.3). The second $90^\circ_{x'}$ pulse rotates the transverse magnetization vector from the y' axis onto the $-z$ axis. According to Tab. 4.1, the phases of S_0 are along the $-y$, $+x$, $+y$ and $-x$ axis, respectively, when the phases of φ are correspondingly along the $+x$, $+y$, $-x$ and $-y$ axis. The time duration of Δ is equal to several microseconds, during which the φ can be set.

Produced by the series 90°, τ, 180°, τ and 90°, S_2 results from every two coupled ^{13}C nuclei and it is of great relevance. In fact, S_2 is the coherence between the two energy levels of $\alpha\alpha$ and $\beta\beta$ of the spin system of two coupled ^{13}C nuclei, hence the name double-quantum coherence. It becomes a detectable signal by the action of the read pulse.

Because the signals transferred from S_2 are weaker than those of S_0 by two orders of magnitude, the suppression of the signals of S_0 is the key to the measurement of the signals transferred from S_2 for the INADEQUATE experiment. The suppression is achieved by the phase cycling as follows.

The phase of the detector, ψ, matches the phase of the read pulse φ. When the phases of φ are along the $+x$, $+y$, $-x$ and $-y$ axis, respectively, the phases of ψ are correspondingly along the $+x$, $-y$, $-x$ and $+y$ axis, which is identical to the phases of the signals transferred from S_2, hence the signals resulting from S_2

Tab. 4.1 The cycling between φ, ψ, and the phases of three types of signals

φ	S_0	S_1	S_2	ψ
$+x$	$-y$	$+x$	$+x$	$+x$
$+y$	$+x$	$+y$	$-y$	$-y$
$-x$	$+y$	$+x$	$-x$	$-x$
$-y$	$-x$	$+y$	$+y$	$+y$

will be detected. On the other hand, the phases of the signals of S_0 are quite different from those transferred from S_2 so that the former is suppressed.

The above-mentioned conclusion can be drawn in another way. When the phases of φ are along the $+x$, $+y$, $-x$ and $-y$ axis sequentially, these four directions form an anti-clockwise rotation if it is observed from the z axis towards the x'–y' plane. Correspondingly, the four directions of the phases of S_0: $-y$, $+x$, $+y$ and $-x$ also form an anti-clockwise rotation. However, ψ and the signals of S_2 rotate clockwise at the same time (see Tab. 4.1), so that the signals of S_2 are detected while those of S_0 are suppressed.

Because the signals of S_2 are much weaker than those of S_0 by two orders of magnitude and the imperfection of the pulses is inevitable, false signals, S_1, will appear, which should be suppressed both by using phase cycling and by the replacement of the 180° pulse in Fig. 4.62 by $180^\circ_{\pm y}$.

The time duration τ is set according to the following equation in order to obtain the best transfer of double-quantum coherence.

$$\tau = (2n + 1)/4\,{}^1J_{CC} \tag{4-38}$$

where n=1, 2, 3, 4... . In general, n is taken as 0 to decrease the influence of transverse relaxation. Because ${}^1J_{CC}$, the coupling constant of two adjacent ^{13}C nuclei, is about 30 Hz, the value of τ is about 8 ms.

By using the pulse sequence in Fig. 4.63, a one-dimensional INADEQUATE spectrum, a ^{13}C spectrum with the coupling splitting, is obtained. Two adjacent carbon atoms can be found if some peak pairs that are separated by a particular space in their two peak sets exist, which means these two carbon atoms have an identical coupling constant.

The one-dimensional INADEQUATE spectrum has disadvantages. As a result, it is replaced by the 2D INADEQUATE experiment, whose pulse sequence is identical to that shown in Fig. 4.63 except that Δ is changed to the time variable t_1.

The 2D INADEQUATE spectrum of δ-vitamin E is shown in Fig. 4.63.

The chemical shift of carbon atoms is shown on the F_2 axis of the 2D INADEQUATE spectrum while the double-quantum frequency, which is the average of the δ values of two coupled carbon atoms, is shown on the F_1 axis. The ^{13}C spectrum of the sample is placed over the 2D INADEQUATE spectrum. In the 2D INADEQUATE spectrum, there is a pseudo-diagonal, which is described as $\omega_1 = 2\omega_2$. The correlated peaks of two coupled (adjacent) carbon atoms are situated on a horizontal line with equal distances from the cross point of the horizontal line and the pseudo-diagonal. Therefore, the entire connection of carbon atoms of an unknown compound can be determined by finding all correlated peaks.

As compound **C4-3** has a side chain with 16 carbon atoms, whose peaks are close in the fairly high field region, it is difficult to determine the connection to the side chain. However, the connection of the phenyl group can be easily determined. It is surprising that the methyl group, which is connected to the phenyl ring, has the lowest δ value.

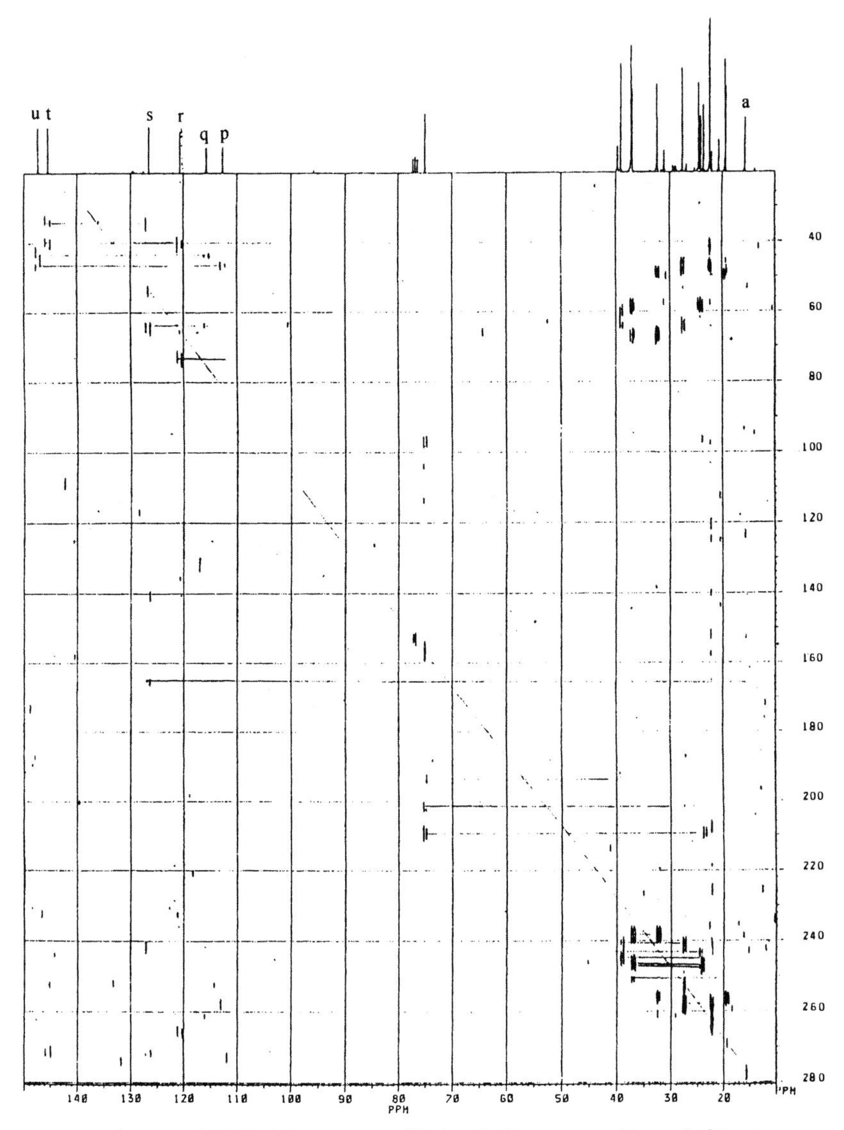

Fig. 4.63 The 2D INADEQUATE spectrum of δ-vitamin E, courtesy of *Journal of Spectroscopy* (Chinese version).

4.9.2 Two-dimensional Double Quantum Spectra of ^1H

The pulse sequence of the two-dimensional double quantum spectra of ^1H is similar to that of 2D INADEQUATE except that the pulse sequence acts on ^1H, the acquisition of ^1H is achieved without decoupling, and the calculation of Δ is performed by $^3J_{HH}$.

The two-dimensional double quantum spectrum of ^{1}H is similar to that of 2D INADEQUATE. The chemical shift of ^{1}H nuclei is shown on the ω_2 axis while the double quantum frequency is shown on the ω_1 axis. Two correlated peaks, which are situated on a horizontal line with equal distances from the cross point of the horizontal line and the pseudo-diagonal, illustrate the coupling of two hydrogen atoms with ^3J, that is, two hydrogen atoms in the structural unit of H–C–C–H.

The two-dimensional double quantum spectrum of ^{1}H has the following advantages:

1. As it has no diagonal peaks, so there is no problem with interference of cross peaks by the diagonal peaks.
2. In addition to the couplings from ^3J, it shows couplings from long-range couplings and so forth [1].

4.10
^{1}H Detected Heteronuclear Correlation Spectra

As discussed in Section 4.5, the heteronuclear correlation spectrum, especially H, C-COSY, is very important for the identification of organic compounds. However, its sensitivity is low because its acquisition is performed for the heteronucleus, which has a very low γ value. Therefore, the experiment requires a fairly large amount of sample and a rather long time for the acquisition in order to obtain a spectrum with a high S/N (signal-to-noise ratio).

According to reference [1], the S/N in one-dimensional experiments is described by the Eq. (4-39).

$$S/N \propto N\gamma_{\text{exc}}\gamma_{\text{det}}^{3/2}B_0^{3/2}(NS)^{1/2}T_2/T \tag{4-39}$$

On the right-hand side of the equation, N is the number of the nuclei to be measured in the effective volume; γ_{exc} is the magnetogyric ratio of the nucleus to be excited; γ_{det} is the magnetogyric ratio of the nucleus to be measured; B_0 is the magnetic induction strength; NS is the number of the scans; T_2 is the transverse relaxation time; and T is the temperature of the experiment. Therefore, S/N, that is, the sensitivity of the experiment, will increase eight times by changing the nucleus from ^{13}C to ^{1}H. Consequently, the accumulation time and/or the amount of sample can be greatly decreased.

According to the number of chemical bonds, which separate the correlated ^{1}H and ^{13}C, such 2D NMR experiments are divided into two types: The first type corresponds to H,C-COSY, that is, establishes the correlation between two directly connected ^{1}H and ^{13}C. The second type corresponds to the long-range H,C-COSY or COLOC, that is establishes the correlation between two coupled nuclei, ^{1}H and ^{13}C through $^nJ_{\text{CH}}$.

No matter what type it is, the acquisition of ^{1}H is carried out and thus these experiments are known as experiments in an inverse mode.

4.10.1
HMQC and HSQC

HMQC is the abbreviation for [(^{1}H-detected) heteronuclear multiple-quantum coherence (experiment)] and HSQC is the abbreviation for [(^{1}H-detected) heteronuclear single-quantum coherence (experiment)]. Both have the function of H,C-COSY.

The pulse sequence of HMQC is shown in Fig. 4.64.

The principle of HMQC should be explained by using the product operator formalism. However, some parts of the pulse sequence (b) can be explained by using the model of the magnetization vector.

1. As HMQC measures the correlation between ^{1}H and ^{13}C, whose isotopic natural abundance is only 1%, the magnetization vectors of ^{1}H, which are connected directly to ^{12}C, should be removed to suppress their signals, which appear as strips parallel to the F_1 axis at their δ values on the F_2 axis. By using the BIRD sequence at the beginning of the pulse sequence in (b), the magnetization vectors of ^{1}H, which are connected directly to ^{12}C, are inverted from the z axis to the $-z$ axis (see Section 4.1.5). The magnetization vectors can de-

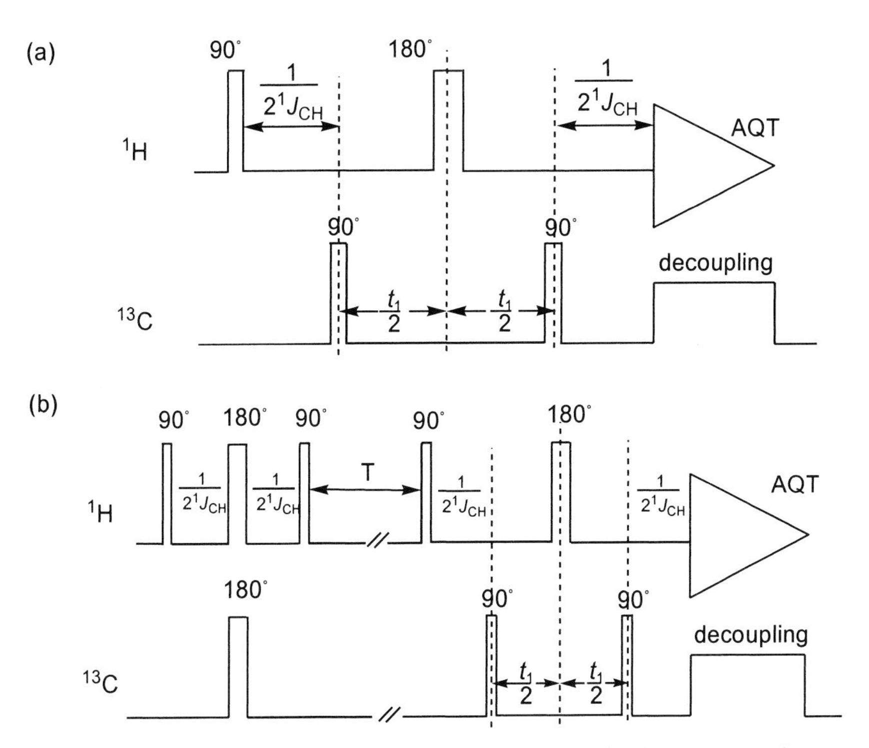

Fig. 4.64 The pulse sequence of HMQC. (a) The basic sequence. (b) The sequence with BIRD.

cay to zero (or close to zero) through the transverse relaxation during T so that their signals can be suppressed. On the other hand, the magnetization vectors of ^1H, which are connected directly to ^{13}C, are always along the z axis during the BIRD sequence, after which the basic pulse sequence of HMQC starts.

2. The sequence of $t_1/2$, 180° and $t_1/2$ takes the role of δ labeling (see Sections 4.1.4 and 4.5.1) to correlate δ_H with δ_C.
3. Because the decoupling to ^{13}C is used for the acquisition of ^1H, the non-split signals of ^1H are obtained.

From the above discussion, it follows that the HMQC spectrum corresponds to that of H,C-COSY. However, the F_2 axis in the HMQC spectrum shows the chemical shift values of ^1H while that in the H,C-COSY shows the chemical shift values of ^{13}C. In the 2D NMR spectrum, the resolution on the F_2 axis is determined by the number of data points in the acquisition while that on the F_1 axis is determined by the number of t_1. Since the number of data points in the acquisition is much greater than the number of t_1, the resolution on the F_2 axis is much better than that on the F_1 axis. In the HMQC spectrum, the F_2 axis with a fairly high resolution shows the chemical shifts of ^1H nuclei, which have a low resolution while the F_1 axis with a rather low resolution shows the chemical shifts of ^{13}C nuclei, which should be described in a high resolution. Therefore, the HMQC spectrum is not satisfactory. Hence when a sufficient amount of the sample to be analyzed is available, the H,C-COSY is preferable to the HMQC in achieving a rather high resolution on the δ_C axis.

In addition, the HMQC spectrum has the disadvantage of showing split cross peaks along the F_1 axis. As a result, HMQC spectra are frequently replaced by those of HSQC.

The pulse sequence of HSQC is shown in Fig. 4.65.

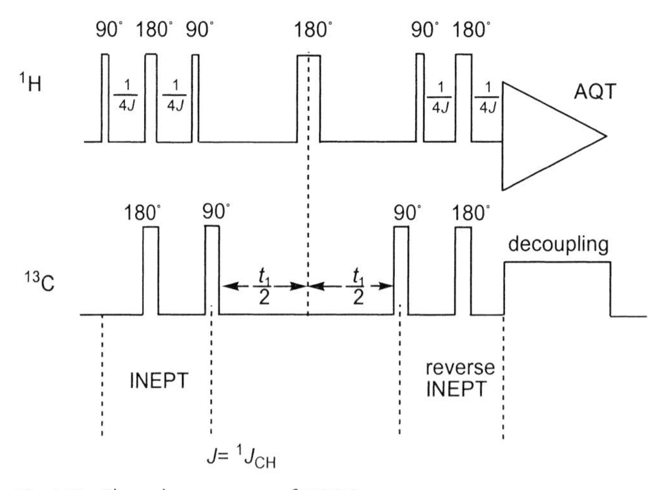

Fig. 4.65 The pulse sequence of HSQC.

The principle of Fig. 4.65 can be illustrated with a model of the magnetization vector.

The beginning of the sequence is the sequence of INEPT, at the end of which the magnetization vectors of ^{13}C are enhanced through polarization transfer. The sequence of $t_1/2$, 180° and $t_1/2$ takes the role of δ labeling, followed by a reverse INEPT, through which the magnetization vectors of ^{13}C transfer to those of ^{1}H. The magnetization vectors of ^{1}H at the beginning of the reverse INEPT are along the $\pm z$ axis. Acted on by the 90° pulse applied to ^{1}H, the pair of opposite magnetization vectors of ^{1}H is now along the $\pm y$ axis. They are refocused through $1/(4J)$, a 180° pulse, which is applied to both ^{1}H and ^{13}C, and another $1/(4J)$. Under these conditions, the acquisition of ^{1}H, whose δ values have been modulated by δ_C during t_1, is carried out.

At the end of the INEPT, the magnetization vectors of ^{1}H return to the $\pm z$ axis. During t_1, there are only the transverse magnetization vectors of ^{13}C in the x'–y' plane so that the cross peaks do not show the splitting by the couplings between ^{1}H and ^{1}H on the F_1 axis after the Fourier transform about t_1.

In the pulse sequence of HMQC, however, since the magnetization vectors of ^{1}H are in the x'–y' plane during t_1, the cross peaks show the splitting by the couplings among the adjacent ^{1}H on the F_1 axis after the Fourier transform about t_1.

The HSQC spectrum of compound **C4-7**, as shown in Fig. 4.66, is taken as an example of HSQC.

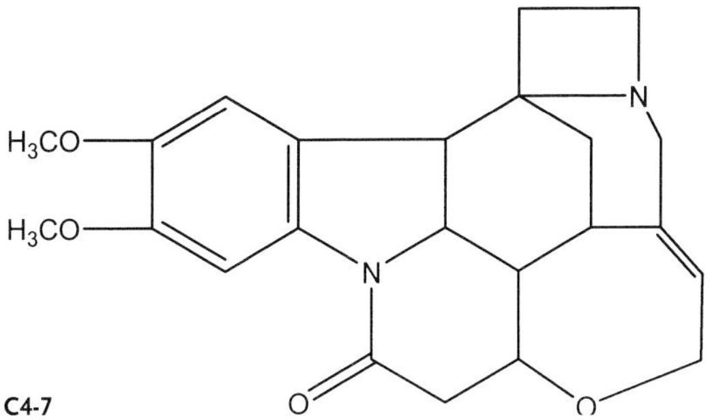

H_3CO
H_3CO
N
N
O
O

C4-7

Compound **C4-7**, brucine, has a structure similar to that of strychnine (except that brucine has two substituents of methoxy groups in the phenyl ring). Fig. 4.66 is a phase-sensitive spectrum, in which cross peaks are distinguished clearly by positive and negative signs.

The HMQC spectrum is similar to that of HSQC except the split in the F_1-axis.

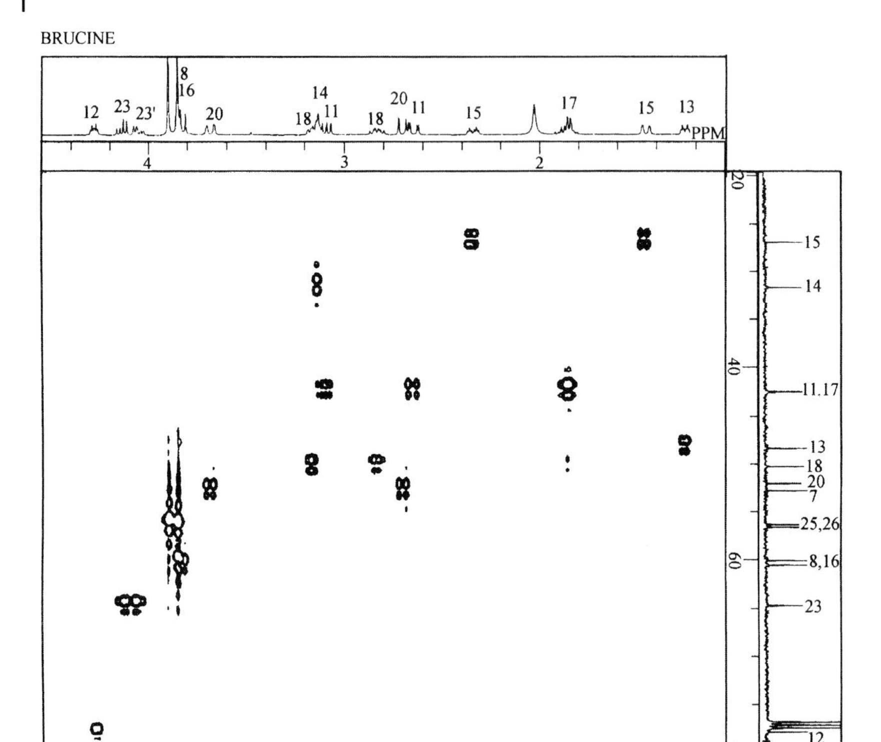

Fig. 4.66 The HSQC spectrum of compound **C4-7**.

4.10.2
HMBC

HMBC is the abbreviation for [(^{1}H-detected) heteronuclear multiple-bond correlation (experiment)], which correlates ^{13}C and ^{1}H separated by several chemical bonds from the ^{13}C. The function of HMBC is the same as that of long-range H,C-COSY or COLOC.

The basic pulse sequence of HMBC is shown in Fig. 4.67.

The principle of Fig. 4.67 should be explained by the product operator formalism. However, the following two points can be explained on the basis of our present knowledge.

1. As has been discussed previously, the sequence of $t_1/2$, 180° and $t_1/2$ has the role of δ labeling.
2. The first half of the pulse sequence is called the low-pass *J*-filter, which is an ingenious design to strongly suppress the magnetization vectors of ^{1}H connected directly to ^{13}C with those of ^{1}H connected indirectly to ^{13}C left [28]. Therefore, the signals of long-range couplings between ^{13}C and ^{1}H are emphasized.

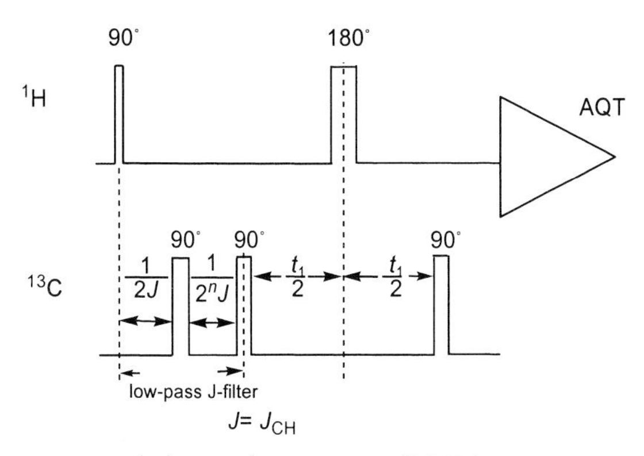

Fig. 4.67 The basic pulse sequence of HMBC.

If a BIRD sequence is added to the beginning of the basic sequence, the long-range correlation can be enhanced, in which case the time spaces of τ in the BIRD sequence are calculated to be $1/(2^n J_{CH})$.

The high field portion of the HMBC spectrum of brucine is shown in Fig. 4.68.

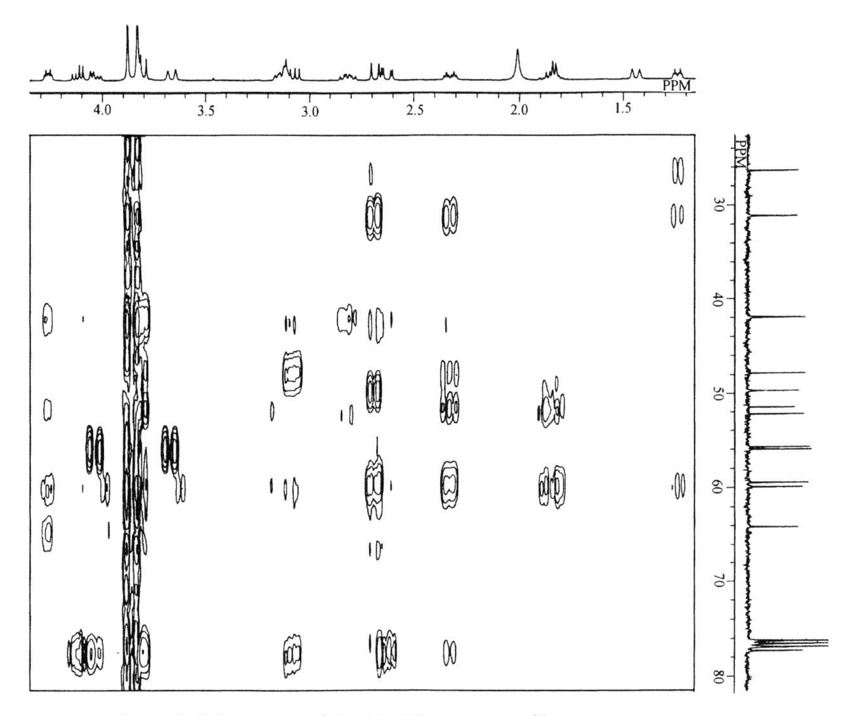

Fig. 4.68 The high field portion of the HMBC spectrum of brucine.

When considering the resolution on the F_1 axis of the spectrum, it is recommended that COLOC should be used to replace HMBC if sufficient sample to be measured is available.

4.11
Combined 2D NMR Spectra

In general, a 2D NMR spectrum determines only one type of correlation (such as a correlation spectrum) or produces one type of result (such as a *J*-resolved spectrum). If a combined pulse sequence is used, two results corresponding to the two pulse sequences of the combined pulse sequence can be obtained at once. This type of 2D NMR spectra is known as the combined 2D NMR spectra.

The relayed H,C-COSY/HOHAHA is taken as an example, the pulse sequence for which is shown in Fig. 4.69.

The principle of Fig. 4.69 can be clearly illustrated with the model of the magnetization vector. The 90° pulse applied to ^1H produces the transverse magnetization vectors. At the end of the pulse, t_1 begins. The 180° pulse applied to ^{13}C at the midpoint of t_1 modulates the δ values of ^1H so that the chemical shifts of the ^1H and those of ^{13}C directly connected to the ^1H are correlated (see Section 4.1.4). An MLEV-17, during which the isotropic mixing and the transfer of magnetization vectors from a particular ^1H to other ^1H in the spin system take place (see Section 4.1.7), is used at the end of t_1. The pulse sequence from the end of τ to the beginning of the 90° pulse applied to ^{13}C forms that of INEPT (because the magnetization vectors of ^1H are already in the x'–y' plane, the first 90° pulse applied to ^1H is not needed). The magnetization vectors of ^{13}C are enhanced by the INEPT. They are rotated from the $\pm z$ axis onto the $\pm y'$ axis by the 90° pulse to ^{13}C. After the period of $\Delta_2 = 1/(3J_{CH})$, the opposite magnetization vectors of ^{13}C close up to give the signals of ^{13}C without any split under the decoupling for ^1H.

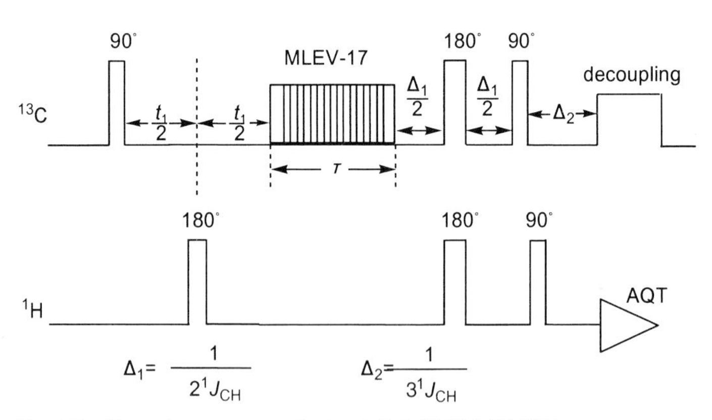

Fig. 4.69 The pulse sequence of relayed H,C-COSY/HOHAHA.

From the above discussion, it follows that the pulse sequence shown in Fig. 4.69 has the functions of both H,C-COSY and HOHAHA. Suppose that a molecule has structure **37**.

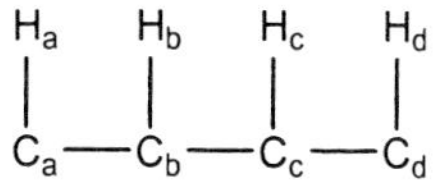

H_a H_b H_c H_d

C_a—C_b—C_c—C_d

37

The correlation of the directly connected ^1H and ^{13}C is shown because of the function of H,C-COSY while the spin system from H_a to H_d is also shown because of the function of HOHAHA. In this 2D NMR spectrum, starting from $\delta(C_a)$ on the F_2 axis, the cross peaks of H_a, H_b, H_c and H_d can be found sequentially, which is obviously very useful in determining the structural unit.

4.12 Three-dimensional NMR Spectra

The progression from ordinary (one-dimensional) NMR spectra to 2D NMR spectra is a milestone in NMR spectroscopy. The 2D NMR methods are more objective and more reliable than those of 1D NMR for the identification of organic compounds. In addition, 2D NMR provides a method to determine the secondary structure of the protein molecule containing less than 80–90 amino acid residues. When a protein molecule has more amino acid residues, the overlaps will be serious because of the augmentation of peaks and the deterioration of spectral resolution. One possible solution is to add a third frequency variable so that an overlapped 2D NMR spectrum is "resolved" into a series of 2D NMR spectra without overlap. Therefore, the 3D NMR spectra (and 4D NMR spectra) are the inevitable consequence of 2D NMR spectra.

4.12.1 Principle of Three-dimensional NMR Spectra

A 3D NMR spectrum is produced by performing the Fourier transform three times with respect to three time variables in the following pulse sequence with the time axis for one-dimensional (A), two-dimensional (B) and three-dimensional (C) NMR experiments, respectively:

A preparation period → excitation → detection period t
B preparation period → evolution period t_1 → mixing period → detection period t_2
C preparation period → evolution period t_1 → mixing period 1 → evolution period t_2 → mixing period 2 → detection period t_3

The pulse sequence of the 3D NMR experiment consists of two pulse sequences of the 2D NMR experiment: the first pulse sequence includes a preparation period, an evolution period and a mixing period, while the second pulse sequence includes an evolution period, a mixing period and a detection period.

According to this concept, the pulse sequence of four- or more dimensional NMR experiments can be constructed.

In a 3D NMR experiment, t_1 and t_2 gradually increase independently. For a certain value of t_1, a series of acquisitions (with gradually increased t_2) are taken. If phase cycling is necessary, the acquisitions will be multiplied further (the pulsed-field gradient can be used to replace phase cycling if the S/N ratio is acceptable). Therefore, the 3D NMR experiment is time-consuming.

4.12.2 Classification of 3D NMR Spectra

The 3D NMR spectra are divided into two categories, homonuclear and heteronuclear 3D NMR spectra.

Homonuclear 3D NMR spectra include 3D NOESY-TOCSY, 3D TOCSY-NOESY, 3D NOESY-NOESY, etc., all of which deal with ^{1}H only.

Heteronuclear 3D NMR spectra have found much wider use than homonuclear 3D NMR spectra, for the following reasons:

1. Heteronuclei (^{13}C and ^{15}N) have a very wide distribution of their δ values so as to bring the advantages of their 3D NMR spectra into full play.
2. The coupling constant between a heteronucleus and a proton has a rather high value so that the transfers of the related magnetization vectors are effective. As a result, related cross peaks can have high signal-to-noise ratios.
3. No additional peaks are produced in heteronuclear 3D NMR spectra. A heteronuclear 3D NMR spectrum is the "arrangement" of related 2D NMR spectra along the frequency axis of the heteronucleus.
4. A biological molecule can be easily labeled with ^{15}N and/or ^{13}C.

4.12.3 Application of 3D NMR Spectra

3D NMR spectra are mainly used in the determination of the three-dimensional structures of bio-molecules in their solution. The prerequisite is the assignment of their NMR spectra.

Wüthrich, the winner of the Nobel Prize in chemistry in 2002, established one method of assigning the sequence of amino acid residues of a protein molecule by the combination of COSY (or TOCSY) and NOESY [29]. In the assignment process, the secondary structure of the protein can be determined.

The development of a new assignment method, which is not related to the secondary structure, is more attractive, which means that the assignment is achieved only through the chemical bonds by which the protein molecule is constructed. Many 3D NMR pulse sequences have been created to determine the connections of the main chain, the connections of the side chains and those between the main chain and the side chains in a protein.

The NOE data are the most important information for determining the three-dimensional structure of a protein molecule in its solution. Semi-quantitative data of NOE are necessary for the determination of the distance ranges of related proton pairs. The NOE intensities are calculated from the NOE cross peaks. Related 3D NMR pulse sequences have been developed.

In addition to the NOE data, it is necessary to obtain homonuclear and heteronuclear coupling constants of the protein molecule for the determination of the three-dimensional structure in solution. Related 3D NMR pulse sequences for this use have been set up.

Based on the assignment of the sequence of the protein, NOE data and J values, the three-dimensional structure of the protein molecule in its solution can be calculated step by step by using special software.

4.13 DOSY

DOSY is the abbreviation for diffusion ordered spectroscopy.

The combination of LC and NMR is a good solution to the analysis of a mixture. However, experimental LC-NMR requires an instrumental set-up of HPLC and NMR, and a connection between the HPLC and NMR. A crucial problem of this experiment is also the suppression of solvent peaks, which are much stronger than the signals of the samples to be analyzed, by several orders of magnitude. In addition, before the experiment with the hyphenated instruments, it is necessary to find the experimental conditions for the separation of the mixture by HPLC. Is there any other method of analyzing a mixture just with an NMR instrument? Can we obtain the NMR spectra of pure components directly from a mixture? The answer is positive, if DOSY is applied.

DOSY is a new type of multi-dimensional NMR spectroscopy [30, 31]. One dimension of DOSY is not an NMR parameter, such as δ or J, but a molecular property: the diffusion coefficients. In a 2D DOSY spectrum, the horizontal direction shows the ^{1}H spectra of pure components of the mixture to be analyzed, which are arranged in the order of the diffusion coefficients of the components along the vertical axis. Similarly, a 3D DOSY is a series of 2D NMR spectra of the components arranged vertically according to the order of the diffusion coefficients of the components. In general, DOSY increases the dimensionality of an NMR experiment by one. From this discussion, it follows that the new dimension is not created by a Fourier transform from a time variable but in another way.

The principle of DOSY is closely related to the application of the pulsed-field gradient, PFG. One of the functions of PFG is the suppression of the solvent peak in an NMR spectrum. Acted on by a pair of PFGs, the magnetization vectors of the sample molecules to be measured can refocus after its defocusing while the magnetization vector of the solvent molecules defocuses further after its defocusing, so that the solvent peak is suppressed. Obviously this result is ideal but not completely true, because the sample molecules cannot be immobile in the solution. The greater diffusion coefficient of a component, the weaker the NMR spectrum it obtains for a given intensity of PFG. On the other hand, the stronger intensity of PFG is applied, the weaker the NMR spectrum a certain component obtains.

We start from 2D DOSY, which produces the ^{1}H spectra of pure components of a mixture to be measured. Some pulse sequences for encoding the diffusion coefficients of components of a mixture are applied. They are longitudinal eddy current delay (LED), bipolar pulse pair-longitudinal eddy current (BPP-LED), bipolar pulse pair stimulated echo, gradient compensated stimulated echo with spin lock, etc. One of these sequences is used for a DOSY experiment. The variation of the gradient amplitude K will result in a change in the signal intensity according to the Stejskal-Tanner equation for a semi-sinusoidal shaped gradient [31]:

$$I(K, \nu) = \Sigma A_n(\nu) \exp[-D_n(\Delta - \delta/3)K^2] \tag{4-40}$$

where $I(K, \nu)$ is the signal intensity, which is a function of K and ν; $K = \gamma g \delta$, where γ is the magnetogyric ratio, and g and δ are the amplitude and duration of the gradient pulse, respectively; ν is a frequency in the ^{1}H spectrum; $A_n(\nu)$ is the signal of the n-th component at ν when g=0; D_n is the diffusion coefficient of the n-th component; Δ is the diffusion time or the diffusion delay, which is the time interval between the two gradient pulses.

Several tens of free induction decays, each associated with a different value of K, are acquired. It should be noted that in every acquisition a high S/N value must be reached even with the strongest K (the weakest signal). After the data collection, the data are transferred into a frequency spectrum through a Fourier transform.

The signal at a certain frequency ν, which is obtained in the transferred spectrum, is the addition of those for all components. Remember that a series of data with different K values has been acquired. The aim of DOSY is to get $A_n(\nu)$, which can be obtained on the basis of the series of data according to Eq. (4-40) through a certain mathematical operation that "resolves" the additional signal into those of components. The mathematical operation is the inverse Laplace transform (ILT). However, because sufficient data are not collected, the ILT is operated under the "ill-condition." Therefore, some prior knowledge is needed. Through the ILT, the signals of pure components at ν are obtained. The ^{1}H spectra of pure components of the mixture to be measured are produced by repeating the ILT at all frequencies. The calculations are performed by special software. In fact, some software can be used to calculate a spectral width of frequencies even in several regions simultaneously. In brief, DOSY "resolves" the additional spectrum into those of pure components according to their diffusion coefficients, which are also obtained in the DOSY experiment.

The principle of 3D DOSY is similar to that of 2D DOSY. In this case, a cross peak produced through the Fourier transform from the acquired data in the

DOSY experiment is considered as the addition of those of pure components. Namely, 3D DOSY "resolves" the addition of 2D NMR spectra into a series of 2D NMR spectra according to diffusion coefficients of components of the mixture to be measured. The 2D NMR spectra may be of the ordinary type, such as COSY, TOCSY, etc.

For all DOSY experiments, it is necessary that the diffusion coefficients of pure components have fairly significant differences. Diffusion coefficient is mainly related to molecular radii as well as to molecular shape.

DOSY is still developing and it will play an increasingly more important role in the analysis of mixtures.

4.14 References

1 R. R. Ernst, G. Bodenhausen, A. Wokaun, *Principles of Nuclear Magnetic Resonance in One and Two Dimensions*, Clarendon, Oxford, **1987**.

2 W. R. Croasmun, K. M. K. Calson, Eds., *Two Dimensional NMR Spectroscopy: Applications for Chemists and Biochemists*, 2nd Edn., VCH, Weinheim, **1994**.

3 H. Kessler, M. Gerke et al., Two-Dimensional NMR Spectroscopy: Background and Overview of the Experiments, *Angew. Chem., Int. Ed. Engl.*, **1988**, *27*, 490–536.

4 A. E. Derome, *Modern NMR Techniques for Chemistry Research*, Pergamon Press, Oxford, **1987**.

5 T. D. W. Clarige, *High-Resolution NMR Techniques in Organic Chemistry*, Pergamon Press, Oxford, **1999**.

6 G. E. Martin, A. S. Zektzer, *Two-Dimensional Methods for Establishing Molecular Connectivity*, VCH, Weinheim, **1988**.

7 K. Naksnish, Ed., *One-Dimensional and Two-Dimensional NMR Spectra by Modern Pulse Techniques*, University Science Books, Mill Valley, CA, **1990**.

8 S. Brann, H. O. Kalinowski, S. Berge, *150 and More Basic NMR Experiments*, VCH, Weinheim, **1998**.

9 A. G. Redfield, S. D. Kunz, *J. Magn. Reson.*, **1975**, *19*, 250–254.

10 G. Bodenhausen, H. Kogler, R. R. Ernst, *J. Magn. Reson.*, **1984**, *58*, 370–388.

11 D. G. Davis, A. Bax, *J. Am. Chem. Soc.*, **1985**, *107*, 2820–2821.

12 A. Bax, D. G. Davis, *J. Magn. Reson.*, **1985**, *65*, 355–360.

13 L. Braunsweiler, R. R. Ernst, *J. Magn. Reson.*, **1983**, *53*, 521–528.

14 J. Stonehause, P. Adell et al., *J. Am. Chem. Soc.*, **1994**, *116*, 6037–6038.

15 C. Le Cocq, J. Y. Lallemand, *J. Chem. Soc., Chem. Commun.*, **1981**, 150–152.

16 S. L. Patt, J. N. Shoolery, *J. Magn. Reson.*, **1982**, *46*, 535–539.

17 J. Homer, M. C. Perry, *J. Chem. Soc., Chem. Commun.*, **1994**, 373–374.

18 J. Homer, M. C. Perry, *J. Chem. Soc., Perkin. Trans.*, **1995**, *2*, 533–536.

19 H. Kessler, C. Griesinger et al., *J. Magn. Reson.*, **1984**, *57*, 331–336.

20 D. J. States, R. A. Habekorn et al., *J. Magn. Reson.*, **1982**, *48*, 286–292.

21 D. Marion, K. Wüthrich, *Biochem. Biophys. Res. Comm.*, **1983**, *113*, 967–974.

22 J. Keeler, D. Neuhaus, *J. Magn. Reson.*, **1985**, *63*, 454–472.

23 R. Freeman, *A Handbook of Nuclear Magnetic Resonance*, John Wiley, New York, **1988**.

24 W. R. Croasmun, K. M. K. Calson, Eds., *Two Dimensional NMR Spectroscopy: Applications for Chemists and Biochemists*, 1st Edn., VCH, Weinheim, 1986, pp. 109–117.

25 D. Newhaus, G. Wagner et al., *Eur. J. Biochem.* (FEBS), **1985**, *151*, 257–273.

26 D. Newhaus, M. Williamson, *The Nuclear Overhauser Effect in Structural and Conformational Analysis*, VCH, **1989**.
27 A. Bax, R. Freeman et al., *J. Am. Chem. Soc.*, **1980**, *102*, 4649–4851.
28 H. Kogler, O.W. Sorensen et al., *J. Magn. Reson.*, **1983**, *55*, 157.
29 K. Wüthrich, *NMR of Proteins and Nucleic Acids*, Wiley, New York, **1986**.
30 K.F. Morris, C.S. Johnson, Jr., *J. Am. Chem. Soc.*, **1992**, *114*, 3139–3141.
31 K.F. Morris, C.S. Johnson, Jr., *J. Am. Chem. Soc.*, **1993**, *115*, 4291–4299.

5
Organic Mass Spectrometry

Mass spectrometry has been an important technique for the structural identification of organic compounds since the late 1950s. This method has two significant advantages over other spectroscopic methods. Firstly, it has much higher sensitivity, and requires the smallest amount of sample. Secondly, it is the only method that can be used to determine the molecular formula of an unknown compound.

The development of mass spectrometry has been incredible. After the award of the Nobel Prize to Dehmelt and Paul in 1989, Fenn and Tanaka shared the Nobel Prize in chemistry in 2002 with Wüthrich because of their contribution to the ESI technique.

Radical changes have taken place in the development of instrumentation. For example, the global annual production of double-focusing mass spectrometers has dropped to ten instruments. On the other hand, by 1996 there were over 200 FT-ICR/MS [1]. Both FT-ICR/MS and ion traps are now commonly used in tandem mass spectrometry (tandem-in-time) as well as in LC-MS systems, which are now being perfected. Furthermore, the time-of-flight (TOF) instruments, including the oa-TOF, are becoming increasingly important because of their high sensitivity and mass detectability (molecular weights as high as $982\,000 \pm 2000$ u have been successfully determined with this type of instrument [2]). In addition, many new soft ionization techniques, as well as computerized retrievals of mass spectra, have been initiated rapidly or improved. Therefore, one entire chapter of this book is devoted to mass spectrometry.

References [3–8] are listed for the reader who wishes to understand the related topics further.

It should be noted that mass spectrometry has had remarkable success in biology, leading to the creation of a new discipline: biological mass spectrometry [9].

Structural Identification of Organic Compounds with Spectroscopic Techniques. Yong-Cheng Ning
Copyright © 2005 WILEY-VCH Verlag GmbH & Co. KGaA, Weinheim
ISBN: 3-527-31240-4

5.1 Fundamentals of Organic Mass Spectrometry

5.1.1 Instruments

An organic mass spectrometer consists of the following units:

1. Inlet system
   A sample to be analyzed is transferred into a mass spectrometer through an inlet system with the vacuum of the mass spectrometer maintained. In an LC-MS combination, the inlet system is replaced by the interface.
2. Ionization and acceleration chamber
   An ionization and acceleration chamber is also known as an ion resource in which molecules are ionized to form ions.
   An acceleration voltage is applied so that the ions arrive at a mass analyzer. Acceleration voltages vary greatly for different types of mass analyzer.
3. Mass analyzer
   A mass analyzer, the key part of a mass spectrometer, differentiates ions according to their mass-to-charge ratios. As each type of mass analyzer has its own principle, function and application area, we will deal with various mass analyzers at length in Section 5.2.
4. Detector
   A detector measures ions with different m/z values.
5. Computer and data system
   A computer controls all processes of a mass spectrometer, including data accumulation, processing and printing, as well as spectrum retrieval (see Section 6.5). Elemental compositions of molecular ions and important fragment ions can be found by the computer of a high resolution mass spectrometer.
6. Vacuum system
   A vacuum system provides the vacuum necessary for the ion source and the mass analyzer. The level of vacuum needed varies with the type of mass analyzer.

5.1.2 Major Specifications

1. Mass range
   The measurable range of mass-to-charge ratios of ions is called the mass range for ions with a single charge. The measurable mass range can be extended when measuring multiply-charged ions.
2. Resolution
   The resolution of a mass spectrometer with a sector magnetic field is defined by Eq. (5-1).

$$R = \frac{M}{\Delta M} \tag{5-1}$$

where M is the average mass of two discernible masses in a mass spectrum and ΔM is the mass difference between the two discernible masses.

Hence resolution is the ability to differentiate two adjacent ion peaks.

Discerning two peaks requires that the height of their valley be 10% of the height of the peak. For clarity, R is denoted as $R_{10\%}$.

Equation (5-1) also applies to FT-ICR and TOF, in which case ΔM denotes the mass difference corresponding to the half-height width.

3. Sensitivity

Sensitivity illustrates the relationship between the intensity of a peak and the amount of sample to be measured.

5.1.3 Mass Spectrum

Ions with different mass-to-charge ratios are separated by a mass analyzer and recorded by a detector as a mass spectrum.

The abscissa of the mass spectrum is the mass-to-charge ratio. In general the direction from left to right denotes an increase in the mass-to-charge ratio. Frequently, recorded ions are singly charged ions, in which case the abscissa illustrates the masses of the ions.

The ordinate of the mass spectrum represents the intensities of the peaks of ions. The highest peak is called the base peak.

5.1.4 Ion Types in Organic Mass Spectrometry

1. Molecular ions

A molecular ion is produced from the ionization of an organic molecule. It is denoted as "$M^{+\bullet}$". The symbol "+" indicates the ionized molecule has lost an electron and "•" expresses the ionized molecule has lost one of its paired electrons, leaving an unpaired electron, which means the ionized molecule is a radical.

The mass-to-charge ratio of the molecular ion is numerically equal to its molecular weight when it is a singly-charged ion.

The molecular ion will produce fragment ions (in a broad sense) as it possesses high internal energy.

2. Quasi-molecular ions

$M+H]^+$ and/or $M–H]^+$ are called quasi-molecular ions. $M+X]^+$, where X denotes a molecule from the medium surrounding the molecules to be analyzed in soft ionization, is called an additional ion. It can also be called a quasi-molecular ion.

The quasi-molecular ion is fairly stable as it does not contain any unpaired electrons.

Although the term quasi-molecular ion has been criticized [10], it is still being used until a more appropriate term is suggested.

3. Fragment ions

Fragment ions in a broad sense are ions that result from fragmentation of molecular ions. Fragment ions (in a narrow sense) are the ions that are produced from the molecular ions through simple fragmentation reactions only (see Section 6.2.2). In this text, "fragment ion" is used in the narrow sense unless otherwise stated.

4. Rearrangement ions

Rearrangement ions result from rearrangement reactions of the molecular ion or other ions (see Section 6.2.3). Therefore their structural units are not those in the original structure.

The above discussion is based on ion structures.

5. Parent ions and daughter ions

If an ion produces another ion, the former is called the parent ion and the latter the daughter ion. A parent ion is also called a precursor ion.

6. Metastable ions

"Metastable" implies "lying between stable and unstable". If an ion is stable, it can move from the ion source to the detector and can be recorded. If an ion is unstable, it decomposes into another ion in the ion source. Metastable ions are formed between the ion source and the detector and they are so important in organic mass spectrometry that a whole section (see Section 5.4) will be dedicated to them.

The above discussion has been conducted from the point of view of ion formation.

7. Odd-electron and even-electron ions

Ions with unpaired electrons are called odd-electron ions. They are highly reactive because of the presence of the unpaired electrons.

Ions without any unpaired electron are called even-electron ions, which are more stable than odd-electron ions.

Odd-electron ions are more important for the interpretation of mass spectra than even-electron ions.

8. Multiply-charged ions

Ions with more than one charge are called multiply-charged ions. As a result, their mass-to-charge ratios decrease correspondingly. These ions are used for measuring the mass of large molecules in some ionization techniques.

The above discussion is based on the electronic structures of ions.

9. Isotopic ions
Molecules that contain an element with a non-unique isotopic composition will produce isotopic ions on ionization. Isotopic ions form isotopic ion clusters.

5.2 Mass Analyzers

The mass analyzer is the key part of a mass spectrometer. Mass spectrometers are classified according to their type of mass analyzer.

5.2.1 Single-focusing or Double-focusing Mass Analyzers

A single-focusing mass analyzer uses a sector magnetic field while a double-focusing mass analyzer uses a combination of an electrostatic field and a sector magnetic field.

Ions formed in an ion source acquire kinetic energy produced by an acceleration voltage, which is denoted by

$$zeV = \frac{1}{2}mv^2 \tag{5-2}$$

where V is the acceleration voltage; ze is the charge carried by the ion (z is a positive integral number and e is the charge unit); m is the mass of the ion; and v is the velocity of the ion after acceleration.

We will first discuss the single-focusing mass analyzer.

Accelerated ions enter the sector magnetic field analyzer (MA), with a direction of motion perpendicular to that of the magnetic lines of force. Moving ions, as with a current, make a circular motion in a plane perpendicular to the magnetic lines of force. The centripetal force for the circular motion is provided by the force exerted on the ions by the magnetic field. Thus we obtain:

$$zevB = \frac{mv^2}{r_\mathrm{m}} \tag{5-3}$$

where B is the applied magnetic field strength and r_m is the radius of the moving ions.

Combining Eqs. (5-2) and (5-3), we have

$$r_\mathrm{m} = \frac{1}{B}\sqrt{2V(m/ze)} \tag{5-4}$$

or

$$m/z = \frac{r_\mathrm{m}^2 B^2 e}{2V} \tag{5-5}$$

As a detector is set at a fixed position, r_m is a constant. Ions can be recorded in the order of their m/z values by varying either B or V. Because mass spectrometers possess a high resolution and sensitivity for a high acceleration voltage, ions are recorded by scanning B.

From Eq. (5-3) it follows that moving ions have different radii according to their m/z values in a constant B, which means that a magnetic field possesses a function for mass dispersion for ions. This is similar to the dispersion of light by a prism.

From theoretical calculation or experiment, it is known that a magnetic field also has the function of directional focusing. A beam of ions with a diverging angle can be focused at a point by a magnetic field, which is called directional focusing. Therefore the mass spectrometer with a sector magnetic field is known as a single-focusing mass spectrometer.

In fact, ions with the same m/z value cannot be focused at one point by a magnetic field because the original kinetic energy of the ions has a certain distribution before their acceleration. Therefore a magnetic field has the function of energy dispersion, which will decrease the resolution of a mass spectrometer.

As with a magnetic field, an electrostatic field has the functions of directional focusing and energy dispersion.

A combination of a magnetic field and an electrostatic field is applied. If the energy dispersion produced by the magnetic field is counteracted by that produced by the electrostatic field, which means they have the same absolute value but opposite signs for energy dispersion, both directional focusing and energy focusing can be realized at the same time. Therefore, mass spectra with a high resolution can be obtained.

In general, an electrostatic analyzer (ESA) is set between the ion source and the MA. An ESA consists of a pair of concentric plates across which a potential difference E is applied. Accelerated ions enter the ESA and make a circular motion in the ESA. The centripetal force for the circular motion is provided by ESA. Thus we obtain:

$$zeE = \frac{mv^2}{r_e} \tag{5-6}$$

where r_e is the radius of the circular motion of ions in ESA.

Combining Eqs. (5-2) and (5-6) leads to

$$2zeV = Ezer_e$$

$$r_e = \frac{2V}{E} \tag{5-7}$$

The mass spectrometers in the order of an ion source, an MA and an ESA are termed reversed geometry instruments, which have specific functions (see Section 5.4).

5.2.2 Quadrupole Mass Analyzers

A quadrupole mass analyzer is also called a quadrupole mass filter, which consists of four parallel rod electrodes, hence the name. In theory, the section of the electrodes should be hyperbolic. In fact, a circular section can be used for general applications. The opposite two electrodes are of equipotential. The two pairs of electrodes have opposite potentials. Both a dc voltage U and an ac voltage $V(\cos \omega t)$ are applied to the electrodes, as shown in Fig. 5.1.

The x and y axes are shown in Fig. 5.1, with the z axis, along which ions fly from the ion source into the detector, being perpendicular to the plane of the book. A potential of several volts between the ion source and electrodes provides kinetic energy for ions to reach the detector. The motion of ions in a quadrupole mass analyzer can be determined precisely because the potential in the analyzer is expressed by the following equation:

$$\Phi = \frac{x^2 - y^2}{2r^2}(U + V \cos \omega t) \tag{5-8}$$

where Φ is the potential at the point (x, y).

We also have

$$F_x = \frac{\partial \Phi}{\partial x} \cdot e \tag{5-9}$$

$$F_y = \frac{\partial \Phi}{\partial y} \cdot e \tag{5-10}$$

$$F = ma \tag{5-11}$$

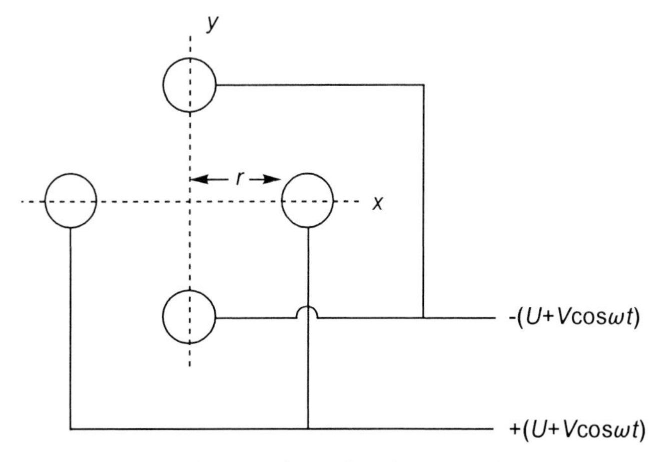

Fig. 5.1 Schematic diagram of a quadrupole mass analyzer.

where F_x and F_y are the electric forces acting on an ion along the x or y axis by the electric field, respectively, m is the mass of the ion and a is the acceleration of the ion. We are assuming that it is a singly-charged ion.

By combining the above four equations, the motion of ions in a quadrupole mass analyzer can be determined. However, the solution is complicated because there are such variables as U, V, ω, r, m and even z when we discuss multiply-charged ions. The reader who is interested in the process of this solution can refer to references [11–13].

The solution can be illustrated by Fig. 5.2.

The trick for Fig. 5.2 is that only two parameters, a and q, are used to characterize the stability of an ion in a quadrupole mass analyzer.

$$a = \frac{8eU}{mr^2\omega^2} \tag{5-12}$$

$$q = \frac{4eV}{mr^2\omega^2} \tag{5-13}$$

The area enclosed by the curved and straight lines is a stable area in which any point represents a stable condition, which means that the ion whose parameters are in the area has a limited trace along the x or y axis, that is, its motion is stable. Therefore it can be detected by the detector through the flight of the ion along the z axis.

Clearly, the region outside the stable area is unstable, which means that any ion specified in the unstable area has an unstable motion, and it will collide with an electrode and hence give no signal.

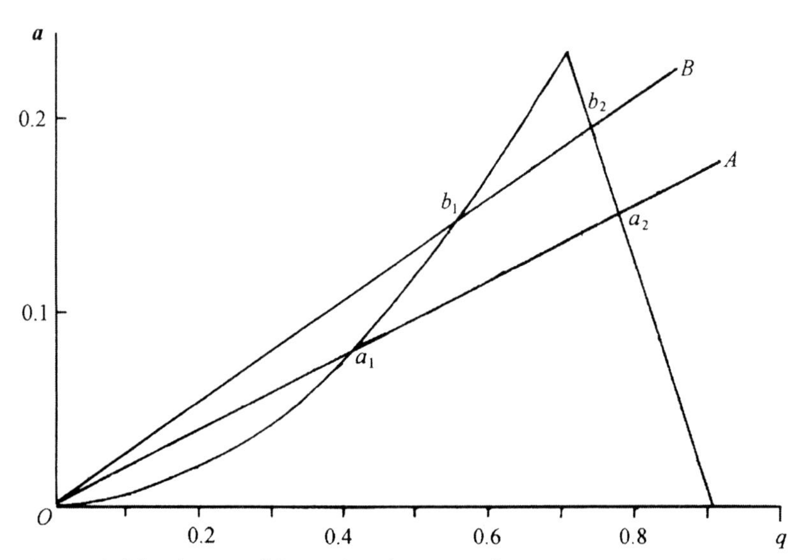

Fig. 5.2 Stability diagram of the quadrupole mass analyzer.

There are three parameters for operation: U, V and ω. In general, ω is fixed and $a/q=2U/V$ is kept constant (U and V both increase at the same time or vice versa). This operation is illustrated as a straight line through the origin, called a scan line. From the two intersection points of a scan line with the stability diagram, the lower limit of mass (from the right intersection point) and the upper limit of mass (from the left), between which ions are recorded, can be calculated. Obviously, the greater the slope of a scan line, the narrower the mass range of the detected ions. The extreme position is the scan line passing through the apex, which corresponds to the highest resolution.

If U and V are increased simultaneously with U/V kept constant, ions pass through a quadrupole mass analyzer from a lower mass-to-charge ratio to a higher mass-to-charge ratio.

Another operation mode is the mere use of V while $U=0$, in which case the scan line is the q axis. All ions except those within a very low mass-to-charge ratio will pass through the quadrupole analyzer, which can be used as the collision chamber in tandem mass spectrometry (see Section 5.5.2).

The specifications of the quadrupole analyzer have been improved with a mass range of 4000 u and a mass resolution of 0.1 u at $m=900$ u.

The quadrupole mass spectrometer is widely used for the following reasons:

1. It is simple in structure, small in size, light in weight, low in price and easy to operate and clean.
2. As it uses no magnetic field, it shows no magnetic hysteresis, leading to a rapid response, for which reason it can be used in combination with chromatographic instruments: GC-MS and LC-MS. It is also convenient to trace rapid chemical reactions, and so forth.
3. The quadrupole mass spectrometer requires a fairly low vacuum, which is another reason for its combination with chromatographic instruments.

The disadvantages of the quadrupole mass spectrometer are as follows:

1. A rather low mass resolution.
2. A mass discrimination effect for high mass-to-charge ions.

We will discuss ion traps in the next section. An ion trap is similar to a quadrupole mass analyzer in principle. Some characteristics, for example β lines, are applicable to quadrupole analyzers.

5.2.3 Ion Trap

Because of their similarity in principle to quadrupole mass analyzers, ion traps are known as quadrupole ion traps, or quadrupole ion storage (QUISTOR), which means the trap can store ions. From these technical terms, other abbreviations can arise, for example, ESQUIRE (external source quistor with resonance ejection).

Studies on the ion trap started at the beginning of the 1950s. It has been used as a mass analyzer in organic mass spectrometry since the middle of the

1980s. Paul and Dehmelt were awarded the Nobel Prize in physics in 1989 for their contributions to the development of the ion trap and its application in atomic physics.

Considering its principle, an ion trap is similar to a quadrupole mass analyzer. It can be understood from Fig. 5.3.

Suppose that a quadrupole mass analyzer has a rotation of 180° about the y axis, a ring-like electrode having an inner cross section of a hyperboloid, known as a ring electrode, is formed from the pair of hyperboloids symmetrical about the x axis. A pair of electrodes having a hyperbolic cross section, whose two sides are symmetrical about the y axis, remains unchanged during the rotation. They are called end cap electrodes. An ion trap consists of a ring electrode and a pair of end cap electrodes. The sample to be ionized is introduced and ions are ejected through the pores, which are drilled into the end cap electrodes. The ring electrode is isolated from the pair of equi-potential end cap electrodes.

It should be noted that in other literature on ion traps, the axis along the pair of end cap electrodes is usually denoted as the z axis, and the direction through the ring electrode r.

There are several ways of applying potentials to electrodes. The usual way is that the ground potential is applied to the end cap electrodes while an RF (radiofrequency) oscillating "drive" potential is applied to the ring electrode.

Although the ion trap is simple in structure, the theory is the most complicated of all types of mass analyzer.

As an ion trap is used not only for the detection of ions with different mass-to-charge ratios, but also in tandem-in-time mass spectrometry because it can store ions with particular mass-to-charge ratios, it is necessary to understand the theory [12–16]. Similar to the quadrupole mass analyzer, ions in the ion trap are affected by a quadrupole field.

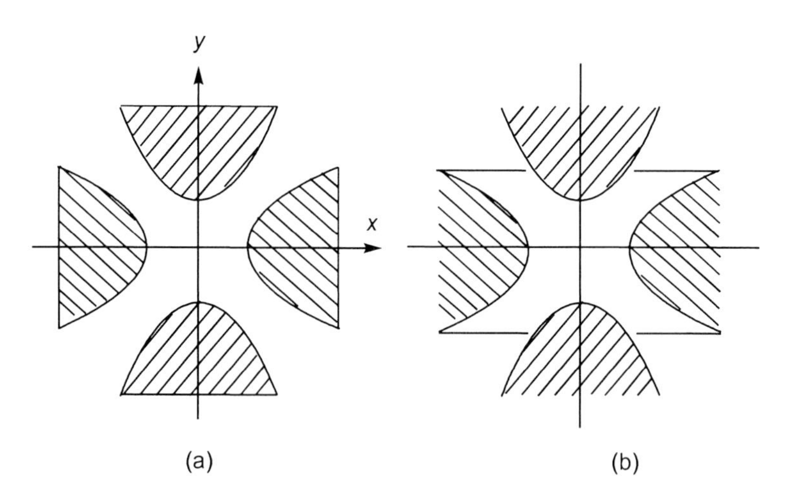

Fig. 5.3 An ion trap is formed by the rotation of a quadrupole mass analyzer about the y axis.

The theoretical treatment of an ion trap is similar to that for a quadrupole mass analyzer [12, 13]. Ion motion in an ion trap is either stable or unstable, which depends on its a_z and q_z, as shown in Fig. 5.4.

$$a_z = -2a_r = -\frac{16eU}{m(r_0^2 + 2z_0^2)\Omega^2} \tag{5-14}$$

$$q_z = -2q_r = \frac{8eV}{m(r_0^2 + 2z_0^2]\Omega^2} \tag{5-15}$$

In the coordinates of a_z and q_z there are several stability regions of which the one with the smallest a_z and q_z is usually used, as shown in Fig. 5.4. The lines crossing the stability region are so-called iso-β lines, which describe the detailed trajectories of the ions characterized by these lines. The stability region is restricted from β_r, $\beta_z=0$ to β_r, $\beta_z=1$. The region outside the stability region is the unstability region. The ions in the stability region have limited trajectories both along the r direction and along the z direction, so that these ions can be stored in an ion trap for a long period of time. On the other hand, the ions with unstable and large trajectories outside the stability regions result in their disappearance because of their collisions with electrodes.

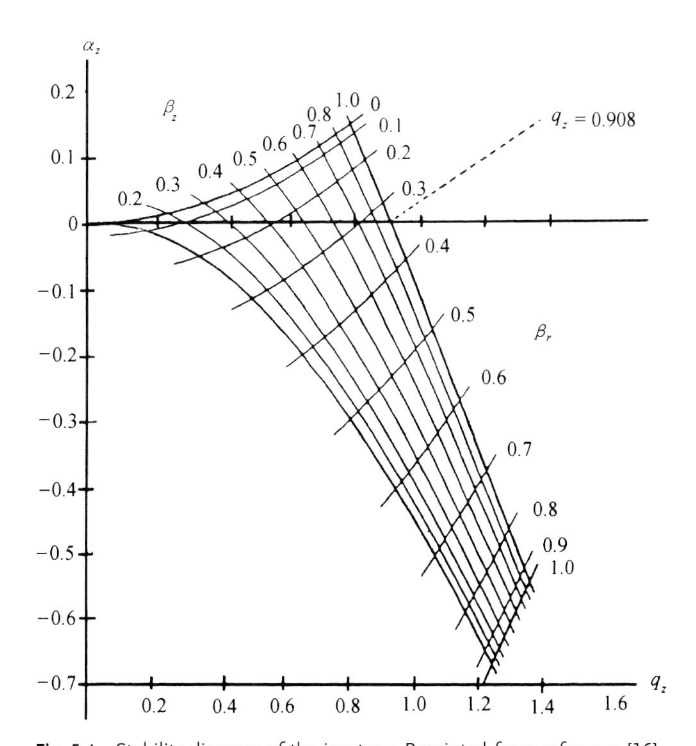

Fig. 5.4 Stability diagram of the ion trap. Reprinted from reference [16], with permission from CRC Press.

β_r and β_z characterize the ion angular frequencies of the radial and axial components, respectively. Related theoretical calculation gives:

$$\omega_{u,n} = +\left(n + \frac{1}{2}\beta_u\right)\Omega \qquad 0 \le n < \infty \tag{5-16}$$

and

$$\omega_{u,n} = -\left(n + \frac{1}{2}\beta_u\right)\Omega \qquad -\infty \le n < 0 \tag{5-17}$$

where n is an integer, $u = r, z$.

When $n = 0$

$$\omega_{u,0} = \frac{1}{2}\beta_u\Omega \tag{5-18}$$

Now we will explain the two equations (5-16) and (5-17). The ion angular frequency in the u (r or z) direction is ω determined by β, ω and n. As n increases, ω_n increases, and in the meanwhile the corresponding motion amplitude decreases rapidly. Therefore, only situations for ω with $n = 0$, ± 1 and ± 2 need to be considered, of which $\omega_{u,0}$, the fundamental frequency, is the most important. Therefore, it becomes evident that the values of $\beta_{u,0}$ ($u = r$, z) are important parameters describing ion motions. Iso-β_z line values increase from the left to the right, and iso-β_r line values from top to bottom. The intersection of the q_z axis with the line with $\beta_z = 1$ is situated at $q_z = 0.908$.

The expression of β_u, containing a_u, q_u and β_u, is very complicated. The β_u value is established gradually using the trial-and-error method by initially setting a numerical value.

The principle of mass scan can be understood from Fig. 5.4 and Eq. (5-15). Assuming $U = 0$ and $\Omega = a$ constant (typically 1.1 MHz or 0.88 MHz), q_z will increase gradually as V increases gradually according to Eq. (5-15). The greater the m/e of the ions, the sooner the ions reach $q_z = 0.908$ and are ejected from the ion trap through the pores of the end cap. Consequently, the mass spectrum is recorded as V increases (e.g., from 0 to 7500 V).

Ion storage in an ion trap can also be comprehended from Fig. 5.4. All ions with q_z values of less than 0.908 can be stored in an ion trap when U equals zero. However, only the ions with a particular m/e value can be selectively stored in an ion trap by the following operation. Firstly, we give the ions a q_z of 0.78, which is the projection of the upper apex of the stability diagram onto the q_z axis, by adjusting V. Then we raise the operating point near the apex vertically by carefully applying a minus voltage to the ring electrode. Consequently, only the ions are selected to be stored in the ion trap. Furthermore, tandem-in-time mass spectrometry can then be carried out.

Filling an ion trap with helium (at about 0.1 Pa) leads to damping of the ion motions, so that all ions concentrate toward the center, contributing to the enhancement of resolution and the extension of the mass range. Ion traps have been used regularly as mass analyzers in organic mass spectrometry only since this improvement [17].

There are several modes for manipulating an ion trap [16].

It is necessary to introduce two techniques in which an RF potential is applied to the end cap electrodes.

The first technique is called axial modulation. A supplementary RF potential whose frequency is about half that of the RF drive potential is applied to the end cap electrodes. According to Eq. (5-15), the frequency is close to the fundamental frequency of the ions that are being ejected at that moment in the scan. Absorbing energy from the supplementary RF field, the ions are excited suddenly, which are then ejected as they are tightly bunched. Therefore, the resolution and sensitivity of the mass spectrum are improved.

The second technique is called resonant ejection, which is a new mass scan mode. In the mass-instability scan mode (with or without the axial modulation), the ions, which are situated at $q_z = 0.908$ and $a_z = 0$ in the stability diagram, are ejected. Now a supplementary RF potential, whose frequency is fairly low, for example, one third the main RF frequency, is applied to the end cap electrodes. Because this potential is rather high, it produces a "hole" in the stability diagram. The hole is situated at the intersecting point between the q-axis and the iso-β_z line, the position of which is determined by the frequency of the supplementary RF potential. The ions that are situated at that point on the q-axis are excited strongly. They are ejected at this point before they arrive at the point of $q_z = 0.908$ and $a_z = 0$.

Resonant ejection can increase the mass range of the mass spectrum recorded by the ion trap. The abscissa of the intersecting point is denoted as q_{eject}, which is less than 0.908. The mass range is increased by a factor of $0.908/q_{eject}$.

An ion trap has the following advantages:

1. Realizing tandem-in-time mass spectrometry, that is, multi-stage $(MS)^n$, by using one mass analyzer, with n now reaching a value greater than 10.
2. A simple structure, which leads to a high value/cost ratio.
3. A high sensitivity, which is 10–1000 times higher than that of a quadrupole MS.
4. A wide mass range from 0 to 6000 u for some commercial instruments.

An ion trap can work in conjunction with a chromatographic instrument as the mass analysis detector.

The disadvantage of the ion trap is that the mass spectrum recorded by it differs slightly from that with other analyzers, which can be overcome by using an external ion source.

5.2.4
Fourier Transform Mass Spectrometer

Strictly speaking, the Fourier transform mass spectrometer should be called the Fourier transform ion cyclotron resonance mass spectrometer, or FT-ICR, which has been developed on the basis of ICR.

The development from ICR to FT-ICR was inspired by FT-NMR [1, 18].

First we will explain what ion cyclotron resonance is.

In a magnetic field B, ion motions are constrained to circular cyclotron orbits in the planes perpendicular to the magnetic field. Reconsider Eq. (5-3):

$$zevB = \frac{mv^2}{r_m} \tag{5-3}$$

Note that $ze = q$ and $\omega = \frac{v}{r}$. After simplification, we have the cyclotron equation:

$$\omega = \frac{qB}{m} \tag{5-19}$$

or

$$f_c = \frac{qB}{2\pi m} \tag{5-20}$$

where f_c is the ion cyclotron frequency in Hz.

An important concept can be introduced from Eq. (5-19) or Eq. (5-20): in a magnetic field, ions with various m/e values circulate at characteristic frequencies that are determined only by their m/e values rather than their kinetic energy.

Assume that there are several types of ions with different mass-to-charge ratios, and that each type rotates at its own characteristic cyclotron frequency between two parallel plates of a capacitor, in which the magnetic field strength is B. An RF (oscillating) potential is applied to the plates, the frequency of which equals the cyclotron frequency for the particular ions, and the ions will then be excited as they absorb energy from the oscillating electric field. Because of the constant cyclotron frequency, the velocity of the ions and the radius of the ion cyclotron orbit will increase, resulting in a spiral path as shown in Fig. 5.5.

Magnetic lines of force are perpendicular to the plane of the paper.

The absorption of energy from the oscillating electric field is called ion cyclotron resonance, and the increase in the ion's orbital radius is known as exciting the cyclotron motion of the ion.

Setting a magnetic field strength B, we can record a mass spectrum by gradually changing the RF frequency to excite the ions according to the order of mass-to-charge ratios. The signal is detected by measuring the power absorption from the oscillating electric field by the ions or by measuring the current, which is formed by the ions striking the plates. No matter which method is used, this is a process like CW NMR: ions are detected in turn.

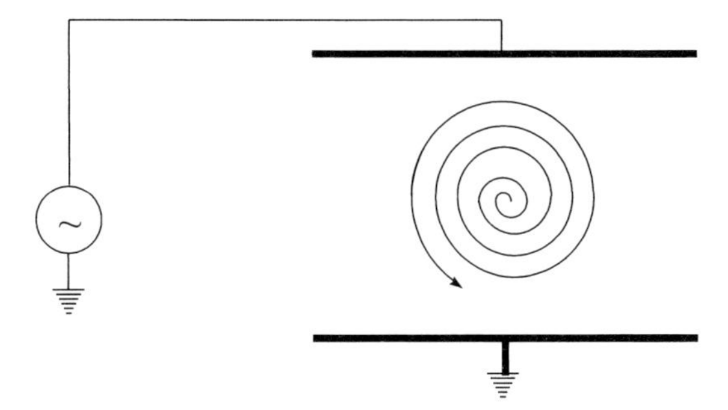

Fig. 5.5 Principle of ion cyclotron resonance.

Exciting all ions with various mass-to-charge ratios, whose cyclotron frequencies cover several orders of magnitude, is more difficult than exciting nuclei in different groups with close resonance frequencies.

It is necessary to discuss the process of FT-ICR further [3, 5, 19]. Before the excitation of ions, the ions with different mass-to-charge ratios have very small motional radii and the ions with the same mass-to-charge ratio have different phases. During the excitation, when an excitation frequency is equal to the cyclotron frequency of the ions with a particular mass-to-charge ratio, the ions form an ion bundle (an ion packet) and they have a coherent phase. The radius and their kinetic energy of the ions increase with the absorption of the energy from the RF field. The relationship between kinetic energy E and their cyclotron radius r of the ions can be deduced easily.

Because

$$E = \frac{1}{2} m v^2 \tag{5-21}$$

and

$$\omega = \frac{v}{r} \tag{5-22}$$

the combination of Eqs. (5-19), (5-21) and (5-22) leads to the following equation:

$$E = \frac{q^2 B^2 r^2}{2m} \tag{5-23}$$

One way to excite ions is “chirp,” that is a rapid frequency-sweep, over a short period of time. The advantages of chirp excitation are its fairly simple implementation and a nearly flat mass range excited if the frequency sweep traverses a constant number of Hertz per second. The disadvantages of chirp excitation are as follows: (1) Rather weak excitation near the low- and high-frequency limits of the sweep leads to incorrect ion intensities. (2) Non-simultaneous excitation, which results in complicated phases in the frequency domain. (3) Non-selective excitation, which prohibits FT-ICR from being used in tandem-in-time MS-MS.

Another way to excite ions is the stored waveform inverse Fourier transform [20], SWIFT, which is more advanced than chirp because of its ability to generate complex broadband excitation waveforms with equal power. Firstly, a mass-domain excitation profile designed by experimental requirements is determined. The profile is converted into a corresponding frequency-domain excitation spectrum. Secondly, the time-domain excitation spectrum, with which the desired mass-domain excitation will be obtained, is produced from the frequency-domain excitation spectrum by an inverse Fourier transform. SWIFT overcomes the three disadvantages of chirp excitation and it can provide a mass spectrum with good quantification for the whole desired mass range. The selection of a given m/e value can be done by using a mass domain excitation profile with a suppression window at the m/e value. All other ions are ejected from the cell. The remaining ions are then dissociated by CID.

After ions with widely different mass-to-charge ratios have been excited in an ICR cell, each ion with its own m/e ratio induces an alternating signal with its cyclotron frequency on the plates of the ICR cell. The signal is a time-domain signal, from which superimposed time-domain signals are formed (this situation is quite similar to that in FT-NMR). Then the Fourier transform is used to convert time-domain signals of FT-ICR into a frequency-domain spectrum, in which the frequency of the peak indicates the ion mass and the area of the peak is proportional to the ion population.

From what has been mentioned above, we know that FT-ICR is similar to FT–NMR not only in principle but also in data processing. The Fourier transform calculation is quite similar in both cases. Zero-fill and window functions are applied in FT-ICR as in FT-NMR.

It is useful to understand the principle of FT-ICR comprehensively. Firstly, the ions with the same mass-to-charge ratio have a coherent cyclotron motion, which means the ions form a "packet," so that their signals have a "phase" similar to that in an FT-NMR signal. Secondly, a potential is applied between the plates in the z direction along the magnetic lines to restrict the ion motion in the z direction, resulting in a cyclical motion with frequency f_z.

Furthermore, a cyclical radial motion, termed the magnetron motion, is also produced which has a frequency f_m. Although there is a relationship for $f_c \gg f_z \gg f_m$, accurate masses cannot be calculated just by Eq. (5-20), because ion motions are complex in an ICR cell. It is necessary to determine accurate masses by calibration on the basis of the known ions. Because of the good reproducibility of FT-ICR, calibration can be carried out independently, which makes the measurement of accurate masses very convenient.

The resolution of a mass spectrum was given by Eq. (5-1). The resolution in FT-ICR is introduced by differentiating then rearranging Eq. (5-20):

$$R = \frac{m}{dm} = -\frac{-qB}{2\pi m d f_c} \tag{5-24}$$

Substituting Eq. (5-20) into Eq. (5-24) gives:

$$R = \frac{m}{\mathrm{d}m} = -\frac{f_c}{\mathrm{d}f_c} \tag{5-25}$$

Thus as shown in Eq. (5-24), the resolution R in FT-ICR is in inverse proportion to the ion mass m and in direct proportion to the magnetic field strength B. The enhancement of B also increases the mass range. FT-ICR instruments with 9.4 T (corresponding to that in the 400 MHz FT-NMR spectrometer) are commercially available.

The key part of an FT-ICR instrument is the analytical cell whose simplest shape is cubic. Cyclotron motions are perpendicular to the z axis. The two orthogonal sets of parallel plates are the receiving and transmitting plates, respectively. Other shapes of the analytical cell are cuboid and cylinder. Ions are produced in the cell or introduced from an external ion source through a special guide tube, which consists of a series of symmetrical electrostatic lenses. The adoption of the external ion source makes the choice of an appropriate ionization possible.

The main advantages of FT-ICR are as follows:

1. It can give an ultrahigh mass resolution, which cannot be achieved with any other type of mass analyzer. Its resolution R can reach 1.5×10^6 when m equals 1000 u in commercially available instruments. This ultrahigh mass resolution is outstanding, also in that the enhancement of resolution does not decrease sensitivity because a high resolution can be achieved by increasing acquisition data.
2. Multiple stage tandem mass spectrometry (tandem-in-time), MS^n, can be accomplished by FT-ICR.
3. Various ionizations can be achieved with external ion sources.
4. A high sensitivity, a large mass range, rapid manipulation and good reliability.

A molecular weight as great as 1.1×10^8 Da has been measured by using ESI-FT-ICR [21].

5.2.5 Time-of-flight (TOF) MS

The key part of the TOF is the ion drift tube. The principle of the TOF is simple. Ions produced by MALDI (see Section 5.3.5) or drawn out by a pulse, enter the drift tube after they are accelerated by an acceleration voltage. As the ions possess the same amount of kinetic energy, the smaller the mass of the ions, the greater the velocity they maintain. Therefore, light ions will arrive at the detector earlier than heavy ones.

We can rewrite Eq. (5-2) as

$$zeV = \frac{1}{2}mv^2 \tag{5-2}$$

Thus

$$v = \left(\frac{2zeV}{m}\right)^{1/2} \qquad (5\text{-}26)$$

Suppose that L is the length of the drift tube and that t is the flight time of an ion, then we have

$$\begin{aligned} t &= \frac{L}{v} = L\left(\frac{m}{2zeV}\right)^{1/2} \\ &= \left(\frac{m}{z}\right)^{1/2} \times L \times \left(\frac{1}{2eV}\right)^{1/2} \end{aligned} \qquad (5\text{-}27)$$

Therefore ion masses can be calculated from their flight times. In general, flight times are in microseconds, and they must be measured with an accuracy of nanoseconds.

As there are distributions in space, in time and in the original kinetic energy of the ions, a distribution of v and t results, which leads to a decrease in mass resolution. There are a few methods for increasing the resolution.

One method is the use of an ion mirror, a reflector, which is set at the end of the drift tube. The ion mirror has a potential that is the same sign as that of the ions. When ions enter the region of the ion mirror, they will be decelerated, stopped and reflected (with a small angle). An ion with a greater velocity will pass through a longer distance in the region of the ion mirror than that with a smaller velocity. Therefore ions with the same m/z but slightly different velocities will focus at the exit of the ion mirror, which is similar to the spin echo in NMR (see Section 4.1.3). By using this method the resolution of TOF can be increased greatly. For this reason, the drift tube has been developed from the "I" type to the "V" type and further to the "W" type. However, increasing the resolution with ion mirrors will decrease sensitivity.

Another method is time-lag focusing (TLF) [22] or delay time [23].

The TOF has the following advantages:

1. It has no upper limit of measured mass, which is particularly suitable for biological research. Monoclonal human immunoglobulin with a molecular weight as great as 982 000 Da has been measured by TOF [2].
2. It is very suitable for matching MALDI.
3. As all ions are measured at the same time, TOF has a high sensitivity, so it can act as the second mass analyzer of a tandem mass spectrometry (see Section 5.5.2).
4. Because of its high scan speed, TOF can be used for the study of rapid processes.

One important disadvantage of TOF is that the mass resolution decreases with the augmentation of the mass to be analyzed.

When TOF is used in the LC–MS combination or in the second mass analyzer of a tandem mass spectrometry, its operation mechanism has to be changed. A new type of TOF, orthogonal acceleration TOF, abbreviated as oa-TOF, has been developed, where ions are drawn out with periodic pulses at an orthogonal angle to the direction of the ion motion. Then the ions fly in a drift tube. In this way, TOF can be used for the measurement of ions that are moving continuously. The oa-TOF uses micro-channel plates, MCP, as the detector.

A technique related to TOF is that of post-source decay, PSD, which is useful for the analysis of product ions produced from precursor ions [24].

5.3 Ionization

In this section, ionization is discussed for mass spectrometry as well as for LC-MS.

5.3.1 Electron Impact Ionization, EI

This used to be called electron impact and now it is known as electron impact ionization or electron ionization.

EI is the most widely used and highly developed method of ionization.

Electrons are accelerated to tens of eV, they have a wavelength of the order of magnitude of the size of a molecule. When the wave passes through or near a molecule, the wave can be decomposed into a series of waves, one of which has the same phase as that of an electron of the molecule to be analyzed. Thus the electron will be expelled so that the molecule will become an ion.

In general, the acceleration voltage is 70 V. Therefore, the energy of electrons to be accelerated is 70 eV. In a high vacuum, the interaction between the electrons and molecules leaves molecular ions with so much energy that most of them break up to form smaller ions as the ionization potential of an organic molecule is in general 7–15 eV. This process is called fragmentation and the smaller ions, fragment ions (in a broad sense).

The absolute majority of ions produced by EI are singly-charged ions whose mass-to-charge values are numerically equal to their mass values.

EI occurs in an ion source. Sample molecules in the vapor phase enter the ion source in a high vacuum through a pinhole. A hot cathode emits electrons, which are accelerated to 70 eV and strike sample molecules. A spiral trace of electrons, which is formed by an auxiliary magnetic field, enhances the probability of collisions between sample molecules and electrons.

The resulting ion beam is drawn out at an orthogonal angle with respect to the direction of the electron beam by a high acceleration voltage (at several thousand volts).

EI has the following advantages:

1. It produces reproducible mass spectra that can be easily used for mass spectral retrieval in a data system.
2. The mass spectra produced by EI contain many peaks of fragment ions (in a broad sense). This is very useful for postulating structures of unknown compounds. In the next chapter we will deal with the interpretation of mass spectra, which is mainly based on mass spectra produced by EI.
3. EI is the most common ionization technique.

Because of these advantages, EI has found wide application.

The disadvantage of EI is generally a low intensity of the molecular ion peaks. Sometimes there is no molecular ion peak when samples to be analyzed are nonvolatile or unstable to heat.

If the electron energy is decreased, the intensity of the molecular ion peak in an EI spectrum will increase while intensities of all other peaks will decrease.

Other ionization techniques, which use lower energy and are termed soft ionization, are used to obtain molecular ion information.

5.3.2
Chemical Ionization, CI

In CI, sample molecules are ionized through chemical reactions, hence the name.

In EI, the ionization takes place in a vacuum of about 1.3×10^{-4} Pa while in CI, the ionization takes place in vacuum of about 1.3×10^{2} Pa because of the presence of the reagent gas. As molecules of the reagent gas are in an overwhelming majority compared with sample molecules, they are ionized by energized electrons, and then a series of complicated chemical reactions take place. When methane acts as the reagent gas, some reactions are as follows:

- $CH_4 + e \rightarrow CH_4^{+\bullet} + 2e$
- $CH_4^{+\bullet} + CH_4 \rightarrow CH_5^{+} + CH_3^{\bullet}$
- $CH_5^{+} + M \rightarrow CH_4 + MH^{+}$

where M is a sample molecule and MH^+ is a quasi-molecular ion, which is the result of a proton transfer reaction.

If ammonia gas is utilized as the reagent gas, it is easy to form $[M+H]^+$ when M is more basic than the reagent gas. On the other hand, it is easy to produce $[M+NH_4]^+$ when M is less basic than the reagent gas. In practice both are termed quasi-molecular ions.

For a long time CI used to produce only positive ions. Later, CI in the negative ion mode has been developed. For example:

$$X^- + M \rightarrow [M - H]^- + HX$$

where M is the molecule to be measured and X is a reagent gas molecule.

Of course, the process producing negative ions is complicated, with the above reaction being only part of it. CI with negative ions is utilized for compounds, which are strongly electrophilic. CI produces quasi-molecular ions.

Besides methane and ammonia gas, isobutane, methanol, etc. can also be used as the reagent gas.

The quasi-molecular ion produced from CI has a small amount of extra energy. In addition, a quasi-molecule is an even electron ion, leading to a fairly high abundance compared with that of $M^{+\bullet}$ produced by EI.

The fragmentation of quasi-molecules is mainly a process of losing small neutral molecules, HY, where Y is a group that exists in the original molecule (before fragmentation). According to proton affinity, PA, the order of loss of HY is:

$$\begin{aligned} &HBr, HCl, HI > H_2O > H_2S > HCN > C_6H_6 > CH_3OH \\ &\qquad > CH_3COOH > NH_3 \end{aligned}$$ [25]

There are far fewer fragment ion peaks in CI spectra. In general, CI spectra and EI spectra are mutually complementary.

Another advantage of the CI spectrum is that it can differentiate isomers better (see Section 9.2).

CI mass spectra can be obtained along with that from EI in the same experiment because CI and EI can be switched rapidly in the ionization source of a commercial mass spectrometer.

The prerequisite to realize CI is that the sample to be analyzed can be vaporized.

5.3.3 Field Ionization and Field Desorption

Vapor sample molecules will be ionized if they pass through a very strong potential gradient produced by a metal needle with a very high potential, because a needle has a high curvature. This type of ionization is termed field ionization.

Field ionization is utilized rarely at present because of its low sensitivity and the requirement of sample molecules to be in the vapor form.

The principle of field desorption is similar to that of field ionization except that sample molecules are deposited on an electrode on whose surface there are a large number of microneedles to increase the yield of the ions. Under the action of an electric field (and even with the help of slight heating), sample molecules form quasi-molecular ions directly without vaporization. Field desorption is suitable for compounds that are difficult to vaporize and unstable to heat, such as peptides, sugars, polymers, salts of organic acids, and so forth.

5.3.4
Fast Atom Bombardment, FAB, and Liquid Secondary Ion Mass Spectrometry, LSIMS

FAB is an effective ionization technique for polar, labile or non-volatile compounds. FAB can be applied to both non-polar and polar compounds, including salts.

FAB uses a beam of fast moving atoms to bombard a target, which is a mixture of a liquid matrix and the sample molecules. The bombarding atom is Xe or Ar. Inert gas atoms are ionized first, and then accelerated by an electric field to produce high kinetic energy. They then become atoms with high kinetic energy through a charge-exchanging reaction, such as:

$$\begin{aligned}&Ar^+ \text{ (with high kinetic energy)} + Ar \text{ (in thermal motion)} \rightarrow \\ &\quad Ar \text{ (with high kinetic energy)} + Ar^+ \text{(in thermal motion)}\end{aligned}$$

Ions in thermal motion are drawn out by an electric field while the beam of atoms with high energy bombards the target.

A sample to be analyzed is mixed with a liquid matrix. Matrixes in common use are glycerol, thioglycerol, 3-nitrobenzylalcohol, polyethyleneglycol, etc. One prerequisite for a matrix is a low vapor pressure so that it does not destroy the vacuum in the system. When the fast moving atom beam strikes a target, its kinetic energy will be dissipated in the mixture of matrix and sample molecules in several ways, which may lead to the sample molecules being vaporized and ionized. Because of the existence of a liquid matrix, sample molecules at the surface can be replaced continuously. In general, the matrix should have good mobility, solubility and electrolytic properties but be low in volatility and chemical reactivity.

There can be several quasi-molecular ion peaks in a mass spectrum from FAB because sample molecules could be associated with matrix molecules, metal atoms (when their salt exists) and hydrogen atoms. There are fragment ion peaks in the FAB spectrum so that the spectrum provides structural information. In the FAB spectrum, there are peaks produced from the matrix to be used, for example, m/z 93, 185, 277, and so forth, when glycerol is used as the matrix. More information will be given in Section 6.6.2.

An FAB spectrum can be changed by another matrix, the work presented in reference [26] being an example.

FAB can produce negative ions, all of which are then detected.

5.3.5
Matrix-assisted Laser Desorption-ionization, MALDI

Although MALDI was developed later than CI, FD and FAB, today it is an important soft ionization technique [27, 28].

The solution of a sample to be analyzed with a concentration of $\mu mol\ L^{-1}$ is mixed with a matrix with a concentration of $mmol\ L^{-1}$. After the vaporization of the solvent used, the mixture becomes crystals or semi-crystals, which are then radiated by laser pulses of a certain wavelength. The matrix transfers the absorbed energy into the sample molecules so that they are vaporized and ionized.

Matrixes in common use are 2,5-dihydrobenzoic acid, sinapinic acid, nicotinic acid, α-cyano-4-hydroxycinnamic acid, etc.

MALDI has the following advantages:

1. Ions containing intact sample molecules can be obtained from samples that are difficult to ionize, such as biological macromolecules: e.g., proteins, nucleic acids.
2. As the ionization takes place instantaneously, MALDI is particularly suitable for matching TOF. Therefore, we frequently encounter the term MALDI–TOFMS.
3. MALDI can also be used for other types of mass analyzer.

A mass spectrum from MALDI can distinguish singly-charged quasi-molecular ions and other quasi-molecular ions. More information will be given in Section 6.6.3. There are rare fragment ions in the mass spectrum from MALDI.

5.3.6 Atmospheric Pressure Ionization, API

API is mainly utilized for the combination of HPLC and MS. It comprises electrospray ionization, ESI, and atmospheric pressure chemical ionization, APCI.

1. Electrospray ionization (ESI)

The solution containing a sample to be analyzed flows out of a capillary tube, at the end of which the solution is almost instantaneously nebulized into very small charged droplets because of the following factors:

1. A high electric potential, positive or negative, of typically 3–8 kV, is applied at the end of the capillary tube.
2. A drying gas (such as N_2) flows along a tube, which is wider and concentric with respect to the capillary tube and terminates at the same location as the capillary tube.
3. A high temperature.

The charged droplets pass through the central orifices of skimmers where the solvent vapor is drawn out continuously by rotary pumps. This region is kept in a modest vacuum. The cylinder in which the charged droplets move is heated slightly to prevent condensation. As the droplets move through this region, the solvent evaporates rapidly from their surface and the droplets get smaller and smaller. As a result, the charge density on the droplet surface increases rapidly, leading to the expulsion of ions into the vapor phase. These ions are singly-charged quasi-molecular ions and/or multiply-charged quasi-molecular ions, which are related to the sample being analyzed. Small molecules are preferable for producing singly-charged quasi-molecular ions. On the other hand, multiply-charged quasi-molecular ions are preferable when a biological sample, that has many acid and basic groups, is analyzed. The multiply-charged quasi-molecular ions form a peak cluster so that the mass range to be measured can be multiplied. This multiplication can be several tens of times, which is very useful for measuring biological macromolecules.

ESI is a type of very soft ionization so that few fragment ions are produced.

2. Atmospheric pressure chemical ionization (APCI)

APCI is similar to ESI in many respects. In APCI, ionization is realized by a corona discharge. For this purpose, a needle with a high potential is set near the exit of a capillary tube from which the solution to be analyzed flows out. As with ESI, the solution is nebulized by a drying gas. First solvent molecules are ionized. Then there are a series of ion–molecule collisions to produce quasi-molecular ions of the sample being analyzed. This is a CI process in which solvent molecules function as the reagent gas in ordinary CI.

Generally speaking, the samples used in APCI are of low molecular weight and less polar than those used in ESI. Overall, the technique of APCI is complimentary to ESI.

Molecular weight information can be easily obtained from quasi-molecular ion peaks in a mass spectrum produced through ESI or APCI. For further structural information, it is necessary to produce fragment ions, which is realized by the application of a potential between skimmers. The fragmentation can be controlled by adjusting the potentials.

The interpretation of the mass spectra from ESI or APCI will be presented in Sections 6.6.4 and 6.6.5, respectively.

5.4
Metastable Ions and their Measurement

A mass spectrum can provide two types of information. From the molecular ion peak (or quasi-molecular ion peak, multiply-charged molecular ion cluster), one can obtain the molecular weight. From fragment ions (in a broad sense), structural units in an unknown compound and their connection sequence can be deduced.

If there are many fragment ion peaks in a mass spectrum, their assignment is sometimes difficult. We often wish to know which fragment ion (in general) is produced from the molecular ion and which ion is subsequently formed from that fragment ion. The knowledge of the "filiations" (formation sequences) of ions is very useful to the deduction of an unknown structure. It is metastable ions that give the information and their study forms a new important branch of organic mass spectrometry. It is therefore necessary to present a brief description of metastable ions in this section.

Our discussion on metastable ions is restricted to the EI spectrum because only EI can produce many fragment ions. We will deal with other ionization

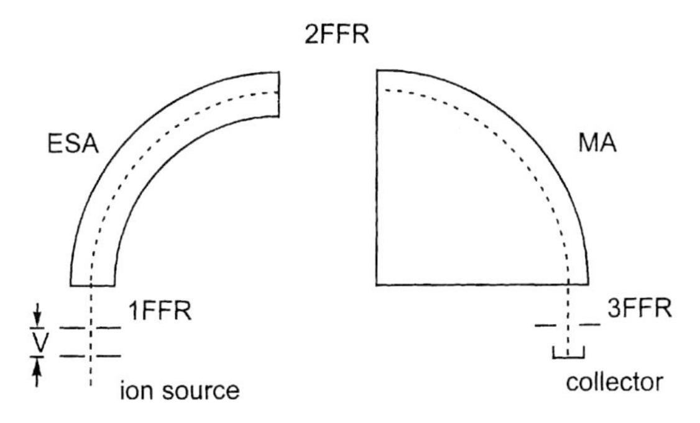

Fig. 5.6 The three field-free regions in a double focusing mass spectrometer (with normal geometry).

techniques in Section 5.5.1 (by CID). The formation and measurement of metastable ions is discussed with respect to a double focusing instrument in this section and with other mass analyzers in Section 5.5.2.

No matter how the sector magnet field analyzer and electrostatic analyzer of a double focusing instrument are set, there are three field-free regions, or FFRs. The first FFR is situated between the ion source and the first mass analyzer, the second between the two mass analyzers and the third between the second mass analyzer and the detector. The three FFRs in a double focusing mass spectrometer (with normal geometry) are shown in Fig. 5.6.

As the probability of forming multiply-charged ions is much less than that of forming singly-charged ions, all equations in this section are deduced on the basis of $z=1$ for simplicity. Consequently the mass-to-charge ratio reduces to m/e.

It is convenient to discuss metastable ions with momentum mv. From Eq. (5-2):

$$eV = \frac{1}{2}mv^2$$

and we have:

$$mv = \sqrt{2eVm} \tag{5-28}$$

We can also obtain an expression from Eq. (5-3):

$$mv = r_\mathrm{m} Be \tag{5-29}$$

It follows from Eqs. (5-28) and (5-29) that for ions with different mass-to-charge ratios the deflection radii in the magnetic field are proportional to their momenta that are in turn proportional to the square roots of the masses.

5.4.1
Metastable Ions Produced in the Second Field-free Region

The time it takes for an ion to travel from the ion source into the detector is of the order of magnitude of 10^{-5} s while the time it takes for an ion to travel in the ion source is of the order of magnitude of 10^{-6} s. When an ion possesses high internal energy, it will decompose quickly in the ion source (in less than 10^{-6} s). A stable ion can arrive at the detector without any decomposition, with a lifetime of longer than 10^{-5} s). Ions whose lifetimes are between 10^{-5} and 10^{-6} s are called metastable ions that decompose frequently after leaving the ion source and before arriving at the detector. We denote this process as:

$$\mathrm{m}_1^+ \rightarrow \mathrm{m}_2^+ + \mathrm{N} \tag{5-30}$$

where m_1^+ is the parent ion; m_2^+ is the daughter ion; and N is a neutral fragment with a mass of m_1–m_2.

Notice that we do not distinguish between odd-electron ions and even-electron ions with "+" and "+•" in this section. Here m_1^+ can be an odd-electron or an even-electron ion.

The reaction (5-30) can take place in any field-free region or field region. If the reaction occurs in a magnetic or electrostatic field region, m_2^+ will be deflected and neutralized. Now we will discuss the measurement of metastable ions produced in the second FFR.

Corresponding to Eq. (5-2), we have:

$$eV = \frac{1}{2} m_1 v_1^2 \tag{5-31}$$

where m_1 is the mass of m_1^+; and v_1 is the velocity of m_1^+ after its acceleration.

Obviously m_2^+ has the same velocity as m_1^+. Namely,

$$v_2 = v_1 \tag{5-32}$$

In the magnetic field m_2^+ takes on a circular motion like the other ions but the radius of the circular motion is different from that for m_1^+.

Corresponding to Eq. (5-3), we have:

$$ev_1 B = \frac{m_2 v_1^2}{r_\mathrm{m}} \tag{5-33}$$

Combining Eqs. (5-31) and (5-33) to eliminate v_1 yields

$$r_\mathrm{m} = \frac{1}{B}\sqrt{2V \frac{m_2^2}{m_1}\frac{1}{e}} \tag{5-34}$$

or

$$\frac{m_2^2}{m_1 e} = \frac{B^2 r_m^2}{2V} \tag{5-35}$$

Comparing (5-35) with (5-5), we know that m_2^+ will be recorded as m_2^2/m_1. Its "apparent mass", which is less than the mass m_2, is denoted as:

$$m^* = \frac{m_2^2}{m_1} \tag{5-36}$$

Because the internal energy of m_1^+ converts partly into kinetic energy, the peaks of metastable ions are not sharp but flattened by 2–3 mass units.

By using Eq. (5-36), one can establish the m_1^+ (parent ion) and m_2^+ (daughter ion) pair. In general, both m_1^+ and m_2^+ have considerable intensities.

5.4.2 Metastable Ions Produced in the First Field-free Region

Because m_1^+ decomposes in the first FFR, the kinetic energy of m_2^+ is equal to that of normal ions multiplied by m_2/m_1. Metastable ions that have much less kinetic energy will be deflected more, thus coming into collision with the analyzer. Two methods can be used to detect metastable ions produced in the first field-free region.

1. High voltage scan (HV method)

In using this method, the voltage at the electrostatic analyzer is kept at the normal value so that ions with normal kinetic energy can pass through the electrostatic analyzer.

If the acceleration voltage for m_2^+ is increased from V_1 to V_1' and

$$V_1' = \frac{m_1}{m_2} V_1 \tag{5-37}$$

The kinetic energy of m_2^+ in this case is

$$\frac{1}{2} m_2 v_2^2 = \frac{m_2}{m_1} e V_1' = \frac{m_2}{m_1} e \frac{m_1}{m_2} V_1 = e V_1 \tag{5-38}$$

Equation (5-38) means that metastable ions produced in the first field-free region can pass through the electrostatic analyzer under these conditions. On the other hand, all normal ions produced in the ion source will strike the electrostatic analyzer because they possess too much kinetic energy in this case. Therefore this method is also called the defocusing method.

Using this mode, one can find all parent ions that produce the same daughter ion.

2. Scan *E* method

It is known that m_2^+ produced from m_1^+ has the kinetic energy of $\frac{m_2}{m_1}eV$. They can pass through the analyzer if the potential applied at the electrostatic analyzer is decreased to $\frac{m_2}{m_1}E$. Obviously all normal ions are defocused under these conditions.

5.4.3 Ion Kinetic Energy Spectrum (IKES)

We in fact discussed the principle of IKES in the above section.

The instrument usually utilized for the measurement of IKES is a mass spectrometer with normal geometry (ESA-MA).

A detector is set at the exit of the electrostatic analyzer. When the acceleration voltage *V* is kept constant, by decreasing *E*, the detector can record all metastable ions produced in the first field-free region. Because the ions have different kinetic energies, the spectrum recorded in this case is a spectrum of the kinetic energy. Therefore it is called an ion kinetic energy spectrum.

5.4.4 Mass-analyzed Ion Kinetic Energy Spectrum (MIKES)

A mass-analyzed ion kinetic spectrum is also called the direct analysis of daughter ions (DADI).

A mass spectrometer with reverse geometry is used for the measurement of MIKES. Its first mass analyzer is the magnetic analyzer through which ions of interest are selected. All metastable ions produced from the ions in the second field-free region pass through the electrostatic analyzer at which the potential is scanned. All metastable ions are recorded at $\frac{m_2'}{m_1}E$, $\frac{m_2''}{m_1}E$, $\frac{m_2'''}{m_1}E$, and so forth, which means that all daughter ions produced from the same parent ion are recorded.

The advantages of MIKES are as follows:

1. Isomers with the same skeleton but different side chains can be distinguished by MIKES if side chains are selected for analysis by MIKES.
2. A mixture can be separated in the magnetic analyzer and then analyzed by the electrostatic analyzer. The separation by the magnetic analyzer is much more rapid than by chromatography.

5.4.5 Linked Scanning

Linked scanning has more advantages than the HV method or scanning *E* method.

There are three parameters for mass spectrometry operation: *B*, *E* and *V*. In linked scanning one parameter is kept constant and the two other parameters are changed according to a particular relationship [7].

1. *B/E* scanning

This is the most frequently applied linked scanning.

Under certain conditions (the acceleration voltage V, the voltage at the electrostatic analyzer E_1 and the magnetic induction field B_1), the normal ion m_1^+ is detected. Thus the ratio between B and E is determined.

In the first field-free region, m_1^+ decomposes into m_2^+. The condition for the passage of m_2^+ through the electrostatic analyzer is

$$E_2 = E_1 \frac{m_2}{m_1} \tag{5-39}$$

Under normal conditions, we have Eq. (5-29):

$$mv = r_m Be$$

For m_1^+, this becomes

$$m_1 v_1 = r_m B_1 e \tag{5-40}$$

For m_2^+ to pass through the magnetic analyzer with the same r_m as m_1^+, the following equation must hold:

$$m_2 v_2 = r_m B_2 e \tag{5-41}$$

Because $v_2 = v_1$, we have

$$m_2 v_1 = r_m B_2 e \tag{5-42}$$

Combining (5-40) and (5-42) yields

$$\frac{B_2}{B_1} = \frac{m_2}{m_1} \tag{5-43}$$

Combining (5-39) and (5-43), we have

$$\frac{B_2}{B_1} = \frac{m_2}{m_1} = \frac{E_2}{E_1}$$

or

$$\frac{B_2}{E_2} = \frac{B_1}{E_1} = \text{constant} \tag{5-44}$$

This constant depends on the value of m_1.

The operation is as follows: For the value of m_1, a ratio of B_1 to E_1 is determined. Then linked scanning is performed. Both B and E are decreased to keep the ratio unchanged. All metastable ions produced from m_1^+ in the first field-free region are recorded, which is useful to the deduction of the structure of m_1^+.

2. B^2/E scanning

When using this scanning, all parent ions that produce the same daughter ion are recorded.

In the first field-free region

$$m_1'^{+} \rightarrow m_2^{+} + N'$$
$$m_1''^{+} \rightarrow m_2^{+} + N'', \text{ and so forth.}$$

For m_2^+ to pass through the electrostatic analyzer, we have

$$eE_2' = \frac{m_2 v_2^2}{r_e} = \frac{m_2 v_1^2}{r_e} \tag{5-45}$$

where v_1 is the velocity of m_1^+.

In the magnetic analyzer, according to Eq. (5-29), we can write

$$r_m B_2' e = m_2 v_2 = m_2 v_1 \tag{5-46}$$

We take

$$(5\text{-}46)^2/(5\text{-}45)$$

After rearrangement, we have

$$\frac{B_2'^2}{E_2'} = \frac{m_2 r_e}{r_m^2 e} \tag{5-47}$$

Because m_2 is given, B_2^2/E_2 is a constant.

In normal operation, for the measurement of m_2^+, V, E_2 and B_2 are determined. Thus the value of B^2/E is obtained. Then linked scanning is performed. Decrease B and E but keep B^2/E constant. Under these conditions, all parent ions that produce the same m_2^+ in the first field-free region are recorded in the order of increasing m_1.

3. The constant neutral loss scan

The scan gives all daughter ions that are characterized by the loss of a constant neutral fragment from different parent ions. For example, by setting the constant neutral fragment at 18 u, all alcohol compounds that lose water molecules can easily be found.

The operation of this mode by a magnetic sector instrument is complicated. If we use tandem quadrupole mass analyzers, the calculation will be greatly simplified (see Section 5.5.2).

5.4.6
Information Provided by Metastable Ions

In this section we present the measurement of metastable ions by a double focusing mass spectrometer. The measurement by other types of mass analyzers will be discussed in Section 5.5. By measuring metastable ions one can identify parent–daughter ion pairs from which the information on ion structures, connection sequences and reaction mechanisms in mass spectrometry can be obtained. The measurement of metastable ions is also useful for mixture analysis. We will present several examples to explain its usefulness.

1. Reaction mechanisms in organic mass spectrometry

In the mass spectrum of structure **38** there is a molecular ion peak at m/z 162 and fragment ions at m/z 147, 119, 105 and 91, and so forth. Obviously ion m/z 147 is produced from M–CH_3. However which CH_3 is the CH_3 in the molecule?

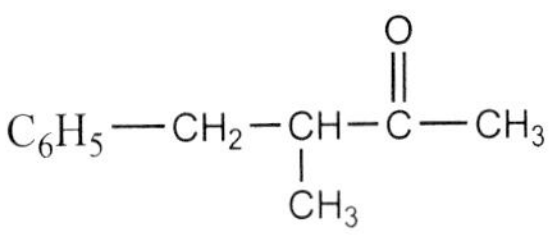

$$C_6H_5—CH_2—CH(CH_3)—C(=O)—CH_3$$

38

From the measurement of its metastable ions we know that there are two fragmentations as follows:

1. m/z 162 → m/z 147 → m/z 119 → m/z 91
2. m/z 162 → m/z 147 → m/z 105

Considering the structure, we can deduce that the fragmentation in 1 is the sequential loss of CH_3, C=O and C_2H_4 from the molecule, finally producing a tropylium ion. Therefore the CH_3 is the one connected to the carbonyl group. Similarly the fragmentation in 2 is the sequential losses of CH_3 (connected to CH) and O=C=CH_2, producing $(C_6H_5–CH_2–CH_2)^+$.

2. Ion structure

If we know the parent ion structure and its fragmentation, then it is certain that we know the structure of its daughter ion, as shown in the above discussion.

Here is another example. The molecular ion peak of a steroid with four rings A, B, C and D is situated at m/z 320. We wish to know the structure of ion m/z 122.

From the measurement of its metastable ions we know that it has two fragmentations, one being the sequential loss of ring D and then rings A and B from the molecular ion to produce 122, and the other is the sequential loss of rings A, B and D. Therefore the ion m/z 122 has a single structure (the C ring).

Information about ion structures can be provided by CID (see Section 5.5.1).

3. Connection sequence of structural units

From the information of the measurement of sequential daughter ions, one can obtain the connection sequence of structure units of an unknown compound.

The last example in Section 6.4 gives the result for a sequence determination of an oligopeptide. However, the result is less reliable than that deduced from metastable ions.

5.4.7
Peak Shapes of Metastable Ions

When a parent ion decomposes into a daughter ion and a neutral fragment, the momentum of the daughter ion can be in any direction while that of the parent ion has only one direction. As a result, metastable ions have a shape that is not sharp but diffuse. The common shapes of metastable ions are shown in Fig. 5.7.

In fact the shape of a metastable ion is related to the release of internal energy during the fragmentation. Readers who are interested in this topic should read reference [7].

5.5
Tandem Mass Spectrometry (MS^n)

We discussed the measurement of metastable ions by a double focusing instrument in Section 5.4. In this section we will deal with the measurement of metastable ions by other types of mass analyzers.

5.5.1
Collision-induced Dissociation (CID)

Collision-induced dissociation is also known as collision-activated dissociation (CAD) or collisional activation (CA).

By soft ionization techniques we can obtain information about the molecular weight. However, since quasi-molecular ions have a very small amount of excessive internal energy and they are even-electron ions, quasi-molecular ions

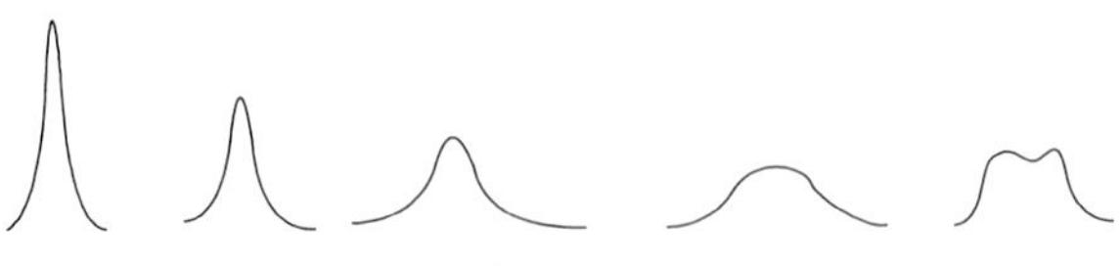

Fig. 5.7 Shapes of metastable ions.

formed by soft ionization give only a few fragment ions (relatively), which is disadvantageous to the deduction of an unknown structure. An ideal result is that we obtain both the information about molecular weight by soft ionization and the structural information by breaking the quasi-molecular ion into fragments. If necessary, some selected fragment ions are broken further, which may be done just by CID.

Some selected ions such as the molecular ion, quasi-molecular ion, fragment ions (in general), collide with atoms (molecules) of an inert gas, thus producing their fragment ions, which is CID.

The device where the collisions take place is called the collision cell. Sometimes a mass analyzer can also perform the function of a collision cell, which will be discussed later.

In CID, the ions before and after the collision are named reactant ions and product ions, respectively.

According to the type of mass analyzer with which CID is applied, CID is divided into two types: CID with high energy and CID with low energy. The first type of CID concerns a double focusing instrument where ions have a kinetic energy of several keV, and the second type is used in a quadrupole, an ion trap or an FT-ICR, where ions have a kinetic energy of less than 100 eV.

Because of the difficulties for studies on CID, some primary conclusions are perhaps not consistent [29]. However, some common conclusions can be drawn:

1. Part of the kinetic energy of the ions is changed into their internal energy, leading to the fragmentation of the ions.
2. The CID spectrum (the mass spectrum is obtained by CID) has some similarities with the EI spectrum, and therefore remedies the defects of soft ionization techniques.
3. The two types of CID have different results.

We will take as an example peptides whose fragment ions have the nomenclature given in structure **39**.

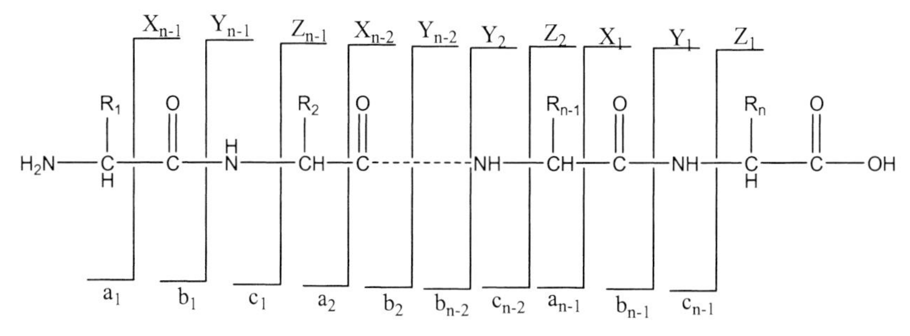

39

N-terminal fragment ions are labeled as a_1, b_1, c_1, a_2... whereas *C*-terminal fragment ions are labeled as x_1, y_1, z_1, x_2...

A peptide containing ten amino acid residues is fragmented by high-energy CID and low-energy CID, respectively (with two types of mass analyzers). The high-energy CID generates all types of fragment ions. In a low-energy CID spectrum, only b- and y-type fragment ions are observed and a- and z-type fragment ions are less frequently observed but c- and x-type fragment ions are not observed. Ions resulting from the loss of water, ammonia or carbon monoxide from the sequence fragment ions are often present in low-energy CID spectra [30]. Notice that b- and y-type fragment ions are useful for mass spectra of peptides.

In high-energy CID, fragmentation takes place easily because ions possess high energy. In low-energy CID, inert gases with greater atomic weight, such as Ar and Xe, are applied to raise the efficiency of CID. If a quadrupole mass analyzer is used as the collision cell, the efficiency of CID is rather high because of the length of the collision cell and the focusing effect produced by the radiofrequency wave. If an FT-ICR or an ion trap is used as the collision cell, the efficiency of CID is also satisfactory because ions move continuously.

CID is applied in tandem mass spectrometry and in LC-MS.

5.5.2 Tandem Mass Spectrometry

Tandem mass spectrometry [31, 32] can also be denoted as MS-MS or MS^n where n denotes the orders of the tandem.

Tandem mass spectrometry is divided into two types: tandem-in-space and tandem-in-time.

Firstly we will discuss tandem-in-space, where two or more mass analyzers are used. IKES, MIKES and linked scanning belong to these operations under certain conditions.

The following experiments can be accomplished by tandem-in-space:

1. The scan of product ions (daughter ions)
   From the discussion in Section 5.4 we know that this scan is the most important scan and it is applied most frequently.
   Ions of interest are selected from the first mass analyzer. In the collision cell they are fragmented by collision with inert gas molecules (He, Ar, Xe, etc.) to produce product ions (daughter ions) that will be analyzed by the second mass analyzer.
2. The scan of precursor ions (parent ions)
   In this mode the first mass analyzer scans over a range of mass-to-charge ratios. In the order of the ratio the selected ions are fragmented sequentially by CID in the collision cell. Only the ions with a certain m/z value are permitted to pass through the second mass analyzer. Therefore all parent ions that produce the same product ion can be recorded by this scan.

3. The scan of constant neutral loss
   Between two mass analyzers a collision cell is set where CID takes place. The two mass analyzers scan at the same time to keep a selected m/z difference. Therefore ion pairs produced from certain neutral fragment losses (loss of H_2O or C=O, etc.) can be recorded, which can be readily accomplished using two quadrupole mass analyzers.

The above-mentioned three scan modes are very useful in the deduction of the structure of an unknown compound. It should be emphasized that these modes can be applied to the analysis of a mixture. The first mass analyzer is used for the separation of the mixture. Compared with chromatographic separation it has the advantage of rapidity. The separation time is much shorter than the retention time in chromatography. If a double focusing instrument is used in the separation, ions with the same nominal mass can be separated. The scan of constant neutral loss is very convenient for analysis of mixtures. With this scan one type of compound can be easily selected. Consequently, the signal-to-noise ratio can be greatly increased. For example, all alcohols can be easily detected because they lose water molecules. By this scan only alcohols are selected and detected.

The three scan modes can be illustrated in Fig. 5.8. We use a circle to illustrate the scan for the entire or a selected range of m/z values. A circle with a dotted line is used to illustrate the product ions with different mass-to-charge ratios produced by CID. An arrow is used to illustrate the detection of ions with a selected m/z value.

For simplicity, the quadrupole mass analyzer, the sector magnetic field mass analyzer and the electrostatic mass analyzer are denoted as Q, B and E, respec-

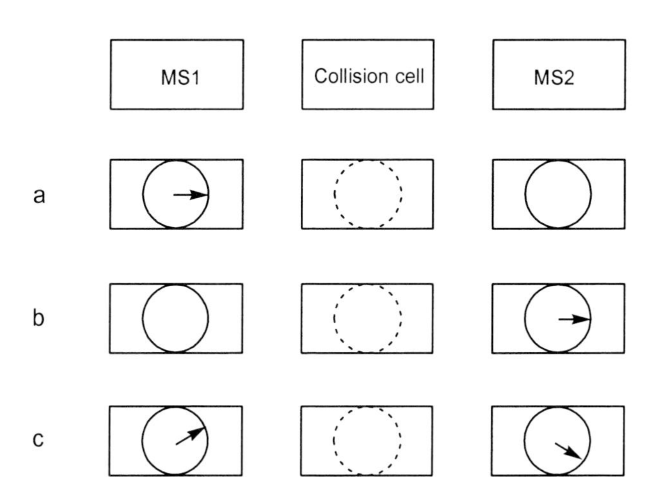

Fig. 5.8 Three scan modes of tandem-in-space mass spectrometry. (a) The scan of product ions. (b) The scan of precursor ions. (c) The scan of constant neutral loss.

tively. There are many possible combinations of Q, B and E. We will present only a few of the combinations most frequently used.

1. The combination of three quadrupole mass analyzers

This combination can be denoted as QQQ or QqQ, where the second Q or q denotes the collision cell. This combination is applied most frequently. It can easily perform all three modes mentioned above. In particular it is easy to scan the constant neutral loss because of the ease of control of an m/z difference between Q_1 and Q_2. On the other hand, the operation of this mode by the combination of E and B is complicated.

When a quadrupole mass analyzer is used as a collision cell, we do not use U, that is $a=0$, so that the scan line coincides with the Oq-axis. In this case all ions except those with smaller m/z ratios can pass through the quadrupole mass analyzer in which CID takes place.

2. Hybrid tandem MS

The term "hybrid" means the combination of B and/or E with other types of mass analyzer, the latter always being Q.

One combination is BEqQ where q means that a quadrupole mass analyzer is used as a collision cell. Three detectors are set after three mass analyzers (B, E and Q), respectively. There is another collision cell between B and E. This combination has two advantages: Firstly, it can perform the function of scanning product ions in several steps. Secondly, it has both high-energy CID and low-energy CID. Therefore, abundant information about ion structure can be obtained. Because the ions passing through B and E possess high kinetic energy, it is necessary to set a deceleration lens before q to decelerate ions.

3. The combination of B and E

One combination is BEEB. B_1 and E_1 can be used as two sequential mass analyzers or as a set of mass analyzers with a high resolution. The function of E_2 and B_2 is similar to that of B_1 and E_1. Between each two mass analyzers, a collision cell can be set, where CIDs are carried out.

4. TOF is used as the second mass analyzer

As the TOF records all ions at the same time, it has the advantage of high sensitivity. A trend in organic mass spectrometry is the use of TOF as the second mass analyzer in tandem mass spectrometry. Because ions pass continuously from the first mass analyzer, the TOF should be an oa-TOF (see Section 5.2.5).

One Q or a set of E and B can be employed as the first mass analyzer. A quadrupole or an octopole is employed as the collision cell.

Now we will discuss tandem-in-time mass spectrometry. In this case only one mass analyzer, an ion trap or an FT-ICR, is used so that the cost could be greatly reduced. In addition, n in the MS^n can be as high as 10, a value for tandem-in-time that is much greater than that in tandem-in-space.

The initial discussion will be on how an ion trap can perform this experiment.

1. According to the stability diagram, the ions of interest with a particular m/z value are selected to be stored in the ion trap by adjusting the appropriate U and V. All other ions are expelled from the ion trap (see Section 5.2.3).
2. Conduct CID. Since there is He gas in the ion trap, CID can be carried out by increasing the kinetic energy of the precursor ions. A tickle potential with a sine wave is applied at the end cap electrodes. When the frequency of the tickle potential is adjusted to the fundamental frequency of ion motion, the ions acquire energy from the tickle potential, which is a resonant excitation process. The excitation is carefully controlled so that the ions are just drawn out from the center of the ion trap to experience effective collisions with inert gas molecules. More excitation will lead to the expulsion of ions from the ion trap. The effective collisions of the ions with inert gas molecules lead to CID.
3. The mass spectrum is obtained by the scan of product ions, which is accomplished by the scan of V (see Section 5.2.3). The three steps form a circle. The operation is performed under the control of a computer. If necessary, one or several more circles can be carried out.

Because ions move continuously in the ion trap, the CID with the ion trap is efficient.

Now we will discuss tandem-in-time mass spectrometry with FT-ICR. The process is as follows:

1. As part of the procedure, the ions of interest with a particular m/z value are stored while all other ions are expelled. If SWIFT is applied, the m/z value is set as the window. All other ions acquire a lot of energy so that they are expelled from the cell. If chirp is applied, the operation is accomplished by two excitations. The first excitation expels all ions with a greater m/z value than the selected one. The second excitation expels all ions with a smaller m/z value than the selected one.
2. The stored ions are excited with the cyclotron frequency of the selected ions. The excitation is so controlled that the motion radius of the ions increases as much as possible (near the wall of the analytical cell). We can then rewrite Eq. (5-23):

   $$E = \frac{q^2 r^2 B^2}{2m} \tag{5-23}$$

   when $r = 3$ cm, $B = 4.7$ T and $m/z = 1000$ u, then $E = 800$ eV.

   Because FT-ICR is operated in a high vacuum, an inert gas is introduced through a pulsed valve to produce CID. After that, the inert gas is expelled and the system returns to a high vacuum.
3. Product ions are excited and detected so that their mass spectrum is obtained.

If necessary, the process is repeated. All processes are controlled by a computer.

Because ions move with cyclotron motion, the CID in FT-ICR is efficient. Obviously tandem-in-time mass spectrometry can only perform the product ion scan. Fortunately this scan is the most important of the three scans.

5.6 Combination of Chromatography and Mass Spectrometry

Being a very effective separation method, chromatography, is widely applied. However, the retention time obtained from chromatography gives no structural information. Mass spectrometry does give structural information with very high sensitivity. Therefore, the combination of chromatography and mass spectrometry makes an ideal method for the separation and identification of the mixtures of unknown compounds, which can perform the function that can not be achieved either by chromatography or mass spectrometry alone.

5.6.1 GC-MS

The separation of mixtures can be performed by capillary gas chromatography in which the carrier gas pressure is low. Therefore the combination of capillary gas chromatography and mass spectrometry is easy to achieve. The two sets of instruments can be connected directly and the carrier gas is withdrawn with a high-speed pump. The components of the mixture to be separated flow out sequentially from the capillary column and they are then ionized and detected by MS. EI and CI can often be applied alternately in the ionization to give both EI spectra and CI spectra.

The acquisition speed of the mass spectrometer must be rapid enough compared with the elution speed of the components so that the mass spectra can illustrate ion abundances correctly. Suppose that an acquisition starts at the beginning of an eluted peak and ends at the apex of the peak. The peaks in the lower m/z region of the mass spectrum will be suppressed because these ions have fairly low concentrations when they are ionized. If a mass spectrum is distorted, it cannot be used for mass spectrum retrieval.

Because the quadrupole mass analyzer has the advantages of rapid scan, reproducible mass spectra and low price, it is used most frequently in combination with GC.

A computer controls the operation. Acquisitions are carried out repeatedly. All data are stored on the hard disk. In addition to control, the computer has two additional important functions:

1. The treatment of the acquired data.
   i. The correction of the originally acquired data by background subtraction.
   ii. It can give a total ion current chromatogram, TIC, which is a chromatogram detected by the mass analyzer (not by an ordinary chromatographic detector).

iii. It can give a mass chromatogram and it can also give the chromatogram of the ions with a selected m/z value. For example, the components containing the benzene ring in a mixture can be seen when ions with an m/z value of 77 are selected.

This function can be extended to detect ions with two or more m/z values.

2. The retrieval of unknown mass spectra.

Please see Section 6.5.

5.6.2
LC-MS and LC-MS^n

The term LC-MS in fact implies HPLC-MS.

A prerequisite for the separation of a mixture by gas chromatography is that the components of the mixture are stable in their vapor phase. Many compounds do not meet this demand. They may not be vaporized or they may decompose during the vaporization. A rough estimation is that 80% of organic compounds are not suitable for separation by gas chromatography. However, they can be separated by liquid chromatography. It should be emphasized that HPLC is an ideal separation method for all organic compounds as well as many biological samples.

The combination of LC and MS is much more difficult than that of GC-MS because there are many more elution solvent molecules in HPLC than carrier gas molecules in GC (with the difference being several orders of magnitude). In spite of the difficulty, LC-MS has been developed as a routine separation-identification method [33]. For this combination, it is necessary to establish an interface between the HPLC and MS. The following items should be realized through the interface:

1. All molecules including those of the solvent and the analyte (the substance to be detected) are vaporized.
2. Solvents should be separated and eluted.
3. The sample molecules are ionized.
4. In some cases CID is carried out.

Some non-volatile compounds can be used as a buffer. These compounds cannot be eluted using the pumps of the combination. Therefore the system of vaporizing the solvents and the analyte can be of various designs, for example, a path in a "Z" configuration, to reduce contamination at the interface.

In fact, the two interfaces employed most frequently in LC-MS are electrospray ionization and atmospheric pressure chemical ionization, both of which were discussed in Section 5.3.

Because electrospray ionization and atmospheric pressure chemical ionization are very soft ionization techniques, quasi-molecular ions or those ions together with the multiply-charged ions including the sample molecule are obtained by LC-MS. It is important to obtain the molecular weight of an unknown compound. If a mass analyzer with a high resolution is used, the molecular formula can be deduced from the exact measured molecular weight.

Because the mass spectrum of LC-MS is obtained after the effective separation by HPLC, it (the mass spectrum) is favored when measuring the exact mass (see Section 6.1.7).

To remedy the deficiency of fragment ions, CID is applied in LC-MS. A potential is applied to the skimmers in the interface. The higher the applied potential, the more fragment ions there are.

Because the quadrupole mass analyzer has the advantages of rapid response and simplicity, it is frequently used in LC-MS, especially for the quantitative analysis of mixtures. Tandem three quadrupole mass analyzers are used in LC–MS–MS. In this case a third quadrupole mass analyzer is used for the detection of product ions produced by CID in the second quadrupole mass analyzer.

The ion trap is being applied more and more widely in LC-MS because it has the capacity for tandem-in-time mass spectrometry.

An oa-TOF can be used in LC-MS-MS as the second mass analyzer.

FT-ICR can be used in LC-MS. As it has the function of tandem-in-time mass spectrometry, the combination can work with LC–MS^n. In addition, the combination can offer data with a high resolution, which is very useful for the application described in Section 6.1.2.

The combination of LC-MS or LC-MS^n is controlled by a computer.

Multiple reaction monitoring (MRM) and selected reaction monitoring (SRM) are applied frequently in LC-MS-MS. Consider QqQ as an example. One type of particular precursor ion is selected from the first quadrupole, CID is carried out in the second quadrupole and product ions are detected through the third quadrupole. Only one type of product ion selected is detected in the SRM mode. Several types of product ions selected are detected in the MRM mode. A component in a mixture can readily be quantified by MRM or SRM.

5.7 References

1 M. B. Comisarow, A. G. Marshall, *J. Mass Spectrom.*, **1996**, *31*, 581–585.

2 R. W. Nelson, D. Dogruel et al., *Rapid Commun. Mass Spectrom.*, **1994**, *8*, 627–631.

3 E. D. Hoffmann, J. Charette, V. Stroobant, *Mass Spectrometry, Principle and Application*, John Wiley & Sons, Chichester, 1996.

4 H. Budzikiewicz, *Massenspektrometrie, Einführung*, 3. erweiterte Auflag, VCH, Weinheim, 1992.

5 J. R. Chapman, *Practical Mass Spectrometry*, 2nd Edn., John Wiley & Sons, Chichester, 1994.

6 M. E. Rose, R. A. W. Johnstone, *Mass Spectrometry for Chemists and Biochemists*, The University Press, Cambridge, 1982.

7 F. W. McLafferty, Ed., *Tandem Mass Spectrometry*, John Wiley & Sons, Chichester, 1983.

8 R. G. Cooks, J. H. Beynon, *Metastable Ions*, Elsevier, Amsterdam, 1973.

9 A. L. Burlingam, R. K. Boyd, S. J. Gaskell, *Anal Chem.*, **1996**, 68, 599R–651R.

10 M. M. Bursey, *Mass Spectrom. Rev.*, **1991**, *10*, 1–2.

11 P. E. Miller, M. B. Denton, *J. Chem. Educ.*, **1986**, 63, 617–622.

12 P. H. Dawson, *Quadrupole Mass Spectrometry and Its Applications*, Elsevier, Amsterdam, 1976.
13 N. W. McLachlan, *Theory and Applications of Mathieu Functions*, Dover Publications, New York, 1964.
14 R. E. March, R. J. Hughes, *Quadrupole Storage Mass Spectrometry*, John Wiley & Sons, Chichester, 1989.
15 J. F. J. Todd, *Mass Spectrom. Rev.*, **1991**, *10*, 3–52.
16 R. E. March, J. F. J. Todd, Eds., *Practical Aspects of Ion Trap Mass Spectrometry* Vols. I, II, CRC Press, Boca Raton, 1995.
17 G. C. Stafford, P. E. Kelly et al., *Int. J. Mass Spectrom. Ion Processes*, **1984**, *60*, 85–98.
18 A. G. Marshall, *Acc. Chem. Res.*, **1985**, *18*, 316–322.
19 B. Asmoto, Ed., FT-ICR/MS; *Analytical Applications of Fourier Transform Ion Cyclotron Resonance Mass Spectrometry*, VCH, Weinheim, 1991.
20 A. G. Marshall, T. C. L. Wang et al., *J. Am. Chem. Soc.*, **1985**, *107*, 7893–7897.
21 R. Chen, X. Cheng et al., *Anal. Chem.*, **1995**, *67*, 1159–1163.
22 R. S. Brown, J. J. Lennon, *Anal. Chem.*, **1995**, *67*, 1998–2003.
23 R. M. Whittal, L. Li, *Anal. Chem.*, **1995**, *67*, 1950–1954.
24 B. Spengler, *J. Mass Spectrom.*, **1997**, *32*, 1019–1036.
25 A. G. Harrison, *Chemical Ionization Mass Spectrometry*, CRC Press, Boca Raton, 1992.
26 C. Dass, *J. Mass Spectrom.*, **1996**, *31*, 77–82.
27 F. Hillenkamp, M. Karas et al., *Anal. Chem.*, **1991**, *63*, 1193A–1203A.
28 M. C. Fitzgerald, L. M. Smith, *Nucleic Acids Rev.*, **1994**, *22*, 117–140.
29 A. K. Shukla, J. H. Futrell, *Mass Spectrom. Rev.*, **1993**, *12*, 211–255.
30 I. A. Papayannopoulos, *Mass Spectrom. Rev.*, **1995**, *14*, 49–73.
31 E. D. Hoffman, *J. Mass Spectrom.*, **1996**, *31*, 129–137.
32 R. K. Boyd, *Mass Spectrom. Rev.*, **1994**, *13*, 359–410.
33 R. Willoughby, E. Sheehan, S. Mittrovich, *A Global View of LC–/MS*, Global View Publishing, Pittsburgh, 1998.

6
Interpretation of Mass Spectra

It is an important fact that a mass spectrum can provide information about both the molecular weight and the elemental composition of an organic compound, which all the other spectroscopic techniques cannot provide. In addition, a mass spectrum can give the structural information about structural units and their connections within the compound, although such information is not definitive. When only trace amounts of a sample are available, its mass spectrum may be the only information provided by spectroscopic techniques.

The information on fragment ions (in a broad sense) is discussed mainly on the basis of EI spectra, for which references [1] and [2], especially the former, are excellent. A few fragment ion peaks in mass spectra produced by soft ionization techniques or CID can be understood by referring to the interpretation of EI spectra.

For the limited space of this treatise, the emphasis of our discussion will be on generality, a knowledge of which is useful for the interpretation of the mass spectra of an unknown compound.

6.1
Determination of Molecular Weight and Elemental Composition

6.1.1
Determination of Molecular Weight by an EI Spectrum

The determination of the molecular weight of an unknown compound is achieved through recognition of the molecular ion peak in the spectrum.

It is difficult to recognize the molecular ion peak in the following cases:

1. The sample cannot be vaporized or it can be vaporized but with decomposition.
2. There are no intact ionized molecules on ionization.
3. Peaks from impurities in the sample may interfere with the recognition of the molecular ion peak, especially when the impurity vaporizes easily or it has a stable molecular ion peak.

Structural Identification of Organic Compounds with Spectroscopic Techniques. Yong-Cheng Ning
Copyright © 2005 WILEY-VCH Verlag GmbH & Co. KGaA, Weinheim
ISBN: 3-527-31240-4

4. When the sample consists of elements with several isotopes, the molecular ion peak is one peak of the isotope cluster.
5. A molecular ion peak may coexist with the peak of $M-1]^+$ or $M+1]^+$.

A molecular ion peak can be recognized with the help of the following facts:

1. The peak with the largest *m/z* could be the molecular ion peak. When it is one peak from an isotope cluster, it can be recognized according to specific rules (refer to Section 6.1.6).
2. A molecular ion peak can be recognized in terms of the relationship between itself and other peaks.
   - The logical loss of neutral fragments is the key to the recognition.
     There are no peaks in the region from M–4 to M–13 because neither four hydrogen atoms nor a CH group can be lost from a molecule without fragmentation. In fact, the peak of M–3 is also very rare. There are larger illogical neutral losses, which are difficult to memorize and limited in use.
   - A molecular ion should have a complete elemental composition. If an elemental composition deduced from any cluster in the mass spectrum is more complete than that of the supposed molecular ion peak, the peak cannot be the molecular ion peak.
   - The total mass of the multiply-charged ions after the number of charges is modified should be less than (or equal to) that of the supposed molecular ion.
3. The nitrogen rule
   The nitrogen rule can be stated as follows:
   If a compound contains an even number or no nitrogen atoms, its molecular weight will be an even number. If a compound contains an odd number of nitrogen atoms, its molecular weight will be an odd number.
   If a compound contains no nitrogen, the peak with the largest mass showing an odd number must not be the molecular ion peak.
   The nitrogen rule can be extended from the molecular ion to all other ions. Its generalized statement is presented as follows: the odd-electron ions containing an even number of nitrogen atoms have an even mass; the even-electron ions containing an even number of nitrogen atoms have an odd mass.
4. The intensity of a molecular ion peak depends on the type of compound structure. The determination of some functional groups by other spectroscopic techniques is helpful in the recognition of a molecular ion peak.
   In EI mass spectra, the intensities of molecular ion peaks can be divided into three classes:
   i. The following compounds give prominent molecular ion peaks whose intensities are arranged in the following order: aromatic compounds > conjugated polyenes > alkyl compounds > short chain alkanes > some sulfur-containing compounds.
   ii. Straight chain compounds, such as ketones, esters, aldehydes, amides, ethers and halogenides in general show molecular ion peaks.

iii. The following compounds with large molecular weights: aliphatic alcohols, amines, nitrates and nitrites have no molecular ion peaks, especially those with highly branched chains.

The above-mentioned three classes are given only as a rough classification.

5. Differentiation of a molecular ion peak from the peak of $M+1]^+$ or $M-1]^+$. Ethers, esters, amines, amides, cyanides, etc. may show a significant $M+1]^+$ peak while aromatic aldehydes, some nitrogen-containing compounds, etc. may have a strong $M-1]^+$ peak. The identification of these functional groups by other spectroscopic techniques can help differentiate the peak of $M^{+\bullet}$ from that of $M-1]^+$ or $M+1]^+$.

Sometimes the analysis of the relationships between fragment ion peaks around a large mass range is also helpful to determine the molecular ion peak.

6.1.2
Determination of the Molecular Weight from a Multiply-charged Ion Cluster in an ESI Spectrum

A compound with a large molecular weight could show a cluster resulting from the multiply-charged ions in a mass spectrum by using ESI (see Section 5.3.6). The cluster is usually situated in the range 500–3000 m/z. The molecular weight measured can reach 1×10^5 u or more.

The cluster of multiply-charged ion peaks in an ESI spectrum is shown in Fig. 6.1.

During ESI, sample molecules will associate with charged species, resulting in an "apparent" mass-to-charge ratio:

$$\frac{M+nX}{n} = \frac{m}{z} \tag{6-1}$$

where M is the molecular weight of the sample; n is the number of the charged species associated with the sample molecules; X is the mass of the charged species; and m/z is the "apparent" mass-to-charge ratio of the associated ions.

Because n can be a series of integers, the associated ions show a cluster in an ESI spectrum. M can be deduced from any two successive peaks.

We now select arbitrarily two successive peaks in the cluster. The right peak corresponds to n_1 while the left peak corresponds to n_2. Thus,

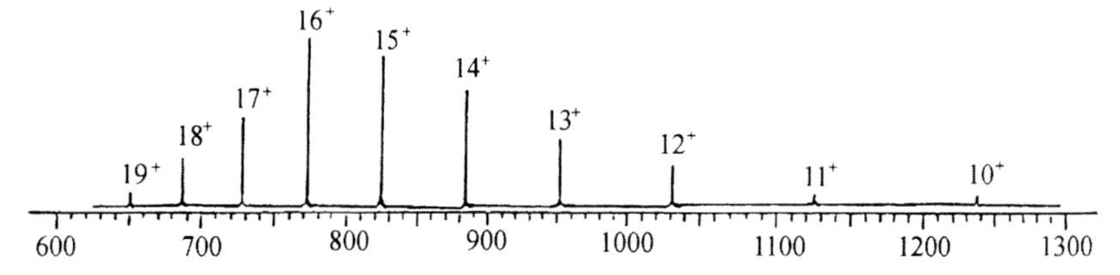

Fig. 6.1 A typical cluster of multiply-charged ion peaks in an ESI spectrum.

$$n_2 = n_1 + 1 \tag{6-2}$$

For simplicity, the m/z of the right peak is represented as m_1 and that of the left peak as m_2. According to Eq. (6-1), we have

$$\frac{M + n_1 X}{n_1} = m_1 \tag{6-3}$$

and

$$\frac{M + n_2 X}{n_2} = m_2 \tag{6-4}$$

The above two equations can be rewritten, respectively, as

$$M + n_1 X = n_1 m_1 \tag{6-5}$$

$$M + n_2 X = n_2 m_2 \tag{6-6}$$

Substituting Eqs. (6-6) and (6-2) into (6-5) results in

$$n_2 m_2 - n_2 X + n_1 X = n_1 m_1$$
$$(n_1 + 1) m_2 - (n_1 + 1) X + n_1 X = n_1 m_1$$

Rearranging, we have

$$n_1 m_1 - n_1 m_2 = m_2 - X$$

That is,

$$n_1 = \frac{m_2 - X}{m_1 - m_2} \tag{6-7}$$

As n is an integer, n_1 calculated from Eq. (6-7) is amended to an integer that is the closest to the calculated value.

M can be calculated from n_1 by using Eq. (6-3):

$$M = n_1 (m_1 - X) \tag{6-8}$$

where X is the mass of the charged species. When the pH value of the sample solution is low, the species are H^+ and $X = 1$.

Special software can be used to calculate any two successive peaks to give a more accurate result.

If two components exist in the solution to be analyzed, the overlap of the two clusters could make the calculation difficult. In this case, FI-ICR/MS has a clear advantage, because it just uses "one peak" of the clusters, which is in fact an isotope cluster in a high resolution spectrum. A compound contains a certain number of carbon atoms. The replacement of ^{12}C by ^{13}C results in an isotope cluster. The charge of an associated ion can be calculated from the isotope cluster. For example, if $\Delta m/z = 0.1$, we have $z = 10$ because $\Delta m = 1$ (a ^{12}C is replaced by a ^{13}C). The molecular weight can be calculated directly from the m/z and z.

6.1.3 Postulation of the Molecular Weight from a Spectrum Obtained Using Soft Ionization Techniques

Soft ionization techniques produce quasi-molecular ions (in positive ion mode or in negative ion mode). The molecular weight of an unknown compound can be postulated from the mass-to-charge ratio of quasi-molecular ion peaks.

Quasi-molecular ions produced by soft ionization techniques are closely related to the technique being used. The mass spectra produced by using CI, FAB, MALDI, ESI and APCI will be discussed in Section 6.6.

6.1.4 Determination of the Molecular Formula from High Resolution MS Data

The atomic weights of the isotopes that are important in organic chemistry are shown in Tab. 6.1.

Mass spectrometers with a high resolution can give exact masses with a high accuracy of mu or even higher. In addition to the limitation of the heteroatom number, the computer attached to the mass spectrometer can easily calculate the elemental compositions of important ions including the molecular ion. The mass accuracy achieved by FT-ICR can be as good as 1 ppm. Thus the determination of elemental compositions is straightforward [3].

Tab. 6.1 The atomic weights of some isotopes

Isotope	*Atomic weight*	*Isotope*	*Atomic weight*
^{1}H	1.00782504	^{31}P	30.9737634
^{2}H	2.01410179	^{32}S	31.9720718
^{13}C	13.0033548	^{34}S	33.9678677
^{14}N	14.0030740	^{35}Cl	34.9688527
^{15}N	15.0001090	^{37}Cl	36.9659026
^{16}O	15.9949146	^{79}Br	78.9183360
^{18}O	17.9991594	^{81}Br	80.9162900
^{19}F	18.9984033	^{127}I	126.904477
^{28}Si	27.9769284		

6.1.5
Peak Matching

High resolution data can be obtained by the peak matching method. For mass spectrometers with a sector magnetic field, remember

$$m/z = (r_m^2 B^2 e)/2V \tag{6-9}$$

To keep B and r_m invariable, we have

$$m_1 : m_2 = V_2 : V_1 \tag{6-10}$$

where m_1 is the exact mass of a known ion; m_2 is the exact mass of an unknown ion; V_1 and V_2 are two accelerating voltages for m_1 and m_2, respectively.

If ions with m_1 or m_2 enter a mass spectrometer continuously, two signals for m_1 and m_2 will be shown on the oscilloscope. By adjusting V_1 and V_2 carefully, the two signals can be superimposed exactly. The value of m_2 can be calculated from m_1, V_1 and V_2 by Eq. (6-10).

6.1.6
Postulation of the Molecular Weight from Low Resolution MS Data

Organic compounds always contain elements that possess more than one isotope so that their molecular ion peaks will show an isotope peak cluster that can be used to postulate elemental compositions.

The relative isotope abundances of common elements are listed in Tab. 6.2 in which the abundance of the most abundant isotope A (the lightest isotope) is set as 100. The isotopes whose atomic weights are heavier than A by 1 and 2 are noted as A+1 and A+2, respectively.

Tab. 6.2 Relative isotope abundances of common elements in organic chemistry

Element	*A*	*A+1*	*A+2*
C	100	1.11	
H	100	0.015	
N	100	0.37	
O	100	0.04	0.20
F	100		
Si	100	5.06	3.36
P	100		
S	100	0.79	4.43
Cl	100		31.99
Br	100		97.28
I	100		

Assume that an element possesses two isotopes and that m atoms of the element exist in a compound, the relative intensities of the cluster peaks can be calculated by the binomial $(a+b)^{\mathrm{m}}$.

$$(a+b)^m = a^m + ma^{m-1}b + \frac{m(m-1)}{2!}a^{m-2}b^2 + \ldots + \frac{m(m-1)\ldots(m-k+1)}{k!}a^{m-k}b^k + \ldots + b^m \qquad (6\text{-}11)$$

where a is the isotopic abundance of the lighter isotope; and b is the isotopic abundance of the heavier isotope.

Substitute the values of a, b and m into the binomial and each term can be calculated. As each term developed from Eq. (6-11) corresponds to a particular mass, the plus sign in the developed equation implies “coexistence” or proportionality. The first term, a^m, represents the peak intensity resulting from the molecules that consist of only the lightest isotope. The term a^{m-1}b represents the peak intensity of the molecules in which an A atom is replaced by an A+1 atom. The other terms can be considered in the same way.

If a compound contains m_i atoms which consist of i elements with two isotopes, the relative intensities of the cluster peaks can be calculated by Eq. (6-12):

$$(a_1+b_1)^{m_1}(a_2+b_2)^{m_2}(a_3+b_3)^{m_3}\ldots(a_i+b_i)^{m_i} \qquad (6\text{-}12)$$

where a_1, b_1 and m_1 correspond to the first element, and so on.

After the development of Eq. (6-12) the terms that correspond to the same mass should be added. The remaining plus sign has the same meaning as before.

The isotope cluster of a compound containing several halogen atoms can be illustrated clearly in a figure drawn in a way analogous to that which describes NMR splitting. An example of a compound containing $-BrCl_2$ is shown in Fig. 6.2.

From the illustration in Fig. 6.2, it is evident that the molecular ion peak may not be the strongest peak in the cluster. However, it will be the strongest peak in the cluster only if the compound does not contain halogen atoms or it contains less than 100 carbon atoms.

The relative intensities of the cluster peaks provide information about the elemental compositions.

1. The number of carbon atoms in a molecular formula can be estimated from the intensity ratio of the M peak to the M+1 peak:

$$n_{\mathrm{C}} \approx \frac{I(\mathrm{M}+1)}{I(\mathrm{M})} \div 1.1 \qquad (6\text{-}13)$$

where n_{C} is the carbon atom number in a molecular formula; $I(\mathrm{M}+1)$ and $I(\mathrm{M})$ are the relative intensities of the M+1 peak and the M peak, respectively.

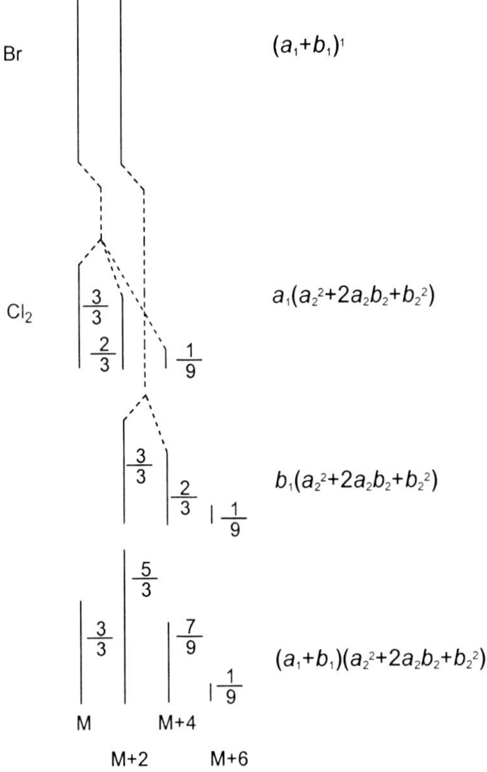

Fig. 6.2 The isotope cluster of a compound containing $-BrCl_2$.

Equation (6-13) should be used with some flexibility. An example is given as follows.

If the M peak has a lower intensity and the M–15 peak has a higher intensity, the intensities of the M–14 and M–15 peaks should be used for Eq. (6-13). Furthermore, one more carbon atom should be added to the calculated result since M–15 is formed from the loss of a methyl from the molecule.

2. The numbers Cl, Br and S atoms can be estimated from the intensity ratio of the M+2 peak to the M peak. However, the following factors should be considered.

From what has been described above, the number of C, Br, Cl and S atoms in a molecular formula can be estimated. The presence or numbers of other elements can be deduced from the following considerations:

i. The presence of fluorine can be verified by the M–20 (M–HF) peak and/or the M–50 ($M-CF_2$) peak.
ii. The presence of iodine can be verified by the M–127 peak. Furthermore, the presence of iodine leads to a smaller value of $I(M+1)/I(M)$.

iii. The presence of m/z 31, 45 and 59 ions indicates the existence of alcohol or ether groups. In addition, a considerable difference between the molecular weight and the combined mass of the known elements could indicate the presence of oxygen atoms.
iv. The number of H atoms can be calculated from the difference between the molecular weight and the mass of all the other elemental compositions.

Therefore, from the formulae, at least a partial elemental composition of an unknown compound can be estimated from the isotope cluster of its molecular ion peak.

6.1.7 Measurement of Exact Masses by a TOF or Quadrupole

The traditional concept is that exact masses can only be measured by a double focusing instrument or by FT-ICR. Since the late 1980s, the measurement of exact masses by other mass spectrometers has been tried with positive results [4].

A high resolution is absolutely necessary to differentiate the packed peaks. However, exact mass measurements can be performed with a relatively low resolution, provided every peak is composed of ions with a single elemental composition. If mass measurement is coupled with HPLC through ESI, this requirement will be satisfied. With a sufficient number of ions and a digital resolution on the m/z scale, exact mass measurement can be performed by using a quadrupole mass spectrometer [4].

A TOF analyzer possesses a higher resolution than the quadrupole mass analyzer. When oa-TOF is used in LC-MS through EPI, exact mass measurements can be achieved with commercial instruments.

6.2 Reactions and their Mechanisms in Organic Mass Spectrometry

6.2.1 Basic Knowledge

1. Labeling of the position of a charge in an ion
   When the charge can be localized on one particular atom, a “+” sign is shown on that atom. If the ion possesses an unpaired electron, it is noted as “+•”, for example, $CH_3–CH_2–CH_2–O^{+\bullet}H$.
   When the charge can not be localized on one particular atom or the labeling of its position is not necessary, square brackets “[ ]” or “]” should be marked at the end of the structural formula of the ion, for example, $[CH_3–CH_2–CH_2–OH]^{+\bullet}$ or $CH_3–CH_2–CH_2–OH]^{+\bullet}$.
2. Representation of the electron transfer
   The “fish hook” representation proposed by Djerassi has been widely adopted [5]. A half-arrow fishhook “⇀” indicates the transfer of a single electron while

a doubly barbed arrow " →" indicates the transfer of a pair of electrons. The cleavage of a chemical bond is classified into the following three types:

- Homolytic cleavage (**40**):

$$R'R''RC-\overset{+\bullet}{O}-H \longrightarrow R^{\bullet} + R'R''C{=}\overset{+}{O}H$$

40

- Heterolytic cleavage (**41**):

$$R-\overset{+\bullet}{O}-R' \longrightarrow R^{+} + {}^{\bullet}OR'$$

41

- Hemiheterolytic cleavage (**42**):

$$R\cdot + R' \longrightarrow R^{\bullet} + R'^{+}$$

42

3. Organic mass spectral reactions are mainly unimolecular
   These are usually initiated by functional groups containing heteroatoms or unsaturated bonds.
4. Primary fragmentation and consecutive fragmentation
   The fragmentation from a molecular ion or a quasi-molecular ion is called primary fragmentation, which is important for the interpretation of the mass spectrum of an unknown compound. The fragment ion (in a broad sense) can be further broken up, a process called consecutive fragmentation, which leads to a complex mass spectrum. The consecutive fragment ions possess less structural information than the primary fragment ions.

6.2.2
Simple Cleavage

A simple cleavage is the process in which only one chemical bond is cleaved so that the products are the structural units of the original molecule.

By a simple cleavage, a molecular ion (an odd-electron ion) produces a free radical and an ion. Therefore, it is inevitable that the ion is an even-electron ion.

Now we will discuss the mass of even-electron ions. If a molecule contains no nitrogen, the molecular weight must be an even number. A free radical produced from the molecule has a composition formula derived from C_nH_{2n+1}, which has a mass with an odd number. As an even number can be the addition of two odd numbers, the mass of the even-electron ion must be an odd number. If the molecule contains a nitrogen atom, the mass of the free radical or the even-electron ion, which contains the nitrogen atom, is an even number, while that of the one which contains no nitrogen atom is an odd number. A molecule containing more nitrogen atoms can be analyzed in a similar manner.

Initiation mechanism of the simple cleavage

According to McLafferty's interpretation [1], the mechanism of a simple cleavage falls into three types:

1. Radical site initiation (α-cleavage)

   This is the most important initiation mechanism, the driving force being the strong tendency of a radical to form electron pairing.

   Some examples are as follows.

   - Compounds containing saturated heteroatoms (**43**):

$$R'\!-\!CR_2\!-\!\overset{+\cdot}{Y}\!-\!R'' \xrightarrow{\alpha} R'^{\cdot} + R_2C\!=\!\overset{+}{Y}R''$$

43

   - Compounds containing unsaturated heteroatoms (**44**):

$$R'\!-\!CR\!=\!\overset{+\cdot}{Y} \xrightarrow{\alpha} R'^{\cdot} + CR\!\equiv\!\overset{+}{Y}$$

44

   - Compounds containing carbon–carbon unsaturated bonds (**45**):

$$R\!-\!CH_2\!-\!CH\!=\!CH_2 \xrightarrow{-e} R\!-\!CH_2\!-\!\overset{+}{C}H\!-\!\dot{C}H_2$$

$$\xrightarrow{\alpha} R^{\cdot} + CH_2\!=\!CH\!-\!\overset{+}{C}H_2 \rightleftharpoons \overset{+}{H_2C}\!-\!CH\!=\!CH_2$$

45

2. Charge-site initiation (inductive cleavage, *i*-cleavage), as in reaction (**46**)

$$R\!-\!\overset{+\cdot}{O}\!-\!R' \xrightarrow{i} \overset{+}{R} + {}^{\cdot}OR$$

46

In general, the importance of the *i*-cleavage is less than that of the *α*-cleavage.

The tendency of the *α*-cleavage generally parallels that of the radical site to donate electrons: N > S, O, *π*, $R^{\bullet}$ > Cl, Br > H, where *π* signifies an unsaturated site and $R^{\bullet}$ an alkyl radical.

In the *i*-cleavage an electron pair transfers to change the charge-site. The tendency of RY to form R^+ is: halogens > O, S >> N, C.

3. *σ*-bond dissociation (*σ*)

If a compound contains neither heteroatoms nor *π* bonds, the only fragmentation is the *σ*-bond dissociation (**47**):

$$R{-}R' \xrightarrow{-e} R{+}R' \xrightarrow{\sigma} R + R'^{+}$$

47

General rules of simple cleavages

Simple cleavages are more regular than rearrangements. Therefore their empirical rules have been summarized, which are very useful for the interpretation of mass spectra of many types of compounds.

1. There are three types of cleavage for compounds containing heteroatoms.
   i. The C–C bond next to the heteroatom is cleaved; or more generally, the bond to be broken is the bond that connects the *α*-carbon atom with a hydrogen atom, or the *β*-carbon atom, or another heteroatom.

   Fragment ions produced from this cleavage are frequently presented in mass spectra. In most instances the ions contain the heteroatom that initiates the fragmentation. The heteroatom can be saturated or unsaturated. The *α*-carbon may be in CH_2, CHR_1R_2 or $CR_1R_2R_3$.

   Some examples are presented in reaction in **48**.

$$R{-}CH(R'){-}\overset{+\bullet}{O}H \longrightarrow R^{\bullet} + R'HC{=}\overset{+}{O}H$$

$$R'{-}CH_2{-}\overset{+\bullet}{O}{-}R' \longrightarrow R'^{\bullet} + H_2C{=}\overset{+}{O}R'$$

$$R'{-}CR_2{-}\overset{+\bullet}{N}H_2 \longrightarrow R'^{\bullet} + R_2C{=}\overset{+}{N}H_2$$

48

$$CH_3-CH_2-\dot{F}^{+\cdot} \longrightarrow CH_3^{\cdot} + H_2C=\overset{+}{F} \longleftrightarrow H_2\overset{+}{C}-F$$

$$R-C(=O^{+\cdot})-R' \longrightarrow R^{\cdot} + \overset{+}{O}\equiv C-R'$$

$$R-C(=O^{+\cdot})-R' \longrightarrow R^{+} + \dot{O}\equiv C-R'$$

$$R-C(=O^{+\cdot})-O-R' \longrightarrow R-C\equiv\overset{+}{O} + \dot{O}R$$

$$R-C(=O^{+\cdot})-O-R' \longrightarrow R^{\cdot} + {}^{+}O\equiv C-OR'$$

$$R-C(=O^{+\cdot})-OH \longrightarrow R^{\cdot} + {}^{+}O\equiv C-OH$$

48 cont.

A particular example of the fragmentation is the formation of $M{-}1]^{+}$, **49**.

$$C_6H_5-C(=O^{+\cdot})-H \xrightarrow{-H^{\cdot}} C_6H_5-C\equiv\overset{+}{O}$$

49

Apart from the cleavage of the C–C bond next to the heteroatom, cleavages further away from the heteroatom also lead to the production of ions, probably through a complicated mechanism. For simplicity such cleavages are regarded as simple cleavages, for example **50**.

This approach is fairly important for carbonyl compounds.

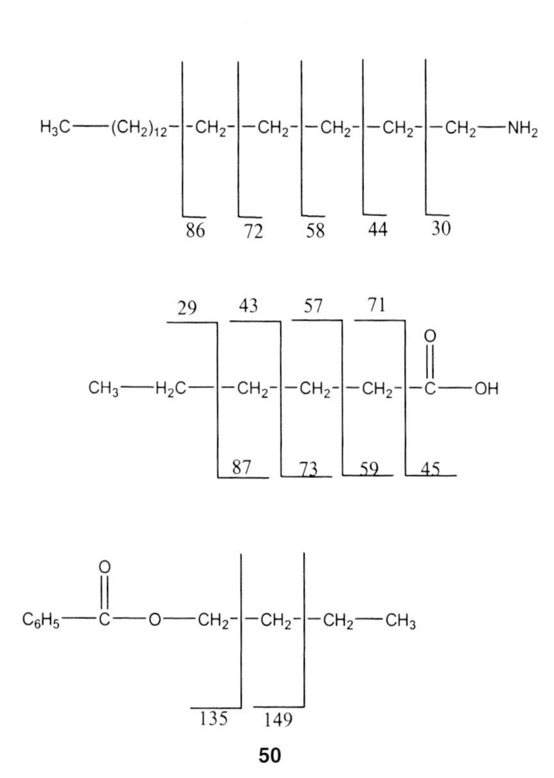

50

ii. The bond between the heteroatom and its α-carbon atom is cleaved and the charge is kept at the site of the alkyl, for example **51**.

$R{-}\overset{+\cdot}{O}{-}R' \longrightarrow R^{+} + {}^{\cdot}OR'$

$R{-}\overset{+\cdot}{S}{-}R' \longrightarrow R^{+} + {}^{\cdot}SR'$

$R{-}\overset{+\cdot}{Cl} \longrightarrow R^{+} + Cl^{\cdot}$

51

iii. The bond between the heteroatom and its α-carbon atom is cleaved and the charge is kept at the site of the heteroatom (**52** and **53**).

$$R{-}Br^{+\bullet} \longrightarrow R^{\bullet} + Br^{+}$$

$$R{-}I^{+\bullet} \longrightarrow R^{\bullet} + I^{+}$$

52

$$R{-}C(=O){-}OR'^{+\bullet} \longrightarrow R{-}\overset{O}{\overset{\|}{C}}{\cdot} + \overset{+}{O}R'$$

53

The cleavage in iii happens less frequently than those in i and ii.

Obviously, unsaturated heteroatoms which connect the α-carbon atom with σ and π bonds can only carry out the cleavage in i. Saturated heteroatoms can carry out reactions from i to iii, but only one reaction is apt to occur.

A If the heteroatom is situated in the upper part of the Periodic Table, especially in upper left part, the cleavage in i is apt to occur. On the other hand, if the heteroatom is situated in the lower-right part of the Periodic Table, the cleavage in iii tends to occur.

B If the heteroatom is connected with H, the cleavage in i usually takes place, while if it is connected with an alkyl, the cleavages in ii and iii are most likely to occur. Thus, $-NH_2$, –NHR and –OH generally only initiate the cleavage in i; –OR initiates those in ii and iii; halides frequently initiate that in iii and so on.

2. The C–C bond next to an unsaturated C=C bond can be easily cleaved, for example **54**.

$$R{-}CH_2{-}CH{=}CH{-}R' \xrightarrow{-e} R{-}CH_2{-}\overset{+\bullet}{CH}{-}CH_3{-}R'$$

$$\downarrow$$

$$\overset{+}{C}H_2{-}CH{=}CH{-}R' \longleftrightarrow CH_2{=}CH{-}\overset{+}{C}H{-}R' + R^{\bullet}$$

54

According to the resonance effect, the more resonant structures of an ion that exist, the greater the stability of the ion.

3. The C–C bond next to a phenyl can be easily cleaved, which is in agreement with 2 (**55**).

55

This is similar to the situation for heteroaromatic compounds (**56**).

56

4. A cleavage is possible at an alkyl-substituted carbon atom. The more of the carbon atom is substituted, the more easily a cleavage takes place, which is a consequence of the following cation stability order: $^{+}CR_3 > {}^{+}CHR_2 > {}^{+}CH_2R > {}^{+}CH_3$.
5. Saturated rings tend to lose alkyl side chains at the branch, which is a special case of 4 (**57**).

m/z 83

57

6. If several alkyls are connected with a carbon atom, the cleavage that results in the loss of the largest subsistent will be the most likely to occur.

With knowledge of the rules mentioned above, one can predict the existence of ions resulting from simple cleavages for a known structural formula, which is very useful for the interpretation of mass spectra.

6.2.3
Rearrangements

1. Characteristics of rearrangements

A rearrangement involves the cleavage of bonds and the formation of new bonds, that is, it requires changes to at least two bonds. A rearrangement produces ions that are not structural units of a precursor ion.

The most common rearrangement leads to the loss of a smaller molecule. As a molecular ion is an odd-electron ion, an ion rearranged from a molecular ion is also an odd-electron ion. If the molecule contains no nitrogen atom, the mass of the rearrangement ion will be an even number, while that of the ions resulting from a simple cleavage will be an odd number. Therefore, the rearranged ion can be differentiated from the ions resulting from a simple cleavage by the odd or even number of the ion mass. For example, the ion m/z 91 is produced from an alkyl benzene through a simple cleavage while the ion m/z 92 by a rearrangement. The mass difference of 1 u involves different mechanisms. Ions produced from compounds containing nitrogen atoms by rearrangements and simple cleavages can be similarly analyzed as above.

During rearrangements, the formation of new bonds partly compensates the energy loss of bond cleavages, and hence rearrangements have lower active energies than simple cleavages. If the bombardment energy in EI is decreased, the relative abundances of rearrangement ions will increase.

Unlike simple cleavages, rearrangements are so diverse and complex that it is difficult to summarize general rules for various functional groups. There are even random rearrangements. In spite of the difficulties, we can still manage to present some rearrangement rules to facilitate the interpretation of mass spectra.

2. McLafferty rearrangement

This rearrangement produces two types of ions. The general reactions are given as in **58**, where D=E is a double or triple bond; C is a carbon atom or a heteroatom; H is a hydrogen atom at the γ-position with respect to the unsaturated bond.

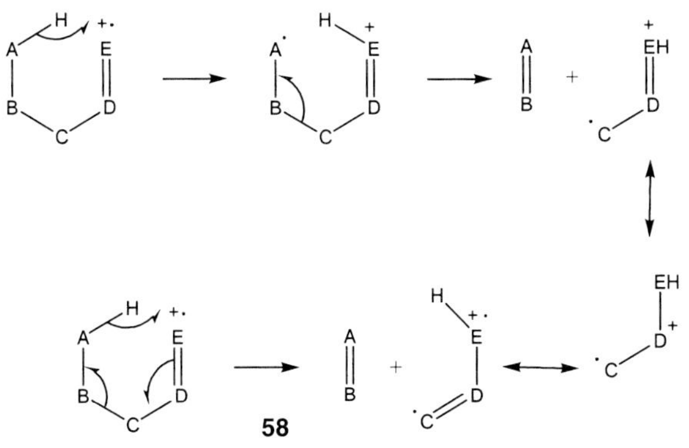

58

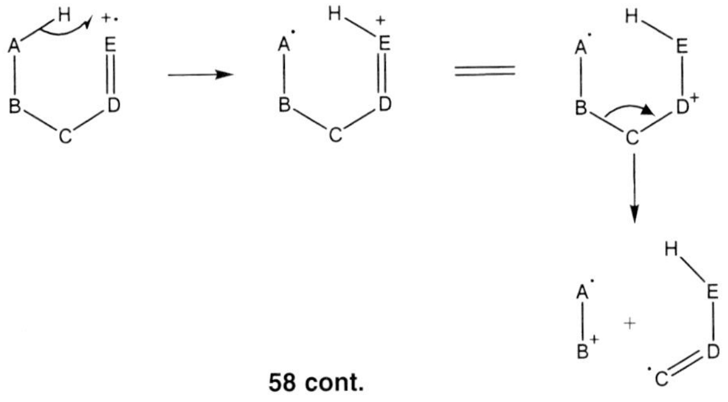

58 cont.

Because the hydrogen atom is close to the unsaturated bond, its transfer through a six-membered ring is favorable. If the specific condition (i.e., the existence of an unsaturated bond and its γ-H) is satisfied, this rearrangement may well take place. This rearrangement can also take place for a second time if the conditions for the rearranged ion are still satisfied.

Although it is possible to produce two types of ions by this rearrangement, the formation of the ion containing the unsaturated bond is slightly more probable.

The ions produced by the McLafferty rearrangement for common functional groups are listed in Tab. 6.3, which is useful for the recognition of such ions.

The masses of the rearrangement ions listed in Tab. 6.3 are those of the smallest rearrangement ions. For homologues, the masses of the rearrangement ions are the addition of $n \times 14$ to the value listed in Tab. 6.3, where n is a positive integer.

3. Retro-Diels-Alder reaction (RDA)

Compounds containing an unsaturated six-membered ring can undergo a retro-Diels-Alder reaction, which is named after the reverse reaction of a Diels-Alder reaction (**59**).

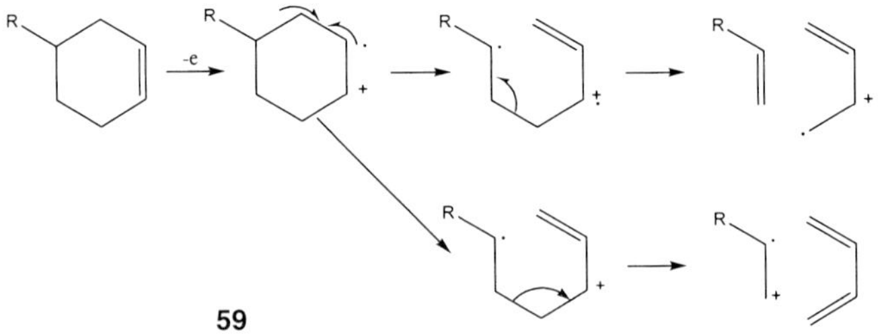

59

The formation of the ion containing the double bond is more likely.

If the conditions for RDA are provided, polycyclic compounds may undergo a retro-Diels-Alder reaction, for example **60**.

Tab. 6.3 The McLafferty rearrangement ions (the smallest mass number)

Compound	*Smallest rearrangement ion*	*Mass number of the smallest rearrangement ion*
Alkenes	$H_2\dot{C}-\overset{+}{C}H-CH_3$	42
Alkyl benzenes	$C_7H_8^{+\bullet}$ (ring, +•)	92
Aldehydes	$H_2\dot{C}-C(=\overset{+}{O}H)-H$	44
Ketones	$H_2\dot{C}-C(=\overset{+}{O}H)-CH_3$	58
Carboxylic acids	$H_2\dot{C}-C(=\overset{+}{O}H)-OH$	60
Carboxylates	$H_2\dot{C}-C(=\overset{+}{O}H)-OCH_3$	74
Amides	$H_2\dot{C}-C(=\overset{+}{O}H)-NH_2$	59
Nitriles	$H_2\dot{C}-C\equiv\overset{+}{N}H$ or $H_2C=C=NH^{+\bullet}$	41

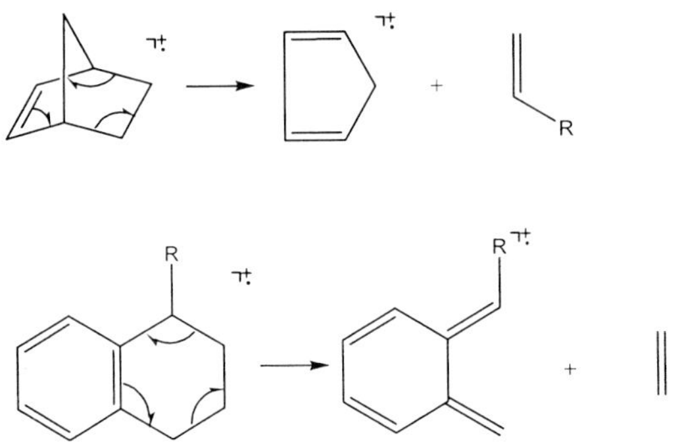

60

If a compound contains another functional group, which can initiate the fragmentation of the molecule, the RDA ions will not be abundant.

4. Compounds containing heteroatoms lose smaller molecules

For example, alcohol loses water or loses water and substituted ethylene simultaneously, **61**.

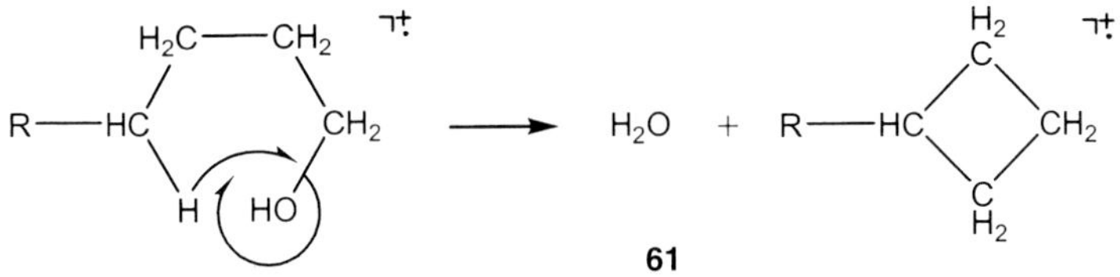

61

An alcohol with a straight chain (carbon number ⩾4) loses a water molecule and a substituted ethylene. Isotopic labeling shows that more than 90% of H_2O is lost through a 1,4-elimination reaction, that is, through the migration of a six membered ring, **62**.

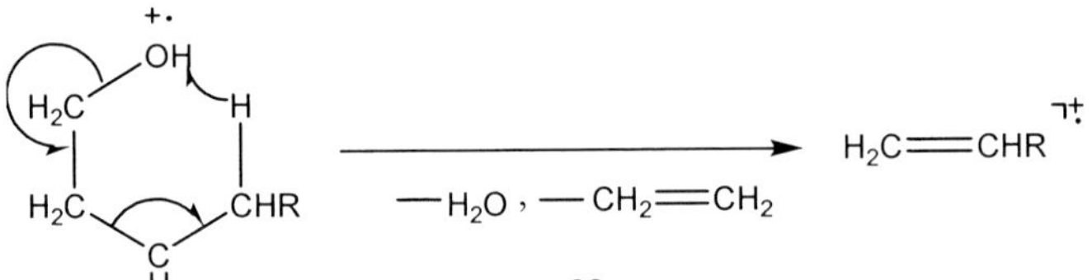

62

Another example is that a halide loses a hydrogen halide. Isotopic labeling also shows that for the loss of HCl from chlorides, 72% of the hydrogen atoms come from C-3 while 18% from C-4; and for the loss of HBr from bromide, 86% of hydrogen atoms come from C-2 to C-4.

Further examples are the loss of HCN from cyanides and the loss of H_2S from thiols. Other small molecules that may be lost include CH_3COOH, CH_3OH, $CH_2{=}C{=}O$, etc.

Aromatic or heteroaromatic rings substituted by a heteroatom may lose a small molecule (see Tab. 6.4). The loss of CO from quinones also belongs to this category, **63**.

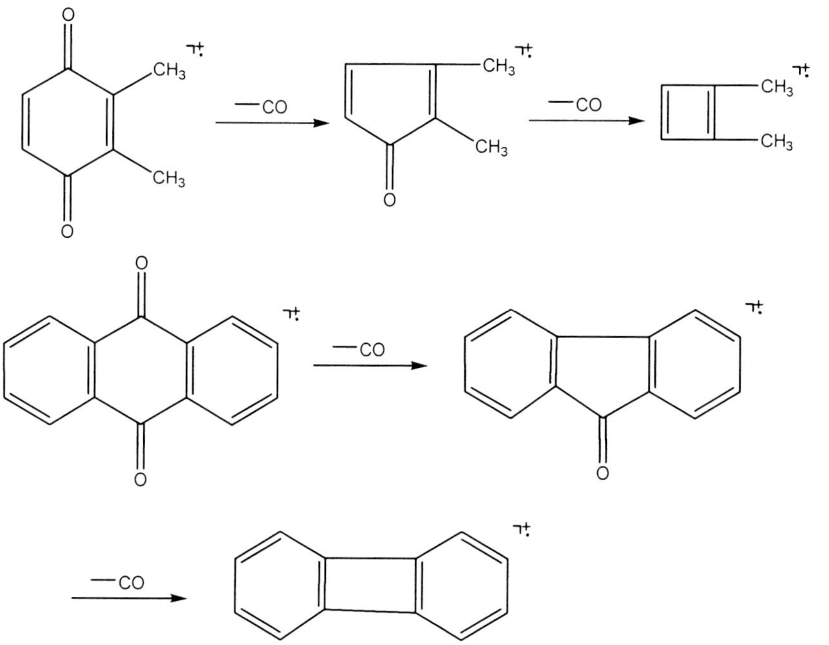

−CO −CO −CO −CO

63

The fragment ion containing carbonyl may lose a carbonyl, **64**:

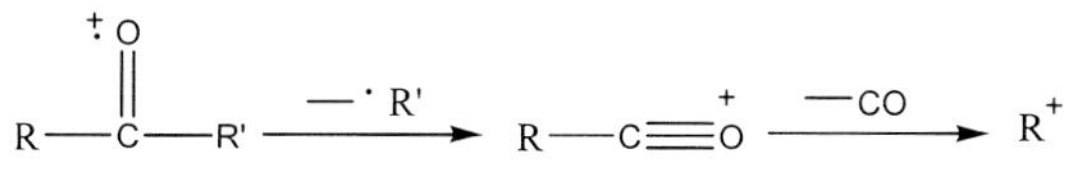

−·R' −CO R^+

64

It should be emphasized that two substituents at adjacent positions of a benzene ring can lose a small molecule. This is called an *ortho*-effect, whose general reaction is as in structure **65**.

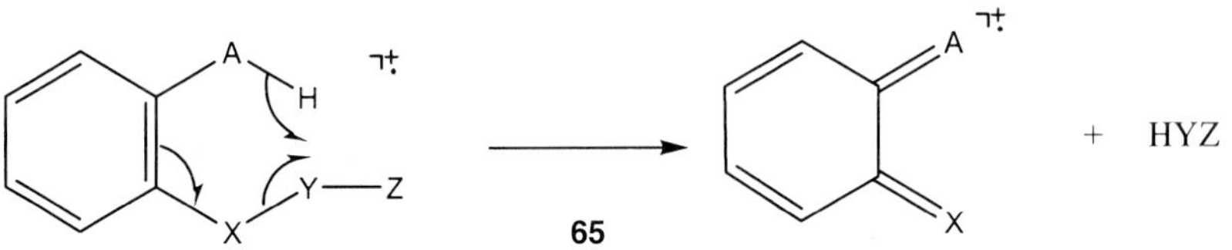

65

An example is shown in structure **66**.

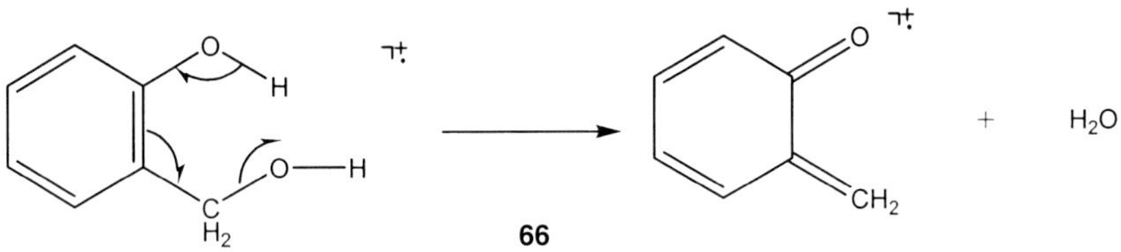

66

Heteroaromatic rings also show the *ortho*-effect.
Two *cis*-substituents of a double bond may undergo similar reactions.

5. Rearrangements through the transfer of a four-membered ring

This rearrangement includes the cleavage of X–C (where X represents a heteroatom) and the formation of X–H. The H migrates from a β-C to the heteroatom through a four-membered ring. Therefore, the alkyl connecting the heteroatom must have at least two carbon atoms; but the longer the chain is, the lower the probability of the rearrangement.

For the rearrangement to take place it is necessary that the compound contains a heteroatom. Although a molecular ion may undergo the rearrangement,

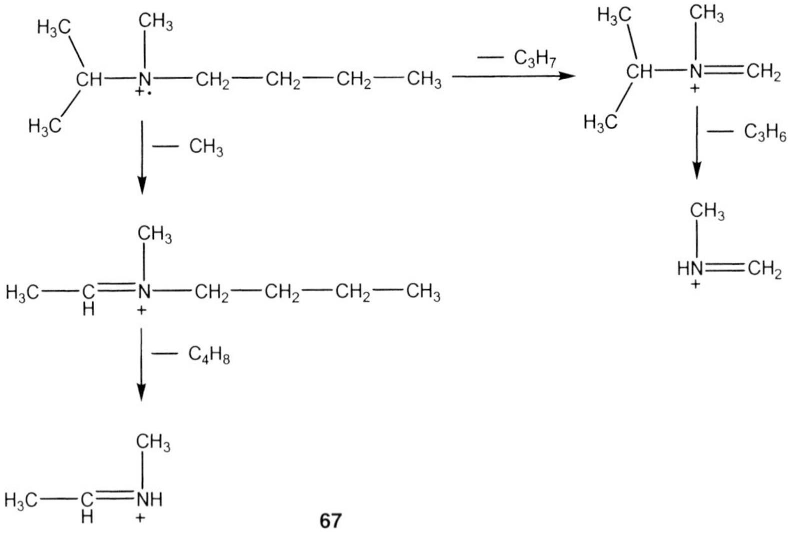

67

the probability is rather low. The rearrangement mainly takes place in fragment ions, which is an important way for the successive fragmentation of even-electron ions containing heteroatoms, **67**.

6. Rearrangement with two hydrogen atoms

Carboxylates, amides, phosphates, carbonates and other compounds containing unsaturated bonds can undergo this rearrangement, **68**.

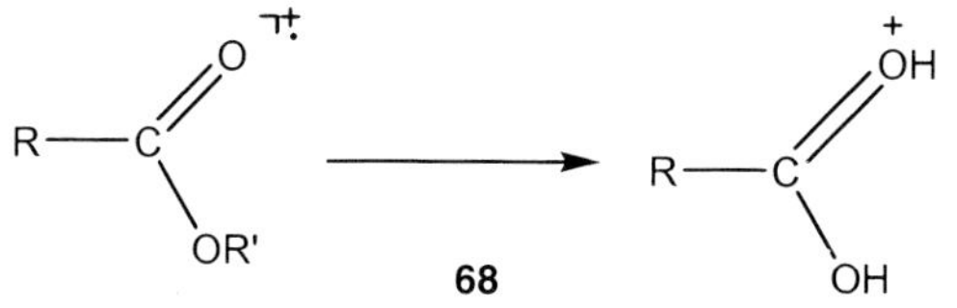

The ion produced by the rearrangement has a mass 2 u more than the ion formed by the corresponding simple cleavage.

7. Other rearrangements

A chain halide ion may form a cyclic ion by a rearrangement, **69**.

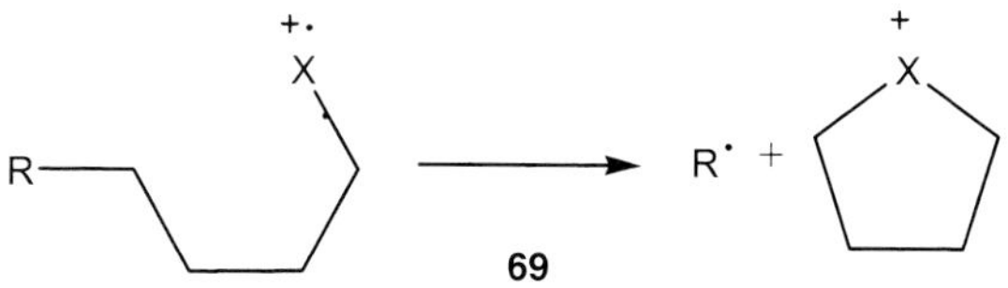

Note that the rearranged ion has a mass with an odd number as the molecular ion has lost a radical.

Chain amines and chain cyanides have similar rearrangements.

An alkyl ion can lose a hydrogen molecule. The mechanism has been proved by the measurement of its metastable ion.

6.2.4 Cleavage of Alicyclic Compounds

An alicyclic molecule must cleave two bonds in order to lose a fragment. This process may include the migration of a hydrogen atom through a six- or a five-membered ring. A typical example is the fragmentation of three isomers of methyl cyclohexanol.

- 2-methyl cyclohexanol, **70**

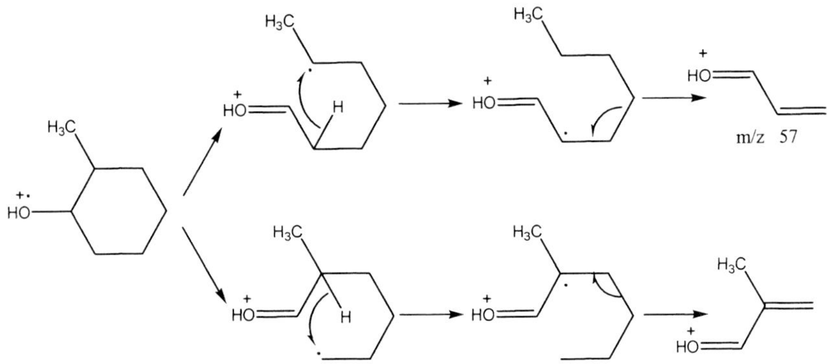

H_3C $HO^{+\bullet}$ → H_3C HO^+ H → H_3C HO^+ → HO^+ m/z 57

H_3C HO^+ H → H_3C HO^+ → H_3C HO^+ m/z 71

70

Because the stability of tertiary carbon radicals is greater than that of secondary carbon radicals, the intensity of the ion *m/z* 57 is greater than that of the ion *m/z* 71.

- 3-methyl cyclohexanol, **71**

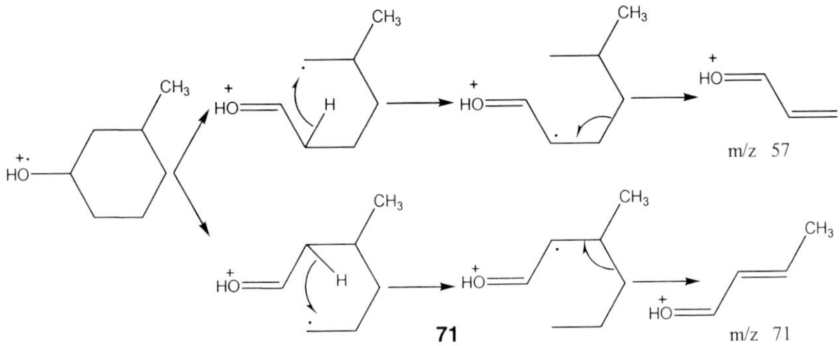

CH_3 $HO^{+\bullet}$ → CH_3 HO^+ H → CH_3 HO^+ → HO^+ m/z 57

CH_3 HO^+ H → CH_3 HO^+ → CH_3 HO^+ m/z 71

71

Because of the superconjugation of the methyl, the intensity of the ion *m/z* 71 is greater than that of the ion *m/z* 57.

- 4-methyl cyclohexanol, **72**
  Because of the symmetry of the molecule, two fragmentation routes are identical.

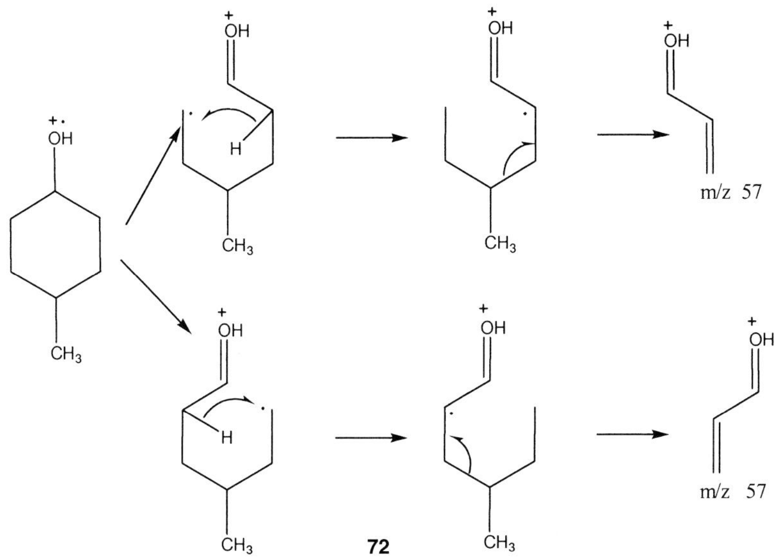

m/z 57

m/z 57

72

6.2.5
Consecutive Decompositions of Primary Fragmentation Ions

Fragment ions or rearrangement ions produced from molecular ions are primary fragmentation ions (in a broad sense), which will decompose consecutively to produce subsequent ions.

The decomposition depends on whether the electrons of the ions are paired or not. Generally, even-electron ions possess higher stabilities and lower activities because of their contracted electron orbits. On the other hand, odd-electron ions possess higher activities and lower stability, so they can expel a radical and form an even-electron ion, or a smaller molecule, thus producing a smaller odd-electron ion. In contrast, an even-electron ion can only expel a small molecule to produce a smaller even-electron ion. If an even-electron ion expelled a radical so as to form an odd electron, energywise the process would be unfavorable.

Note that primary fragmentation ions have more important structural information than consecutive fragmentation ions.

6.2.6
Stevenson-Audier's Rule

Stevenson-Audier's rule can be used to predict the charge retention during the decomposition of an odd-electron ion. The fragment with higher ionization energy is apt to retain the unpaired electron and thus become the neutral product, while the fragment with lower ionization energy is apt to form the ion. Note that the ionization energy is that of the radicals.

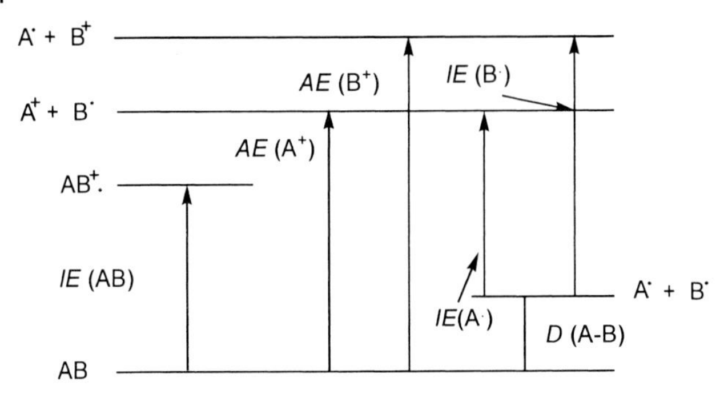

Fig. 6.3 Stevenson-Audier's rule.

The rule can be understood through consideration of the energy (Fig. 6.3). In the reaction $AB^{+\bullet} \rightarrow A^{+} + B^{\bullet}$ or $A^{\bullet} + B^{+}$, if the activation energy of the reverse reaction can be neglected and $IE(A^{\bullet}) < IE(B^{\bullet})$, the dominant products will be A^{+} and $B^{\bullet}$, where D(A–B) is the dissociation energy of the bond A–B.

Two examples of its application are as follows:

1. Reactions **73**

OH → + OH

IE=9.0eV IE=9.4eV

OH → OH +

IE=8.7eV IE=9.0eV

73

From the rule, it is easy to understand that RDA may produce different ions with the substitutions.

2. Reactions **74**

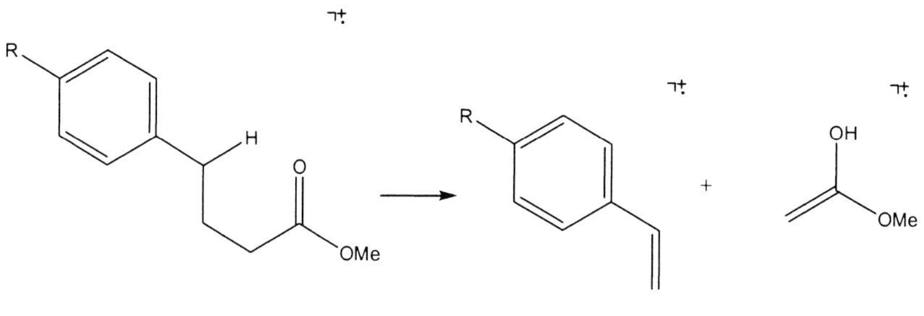

74 m/z=M-74 m/z 74

When R=NO_2, the ion m/z 74 is the base peak.
When R=OCH_3, the ion M–m/z 74 is the base peak.
When R=H, each ion has a relative intensity of 50%.
These reactions can be explained by the decrease in the ionization energy of M–74 with the increase in the electron-donating capacity of R.

This rule can also be applied to the reaction: $CD^{+\bullet} \rightarrow C^{+\bullet} + D$ or $C + D^{+\bullet}$.

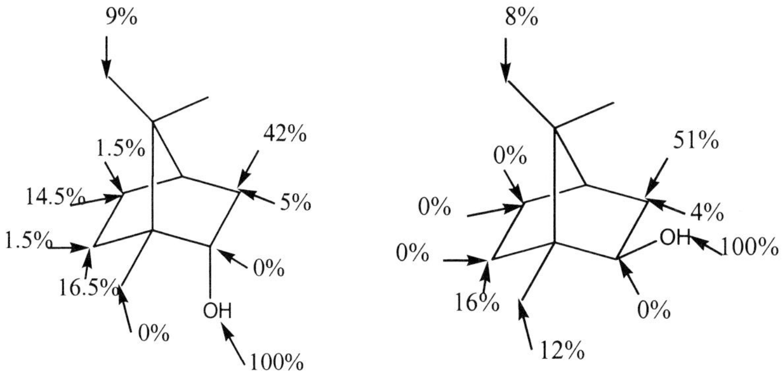

75

6.2.7
Methods to Study Reaction Mechanisms of Organic Mass Spectrometry

The most common methods concern metastable ions and isotopic labeling. As metastable ions were discussed in Section 5.4 in detail, isotopic labeling will be presented briefly in this section.

When compounds are synthesized with deuterium labeling at particular sites and their mass spectra measured, some fragment ions can be easily recognized by the comparison of such compounds with the unlabelled compound to find a mass increment of 1 u. The mechanism can be deduced from the ions containing deuterium atoms.

1. Removing water from a chain alcohol
   Experimentally, the removed water comes from the hydroxyl and the deuterium atom at position 3, 4 or 5 (mostly from position 4).
2. Removing water from monoterpenols
   The result (reaction 77) comes from two monoterpenols labeled with deuterium, which shows the influence of the steric structure on mass spectrometry reactions.

6.3 Mass Spectrum Patterns of Common Functional Groups

The interpretation of mass spectra is based mainly on empirical rules summarized from the mass spectra of known compounds. Therefore, it is useful to discuss mass spectra according to the functional groups in the sample. Because of the limited space we will interpret mass spectra according to the generalized structural classes. With this approach, along with the rules of simple cleavage and rearrangement, readers will be capable of interpreting general mass spectra.

6.3.1 Alkanes

Alkanes are very common organic compounds, with mass spectrum patterns that can be used to interpret the mass spectra of the compounds containing such groups.

Straight chain alkane

A typical mass spectrum of a straight chain alkane is shown in Fig. 6.4.

Its characteristics are as follows:

1. The straight chain alkane produces a weak molecular ion peak.
2. The mass spectrum consists of a series of clusters of peaks separated by gaps of 14 u. The elemental composition of the strongest peak in a cluster is C_nH_{2n+1}, the other peaks have elemental compositions of C_nH_{2n}, C_nH_{2n-1}, and so forth. C_nH_{2n-1} is produced by removing H_2 from C_nH_{2n+1}, which has been proved by its metastable ion.
3. The tops of the clusters form a smooth curve with the highest point at C_3 or C_4. Because every C–C bond has a certain cleavage probability and the produced ions can undergo consecutive cleavages, the C_3 or C_4 ion has the greatest abundance.

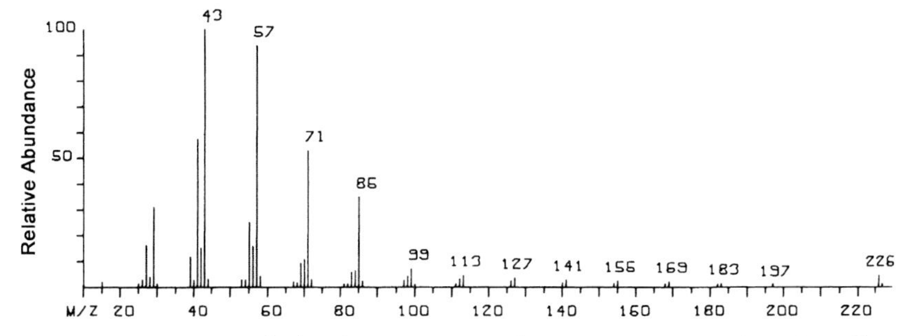

Fig. 6.4 Mass spectrum of n-hexadecane, courtesy of University Science Books, in McLafferty's *Interpretation of Mass Spectra*.

4. The top of the cluster next to the molecular ion peak is M–29, which is a very important criterion by which to distinguish a straight chain alkane from a branched one.

Branched alkanes

The mass spectrum of 5-methyl pentadecane is shown in Fig. 6.5.

Although mass spectra of branched alkanes are very similar to those of straight chain alkanes, they have the following differences:

1. Branched alkanes produce a weaker molecular ion peak than straight chain alkanes.
2. Because cleavages are likely to occur at a branched chain, the abundance of related ions will disturb the smooth curve, for example, the ion m/z 85 ($M{-}C_{10}$) in Fig. 6.5.
3. Noticeable C_nH_{2n} ions arise at a branched chain, whose intensities may sometimes be larger than those of C_nH_{2n+1}. They are produced from C_nH_{2n+1} with the loss of an H, for example, the ions m/z 168 and 140 in Fig. 6.5.
4. The branched methyl produces the ion M–15.
   Generally, the more branches a molecule has, the lower intensity the molecular ion possesses.

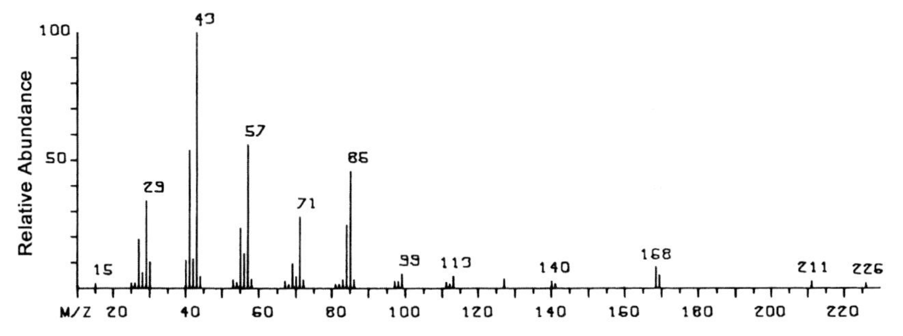

Fig. 6.5 The mass spectrum of the 5-methyl pentadecane, courtesy of University Science Books, in McLafferty's *Interpretation of Mass Spectra*.

Cycloalkanes

The mass spectra of cycloalkanes have the following characteristics:

1. Their molecular ion peak has a higher intensity than that of straight chain alkanes.
2. Cleavage at a side chain is common, which leads to the production of ion C_nH_{2n-1}. The loss of a hydrogen atom from C_nH_{2n-1} results in the production of the ion C_nH_{2n-2}.
3. The consecutive loss of C_2H_4 (or sometimes C_2H_5) is common.

6.3.2 Unsaturated Hydrocarbons

Unsaturated straight chain hydrocarbons

The mass spectra of unsaturated hydrocarbons have the following characteristics:

1. The molecular ion peak has a higher intensity owing to the existence of the double bond.
2. As with saturated hydrocarbons, there exist a series of peak clusters separated by 14 u, but the largest peaks in the clusters have the elemental composition of C_nH_{2n-1}, which results from the migration of the double bond along the chain.
3. The tops of the clusters form a smooth curve.
4. If a γ-H to an unsaturated group exists, the McLafferty rearrangement will occur.
5. *Cis-* and *trans-*isomers have similar mass spectra. The mass spectrum of 1-hexadecene is shown in Fig. 6.6.

Unsaturated and branched hydrocarbons

The mass spectra of unsaturated and branched hydrocarbons are similar to those of unsaturated straight chain hydrocarbons with two exceptions as follows:

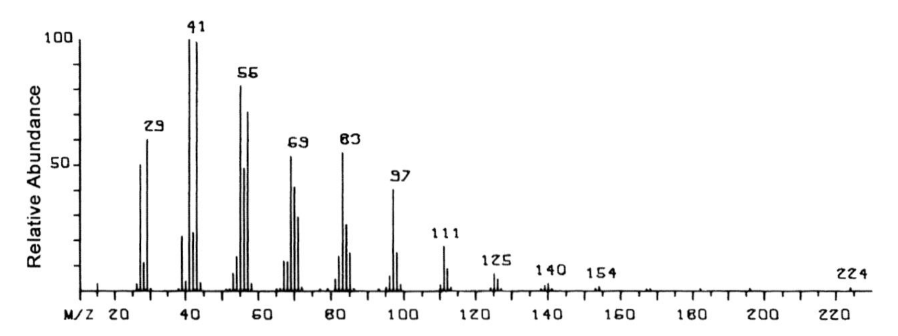

Fig. 6.6 The mass spectrum of 1-hexadecene, courtesy of University Science Books, in McLafferty's *Interpretation of Mass Spectra*.

1. Some tops of the clusters disturb the smooth curve because of the possible cleavages at branched chains.
2. If conjugated double bonds exist, they can not migrate along the chain.

Unsaturated alicyclic hydrocarbons

1. RDA will occur if the conditions are satisfied. RDA plays an important role in the mass spectra.
2. Cleavages at its side chains are probable.

6.3.3 Aliphatic Compounds Containing Saturated Heteroatoms

Their general formula is RX, where X is NH_2, NHR′, NR′R″, OH, OR′, SH, SR′, F, Cl, Br or I.

These compounds have a low or even no molecular ion peak. As the compounds contain an alkyl group, clusters of peaks separated by 14 u exist in the mass spectra. The heteroatom will initiate some fragmentations, as discussed in Section 6.2. Generalizations about these compounds will now be presented.

Simple cleavages

1. The C–C bond next to the heteroatom is cleaved with the charge kept at the site of the heteroatom. The general reaction is as in **76**, and for example **77**.

$$R''-C(R')(R''')-X^{+\cdot} \longrightarrow R'^{\cdot} + R''-C(R''')=X^{+}$$

76

$R-CH_2-NH_2$ (30) $R-CH(CH_3)-NH_2$ (44; M-15)

77

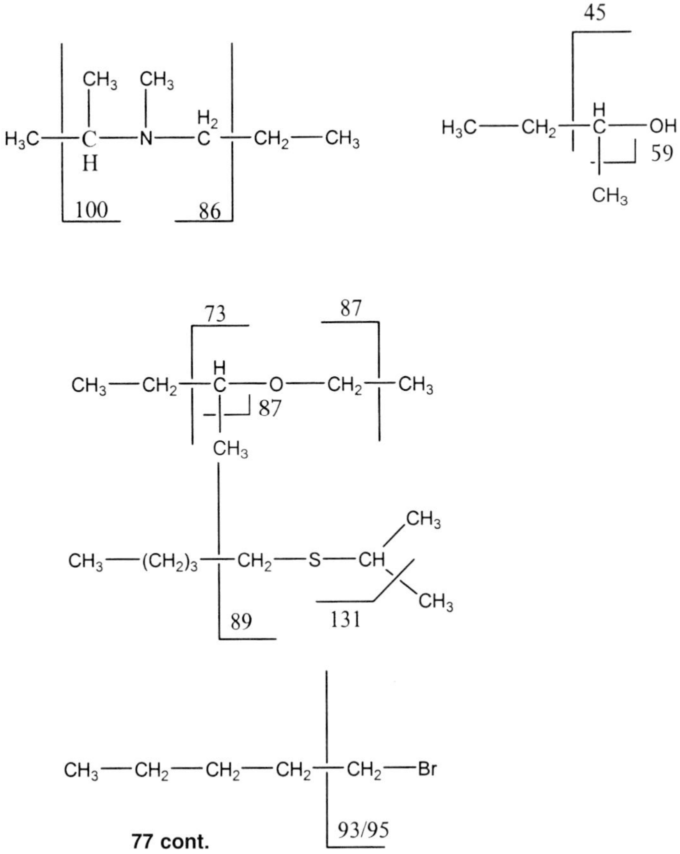

$H_3C—CH(CH_3)—N(CH_3)—CH_2—CH_2—CH_3$ (100, 86)

$H_3C—CH_2—CH(CH_3)—OH$ (45, 59)

$CH_3—CH_2—CH(CH_3)—O—CH_2—CH_3$ (73, 87, 87)

$CH_3—(CH_2)_3—CH_2—S—CH(CH_3)_2$ (89, 131)

$CH_3—CH_2—CH_2—CH_2—CH_2—Br$ (93/95)

77 cont.

As discussed in Section 6.2.2, this cleavage has a high probability. Therefore, amines produce $30+n\times14$ ions; alcohols or ethers $31+n\times14$ ions; and thiols or thioethers $47+n\times14$ ions, where n is an integer, whose value is determined by three factors: the number of carbon atoms that are connected to the heteroatom, the order and the substitution of the carbon atoms.

The ions produced by the cleavage between the α-C and β-C can undergo the four-membered rearrangement (see 5 in Section 6.2.3), giving rise to other ions. For example, a secondary alcohol can produce m/z 59 and 31 ions. The cleavages far away from the heteroatom can also lead to the production of other ions.

2. The bond between the heteroatom and its α-carbon atom is cleaved with the charge remaining at the alkyl site. The general reaction is **78**, and for example **79**.

$$R—X^{+\cdot} \longrightarrow R^{+} + X^{\cdot}$$

78

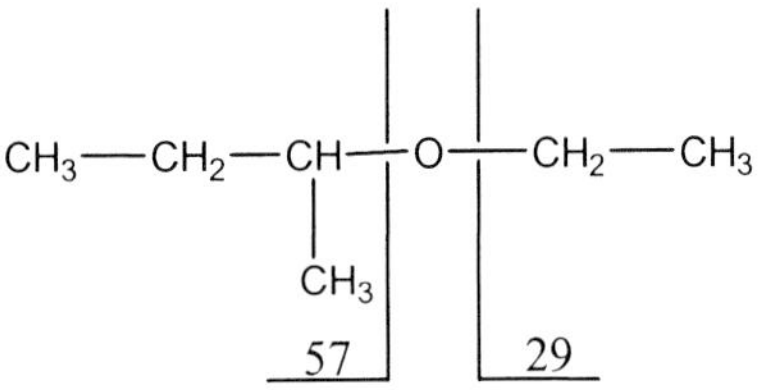

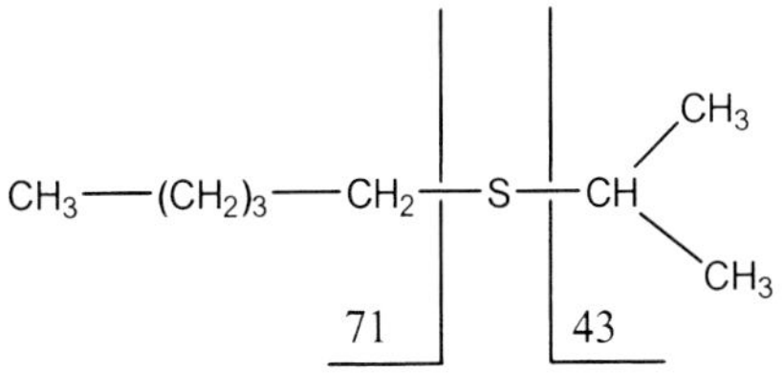

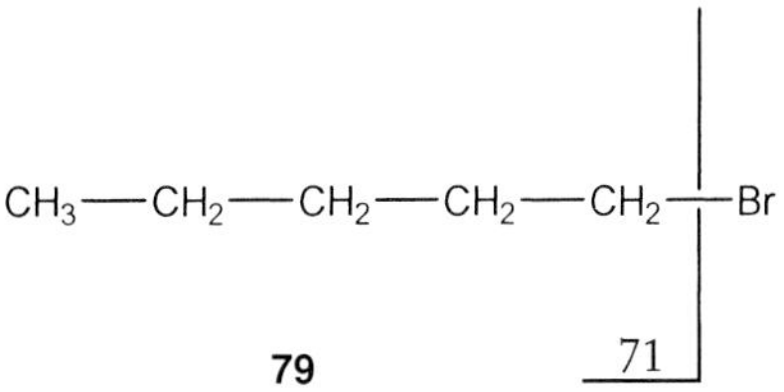

$CH_3—CH_2—CH(CH_3)—O—CH_2—CH_3$

57 29

$CH_3—(CH_2)_3—CH_2—S—CH(CH_3)_2$

71 43

$CH_3—CH_2—CH_2—CH_2—CH_2—Br$

79 71

3. The cleavage is the same as in point 2, but the charge is kept at the heteroatom site. The general reaction is **80**, and for example **81**.

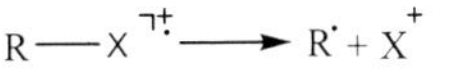

$R—X^{+\cdot} \longrightarrow R^{\cdot} + X^{+}$

80

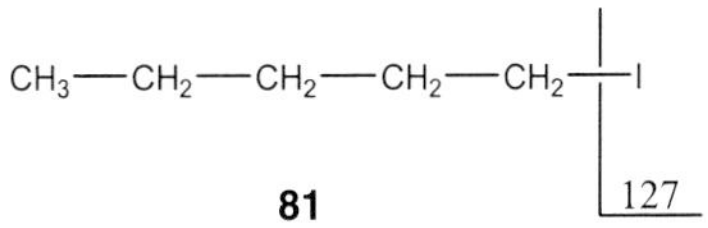

$CH_3—CH_2—CH_2—CH_2—CH_2—I$

81 127

As mentioned before, 3 is less common than 1 or 2.

Rearrangement

1. Rearrangement to lose a smaller molecule
   This rearrangement is very common for compounds containing saturated heteroatoms. For example: alcohols lose water or water and ethylene; thiols lose H_2S or H_2S and ethylene; halides lose HX (and H_2X in small amounts). However, primary amines are very likely to undergo the cleavage between α-C and β-C, and it is rare to lose NH_3 from primary amines.
2. Four-membered rearrangements
   These rearrangements usually start from ions produced by simple cleavages and occur consecutively.

3. Cyclization
   Chlorides and bromides with an alkyl longer than C_6 may undergo the cyclization. The general reaction is that in **82**.

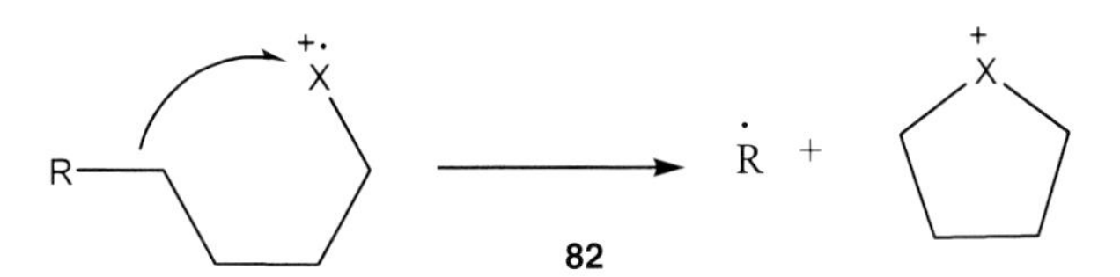

Using the above-mentioned rules, readers will be able to interpret important peaks of common mass spectra.

Now we will briefly summarize the mass spectra of cycloalkanes substituted by a heteroatom. Some fragmentations can be predicted as follows.

1. If the heteroatom is situated at a side chain, the fragmentation will be similar to that discussed under Simple Cleavages earlier in this section.
2. If the heteroatom is situated at the ring or connected to the ring, the cleavage between the α-C and β-C leads to rupture of the ring. The unpaired electron at β-C may initiate the loss of two carbon atom fragments from the ring.
3. If a carbon atom with an unpaired electron has a hydrogen atom γ- (or β-, δ-) next to it, the hydrogen atom usually migrates to combine with the unpaired electron and to leave another unpaired electron at its connected carbon atom. Consequently, the newly produced unpaired electron initiates a cleavage, which leads to the loss of a fragment from the ring.

6.3.4
Aliphatic Compounds Containing Unsaturated Heteroatoms

In these compounds the heteroatom is connected to adjacent carbon atoms by σ and π bonds. The general formulae are R′–C(R)=X, R–C≡X, and so forth, where R or R′ is an alkyl or another functional group.

The molecular ion peaks normally exist in the mass spectra for these compounds. Because of the connection between the unsaturated heteroatom and its α-carbon atom, cleavages at other bonds and some rearrangements may occur. The main reactions are listed below.

1. Reaction **83**, and sometimes the fragmentation leads to the production of an alkyl ion, for example, **84**.

$$\mathrm{R{-}C({=}X){-}R'} \longrightarrow \mathrm{R{-}C{\equiv}\overset{+}{X}} \quad \text{or} \quad \mathrm{\overset{+}{X}{\equiv}C{-}R'}$$

83

$$R-\overset{O}{\overset{\|}{C}}-R' \longrightarrow \overset{+}{R} + \cdot\overset{O}{\overset{|||}{C}}-R'$$

84

2. Cleavages that occur at bonds further away from the heteroatom (see Section 6.3.2).
3. McLafferty rearrangement.
4. Rearrangements that lead to the loss of a smaller molecule, such as those in which an aldehyde loses H_2O or CO, acetates lose the acetic acid, and so forth. The reactions mentioned above are common for these compounds.
5. Rearrangement of two hydrogen atoms, which occurs mainly among carboxylates.

Readers can interpret common mass spectra for this type of compounds by using these summarized rules. In this way the most important peaks can be assigned.

6.3.5
Alkyl Benzenes

The mass spectra of alkyl benzenes are characterized as follows:

1. A significant molecular ion peak.
2. A simple cleavage resulting in the production of a tropylium ion.
   If the benzene ring is connected to a CH_2 group, the peak of m/z 91 will be strong; and if the benzene ring is connected to a CH group, the peak will be discernable, **85**.

H_2C-R ┐+· → CH_2^+ → ⊕

85

3. McLafferty rearrangement, **86**

H_2C — CH_2 — CH (H, R) ┐+· → CH_2, H, H ┐+·

86

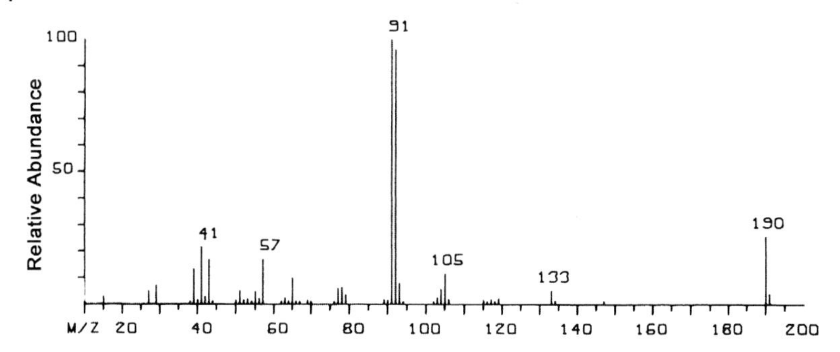

Fig. 6.7 The mass spectrum of n-octyl benzene, courtesy of University Science Books, in McLafferty's *Interpretation of Mass Spectra*.

If a γ-H relative to the benzene ring is present, the peak of m/z 92 will be considerable.

4. The fragment ions containing the benzene ring can successively lose C_2H_2.
   $m/z\ 91 \rightarrow m/z\ 65 \rightarrow m/z\ 39$
   $m/z\ 77 \rightarrow m/z\ 51$ (metastable ion m^*: 33.8)
   Thus, the mass spectra of phenyl derivatives usually show the peaks of m/z 39, 51, 65, 77, etc., as shown in Fig. 6.7.

The mass spectra of multi-ring aromatics are similar to those of alkyl benzenes except that the former have a stronger molecular ion peak.

6.3.6
Aromatic Compounds with Heteroatom Substitutions

Their mass spectra can be interpreted according to the substituted position.

1. The heteroatom substitutes the compound at its side chain.
   In this case, the mass spectrum is formed by superimposing that of an alkyl benzene on that of the substituted side chain.
2. The heteroatom is connected to the benzene ring directly.
   Here the most common reaction is the loss of neutral fragments that are, in general, produced by rearrangements. Tab. 6.4 is useful for finding the substituent.
   The fragments produced by simple cleavages are indicated in parentheses.
3. Multiple substitutions
   The *ortho*-effect resulting from two substituents on an aromatic ring (see Section 6.3.3) can be used to differentiate *ortho*-isomers from the *para*- and *meta*-isomers. The mass spectra of the *para*- and *meta*-isomers can be roughly predicted from Tab. 6.4.

Tab. 6.4 Lost neutral fragments from phenyl derivatives by rearrangements [6]

Substituent	*Lost neutral fragments*
$-NO_2$	NO, CO, (NO_2)
$-NH_2$	HCN
$-NHCOCH_3$	C_2H_2O, HCN
–CN	HCN
–F	C_2H_2
$-OCH_3$	CH_2O, CHO, (CH_3)
–OH	CO, CHO
–SH	CS, CHS, (SH)
$-SCH_3$	CS, CH_2S, SH, (CH_3)

6.3.7
Heteroaromatic Compounds and their Derivatives

The mass spectra of these compounds show their significant molecular ion peaks. Pyridine, pyrrole, furan and thiophene may lose the fragment containing the heteroatom so as to produce the ion *m/z* 39, which is identical to *m/z* 39 produced from benzene. Five-membered heteroaromatics also produce the fragment:

where X represents NH, O or S.

The mass spectra of the substituted heteroaromatics can be interpreted in the same manner as those of the substituted benzene.

6.4
Interpretation of Mass Spectra

The purpose of this treatise is to deal with the structural identification of organic compounds mainly by NMR, and the aim of this section is to show how mass spectra can be used to obtain as much structural information as possible. If a compound is fairly simple, it is possible to find several probable structures from its mass spectrum.

The discussion in this section is aimed mainly at EI spectra that have the most abundant fragment ions and thus are useful for deducing an unknown structure. A few fragment ions in mass spectra obtained by soft ionization techniques can be interpreted by those in EI spectra.

6.4.1
Steps of the Interpretation

1. Determination of the molecular ion peak
   Refer to Section 6.1.1. If there is no molecular ion peak in an EI spectrum, data from soft ionization are necessary.
2. A glance at the mass spectrum
   The structural type of the unknown compound may be judged from the intensity of the molecular ion peak, the number of fragment ion peaks and the ion series in the lower *m/z* region. For example, an aromatic compound has a significant molecular ion peak, a small number of fragment ion peaks and the ion series of *m/z* 39, 51, 65, 77, etc.
3. Determination of the molecular formula
   See Section 6.1.6.
   The index of hydrogen deficiency can be calculated immediately from the molecular formula.
4. Interpretation of important ions
   i. Ions close to the high mass edge
      The importance of these ions is much greater than that of ions with smaller masses because the former have a closer relationship with the molecular ion than the latter. Whether these ions are produced by simple cleavage or rearrangement, they possess some structural information about the unknown compound.
   ii. Rearrangement ions
      All discernible rearrangement ions show more structural characteristics. Thus, interpreting rearrangement ions is an important way of deducing an unknown structure.
      Most rearrangement ions are odd-electron ions, which are distinguishable from ions produced by simple cleavages.
   iii. Metastable ions
      The parent ion and its daughter ion can be found from the related metastable ion, which is very important for deducing an unknown structure. The metastable ion of the molecular ion is even more important.
   iv. Characteristic ions
      These ions show the characteristics of unknown compounds. For example, all phthalates produce a strong peak at *m/z* 149, **87**.

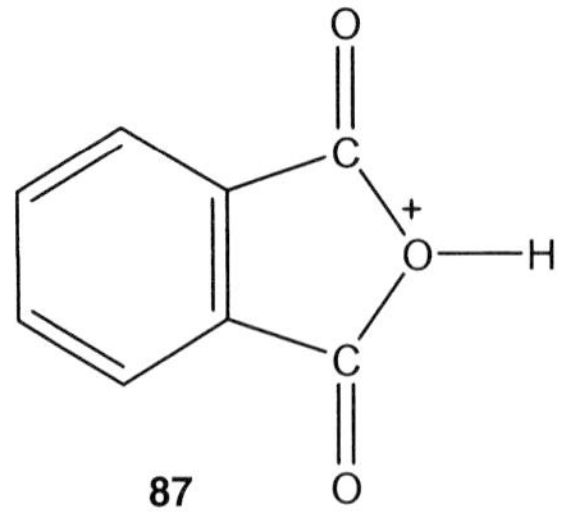

O
C
+
O—H
C
O
87

5. Deduction of the structure of the unknown compound
   We begin with some structural units and then extend them as far as possible. Some examples will follow.
6. Assignment
   At least most of the important peaks in the mass spectrum (the base peak, peaks close to the high mass edge, rearrangement ion peaks and strong peaks) should be interpreted reasonably.

   However, the peaks predicted from the deduced structure should exist in the mass spectrum. Reasonable assignment increases the probability of the correctness of the deduced structure.

6.4.2
Examples

Example 1
The molecular formula for an unknown compound is $C_4H_{11}N$. Try to postulate seven isomeric structures from the relative abundances of the isomers shown in Tab. 6-5.

Solution
It is possible to draw all the eight isomeric structures according to the formula. For simplicity, only the skeletons are drawn in **88**.

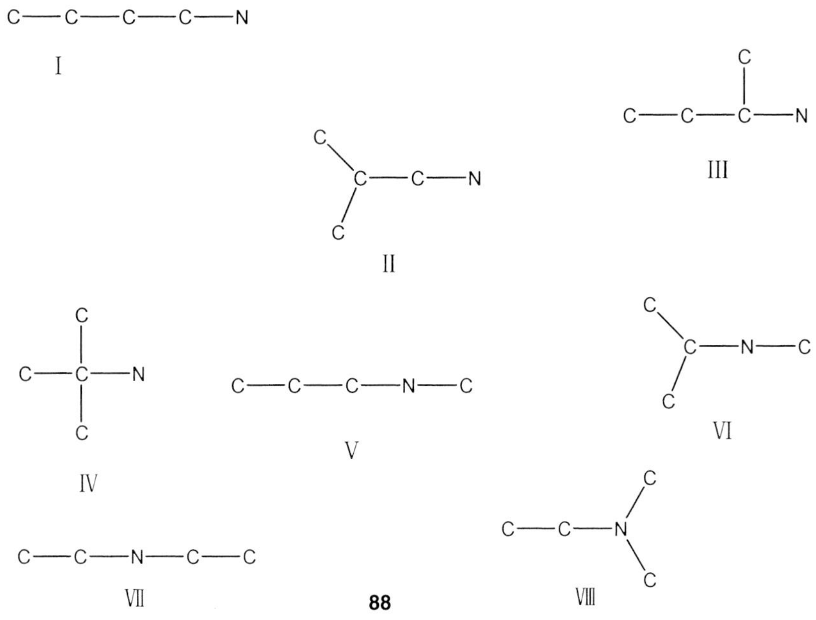

C—C—C—C—N
I

C(C)—C—C—N
II

C—C—C(C)—N
III

C—C(C)(C)—N
IV

C—C—C—N—C
V

C(C)—C—N—C
VI

C—C—N—C—C
VII

C—C—N(C)—C
VIII

88

Tab. 6.5 Relative abundance of isomers

m/z	*A*	*B*	*C*	*D*	*E*	*F*	*G*
30	13.7	73.6	29.2	10.9	100	100	11.6
43	7.1	3.6	7.3	4.1	<2	<3	4.3
44	24.9	29.4	9.4	100	<2	<3	<1
58	100	100	100	10.7	<2	0	100
72	19.3	19.1	10.5	2.8	<2	<2	0
73	23.4	31.5	11.4	1.0	4.9	10.1	0

Each set of data can be correlated with an isomer.

This example is a logical exercise for interpreting the mass spectra of the compounds containing saturated heteroatoms. For these isomers, the nitrogen atom will initiate fragmentation. Therefore one bond of the α-carbon atom except the bond of C–N will be broken off, thus forming a stable even-electron ion by losing any of CH_3, C_2H_5 and C_3H_7, or even an H atom. Because an even-electron ion can not produce an odd-electron ion by losing a radical group, the successive fragmentations must be the four-membered rearrangements.

As the molecular weight is 73, the mass discrimination effect is not serious.

The assignment starts from the base peak, followed by other important peaks.

From the data, it follows that four isomers (A, B, C and G) have the base peak at *m/z* 58, and that five structures (III, IV, VI, VII and VIII) may lose a methyl so as to produce the ion *m/z* 58. However, III will also produce the abundant ion *m/z* 44, which does not agree with any of the sets of data. Thus III can be eliminated at this step.

It is easy to assign IV as G because the carbon atom adjacent to the nitrogen atom is connected to three methyls, and IV loses a methyl easily so as to produce a strong peak at *m/z* 58 with its molecular peak missing.

In the three remaining sets of data (A, B and C), the abundances of *m/z* 30 which show the probability of undergoing four-membered ring rearrangements, can be used for the assignment.

VII can undergo the rearrangement at both sides of the nitrogen atom. Each of the two rearrangements leads to the production of the ion *m/z* 30. Thus VII can be assigned as B, which has the highest abundance of the ion *m/z* 30.

VI can produce the ion *m/z* 30 by two successive rearrangements. Firstly it rearranges at the right side to produce the ion *M*–1, which rearranges at its left side so as to produce the ion *m/z* 30. From the medium intensity of the ion *m/z* 30 of isomer C, VI can be assigned as C.

No rearrangements of VIII can lead to the production of the ion *m/z* 30. Instead, VIII can undergo two successive rearrangements to form the ion *m/z* 44, which suggests the assignment of VIII as A.

Only isomer D has the base peak at *m/z* 44, which agrees with III and V. As V can not produce the ion *m/z* 58, only III satisfies D (the base peak *m/z* 44 and a peak of *m/z* 58).

Finally, only isomers E and F are left, both of which are characterized by the base peak at *m/z* 30. The two remaining structures I and II agree with the base peak. Because II has a methyl branch, it must have a lower molecular ion intensity. Therefore, II can be assigned as E and I as F.

Example 2

Try to deduce the structure of an unknown compound from Fig. 6.8 and the following data: *m/z* 222, 223 and 224 with relative abundances 3.0, 0.4 and 0.04, respectively.

Solution

It can be concluded that the peak at *m/z* 222 meets all the requirements for the molecular ion peak.

From the intensities of the molecular peak cluster, S, Cl and Br are excluded from the molecular formula, and the carbon atom number is 12. The left mass suggests that the unknown has four oxygen atoms. Therefore, the molecular formula is $C_{12}H_{14}O_4$, and it can be established that $\Omega = 6$.

From the ion series of phenyl (*m/z* 39, 65, 77, etc.), the value of Ω, and a relatively small number of fragment ion peaks in the spectrum, it follows that the compound must contain a benzene ring.

The key to interpreting this example lies in the base peak at *m/z* 149, which is the characteristic peak (base peak) of all phthalates, **89**.

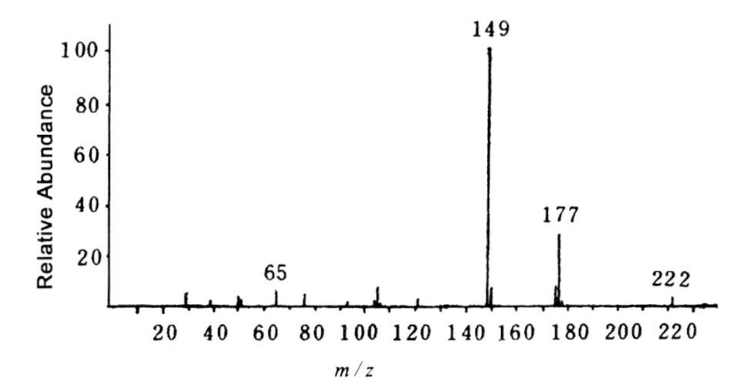

Fig. 6.8 The mass spectrum of an unknown compound.

O
C
+
O—H
C
O
89

Tab. 6.6 Three sets of mass spectral data

$M^{+\bullet}$ (Intensity %)	Base peak m/z	The second strongest peak m/z	The third strongest peak m/z	The fourth strongest peak m/z	The fifth strongest peak m/z
1 154 (12.8)	84	139	93	83	41
2 154 (0)	121	93	95	43	136
3 154 (24.4)	112	69	41	53	139

If its molecular weight is also considered, we can deduce the structure as diethyl phthalate.

Example 3

Try to correlate the following three structures with three sets of mass spectrum data (Tab. 6.6 and structures **90**).

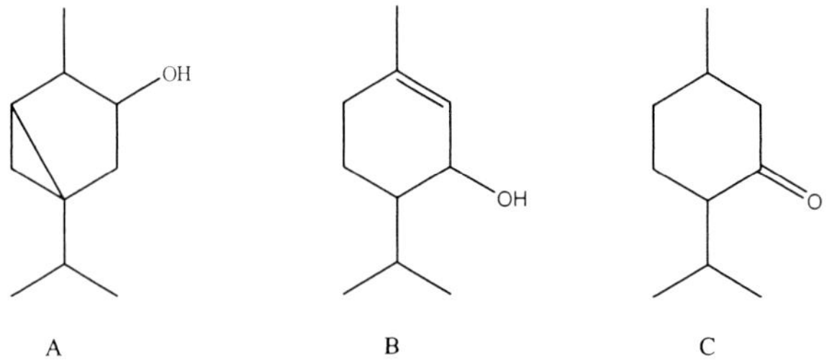

A B C

90

Solution

The first set of data corresponds to structure B. The RDA reaction **91** leads to the production of the base peak at m/z 84. In addition, a medium intensity of the molecular ion peak agrees with the reaction **91**.

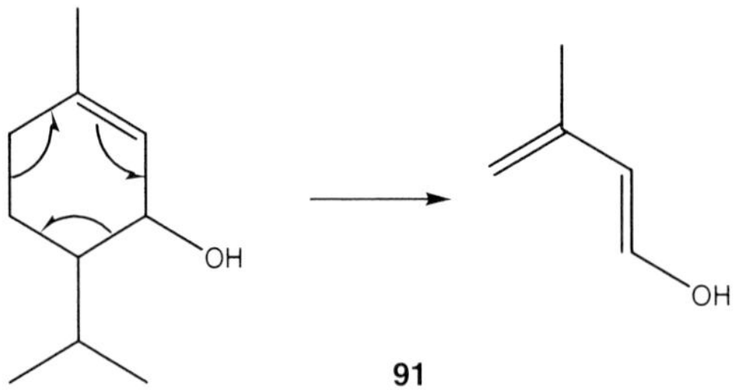

91

The second set of data corresponds to structure A, which has no molecular ion peak. M–H_2O–CH_3 produces the base peak and M–H_2O–C_3H_7 (the side chain) forms *m/z* 93.

The third set of data corresponds to structure C, because a ketone possesses a stronger molecular ion peak than an alcohol. Reaction **92** also helps draw the same conclusion.

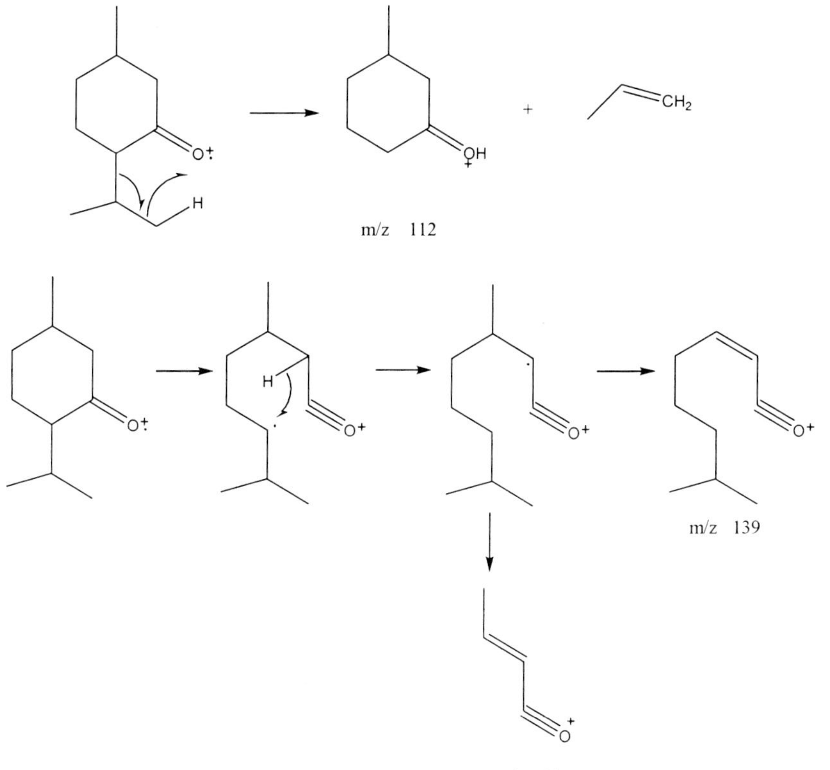

CH_2

m/z 112

m/z 139

m/z 69

92

Example 4

Try to explain Fig. 6.9.

Solution

The compound has its main fragmentations as in sequence **93**.

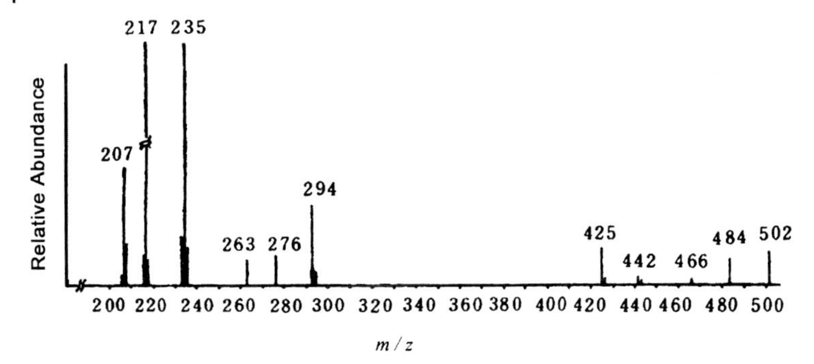

Fig. 6.9 The mass spectrum of an unknown compound.

93

Example 5

A triterpenoid with a known structure has its mass spectrum shown in Fig. 6.10a. Another triterpenoid with an unknown structure has its mass spectrum shown in Fig. 6.10b. Try to determine the unknown compound by comparing the two mass spectra.

Solution

The main fragmentations of the known triterpenoid are as shown in sequence 94.

The molecular ion peak in Fig. 6.10b is situated at m/z 458, which is greater than that in Fig. 6.10a by 32 u. Therefore, the unknown triterpenoid has two more oxygen atoms than the known triterpenoid.

The base peak in Fig. 6.10b is situated at m/z 234, which is greater than that in Fig. 6.10a by 16 u. This can lead to two deductions. Firstly, the two triterpenoids have similar skeletons and more importantly, the positions of the double bond (which initiates the RDA reaction) in the two structures are the same. Secondly, the ion produced from the unknown triterpenoid has one more oxygen atom than that from the known triterpenoid, which indicates the substi-

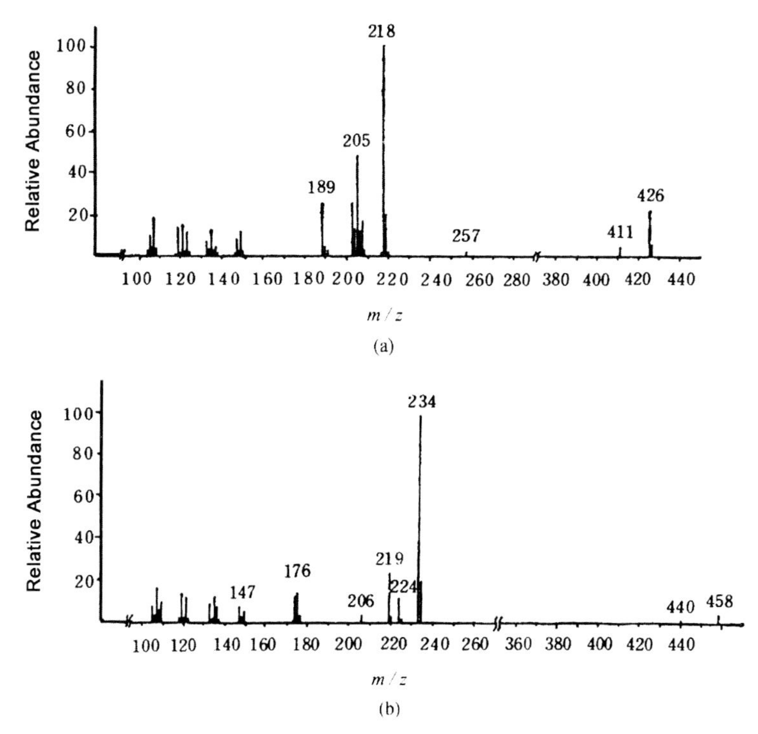

Fig. 6.10 Mass spectra of (a) a triterpenoid with a known structure, and (b) another triterpenoid with an unknown structure.

RDA
m* ≈ 111
HO
m/z 426
m* ≈ 396 —$\dot{C}H_3$
m/z 411
m* ≈ 376 —H_2O
m/z 393
—$\dot{C}_2H_5$
m/z 189
m/z 218
m* ≈ 164 —$\dot{C}H_3$
m/z 203

94

tution of H by a hydroxyl group. As the molecular weight of the unknown triterpenoid is 32 u greater than that of the known triterpenoid, the neutral fragment of the former must have a substitution of H by another hydroxyl group. This deduction method is called the mass shift method.

In fact, the two hydroxyl groups substitute H separately on A and E rings.

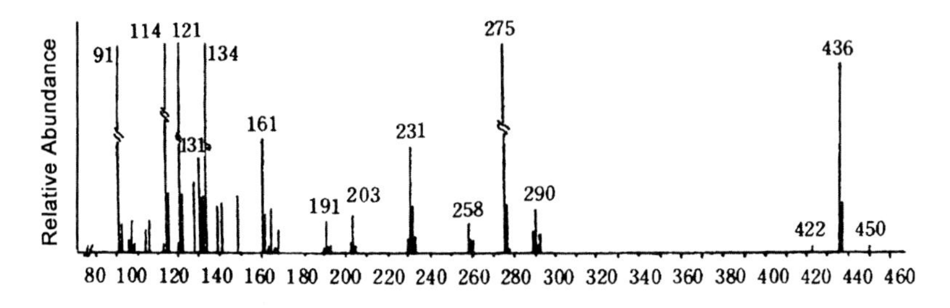

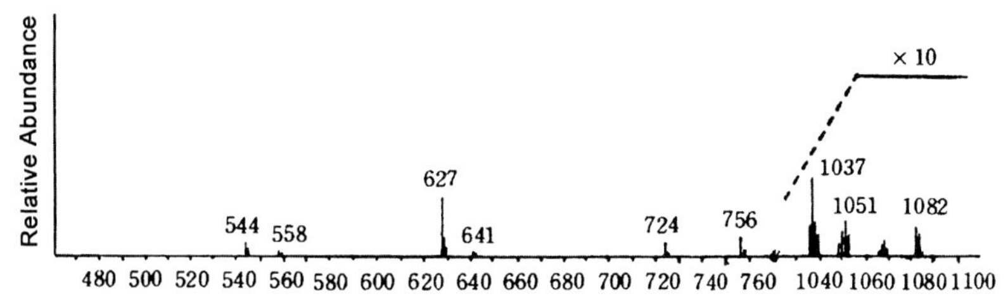

Fig. 6.11 The mass spectrum of an oligopeptide.

Example 6

Try to determine the sequence of an oligopeptide from Fig. 6.11.

Solution

In order to facilitate the vaporization of the sample, the oligopeptide is derived by acetylation followed by a reaction with iodomethane. Under these conditions, a primary amine $-NH_2$ is changed into $CH_3-N-CO-CH_3$; a secondary amine $-NH$ into $-N-CH_3$; and $-OH$ into $-OCH_3$.

There are significant peaks at m/z 91, 114, 121, 134, 161, 231, 275, 436, 627, 756 and 1082. The last should be the molecular ion peak.

Firstly, we will consider the ion of the first amino acid residue in the sequence, which should be one of the ions m/z 114, 121 or 134. Thus the mass of every amino acid residue under the above-mentioned conditions can be calculated. The derived glycine residue (**95**) just agrees with m/z 114.

$H_3C-C(=O)-N(CH_3)-CH_2-C(=O)-$

95

Now we will look for the second residue. The significant peaks after m/z 114 are m/z 161, 231 and 275. The mass difference between 161 and 114 is too small for a residue to exist. The difference between 231 and 114 does not match any derived residue. Compared with m/z 114, the peak m/z 275 has a mass increment of 161 u, which just corresponds to a derived phenyl alanine, **96**.

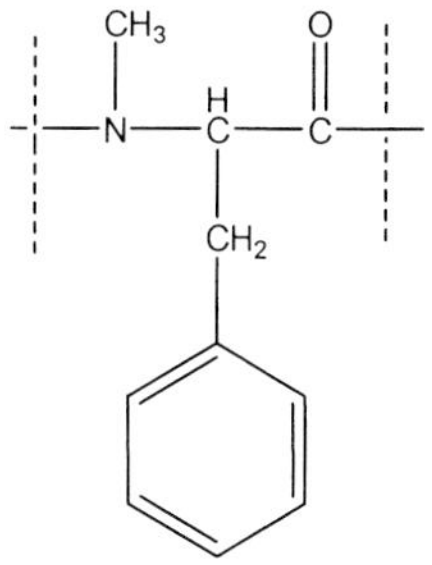

161 u **96**

Similarly, the third and fourth derived residues can be determined from a phenyl alanine and a tyrosine by the peaks m/z 436 and 627, **97**.

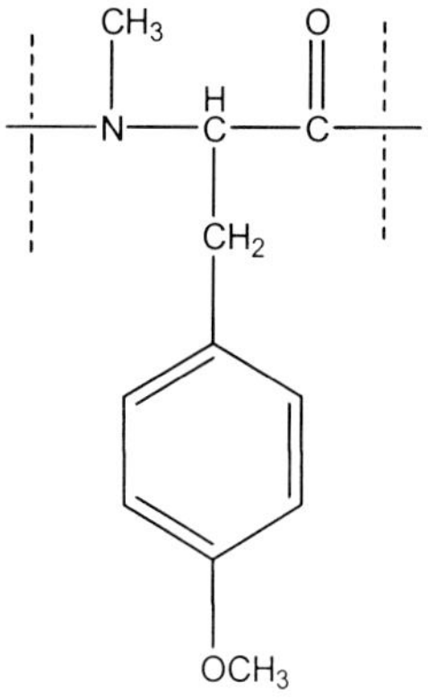

191 u **97**

Finally, we can find the derived oligopeptides, **98**.

Note that the fragment ion produced at proline usually can not be seen. The oligopeptides sequence is shown in **99**, that is, glycine, phenyl alanine, phenyl alanine, tyrosine, threonine, proline and lysine.

Because of the lack of proof from metastable ions, this conclusion is only a tentative one. The definitive conclusion should be obtained by using tandem mass spectrometry.

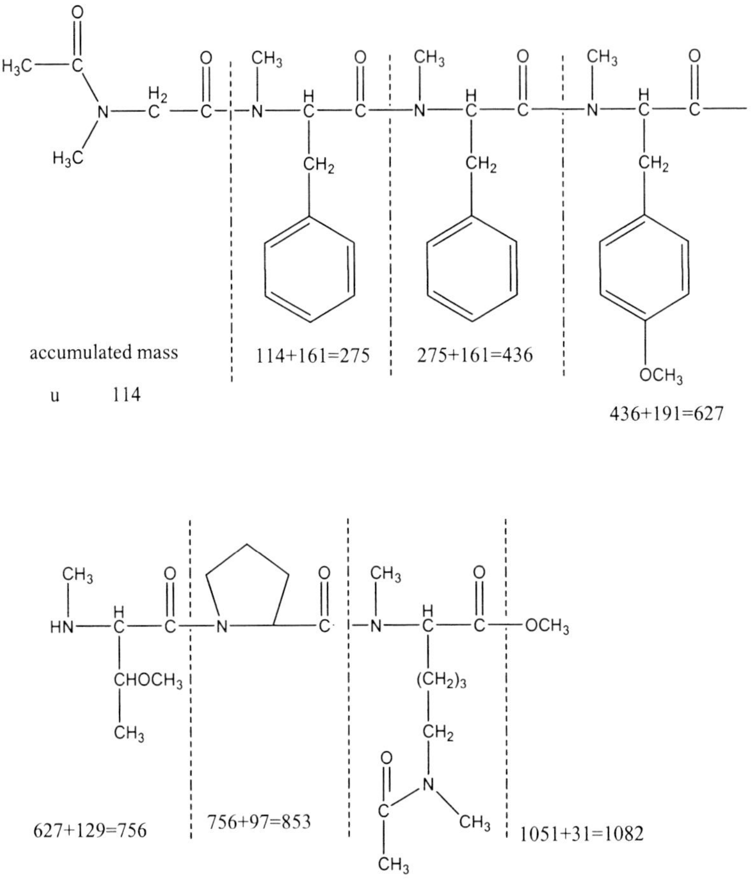

O
H_3C—C
O CH$_3$ O CH$_3$ O CH$_3$ O
H$_2$ H H H
N—C—C—N—C—C—N—C—C—N—C—C—
H_3C
CH$_2$ CH$_2$ CH$_2$
OCH$_3$

accumulated mass

u 114

114+161=275

275+161=436

436+191=627

CH$_3$ O O CH$_3$ O
H H
HN—C—C—N—C—N—C—C—OCH$_3$
CHOCH$_3$ (CH$_2$)$_3$
CH$_3$ CH$_2$
O N
C CH$_3$
CH$_3$

627+129=756

756+97=853

893+198=1051

1051+31=1082

98

O O O O O O O O
H$_2$ H H H H H
H$_2$N—C—C—NH—C—C—NH—C—C—NH—C—C—NH—C—C—N—C—NH—C—C—OH
CH$_2$ CH$_2$ CH$_2$ HC—OH (CH$_2$)$_3$
CH$_3$ CH$_2$
NH$_2$
OH

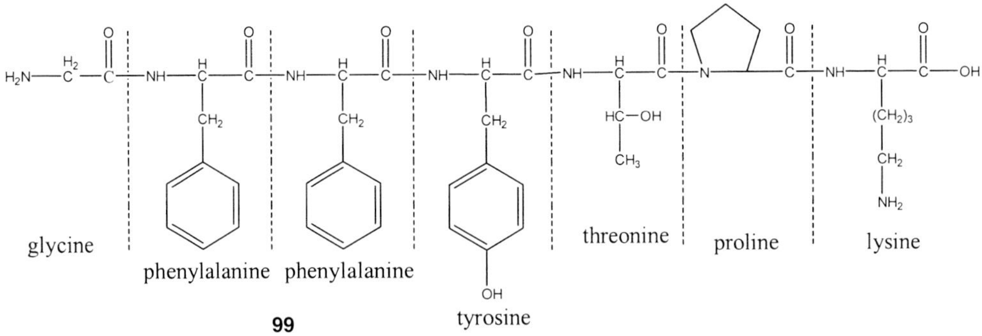

glycine

phenylalanine

phenylalanine

tyrosine

threonine

proline

lysine

99

6.5 Library Retrieval of Mass Spectra

All commercial mass spectrometers have their own computer systems for the instrument operations, spectral storage and spectral library retrieval. When an unknown compound is measured, its mass spectrum will be compared by the computer system with the data system in which tens of thousand of mass spectra of known compounds are stored. Several mass spectra that are most similar to the unknown spectrum will be selected. The similarities between the unknown spectrum and the selected spectra are given in terms of percentages (%) or in term of ‰. Under good conditions, including a complete separation by chromatography, sufficient ion abundances and the existence of similar spectra in the data system, the library retrieval result is reliable. At least the result is helpful for the deduction of the unknown structure.

There are three retrieval systems: INCOS (used by e.g., Finnigan), NIST (used by e.g., Micromass and Varian) and PBM (used by e.g., Hewlett Packard). In principle, INCOS and NIST are similar to each other.

No matter which system is used, both the unknown spectrum and known spectra have to be simplified for the comparison, and the calculations have similarities.

We will start with INCOS, whose mathematical model is fairly understandable [7]. The steps in the library retrieval are as follows:

1. Simplification of the unknown mass spectrum

i. Delete all the peaks whose m/z values are less than 33.
ii. Replace the intensity of each peak I with $\sqrt{mI}$, where m is the mass of the peak.
iii. Select the 16 strongest peaks from all the weighted peaks. At most only one peak will be selected from a peak cluster.

2. Pre-retrieval of the simplified unknown spectrum

i. Firstly, mass spectra that have the molecular weights within our range of interest are chosen. Secondly, every selected known spectrum is compared one by one with the unknown spectrum. The eight strongest peaks in each known spectrum are used for the comparison. Thus some known spectra are selected.
ii. Some factors including the temperature in the ion source, the ionization voltage, and the acquisition time related to chromatographic peaks affect ion abundances. Therefore, some ranges of the known spectra should be modified by multiplying them by a factor (from 1/2 to 2), a procedure known as local normalization, whereas global normalization means multiplying a whole known spectrum by a factor, which is determined by the ratio of the total peak intensity of the known spectrum to that of the unknown spectrum.

3. Calculation of purity, FIT and reverse FIT (RFIT)
These calculations are the key to library retrieval.

The principle of the calculation is as follows: Every peak in a simplified mass spectrum forms one component vector in a multi-dimensional vector space. The weighted intensity $\sqrt{mI}$ of a particular peak determines the magnitude of the component vector. A mass spectrum consists of peaks with m/z values. In a multi-dimensional vector space, the mass spectrum is the resultant vector of its component vectors. The degree of similarity between an unknown mass spectrum and a known mass spectrum is measured by the angle between the two resultant vectors. Now we can define the following three functions

$$\mathrm{UTOTAL} = \Sigma U(m) \cdot U(m) \tag{6-14}$$

where $U(m)$ is the square root of the simplified and weighted peak intensity of an unknown mass spectrum; and UTOTAL is the simplified and weighted total ion current of an unknown mass spectrum.

$$\mathrm{LTOTAL} = \Sigma L(m) \cdot L(m) \tag{6-15}$$

where $L(m)$ is the square root of the simplified and weighted peak intensity of a known mass spectrum; and LTOTAL is the simplified and weighted total ion current of a known mass spectrum.

$$\mathrm{ULTOTAL} = \Sigma U(m) \cdot L(m) \tag{6-16}$$

where ULTOTAL is the sum of dot products of corresponding components of the unknown and known spectra.

Define

$$\mathrm{Purity} = (\boldsymbol{U} \cdot \boldsymbol{L}) = 1000 \times \cos^2 \theta \tag{6-17}$$

where $\boldsymbol{U}$ and $\boldsymbol{L}$ are the resultant vectors of the unknown and known mass spectra, respectively; and θ is the angle formed by $\boldsymbol{U}$ and $\boldsymbol{L}$.

From the calculation rule of vectors, we have

$$\mathrm{Purity}(\boldsymbol{U} \cdot \boldsymbol{L}) = \frac{1000\,(ULTOTAL)^2}{(UTOTAL)(LTOTAL)} \tag{6-18}$$

The purity shows the similarity between the unknown mass spectrum and the known mass spectrum. If the two spectra are identical, $\boldsymbol{U}$ and $\boldsymbol{L}$ will agree with each other well and the purity will be 1000.

The FIT shows the extent to which the known mass spectrum is included in the unknown mass spectrum. FIT=1000 means that the unknown spectrum contains all the peaks of the known spectrum and that all the corresponding peak pairs have the same intensity ratio. In the calculation of FIT, only the peaks of the unknown spectrum that exist in the compared known spectrum are considered. In this case, Eq. (6-16) becomes

$$(\mathrm{UTOTAL})' = \Sigma U'(m) \times U'(m) \tag{6-19}$$

where $U'(m)$ is the square root of intensity of a peak of the unknown spectra that also exists in the compared known spectra.

$$\mathrm{FIT} = 1000 \frac{(ULTOTAL)^2}{(UTOTAL)'(LTOTAL)} \tag{6-20}$$

If the FIT is 1000 but the purity is less than 1000, it can be deduced that the measured sample has the identical structure to that of the compared known compound but the sample is contaminated by impurities.

The RFIT shows the extent to which the unknown mass spectrum is included in the known mass spectrum. In the calculation of the RFIT, only the peaks of the known spectrum, which exist in the compared unknown spectrum, are considered.

As with Eq. (6-20), we have

$$\mathrm{RFIT} = 1000 \frac{(ULTOTAL)^2}{(UTOTAL)(LTOTAL)'} \tag{6-21}$$

Note that the right sides of the three equations above (6-18), (6-20), (6-21) have the same numerator.

A set of data is listed in Tab. 6.7 to facilitate the intuitive understanding of the calculations.

For simplicity the two mass spectra are described as only having two or three peaks.

Again for simplicity, these intensities are not weighted by $\sqrt{m}$.

The unknown spectrum $\boldsymbol{U}$= (10,10,10).

The known spectrum $\boldsymbol{L}$= (12, 15,0).

The two sets of data are shown in Fig. 6.12.

In general, several compounds are selected in the pre-retrieval, and their purities, FITs and RFITs, are calculated by the computer system. Now just one set of data is calculated.

Tab. 6.7 Data to facilitate understanding of the calculations

m/z	*Intensities (unknown)*	*Intensities (known)*
43	100	144
57	100	225
81	100	0

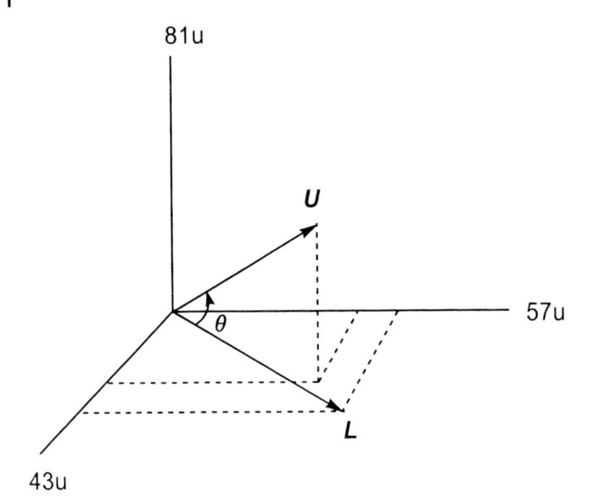

Fig. 6.12 The unknown and known spectra in the multi-dimensional vector space.

$$\begin{aligned}\text{PURITY} &= 1000\frac{(UTOTAL)^2}{(UTOTAL)(LTOTAL)}\\ &= 1000\frac{(10\times 12+10\times 15+10\times 0)^2}{(10\times 10+10\times 10+10\times 10)(12\times 12+15\times 15+0\times 0)}\\ &= 659\end{aligned}$$

By using Eqs. (6-20) and (6-21), we can obtain: FIT=988 and RFIT=659.

The values of FIT, RFIT and PURITY show quantitatively the degree of similarity between the unknown mass spectrum and known mass spectra selected from the data system. The greater the values are, the more similar the unknown and known spectra will be. This is an important step in the process of mass spectrum interpretation.

The NIST method is similar to the INCOS, both of which consider a mass spectrum as a resultant vector in a multi-dimensional vector space. The calculation of the MF (match factor) in NIST is also based on Eq. (6-18), that is, the MF in NIST corresponds to the PURITY in INCOS. The NIST method can also be used to postulate the substructures of an unknown [8].

Next we will present the PBM method (probability based matching) [9, 10], which is based on the theory that the total probability is the product of all the individual probabilities. If the probability of the appearance of the peak m_1/z is p_1 and that of the peak m_2/z is p_2, then the probability of the simultaneous appearance of the two peaks is $p_1 \times p_2$. In logarithms, multiplication becomes addition.

In the PBM system, the probability that the mass spectrum to be examined is due to the compound being sought has been expressed as the confidence index K. The value of K shows the reliability of the retrieval. The probability that the mass spectrum could arise from a compound selected at random is $1/2^k$.

PBM is a reverse retrieval in which the unknown mass spectrum is compared with known mass spectra one by one. Several known mass spectra are selected according to their m/z values and intensities, which are similar to those of the unknown mass spectrum. The selected peaks are called reference peaks. Suppose the k value of the j-th reference peak is k_j, we have

$$K = \Sigma k_j \tag{6-22}$$

where K is the total confidence index.

For the calculation of k_i, it follows from the theory of PBM that

$$k_j = (U_j - A_j - D + W_j) \tag{6-23}$$

where k_j is the confidence index of the j-th peak; U_j is the uniqueness of the j-th peak; A_j is the abundance of the j-th peak; D is the dilution factor; and W_j is the window tolerance of the j-th peak.

Peaks which have larger k_j values should be selected so as to produce a larger total confidence index K.

Now we will discuss the factors mentioned above in detail.

U_j is the most important factor, which shows the characteristics of a selected peak. The rarer the appearance in a mass spectrum of a peak, the larger U_j value it has. In general, U_j increases with the m/z value.

A_j is a modification of uniqueness. When a peak has a small abundance, its U_j value should be lowered. Therefore, the selected reference peaks should have large U_j values as well as high abundances.

If the target compound is pure or its content is over 50%, its D value will be zero. Otherwise, D will lower the value of k_j.

W_j is a modification term concerning the abundance agreement between the unknown spectrum and the selected spectrum. If the two abundances are close, then W_j has a larger value, that is, its k_j value increases.

On the basis of the K value, together with other parameters, the quality of the match can be calculated in percentage terms, which indicates the probability that the selected mass spectrum is correctly identified as the unknown mass spectrum.

No matter what retrieval system is used, a final judgment by the researcher is absolutely necessary. Firstly, one or two peaks may be very important for the identification, but this fact could be neglected by any retrieval system. Secondly, there are some complicated or difficult factors, such as insufficient known spectra, different ionizations, and so forth. Then it is inevitable that our ability to make spectral interpretations should improve continuously, as the final decisions in the interpretations can never be made by a computer.

6.6 Interpretation of the Mass Spectra from Soft Ionization

Two subjects should be considered. Firstly, how can we determine the molecular weight of an unknown compound from a mass spectrum from soft ionization techniques? Secondly, how can we interpret peaks in the mass spectra from soft ionization? This section is dedicated to these two subjects.

References [11–13] are recommended for this section.

6.6.1 Mass Spectra from CI

As EI and CI can be switched rapidly in an ionization source in a commercial mass spectrometer, both the CI mass spectrum and the EI mass spectrum of an unknown compound can be obtained in one experiment.

This section mainly focuses on the positive ion CI because it is more frequently applied than the negative ion CI.

The CI mass spectrum gives two types of information: molecular weight and structural information. These two subjects are closely related to the reagent gas used in the CI process.

When the molecular ion of a sample is decomposed easily by EI, CI can give the information about its molecular weight. Quasi-molecular ions, or more precisely the adduct ions which contain the intact molecule of the sample measured, can be found in a CI spectrum.

The formation of quasi-molecule ions is related to the reagent gas being used. For example, if NH_3 is used as the reagent gas, $[M+H]^+$ is likely to be produced when the sample molecules are more basic than NH_3, while $[M–H]^+$ is likely to be produced when they are more acid.

The adduct ions which appear most frequently in CI mass spectra are $[M+H]^+$, protonated molecules. As they are even-electron ions, which are more stable than odd-electron ions, and the ions produced in CI possess much lower internal energy than parent ions produced in EI, the peak of $[M+H]^+$ is usually distinguished in a CI mass spectrum. Therefore, the molecular weight of the unknown compound can be postulated from the m/z value minus 1. However, the existence of other quasi-molecular ions can complicate this situation. Three possibilities should be considered.

Firstly, in addition to $[M+H]^+$, there are other adduct ions that consist of the analyte molecule and an ion produced from the reagent gas. For example, $[M+H]^+$, $[M+C_2H_5]^+$ and $[M+C_3H_5]^+$ coexist in a CI mass spectrum when methane is used as the reagent gas. If isobutane is used as the reagent gas, the peak of $[M+C_4H_9]^+$ can appear.

Secondly, the ionization process in CI is closely related to the reagent gas being used. In general, the proton affinity of an analyte is greater than that of the reagent gases. In this case, the formation of $[M+H]^+$ is the main result during the CI process. If the proton affinities of the analyte and the reagent gas are nearly equal, adduct ions, formed by the association of the reagent ion with the analyte molecule, could be observed instead of the protonated molecule. For example, the ions of $[M+NH_4]^+$ can be formed when ammonia is used as the reagent gas.

Thirdly, when a hydrocarbon sample is analyzed, the ions of $[M-H]^+$ are formed because the proton affinities of hydrocarbons are less than those of reagent gases. This phenomenon can be observed for other types of sample, for example, some alcohols, esters, etc.

In any case the molecular weight of an unknown compound can be determined from its CI mass spectrum. If methane is used as the reagent gas, the molecular weight can be determined from the three peaks which have an m/z difference of 28 (between the peaks of $[M+H]^+$ and $[M+C_2H_5]^+$) and 40 (between the peaks of $[M+H]^+$ and $[M+C_3H_5]^+$). As the reagent gas is known, the molecular weight can be deduced from the ions of $[M+NH_4]^+$ when ammonia is used as the reagent gas.

Besides information about the molecular weight of an unknown compound, information about the structure or even about the stereo-structure of the compound can be found from CI mass spectra. Isomers can show distinguishable CI mass spectra. A discussion about the discernment of stereo-isomers will be presented in Section 9.2.2.

The fragmentation in CI is related closely to the reagent gas being used. In general the proton affinity of an analyte is greater than that of the reagent gases. The greater the difference in proton affinity between an analyte and the reagent gas to be used in CI, the more energy will be transferred to the protonated molecule of the analyte, the greater the fragmentation of the analyte and the more fragment ions will be present in the CI spectrum. As the order of proton affinities of reagent gases is ammonia > isobutane > methane, the order of the number of fragment ion peaks of an analyte with reagent gases is methane > isobutane > ammonia in CI mass spectra.

The fragmentation reactions that occur frequently in CI are the elimination of HX from the protonated molecule of the analyte where X is a heteroatom or a functional group existing in the molecule. These reactions are different from those in EI because they start from an even-electron ion while the reactions in EI start from an odd-electron ion.

In the negative ion mode in CI, the quasi-molecular ion is most frequently $[M-H]^-$.

6.6.2 Mass Spectra from FAB

The mass spectra from FAB clearly differ in appearance from those from EI or CI. Many intense peaks produced from the matrix exist in the mass spectrum from FAB. In the positive ion mode of FAB, these peaks represent X_nH^+ where X is the matrix molecule, and dehydration products of X_nH^+. The ions produced from the matrix are also present in the mass spectrum from FAB. Therefore, these ions dominate the mass spectrum from FAB. If glycerol is used as the matrix, there are peaks at *m/z* 45, 57, 75, 93, 185, 277, 369, 461, 553, and so forth. Similarly, there are peaks at *m/z* 91, 183, 275, 367, 459, 551, and so forth in the mass spectrum in the negative ion mode.

The adduct ions of the matrix molecule and Na^+ and/or K^+ which are added to the sample or are present as contaminants can also appear in the mass spectrum from FAB. For example, the peaks of *m/z* 115 $[glycerol + Na]^+$, 131 $[glycerol + K]^+$, 207 $[2\ glycerol + Na]^+$, 223 $[2\ glycerol + K]^+$, etc., could be present in the mass spectrum in the positive ion mode.

In addition to these peaks, randomly fragmented pieces mainly from the matrix appear as background peaks. Sometimes this background is referred to as "grass".

The information about the molecular weight of an unknown compound can be postulated from adduct ions that appear in the region of high mass-to-charge ratios in a mass spectrum from FAB. When the positive ion mode is operated, the protonated sample molecule, $[M + H]^+$, and the cationized molecules, $[M + Na]^+$ and $[M + K]^+$, frequently exist in the spectrum. From the differences in the region of high mass-to-charge ratios of 22 (between $[M + Na]^+$ and $[M + H]^+$) and 16 (between $[M + K]^+$ and $[M + Na]^+$), the molecular weight of the compound can be deduced. The intensities of the peaks of $[M + K]^+$ and $[M + Na]^+$ can be enhanced by adding sodium and potassium salts to the mixture of the matrix and analyte intentionally. For an intense $[M + H]^+$ ion signal, the matrix should be more acidic than the sample molecule.

Other adduct ions can appear in the region of high mass-to-charge ratios. If the analyte has basic sites, species such as $[M{-}H + 2Na]^+$, $[M{-}H + 2K]^+$, and so forth will be observed. Sometimes cluster ions of sample molecules, such as $[2M + H]^+$, can appear in the mass spectrum.

If the negative ion mode is operated, the quasi-molecular ion is most frequently $[M{-}H]^-$. For a good signal, the matrix should be more basic than the analyte molecule.

There are fragment ions in the mass spectra from FAB. They can be used to postulate the structure of the unknown compound. The degree of the fragmentation in FAB is related to the matrix being used.

If an ionic compound, A^+B^-, is studied by FAB, A^+ will be detected in the positive ion mode and B^- in the negative ion mode.

6.6.3
Mass Spectra from MALDI

MALDI is mainly applied for TOF. This ionization is appropriate for the mass analysis of molecules with a high molecular weight. The low-mass region (< 500 *m/z*) in MALDI mass spectra is obscured by the noise produced from the matrices used. However, this disadvantage does not affect the application of MALDI mass spectra for the high-mass region.

MALDI only gives information about the molecular weight of the sample because few fragment ions appear in MALDI mass spectra. Therefore, it is appropriate for the analysis of mixtures.

In the MALDI mass spectrum, the peak of $[M+H]^+$, the protonated molecule, is significant. Other peaks which are produced from the analyte molecule could be present, such as $[M+2H]^{2+}$, $[M+3H]^{3+}$, $[2M+H]^+$, and so forth. The peak intensity of $[2M+H]^+$ increases with the enhancement of its concentration in the sample. The peak of $[M+Met]^+$, where Met denotes a metal atom, can be found in a MALDI mass spectrum.

6.6.4 Mass Spectra from ESI

The mass spectra from ESI are different to those from FAB or MALDI. There are no background peaks in ESI mass spectra, which is an important advantage for the interpretation the spectra.

An attractive advantage of ESI mass spectra is the formation of multiply-charged ions, which realizes of the detection of macromolecules by mass spectrometry with reduced values of m/z. Therefore, macromolecules can be analyzed by mass spectrometry using inexpensive mass analyzers, for example, a quadrupole which has a mass-to-charge range of less than 4000.

Multiply charged ions form a cluster of peaks, which are situated at some m/z values for a range of mass-to-charge ratios. These peaks result from $[M+nH]^{n+}$, or $[M+nNa]^{n+}$ in the positive ion mode or $[M-nH]^{n-}$ in the negative ion mode. The determination of the molecular weight of an unknown compound from the cluster peaks is presented in Section 6.1.2. It is necessary to calculate the parameter n of a selected peak first. If a mixture is analyzed, the overlap of related peak clusters makes the calculation difficult, in which case the use of an ESI mass spectrum with a high resolution is a good solution (see Section 6.6.2).

The ESI technique can also be used for the mass analysis of small molecules. In positive ion ESI mass spectra, there are the peaks produced from intact analyte molecules which may include $[M+H]^+$, $[M+Met]^+$, $[M+NH_4]^+$, and so forth, where Met denotes a metal atom. In negative ion ESI mass spectra, the main quasi-molecular ion is $[M-H]^-$.

The tendency to form $[2M+H]^+$ or $[2M+Na]^+$ increases with analyte concentration, which is common to all soft ionization techniques.

The ESI produces ions with low internal energies. Peaks of fragment ions are rare in ESI mass spectra. The structural information from fragment ions is obtained mainly by using CID (see Section 5.5.1). It should be noted that the fragmentation pathways from an even-electron ion are different from those from an odd-electron ion. They have different fragmentation pathways.

6.6.5
Mass Spectra from APCI

The principle of APCI is similar to that of CI. However, they are different in two respects. Firstly, the pressure in the ionization chamber for CI is approximately 100 Pa but that for APCI 100 kPa. Therefore, the ionization efficiency in APCI is much higher than that in CI. Secondly, the ionization in CI begins with the electrons being emitted from a filament while the ionization in APCI begins from a corona discharge.

The mass spectrum from APCI is similar to that from CI except that there are peaks of $[(H_2O)_nH]^+$ in the APCI mass spectrum, because H_2O has the highest proton affinity of those gases normally present in the ionization.

6.7
References

1 F.W. McLafferty, F. Turecek, *Interpretation of Mass Spectra*, 4th Ed, University Science Books, Mill Valley, 1994.

2 R.M. Silverstein, G.C. Bassler, *Spectrometric Identification of Organic Compounds*, 6th Ed, John Wiley, Chichester, 1998.

3 B. Asamoto, Ed., FT-ICR/MS: *Analytical Applications of Fourier Transform Ion Cyclotron Mass Spectrometry*, Chapter 7, VCH, Weinheim, 1991.

4 A.N. Tyler, E. Clayton et al., *Anal. Chem.*, **1996**, *68*, 3561–3569.

5 H. Budzikiewicz, C. Djerassi, D.H. Williams, *Mass Spectrometry of Organic Compounds*, Holden-Day, San Francisco, 1967.

6 M.E. Rose, R.A.W. Johnstone, *Mass Spectrometry for Chemists and Biochemists*, Cambridge University Press, Cambridge, 1982.

7 S. Sokolow, J. Karnofsky et al., *Finnigan Application Report*, **1978**, No. 2.

8 S.E. Stein, *J. Am. Soc. Mass Spectrom.*, **1995**, *6*, 644–655.

9 G.M. Pesyna, R. Venkataraghavan et al., *Anal Chem.*, **1976**, *48*, 1362–1368.

10 F.W. McLafferty, R.H. Hertel et al., *Org. Mass Spectrom.*, **1974**, *9*, 690–702.

11 A.E. Ashcroft, *Ionization Methods in Organic Mass Spectrometry*, The Royal Society of Chemistry, Cambridge, 1997.

12 B.N. Pramanik, A.K. Gangyly, M.L. Gross, *Applied Electrospray Mass Spectrometry*, Marcel Dekker, New York, 2002.

13 J.T. Watson, *Introduction to Mass Spectrometry*, Lippincott-Raven, 1997.

7
Infrared Spectroscopy and Raman Spectroscopy

This chapter discusses molecular vibration spectroscopy, including infrared absorption spectroscopy and Raman scattering spectroscopy, with emphasis on the former for it is more widely applied in laboratories and more spectra have been accumulated than for Raman spectroscopy.

Infrared spectroscopy has found wide application because of the following:

1. Any sample in a solid, liquid or gaseous phase can be measured by infrared spectroscopy.
2. Every compound has its own infrared absorption bands. The absorption bands in the region around the functional group shows the functional groups of the compound. The absorption bands in the fingerprint region are the "fingerprint" of the compound.
3. A commercial infrared spectrometer is much cheaper than spectrometers for NMR or MS.
4. The required amount of a sample to be analyzed is small. Detection techniques for specific samples have been developed.

7.1
General Information on Infrared Spectroscopy

7.1.1
Wavelength and Wavenumber

The propagation of electromagnetic waves can be described by Eq. (7-1).

$$C = \nu\lambda \tag{7-1}$$

where C is the propagation velocity of electromagnetic wave, that is, the velocity of light, 2.998×10^{10} cm s^{-1}; ν is the frequency, in cps or Hz; and λ is the wavelength, in cm.

Equation (7-1) can be rewritten as:

Structural Identification of Organic Compounds with Spectroscopic Techniques. Yong-Cheng Ning
Copyright © 2005 WILEY-VCH Verlag GmbH & Co. KGaA, Weinheim
ISBN: 3-527-31240-4

$$\nu = \frac{C}{\lambda} \tag{7-2}$$

Now define:

$$\bar{\nu} = \frac{1}{\lambda} = \frac{\nu}{C} \tag{7-3}$$

where $\bar{\nu}$ is the number of waves, that is the wavenumber, which represents the number of vibrations per unit length. As λ is expressed in cm, $\bar{\nu}$ is expressed in cm^{-1}.

7.1.2
Near, Medium and Far Infrared Rays

Infrared rays are classified into near, medium and far infrared rays according to their wavelengths.

In organic analysis the medium infrared is the most important infrared region. Commercial infrared spectrometers usually cover 4000–650 cm^{-1} or 4000–400 cm^{-1}. Far IR absorption bands show the stretching vibrations of heavy atoms and bending vibrations of some functional groups. Near IR absorption bands can be used for quantitative analysis of O–H, N–H, C–H, etc.

The abscissa with a linear scale in wavenumber is commonly used in IR spectra in recent textbooks and literature.

7.1.3
The Ordinate of IR Spectra

IR absorption intensities are presented on the ordinate of IR spectra, which use either transmittance or absorbance. The intensity measurement relates to the slit width of the spectrometer used but the slit cannot usually be set to a very narrow width, which is necessary for precise measurements. In addition, other variable experimental parameters are difficult to fix. Accordingly, IR absorption intensities are usually expressed approximately as very strong ($\varepsilon > 200$), strong ($\varepsilon = 75$–200), medium ($\varepsilon = 25-75$), weak ($\varepsilon = 5-25$) and very weak ($\varepsilon < 5$), where ε is the apparent molar absorptivity.

7.2
Basic Theory of IR Spectroscopy

7.2.1
IR Absorption Frequencies of a Diatomic Molecule

1. Treatment with classical mechanics

In classical mechanics, a harmonic oscillator model is used to study the vibrations of a diatomic molecule. In the model, two rigid balls are connected to-

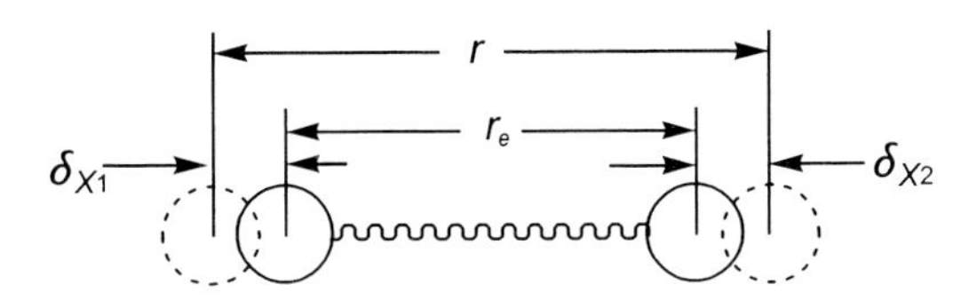

Fig. 7.1 Atomic displacements during a diatomic vibration.

gether by a spring, which has no mass. The two atoms correspond to the two balls, and the chemical bond to the spring.

The displacement of the two atoms during the vibration are illustrated in Fig. 7.1, where r_e is the equilibrium distance between the two atoms; and r is the instantaneous distance between the two atoms during the vibration.

From Huck's law, it follows that:

$$F = -k(r - r_e) = -k(\delta x_2 - \delta x_1) = -kq \tag{7-4}$$

where k is the force constant of the chemical bond; δx_2 and δx_1 are the displacements of atoms 2 and 1 on the x axis, respectively; and q is the vibrational coordinate.

From Eq. (7-4),

$$q = r - r_e = \delta x_2 - \delta x_1 \tag{7-5}$$

For atom 1, the following equation holds:

$$F = m_1 \frac{d^2}{dt^2}(\delta x_1) \tag{7-6}$$

and for the atom 2:

$$F = m_2 \frac{d^2}{dt^2}(\delta x_2) \tag{7-7}$$

If the vibration without any change in the molecular mass center is just discussed, then:

$$-m_1 \delta x_1 = m_2 \delta x_2 \tag{7-8}$$

Combining Eqs. (7-5) and (7-8) gives a new equation:

$$-\delta x_1 = \frac{m_2}{m_1 + m_2}(r - r_e) \tag{7-9}$$

Similarly,

$$\delta x_2 = \frac{m_1}{m_1 + m_2}(r - r_e) \tag{7-10}$$

Substituting Eq. (7-9) into Eq. (7-6) and then combining with Eq. (7-4) leads to

$$\left(\frac{m_1 m_2}{m_1 + m_2}\right)\left(\frac{\mathrm{d}^2(r - r_\mathrm{e})}{\mathrm{d}t^2}\right) = -k(r - r_\mathrm{e}) \tag{7-11}$$

Defining

$$\mu = \frac{m_1 m_2}{m_1 + m_2} \quad \left(\text{or } \frac{1}{\mu} = \frac{1}{m_1} + \frac{1}{m_2}\right) \tag{7-12}$$

where μ is the reduced mass.

Substituting Eqs. (7-12) and (7-5) into Eq. (7-11) yields

$$\frac{\mathrm{d}^2 q}{\mathrm{d}t^2} = -\frac{k}{\mu} q \tag{7-13}$$

The solution to the differential equation is

$$q = q_0 \cos 2\pi\nu t \tag{7-14}$$

where q_0 is a constant, which represents the amplitude of the vibration; and ν is the vibrational frequency.

Differentiating ν twice with respect to t produces the following equation:

$$\nu = \frac{1}{2\pi}\sqrt{\frac{k}{\mu}} \tag{7-15}$$

Or combining with Eq. (7-3), it holds that

$$\bar{\nu} = \frac{1}{2\pi \mathrm{C}}\sqrt{\frac{k}{\mu}} \tag{7-16}$$

Based on classical mechanics, if the diatomic molecule possesses a dipolar moment, the vibration of the dipole produces an electromagnetic wave, which can act on the incident electromagnetic wave, leading to an absorption band at the vibrational frequency.

2. Treatment with quantum mechanics

Schrodinger's equation is

$$H\Psi = E\Psi \tag{7-17}$$

where H is the energy operator or Hamiltonian; Ψ is a wave function; and E is the eigenvalue of Ψ.

The Hamiltonian for the vibration of a diatomic molecule is

$$H = \left(\frac{-h^2}{8\pi^2\mu}\right)\left(\frac{d^2}{dq^2}\right) + V \tag{7-18}$$

where V is the internal energy of the system; μ is the reduced mass; q is the vibrational coordinate; and h is Planck's constant.

Using the approximation of a harmonic oscillator,

$$V = \frac{1}{2}kq^2 \tag{7-19}$$

then substituting Eq. (7-19) into Eq. (7-18), one has

$$H = -\frac{h^2}{8\pi^2\mu} \cdot \frac{d^2}{dq^2} + \frac{1}{2}kq^2 \tag{7-20}$$

Substituting Eq. (7-20) into Eq. (7-17) yields

$$-\frac{h^2}{8\pi^2\mu} \cdot \frac{d^2\Psi}{dq^2} + \frac{1}{2}kq^2\Psi = E\Psi \tag{7-21}$$

The solution to Eq. (7-21) is

$$E_\mathrm{v} = \left(\mathrm{v} + \frac{1}{2}\right)\frac{h}{2\pi}\sqrt{\frac{k}{\mu}} \quad (\mathrm{v} = 0, 1, 2, 3\ldots) \tag{7-22}$$

where v is the vibrational quantum number; E_v is the energy of the system at energy level v.

Using Eq. (7-15), Eq. (7-22) can be rewritten as:

$$E_\mathrm{v} = \left(\mathrm{v} + \frac{1}{2}\right)h\nu \quad (\mathrm{v} = 0, 1, 2, 3\ldots) \tag{7-23}$$

It follows that $E_\mathrm{v}>0$ even when $\mathrm{v}=0$, so E_ν is known as the zero-point vibrational energy. The selection rule for transitions is $\Delta\mathrm{v}=\pm1$.

For the transition from $\mathrm{v}=0$ (ground state) to $\mathrm{v}=1$ (first excited state), the energy difference ΔE is $h\nu$. On the other hand, the absorption frequency can be calculated from Eq. (7-22):

$$\nu = \frac{\Delta E}{h} = \frac{1}{2\pi}\sqrt{\frac{k}{\mu}}$$

This is the same result as in classical mechanics, and the frequency is known as the fundamental frequency.

The equations mentioned above are deduced on the basis of the hypothesis that the vibration of a diatomic molecule is a harmonic oscillator, whose poten-

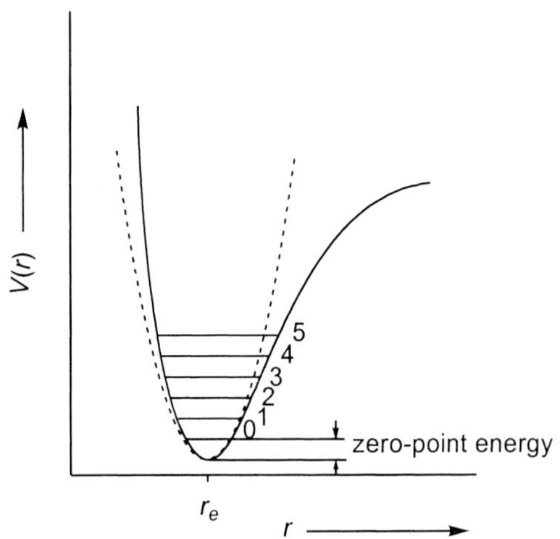

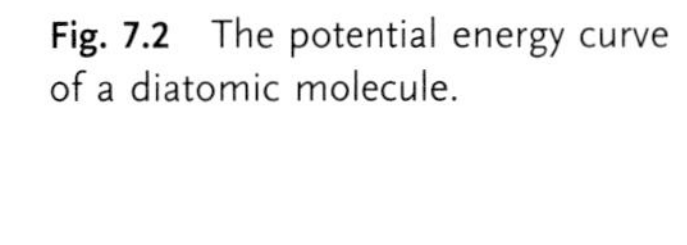
Fig. 7.2 The potential energy curve of a diatomic molecule.

tial energy curve is shown by the dotted line in Fig. 7.2. However, the actual potential energy curve is shown by the solid line in Fig. 7.2. Therefore, the internal energy function should be corrected for anharmonicity, which leads to:

$$E_{\mathrm{v}} = \left(\mathrm{v} + \frac{1}{2}\right)h\nu - \left(\mathrm{v} + \frac{1}{2}\right)^2 \chi h\nu + \text{higher power terms} \qquad (7\text{-}24)$$

where χ is the anharmonicity coefficient.

Because of the vibrational anharmonicity, the transitions are not strictly confined to the condition $\Delta\mathrm{v} = \pm 1$; for example, the transition of $\Delta\mathrm{v} = \pm 2$ can also occur, though with a lower probability. According to Eq. (7-23), this transition possesses a frequency of 2ν, which is termed the overtone. In fact, the frequency is slightly lower than 2ν because of the anharmonicity [Eq. (7-24)].

7.2.2
IR Absorption Frequencies of a Polyatomic Molecule

A diatomic molecule possesses only one vibrational mode while a polyatomic molecule possesses more than one. Because an atom has three coordinates, a molecule, which consists of n atoms, requires $3n$ coordinates to define its position, that is, a polyatomic molecule has $3n$ degrees of freedom. As a polyatomic molecule is considered as a whole, the $3n$ degrees of freedom can be decomposed into those of the mass center, rotations and vibrations of the molecule.

The mass center has 3 degrees of freedom. A nonlinear molecule possesses 3 rotational degrees of freedom while a linear molecule possesses only two. Therefore, a nonlinear molecule has $3n$–6 vibrational degrees of freedom while a linear molecule has $3n$–5 vibrational degrees of freedom. These vibrations are termed normal vibrations because they are concerned with neither the motion of the mass center nor the rotations of the molecule.

A polyatomic molecule has many modes of vibrations. Even for a CH_2 group, there are various vibrational modes, as shown in Fig. 7.3.

The vibrations can be classified into two types: stretching and bending vibrations. The motions along chemical bonds are stretching vibrations while all the other vibrations are bending vibrations, which have lower frequencies than the stretching ones.

Although a polyatomic molecule has $3n$–6 normal vibrations, a polyatomic molecule usually has far fewer than $3n$–6 IR absorption bands for the following reasons:

1. Some vibrations could be inactive hence they do not produce absorption bands.
2. Some absorption bands are outside of the medium IR region.
3. Some vibrations are degenerate.
4. Some absorption bands are too close to be resolved, etc.

According to the nature of the IR absorption frequencies, they can be classified as:

1. Fundamental frequency.
2. Overtone.
3. Combination tone: sum or difference of the fundamental frequencies.
4. Coupling frequencies: When two groups with close fundamental frequencies are connected to each other, the coupling of the vibrations leads to displace-

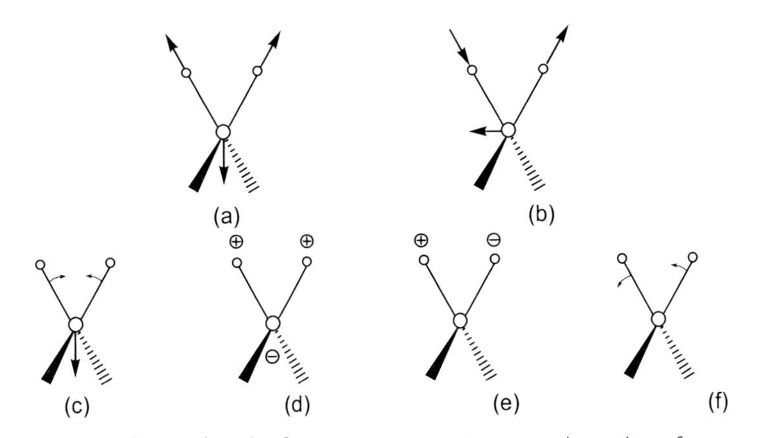

Fig. 7.3 Vibrational modes for a CH_2 group. (a) Symmetrical stretching, frequency ν_s; (b) asymmetrical stretching, frequency ν_{as}; (c) in-plane bending or scissoring; (d) out-of-plane bending or wagging; (e) out-of-plane bending or twisting; (f) in-plane bending or rocking. ⊕ and ◯ denote the motions perpendicular to the plane of the book.

ments of the fundamental frequencies: one becomes higher and the other lower.

5. Fermi resonant frequencies: When an overtone or a combination tone is close to another fundamental frequency, their interaction will produce a splitting of an absorption band.

7.2.3 IR Absorption Intensities

IR absorption intensities are determined by the probabilities of IR transitions. In theory,

$$\text{transition probability} \propto |\mu_{ab}| E_0^2 \qquad (7\text{-}25)$$

where E_0 is the electrical component of the IR rays; and μ_{ab} is the dipolar moment of the transition, which shows the variation in the dipolar moment during the vibration.

According to Eq. (7-25), functional groups containing heteroatoms usually have strong IR absorptions. On the contrary, carbon–carbon bonds with similar substituents at each side have weak IR absorptions.

7.3 Characteristic Frequencies of Functional Groups

7.3.1 Functional Groups Possessing Characteristic Frequencies

A polyatomic molecule may have many absorption bands. Consequently the assignment of the absorption bands could be difficult. However, a particular functional group in various compounds may have an approximate absorption frequency. Having measured many standard samples, spectrochemists have generalized the characteristic frequencies of functional groups, which are very useful for the postulation of unknown compounds.

It can be explained theoretically that functional groups have characteristic frequencies. If functional groups contain hydrogen atoms, or double or triple bonds, they have higher absorption frequencies than other functional groups because of their smaller reduced masses or the larger force constants of the bonds. In addition, these higher frequencies are just slightly affected by adjacent functional groups. Consequently, these functional groups show characteristic frequencies.

Characteristic frequencies are important for the determination of unknown functional groups. However, band shapes and absorption intensities must also be examined.

To assign an absorption band of a functional group, it is necessary that all of the three factors of the absorption band are in accord with the known data.

7.3.2
Factors Affecting Absorption Frequencies

The discussion here will focus mainly on structural factors. The absorption bands of carbonyls are sensitive to structural changes. Therefore they will be discussed as an example.

1. Electron effect

A carbonyl has a double bond producing an absorption band at a greater wavenumber. If a structural change induces it from (δ+) C=O (δ–) towards (+) C–O (–), this can displace the absorption band towards a smaller wavenumber.

i. Induction effect
   The normal absorption band of the aliphatic ketone is found at about 1715 cm^{-1}. The substitution of a carbonyl by a halogen atom increases the double bond tendency of the carbonyl (it has more difficulty in changing towards C–O) so that the carbonyl will have an absorption band at a larger wavenumber.
ii. Mesomeric effect
   The mesomeric effect is also known as the resonance effect. The absorption of amides is a typical example. All primary, secondary and tertiary amides possess an absorption band of less than 1690 cm, which results from reaction **100**.

$$R-\overset{\overset{\large O}{\|}}{C}-NH_2 \longleftrightarrow R-\overset{\overset{\large O^-}{|}}{C}=\overset{+}{N}H_2$$

100

   Resonance reduces the double bond tendency of the amide, so its absorption band is displaced towards a rather smaller wavenumber.
iii. Conjugation effect
   If a carbonyl is conjugated with another double bond, the delocalization of electrons of the carbonyl decreases the bond order of the carbonyl double bond. Therefore, its absorption band is displaced towards a rather smaller wavenumber. The absorption bands of unsaturated ketones and those of aromatic ketones are situated at about 1675 and 1690 cm^{-1}, respectively. Both values are smaller than those of aliphatic ketones.

A substitution may take effect in several respects, so it has an overall influence. The above discussion is focused only on the effect that plays a dominant role in the structural change.

2. Steric effect

i. Cyclic tension

In general, the stronger tension a cycle has, the larger wavenumber its absorption band shifts towards. For example, the absorption bands of cyclopropane can reach 3050 cm^{-1}.

ii. Steric hindrance

A conjugated system has to maintain coplanarity. If the coplanarity is deviated or distorted by a structural change, for example, by a substitution, the absorption band of the system will shift to a larger wavenumber.

3. The influence of hydrogen bonds

Both inter- and intra-hydrogen bonds weaken the bond that participates in the formation of the hydrogen bond. As a result, the absorption band of the bond shifts to a smaller wavenumber. However, the dipolar moment change of the bond affected by the hydrogen bond increases, so that its absorption intensity also increases. The absorption bands of alcohols can be listed as an example: isolated 3610–3640 cm^{-1}; dimer 3500–3600 cm^{-1}; and polymer 3200–3400 cm^{-1}.

The strong hydrogen bonds between carboxylic acid molecules can shift the hydroxyl absorption band to about 3000 cm^{-1} with its tail at about 2500 cm^{-1}.

4. Mass effect (deuteration effect)

If a group containing hydrogen atoms is deuterated, its absorption frequency will decrease because of the augmentation of its reduced mass.

On the basis of Eq. (7-15), an approximate result can be deduced:

$$\frac{\nu_{X-H}}{\nu_{X-D}} \cong \sqrt{2} \tag{7-26}$$

A compound in a solid, liquid or gaseous phase may have different IR spectra. Usually, a pressed disc containing ground powder of KBr crystals and the sample to be measured is used. IR spectra measured in this way are, in general, reproducible. On the other hand, IR spectra measured by solution may be influenced by the polarity of the solvent, the concentration of the solution, temperature, and so forth. Sometimes isomerization in solution may change the IR spectra. Therefore, a comparison of the IR spectrum of an unknown compound with that of a known compound should be made under identical experimental conditions.

7.3.3 Characteristic Frequencies of Common Functional Groups

The characteristic frequencies of common functional groups are listed in Appendix 2.

7.4 Interpretation of IR Spectra

7.4.1 Wavenumber Regions of IR Absorption Bands

An IR spectrum can be divided into the following six regions.

1. 4000–2500 cm^{-1}

This is the frequency region for stretching vibrations of the functional groups containing hydrogen atoms, that is the group of X–H (X=C, H, O, S, etc.).

i. Hydroxyl group
The absorption bands of hydroxyl groups are situated in the 3200–3650 cm^{-1} region. Inter- or intra-hydrogen bonds between hydroxyl groups have a great influence on the position, shape and intensity of hydroxyl absorption bands. Isolated hydroxyl groups, which are present in solutions of low polar solvents with low concentrations, or in the gaseous phase, have absorption bands at the large wavenumber end (3640–3610 cm^{-1}) with a sharp shape. An associated hydroxyl group has an absorption band at about 3300 cm^{-1} with a broad, blunt shape. If isolated and associated hydroxyl groups coexist, two related absorption bands, a sharp one with a rather large wavenumber and a blunt one with a rather small wavenumber, will appear in the IR spectrum. Because of the influence of strong hydrogen bonds, the strong absorption bands of carboxylic acids are situated at about 3000 cm^{-1}, which are overlapped by the absorption bands of C–H stretching vibrations (see below).
Trace moisture in the KBr could produce a small band at about 3300 cm^{-1}. If a sample contains isolated water, it shows an absorption band at 1630 cm^{-1} and an overtone at about 3300 cm^{-1}, which may interfere with the detection of hydroxyls. Therefore, it is necessary to determine the existence of a hydroxyl group by checking the hydroxyl absorption bands in the fingerprint region.

ii. Amino groups
The absorption bands of isolated primary amino groups are situated in the 3300–3500 cm^{-1} region. Their association decreases the frequency by about 100 cm^{-1}.
A primary amino has two absorption bands because of the symmetrical and asymmetrical stretching vibrations of NH_2. Primary amino groups are differentiated from hydroxyl groups distinctly by the absorption band shape, although their absorption frequencies are close.
A secondary amino group has only one absorption band because it has only one stretching vibration mode. The absorption band of the associated secondary amino is situated at about 3300 cm^{-1} with a band shape less blunt than that of the associated hydroxyl. The aromatic secondary amino has an absorption band at a larger wavenumber with a stronger intensity than that of the aliphatic secondary amino.

A tertiary amino has no absorption band in this region because of the absence of hydrogen atoms.

iii. Alkyls

The boundary between saturated and unsaturated C–H stretching absorption is 3000 cm^{-1} except for the cyclopropyl group. All saturated C–H stretching absorption bands are situated below 3000 cm^{-1}, which is very useful for differentiating the saturated C–H from the unsaturated C–H. Because of low intensity, the absorption bands of unsaturated C–H bonds usually appear to be "shoulder peaks", which are small peaks overlapping a strong band.

The absorption bands of C≡C–H groups are situated at about 3300 cm^{-1} with a sharp shape and distinguishable from other bands of other functional groups at about 3300 cm^{-1}.

Four absorption bands at ≈2960, ≈2870, ≈2925, ≈2850 cm^{-1}, respectively, are assigned as saturated C–H stretching vibrations. The first two result from CH_3 groups which are assigned as ν_{as} and ν_s of the CH_3 group, respectively, while the last two result from CH_2 groups which are assigned as ν_{as} and ν_s of the CH_2 group, respectively. Therefore, the quantitative ratio between CH_3 and CH_2 can be estimated approximately from the intensities of the four bands.

Both CH_3 and CH_2, when connected to an oxygen atom or other electronegative atom, have lower absorption frequencies than those listed above.

The aldehyde group has two absorption bands at about 2820 cm^{-1} and at 2720 cm^{-1}, respectively, which result from the Fermi resonance of ν_{C-H} and the overtone of δ_{C-H}.

The bands at about 3000 cm^{-1} can be used for the differentiation of an organic from an inorganic compound.

2. 2500–2000 cm^{-1}

This is the region for triple bonds (–C≡C, –C≡N) and accumulated double bonds (–C=C=C–, –N=C=O, etc.). Except for the two bands resulting from a CO_2 background (≈2365, 2335 cm^{-1}), all the other bands in this region should not be ignored even when their intensities are small.

3. 2000–1500 cm^{-1}

This is the region for double bonds, which is important in the interpretation of IR spectra.

A carbonyl usually shows a strong band at 1650–1900 cm^{-1}. Except for carboxylic acid salts, the carbonyl group shows a strong band with a sharp or slightly broad shape. This strong band is usually the strongest or the second strongest band in the IR spectrum.

The absorption bands of stretching vibrations of carbon–carbon double bonds appear in this region with low intensities.

The skeleton vibrations of the phenyl group have absorption bands at about 1450, 1500, 1580 and 1600 cm^{-1}. The band near 1450 cm^{-1} may not be distinct from the absorption bands for CH_2 and CH_3, while the other three bands could indicate the presence of a phenyl group.

Heteroaromatic groups have absorption bands similar to those of the phenyl ring. For example, pyran has absorption bands at about 1600, 1500 and 1400 cm^{-1} and pyridine at about 1600, 1570, 1500 and 1435 cm^{-1}.

In addition to the above-mentioned absorption bands, there are the absorption bands of C=N, N=O, etc., in this region. The $-NO_2$ group has two absorption bands because two oxygen atoms connect to one nitrogen atom. Its asymmetric stretching absorption band is situated in this region while the symmetric stretching absorption band appears at about 1350 cm^{-1}.

4. 1500–1300 cm^{-1}

In addition to the above-mentioned bands in the region, attention should be paid to the absorption bands of CH_3 and CH_2.

A methyl group has two absorption bands at about 1380 and 1460 cm^{-1}, respectively. Branching of the band at about 1380 cm^{-1} shows that two or three methyl groups connect to the same carbon atom (e.g., *t*-butyl, isopropyl groups).

A CH_2 group has an absorption band at about 1470 cm^{-1}.

5. 1300–910 cm^{-1}

Many bands appear in this region (see the next section).

6. Below 910 cm^{-1}

The bands resulting from substitutions of a phenyl ring appear in this region. Therefore, these bands offer important structural information (see Appendix 2). However, if the substituents are strong polar groups, the substitutions can not be determined by the absorption bands in this region.

The absorption bands of carbon–hydrogen bending vibrations of alkenes appear here and in the preceding region (1500–1300 cm^{-1}).

7.4.2 Fingerprint and Functional Group Regions

The absorption bands from the first to the fourth region (i.e., 4000–1300 cm^{-1}) have one thing in common, that is, every band in this region reveals the existence of a particular functional group. Therefore, this region, from 4000 to 1300 cm^{-1}, is called the functional region. In fact, the boundary of the functional region at 1300 cm^{-1} is generalized by experiments.

The region below 1300 cm^{-1} is quite different from the previous region. Although there are bands in this region that also reveal the existence of some functional groups, many bands show only an overall characteristic of a compound, just like its fingerprint. Therefore, this region is called the fingerprint region.

Numerous bands are present in the fingerprint region because of the following factors.

1. The stretching vibrations of the functional groups containing neither hydrogen atoms nor double or triple bonds have rather low vibrational frequencies.

2. All bending vibrations have lower frequencies than stretching vibrations.
3. Many absorption bands are produced by couplings of vibrations.
4. Skeleton vibrations of compounds appear in this region.

Although some absorption bands in the fingerprint region can be assigned, numerous absorption bands are difficult to assign. All the absorption bands in this region form a fingerprint for the compound to be measured. However, homologues could have a similar fingerprint. On the other hand, the fingerprint of a compound can change slightly during the preparation of the sample.

From 650 to 910 cm^{-1} in the fingerprint region is also called the substitution region of phenyl rings, which can show the substituent positions on a phenyl ring.

The IR spectrum consisting of a functional group region and a fingerprint region is ideal for structural identification. Functional groups of an unknown compound can be established from absorption bands in the functional group region. The absorption bands in the fingerprint region are very useful for comparing an unknown compound with a known compound, so that a conclusion as to whether the unknown compound is identical with the known compound or not, can be drawn. These two regions are only mutually complementary in function.

7.4.3 Key Points for the Interpretation of IR Spectra

1. Examination of the three characteristics of the absorption band in an IR spectrum: frequency, intensity and band shape [1]

An IR spectrum consists of IR absorption bands. An unknown absorption band can be assigned only if all of the three characteristics of the absorption band coincide with those of a known absorption band. Of course, the frequency of the absorption band is the most important parameter. However, the intensity and the band shape must also be examined.

For example, an absorption band at 1680–1780 cm^{-1} (this is the typical region for carbonyl groups) with a low intensity shows the existence of a carbonyl as an impurity. Take another example. All of the associated primary amino groups, associated hydroxyl groups of alcohols and alkyne hydrogen atoms can have absorption bands at about 3300 cm^{-1}. However, they may differ greatly in band shape: the associated hydroxyl groups show a broad, blunt band; the associated primary amino groups have a band with branching; and the $-C\equiv C-H$ presents a sharp band.

To sum up, only when these three characteristics have been examined, can a safe conclusion be drawn.

2. Examination of the co-existence of all related absorption bands of a functional group

A functional group has several vibration modes (stretching modes and bending modes), so it shows several absorption bands in an IR spectrum. These bands are called related bands of the functional group.

A functional group can be determined only if all related absorption bands appear and each of them coincides with the corresponding known band in the three characteristics: frequency, intensity and band shape. For example, a methyl group can be determined only by the bands at about 2960, 2870 and 1380 cm^{-1} simultaneously and by the appropriate intensities and band shapes.

3. Steps for the interpretation of an IR spectrum

The interpretation can begin with the functional group region to determine the functional groups in the unknown compound. Some absorption bands in the fingerprint region, which are bands related to the functional groups, should also be assigned. Then try to interpret the other bands in the fingerprint region. If the unknown compound is an aromatic compound, its substitution positions can be deduced from the bands at 910–650 cm^{-1}.

If the unknown spectrum is deduced as being a particular compound, the two spectra of the unknown and the known should be compared carefully, especially in the fingerprint region.

4. Application of standard IR spectra

Standard IR spectra, for example, the Sadtler standard spectra, have been widely used for comparisons with measured spectra.

7.4.4 Examples of IR Spectrum Interpretation

Example 1

An unknown compound has a molecular formula of C_8H_{16}. Its IR spectrum is shown in Fig. 7.4. Try to deduce its structure.

Solution

Based on the formula, the degree of unsaturation of the unknown compound is calculated to be 1.

The absorption band at 3079 cm^{-1} indicates the presence of the hydrogen atoms attached to unsaturated carbon atoms. Therefore, the unknown compound should belong to an alkene, which is further confirmed by the existence of the absorption band of C=C stretching vibration at 1642 cm^{-1}.

The absorption bands at 910 and 993 cm^{-1} indicate the presence of $-C{=}CH_2$ (see Appendix 2). The absorption band at 1823 cm^{-1} is that of the overtone of 910 cm^{-1}.

There are many more CH_2 than CH_3 because the absorption bands at 2928 and 1462 cm^{-1} are much stronger than those at 2951 and 1379 cm^{-1}.

To sum up, the unknown compound (at least its main part) is 1-octylene.

Example 2

An unknown compound has a molecular formula of C_3H_6O. Its IR spectrum is shown in Fig. 7.5. Try to deduce its structure.

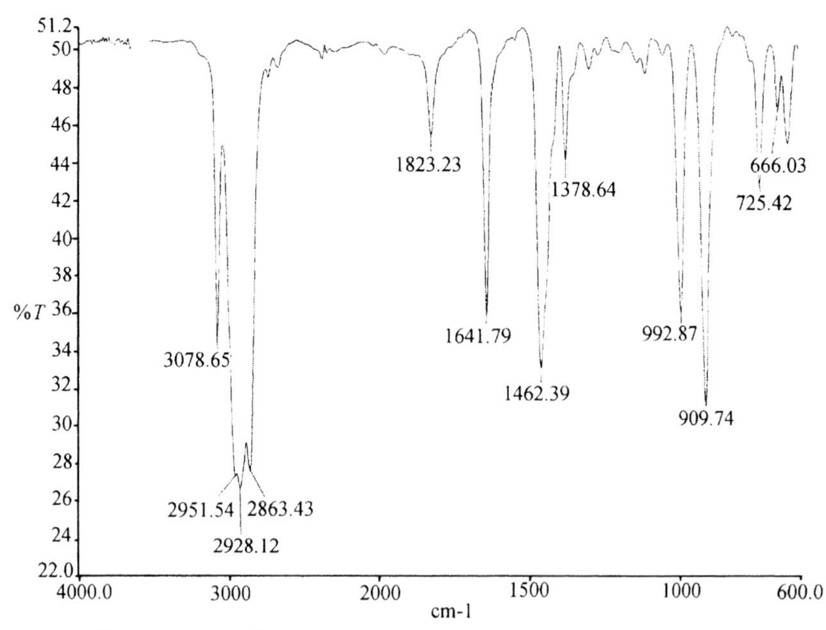

Fig. 7.4 The IR spectrum of an unknown compound, Example 1.

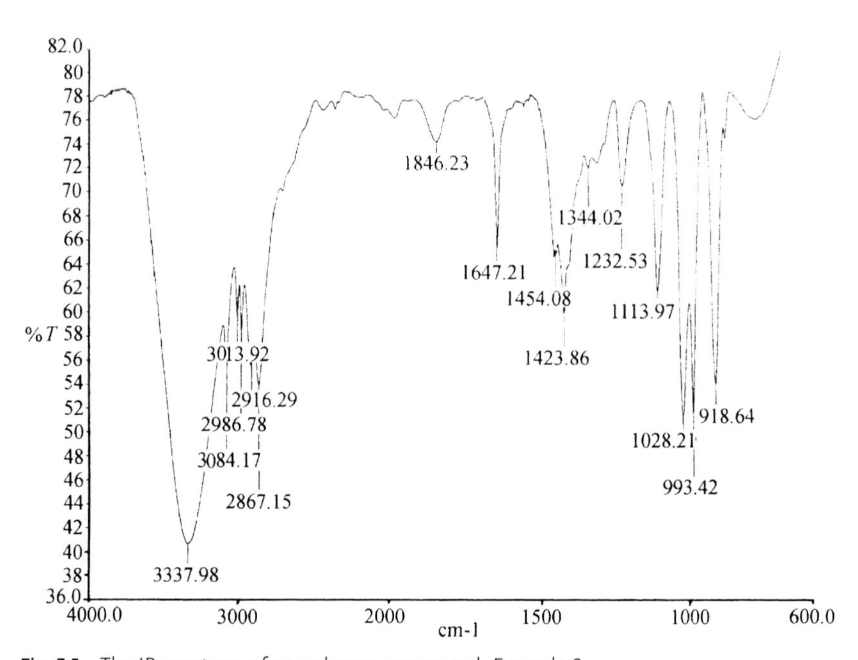

Fig. 7.5 The IR spectrum of an unknown compound, Example 2.

Solution

The degree of unsaturation of the unknown compound is calculated to be 1, from its molecular formula.

As with Example 1, an ethylene substituted at one end can be deduced from the bands at 3084, 3014, 1647, 993 and 919 cm^{-1}.

Because the compound contains an oxygen atom and the absorption band at 3338 is blunt and broad, the unknown compound should be an alcohol. This can be further proved to be a primary alcohol from the band at 1028 cm^{-1}, whose value is less than the normal value (about 1050 cm^{-1}) because the $–CH_2OH$ is connected to a double bond.

Therefore, the unknown compound is $CH_2=CH–CH_2–OH$.

The assignments of the wavenumbers of other absorption bands are as follows:

- 2987 cm^{-1}: an absorption band of $=CH_2$ (another absorption band is situated at 3084 cm^{-1}).
- 2916 and 2967 cm^{-1}: stretching vibration of $–CH_2–$.
- 1846 cm^{-1}: the overtone of 919 cm^{-1}.
- 1423 cm^{-1}: the bending vibration of CH_2. This absorption band is displaced towards a smaller wavenumber because of the connection of the CH_2 to –OH.

Example 3

An unknown compound has a molecular formula of $C_{12}H_{24}O_2$. Its IR spectrum is shown in Fig. 7.6. Try to deduce its structure.

Solution

The degree of unsaturation of the unknown compound is calculated to be 1, from its molecular formula.

A carbonyl can be determined from the strongest absorption band of the spectrum at 1703 cm^{-1}.

As the bands at 2920 and 2851 cm^{-1} are much stronger than those at 2956 and 2866 cm^{-1}, there must be many more CH_2 than CH_3 moieties that can be confirmed by the absorption band at 723 cm^{-1}, which is usually a weak absorption band.

The absorption bands at 2955 and 2851 cm^{-1} overlap on a broad band, which is characteristic of the hydroxyl group of a carboxylic acid group.

A carboxylic acid group can be determined from the absorption bands at 940, 1305 and 1412 cm^{-1}, and particularly with the absorption band of the carbonyl at 1703 cm^{-1}.

From the discussion above, it follows that the unknown compound is $CH_3–(CH_2)_{10}–COOH$.

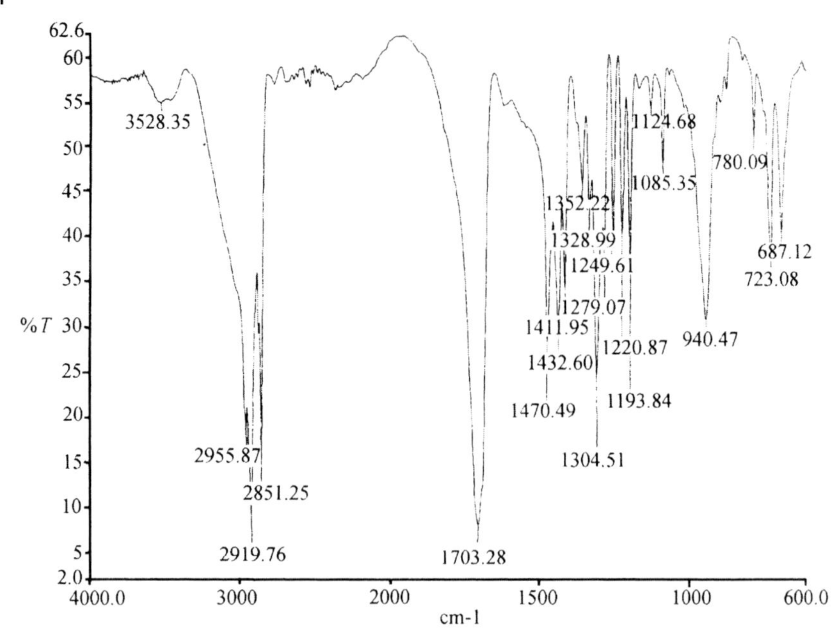

Fig. 7.6 The IR spectrum of an unknown compound, Example 3.

Example 4

An unknown compound has a molecular formula of $C_8H_8N_2$. Its IR spectrum is shown in Fig. 7.7. Try to deduce its structure.

Solution

The degree of unsaturation of the unknown compound is calculated to be 4 according to its molecular formula. Thus the compound probably contains a phenyl group, which is confirmed by the absorption bands at 3030, 1593 and 1502 cm^{-1}. The absorption band at 750 cm^{-1} shows an *ortho*-substitution of the phenyl ring.

The absorption bands at 3285 and 3193 cm^{-1} are characteristic of the associated primary amino group.

From the above-mentioned results it follows that the compound should be *ortho*-diaminobenzene.

The assignments of the wavenumbers of other absorption bands are as follows:

- 3387–3366 cm^{-1}: stretching vibrations of the primary amino group
- 1634 cm^{-1}: bending vibration of NH_2
- 1274 cm^{-1}: stretching vibration of C–N.

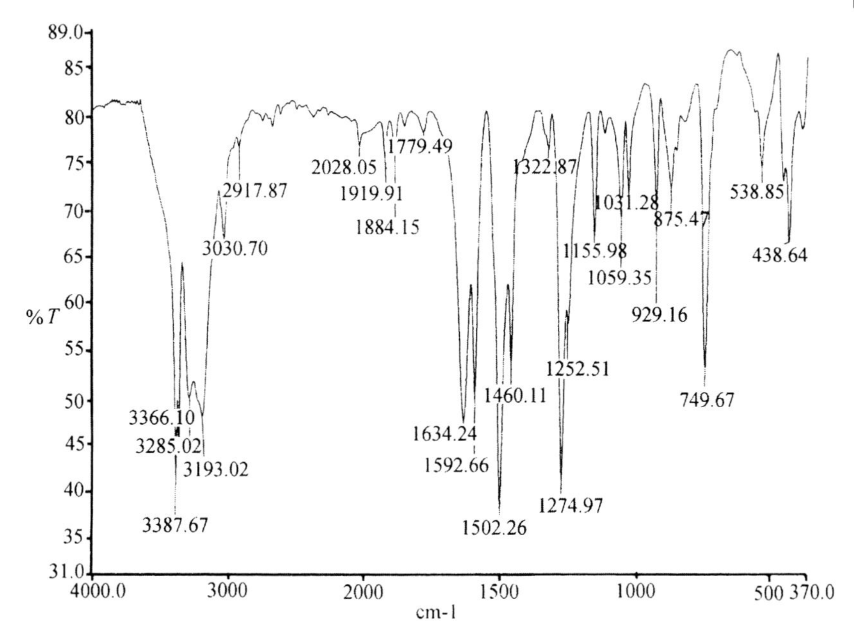

Fig. 7.7 The IR spectrum of an unknown compound, Example 4.

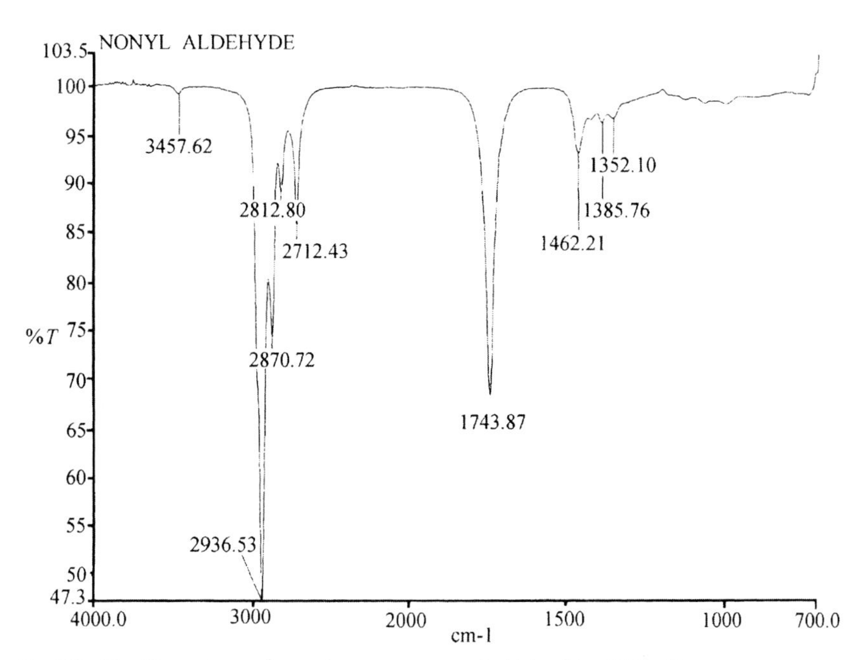

Fig. 7.8 The IR spectrum of an unknown component obtained by GC-FT-IR.

Example 5

Fig. 7.8 is the IR spectrum of an unknown component obtained by GC-FT-IR (see Section 7.5.6). From GC–MS, its molecular formula is calculated as being $C_9H_{18}O$. Try to deduce its structure.

Solution

The IR spectrum measured in the gaseous phase differs from that measured in a condensed phase in absorption position and intensity. However, the two spectra have some similarities.

The absorption band at 2936 cm^{-1} shows the presence of many CH_2 groups. The absorption band at 1744 cm^{-1} is the absorption of a carbonyl group. The absorption bands at 2812 and 2712 cm^{-1} are characteristic of the aldehyde group.

Thus, the compound proves to be a normal aldehyde, nonyl aldehyde.

7.5
Recent Developments in Infrared Spectroscopy

Infrared spectroscopy is undergoing rapid all-round development. However, because of the limited space, only several important topics are presented in this section.

7.5.1
Step Scan

Step scan is one mode of operation of FT-IR, and continuous scan, also known as rapid scan, is another mode. Rapid scan has the distinct advantages of being a rapid process, high signal-to-noise ratios, and so forth, hence since the late 1960s it has become the main mode of FT-IR. However, photo-acoustic spectroscopy (PAS), time-resolved spectroscopy (TRS), etc., require particular acquisitions, which can hardly be achieved by rapid scan. Therefore, step scan came into the spotlight in the early 1980s and has developed rapidly in recent years [2, 3].

It is necessary to understand rapid scan before one comprehends step scan.

The key part of an FT-IR spectrometer is a Michelson interferometer, as shown in Fig. 7.9.

After collimation, parallel infrared rays from the source are directed towards a beam splitter. One half of the rays are reflected by the beam splitter to a mirror, called the fixed mirror, which in turn reflects the rays back towards the beam splitter. Then this half of the rays pass through the beam splitter. The other half of the parallel infrared rays from the source pass directly through the beam splitter and strike a mirror, called the moving mirror. The half of the rays reflected from the moving mirror are then reflected again by the beam splitter. The moving mirror is moved uniformly from a starting point to a terminal point and then quickly returned back to the starting point to complete a scan. Repeated scans are accumulated to achieve an appropriate signal-to-noise ratio.

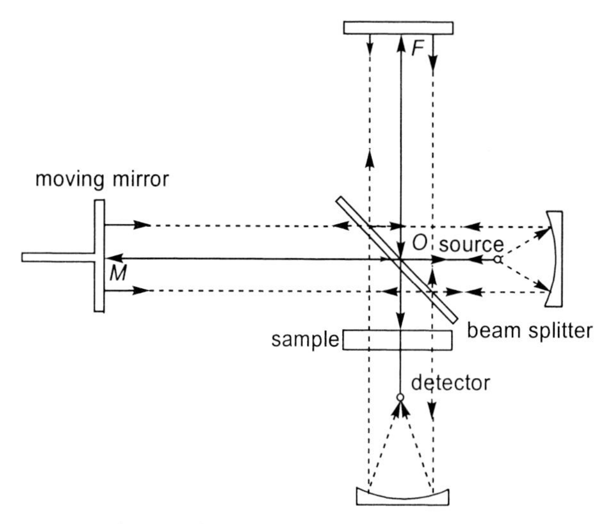

Fig. 7.9 Schematic diagram of a Michelson interferometer.

Because the two portions of the rays travel different distances, ray interference will occur. Interferential rays pass through the sample and arrive at the detector.

For simplicity, monochromatic infrared rays are discussed first.

Because the fixed mirror is motionless while the moving mirror is in motion, the two portions of rays have regular variable optical path differences, that is optical retardation, when they return to the beam splitter after their reflections by the mirrors, respectively. Suppose that the wavelength of the monochromatic rays is λ. In an extreme case, two portions of rays are in-phase when their optical retardation is zero or an even multiple of $\lambda/2$, which leads to constructive interference.

The other extreme case is that two portions of rays are out-of-phase when their optical retardation is an odd multiple of $\lambda/2$, which leads to destructive interference. A general equation can be deduced:

$$I(t) = 0.5I(\tilde{\nu}) \cos 2\pi[\tilde{\nu} \cdot 2vt] \tag{7-27}$$

where $I(t)$ is the alternating intensity measured by the detector; $I(\tilde{\nu})$ is the intensity of incident rays, which is the function of $\tilde{\nu}$; v is the velocity of the motion of the moving mirror; and t is the time of the moving mirror motion.

Suppose $f = \tilde{\nu} \times 2v$ and Eq. (7-27) becomes

$$I(t) = 0.5I(\tilde{\nu}) \cos 2\pi f \tag{7-28}$$

From Eq. (7-28), it can be understood that the measured frequency is modulated by the moving mirror instead of the original infrared frequency, $\tilde{\nu}$. The modulated frequency f is usually in the acoustic frequency range.

When monochromatic rays are scanned, the signal measured by the detector is an interferogram, a cosine curve, which is described by Eq. (7-28). If the incident infrared rays containing several frequencies are used, the signal measured by the detector will be the addition of the interferograms, which correspond to every $I(t)$ and $I(\tilde{\nu})$ for every frequency.

As a practical infrared source possesses a continuous distribution of frequencies, the measured signal is an integral over a frequency range according to Eq. (7-28), which is an overall interferogram. Its abscissa is the time or the distance of the moving mirror while the ordinate is the intensity of the overall interferential rays. The overall interferogram has a maximum, center burst, at zero optical retardation, because every monochromatic ray has constructive interference at this point. The overall signal intensity decreases rapidly with deviation from the zero optical retardation.

As in FT-NMR, the time domain signal is transformed into a frequency domain signal, an infrared spectrum, through a Fourier transform by a computer (see Fig. 1.12).

This type of scan, in which the moving mirror is moved uniformly while the other mirror is fixed, is called rapid scan because the cycle of the scan is short. The signal-to-noise ratio can be improved rapidly by scan accumulations, which is very convenient for conventional measurements. Therefore, rapid scan has been widely adopted in commercial infrared spectrometers since it came into being.

However, rapid scan shows disadvantages in some special measurements, which will be discussed in the following sections.

The difference between rapid scan and step scan is the motional mode of the two mirrors, which leads to the different relationships between optical retardation and t (the motion time of the moving mirror), as shown in Fig. 7.10. Optical retardation varies linearly with t for rapid scan while it varies stepwise for step scan.

How is step scan realized? Different instrument manufacturers have various designs, one of which is presented here [4].

Step scan is achieved by a combination of the motions of the moving mirror and the fixed mirror. The moving mirror is translated backward, while the "fixed mirror" is pulled back by a piezoelectric actuator. These two motions result in a constant optical retardation. After a period of time, which is deter-

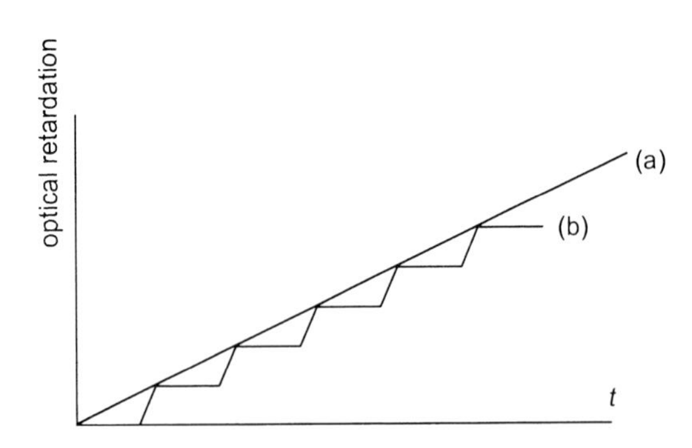

Fig. 7.10 The difference between continuous (rapid) scan (a) and step scan (b), courtesy of Digilab Company.

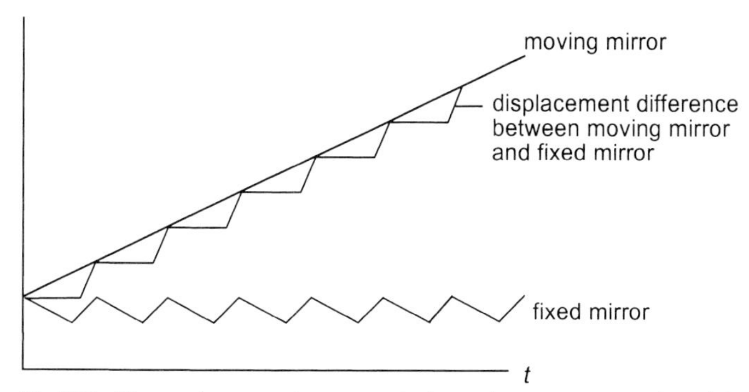

Fig. 7.11 The resultant step-by-step optical retardation, courtesy of Digilab Company.

mined by the step scan rate, the "fixed mirror" is rapidly moved forward by the piezoelectric actuator, resulting in a step in the optical retardation. This process repeats cyclically with discontinuous increasing optical retardations. The constant motion of the moving mirror and a "saw-tooth type" motion of the "fixed mirror" result in step-by-step optical retardation, which is shown in Fig. 7.11.

In the step scan mode, every acquisition is achieved at a fixed optical retardation. The acquisition starts near the beginning of the optical retardation, just after a settling time that is as short as possible. Because the acquisition is carried out over a period of time, a reliable signal-to-noise ratio can be obtained by the accumulation during this time.

From Eq. (7-28), it can be seen that the modulating frequency varies continuously in rapid scan. However, the modulation frequency in step scan is fixed, for example, 8 Hz when data are acquired in eight steps per second.

Step scan is an acquisition mode. Its Fourier transform of the data has the same calculation process as with rapid scan. Some commercial infrared spectrometers can be operated both in rapid scan and in step scan.

The advantages of step scan will be described in the following sections.

7.5.2
Photo-acoustic Spectroscopy

Photo-acoustic spectroscopy (PAS) is a particular absorption spectroscopic technique. In this section we restrict our discussion to infrared photo-acoustic spectroscopy.

The sample to be measured is put into an air-tight photo-acoustic cell. The modulated infrared rays pass through a transparent window of the cell to strike the sample. The infrared rays absorbed by the sample are transformed into heat, which is conducted from inside to the surface of the sample and finally into the gas around the sample in the cell. Because the cell is air-tight, the heat produces gaseous pressure fluctuations. As intensity of the modulated infrared rays change cyclically, the gaseous pressure also changes cyclically, thus producing a sound which is transferred as an electrical signal by a microphone. This signal is amplified by a preamplifier and an amplifier successively. Finally, it is transferred into a spectrum by the Fourier transform.

One advantage of PAS-FT-IR is that a sample can be measured directly without either sample preparation or any damage to the sample, which is very useful for special samples. Another important advantage is that PAS can achieve a depth profile analysis of a sample, which is very useful for the study of some samples, for example, composite materials. The depth profile analysis is based on the following equation:

$$\mu_s = \left(\frac{a}{\pi f}\right)^{1/2} \tag{7-29}$$

where μ_s is the solid thermal diffusion depth of the sample to be measured; a is the thermal diffusivity; and f is the modulating frequency of infrared rays.

From Eq. (7-29), it can be seen that the depth of heat transmission is determined by f. The higher frequency the modulated infrared rays have, the shallower layer the modulated infrared rays or the heat can penetrate through. The structural detail of the deep layer will be stressed if the infrared rays with a fairly low modulated frequency are used. Therefore, depth profile analysis of a sample can be achieved by using PAS with different modulation frequencies in the step scan mode.

On the other hand, the rapid scan mode cannot be used either for depth profile analysis or for analysis with a fixed depth in PAS-FT-IR.

Sowa and Mantsch have carried out a depth profile analysis of a tooth by using step scan [5].

Depth profile analyses by PAS-FT-IR can be achieved by measuring the phase angles of photo-acoustic signals [5–8]. If infrared rays are reflected in a shallow layer of the sample to be measured, the phase angle of the signal is only slightly changed because of the short time duration. However, if the infrared rays are reflected in a fairly deep layer of the sample, the photo-acoustic signal will have a phase lag. Therefore, the change of the phase angles corresponds to different depths.

The phase angle is measured by the following equation:

$$\Phi = \tan^{-1}\frac{Q}{I} \tag{7-30}$$

where Φ is the phase angle; I is the in-phase component, that is, the signal without a lag; and Q is the in-quadrature component, that is, the signal with the phase angle of 90°.

An important development of PAS-FT-IR is DSP (digital signal processing; digital signal processor) [8, 9]. If a modulation frequency of 400 Hz is applied, other higher frequencies can be present because a rectangular wave with a frequency f can be decomposed into a series of harmonic oscillations at discrete frequencies:

$$\sin(f) + \frac{1}{3}\sin(3f) + \frac{1}{5}\sin(5f) + \frac{1}{7}\sin(7f) + \frac{1}{9}\sin(9f) + \ldots$$

Thus, a sample can in fact be acted on by several frequency components if DSP is applied.

7.5.3 Time-resolved Spectroscopy

Time-resolved spectroscopy, TRS, is a technique for studying transient changes including dynamic physical or chemical changes. The time resolution can be in ms, μs or even ns.

In general, cyclic changes are studied by TRS. Let T be the time of the cycle. So the object to be studied is $F(t, T)$, which is a function of t and T. Firstly, the Fourier transform with respect to t at a selected T is carried out:

$$F(t, T) \rightarrow F(\omega, T)$$

Then T is changed so that information about different values of T is obtained.

When the moving mirror moves, both t and T vary, and therefore specific acquisition methods must be adopted to differentiate t from T.

If the rapid scan is used for TRS, a complicated method, the time-sorting method, has to be used [9]. The method is very rigorous. The reaction system, the spectrometer, and so forth, must have high reproducibility.

With the step scan mode, a set of data is collected at a particular optical retardation as a column in Fig. 7.12. All data are regrouped according to T_i. Regrouped data are transformed into a time-resolved spectrum. Therefore, TRS

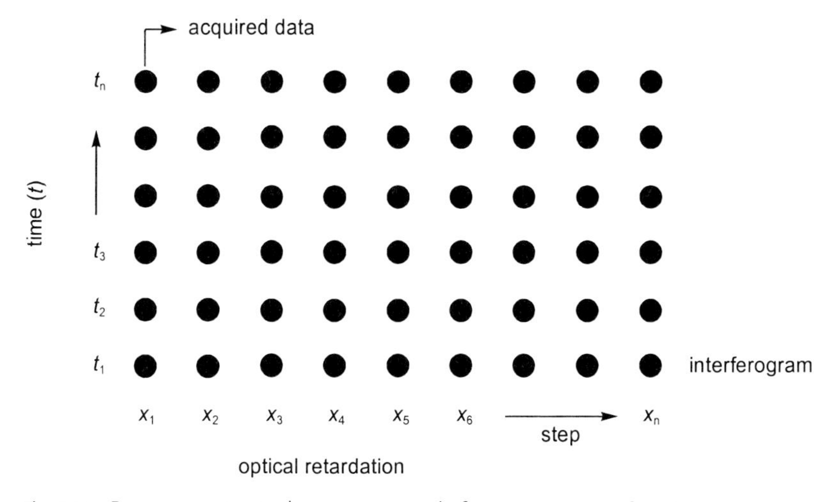

Fig. 7.12 Data regrouping in the step scan mode for TRS, courtesy of Digilab Company.

can be easily achieved. So far a time resolution of 5 ns has been achieved by using advanced spectrometers.

TRS can also be used in the study of non-reversible reactions. In this case, an effective circulation system is used to remove all substances from the reaction system before the reactants are introduced and then the reaction is triggered.

7.5.4
Two-dimensional Infrared Spectroscopy

The concept of two-dimensional infrared spectroscopy (2D IR) was proposed by Noda et al. [10–12]. The development of 2D IR originates from 2D NMR (the COSY type) so that the former is similar to the latter not only in appearance but also in interpretation.

1. The principle of 2D IR

Because the time scale for IR is in the range of 10^{-12}–10^{-13} s, which is much shorter than that for NMR, 2D IR spectra have to be produced by using an alternative appropriate method.

Some perturbation is required to excite the measured molecules. In the early period of studies on 2D IR, the applied perturbation was periodic and its detection was based on the time-resolved technique. Therefore, the instrumental specifications for 2D IR were strict and the applications limited. The general method of 2D IR was proposed by Noda et al. in 1993 [13]. The detection of 2D IR now no longer uses the time-resolved technique. A normal IR spectrometer can be used to studying 2D IR. Furthermore, the perturbation can be mechanical, optical, electrical, thermal, electromagnetic, chemical, etc. Hence, 2D IR has been applied much more widely and it is developing rapidly.

- Dynamic spectrum

  Under the action of the applied perturbation, an infrared spectrum will be changed. The dynamic spectrum $\tilde{y}(v,t)$ is defined as

$$\tilde{y}(v,t) = y(v,t) - \bar{y}(v) \tag{7-31}$$

where v is the abscissa of the infrared spectrum, i.e., wavenumber; and t is the applied perturbation. It can be the chronological time, or any physical quantity, such as temperature, pressure, concentration or voltage. The range of the action of t is:

$$T_{\min} \le t \le T_{\max}$$

$y(v,t)$ is the infrared spectrum measured under the action of t; and $\bar{y}(v)$ is a reference spectrum of the measured system.

The reference spectrum can often be set as an averaged spectrum defined by:

$$\bar{y}(\nu) = \frac{1}{T_{max} - T_{min}} \int_{T_{min}}^{T_{max}} y(\nu, t) dt \tag{7-32}$$

The reference spectrum can also be set as the infrared spectrum measured at a specific reference point ($t = T_{ref}$).

The value of $\tilde{y}(\nu, t)$ is set to zero when t is outside the boundary to be set.

- 2D IR correlation function

The 2D IR correlation function shows the correlation between the absorptions of two wavenumbers in an IR spectrum under the action of the applied perturbation.

Equation (7-31) shows the relationship between an infrared absorption and an applied perturbation. For a selected wavenumber $\nu_1, \tilde{y}(\nu_1, t)$ is obtained correspondingly. Its Fourier transform for t gives

$$\begin{aligned} \tilde{Y}_1(\omega) &= \int_{-\infty}^{\infty} \tilde{y}(\nu_1, t) e^{-i\omega t} dt \\ &= \tilde{Y}_1^{Re}(\omega) + i\tilde{Y}_1^{Im}(\omega) \end{aligned} \tag{7-33}$$

$\tilde{Y}_1(\omega)$ is the result of the Fourier transform of $\tilde{Y}(\nu_1, t)$. $\tilde{Y}_1(\omega)$, which consists of a real and an imaginary part, represents the variation of the infrared absorption at ν_1 under the action of the perturbation t. When the change in t is not continuous, the integration in Eq. (7-32) should be replaced correspondingly by the summation at discrete points.

Similarly, for another selected wavenumber ν_2, one has $\tilde{y}(\nu_2, t)$. The conjugate of its Fourier transform for t is given by the following expression:

$$\begin{aligned} \tilde{Y}_2^*(\omega) &= \int_{-\infty}^{\infty} \tilde{y}(\nu_2, t) e^{i\omega t} dt \\ &= \tilde{Y}_2^{Re}(\omega) - i\tilde{Y}_2^{Im}(\omega) \end{aligned} \tag{7-34}$$

The correlation function is $X(\nu_1, \nu_2)$ defined as

$$\begin{aligned} X(\nu_1, \nu_2) &= \frac{1}{\pi(T_{max} - T_{min})} \int \tilde{Y}_1(\omega) \cdot \tilde{Y}_2^*(\omega) d\omega \\ &= \Phi(\nu_1, \nu_2) + i\Psi(\nu_1, \nu_2) \end{aligned} \tag{7-35}$$

where $\Phi(\nu_1, \nu_2)$ is the synchronous correlation; and $\Psi(\nu_1, \nu_2)$ is the asynchronous correlation.

Synchronous correlation $\Phi(\nu_1, \nu_2)$ is the real component of Eq. (7-35), which shows a similarity in absorption intensity fluctuation at the two wavenumbers under the perturbation. Asynchronous correlation $\Phi(\nu_1, \nu_2)$ is the imaginary component of Eq. (7-35), which shows the difference in absorption intensity variation at the two wavenumbers under the perturbation.

From Eqs. (7-33), (7-34) and (7-35), it follows that

$$\Phi(\nu_1, \nu_2) = \frac{1}{2\pi(T_{\max} - T_{\min})} \int_0^{\infty} [\tilde{Y}_1^{\mathrm{Re}}(\omega)\tilde{Y}_2^{\mathrm{Re}}(\omega) + \tilde{Y}_1^{\mathrm{Im}}(\omega)\tilde{Y}_2^{\mathrm{Im}}(\omega)]\mathrm{d}\omega \tag{7-36}$$

and

$$\Psi(\nu_1, \nu_2) = \frac{1}{2\pi(T_{\max} - T_{\min})} \int_0^{\infty} [\tilde{Y}_1^{\mathrm{Re}}(\omega)\tilde{Y}_2^{\mathrm{Re}}(\omega) - \tilde{Y}_1^{\mathrm{Im}}(\omega)\tilde{Y}_2^{\mathrm{Im}}(\omega)]\mathrm{d}\omega \tag{7-37}$$

If the perturbation is the chronological time, Eq. (7-35) becomes:

$$X(\nu_1, \nu_2) = \frac{1}{\pi T} \int_{-\frac{T}{2}}^{\frac{T}{2}} \tilde{y}(\nu_1, t) \cdot \tilde{y}(\nu_2, t + \tau)\mathrm{d}t \tag{7-38}$$

where T is the cycle observed.

The generalized 2D IR treatment can be used for two different types of spectroscopy, such as IR and Raman. It is termed hetero-correlation spectroscopy.

2. Appearance and interpretation of 2D IR correlation spectra

- Synchronous 2D IR correlation spectrum

  Both the appearance and interpretation of a synchronous 2D IR spectrum are identical to those of COSY. The synchronous 2D IR correlation spectrum consists of diagonal peaks and off-diagonal peaks within a square boundary. Diagonal peaks, also called auto-peaks, have the same band coordinates as a typical IR spectrum. Important information can be obtained from off-diagonal peaks, also called cross peaks, each corresponding to the correlation between two related diagonal peaks (two IR absorption bands) as with COSY (see Section 4.6.1). The correlation means the presence of inter- or intra-molecular interactions between related functional groups.
- Asynchronous 2D IR correlation spectra

  The asynchronous 2D IR correlation spectrum consists only of off-diagonal peaks, that is, cross peaks, within a square boundary. Every cross peak illustrates that the two related functional groups are independent.

3. Applications of 2D IR correlation spectra

Although it is only a short time since 2D IR spectroscopy appeared, it has been successful in many areas, such as liquid crystalline materials [14].

Generalized 2D IR can be used much more widely. For example, Chinese traditional herbal medicines and their products can be appraised correctly using 2D IR spectra [15].

It is worthwhile emphasizing that the resolution of IR absorption bands can be improved considerably using 2D IR. Because of the broad shape of the IR absorption bands for a sample in a condensed phase, they could overlap to a great extent, which can not be overcome by enhancing the resolution of the IR spectrometer. One commonly used method for resolving bands, for example, deconvolution, usually depends on specific models and/or assumptions. However, the method for resolving overlapped bands by 2D IR (the synchronous correlation spectrum and the asynchronous correlation spectrum, especially the latter) has proved to be objective and effective.

As the wave shape of the perturbation has been developed into any arbitrary wave shape and the perturbation has been extended to an acyclic one in generalized 2D IR [16], it can be anticipated that 2D IR will have bright prospects for applications in the future.

7.5.5 Infrared Microscope and Chemical Imaging

An infrared microscope is an important optional accessory to an IR spectrometer. An IR beam with high flux is focused on a tiny area of a sample to be measured. The structural information of the sample can be obtained from the IR beam penetrating the sample (when it is transparent) or from the IR beam reflected by the sample (when it is opaque). The detectable amount can reach the order of magnitude of ng, or even pg. Note that this analysis can be non-destructive.

A visible beam passes along the same path as the IR beam. The area to be analyzed, which has a diameter of as short as 10 μm, can be found using the light of the visible beam by means of a microscope. It is possible to take a photograph or to record it on a video tape.

We studied gun shot residues (GSR) with an IR microscope. Before shooting, we stuck a piece of double-sided adhesive tape onto the shooting hand.

After shooting, particles were picked up to be analyzed by the IR microscope. One of the IR spectra is shown in Fig. 7.13, which was obtained by transmittance.

Although an extremely small amount of sample is used, the IR spectrum is perfect. The strong absorption bands at 1426, 1032 and 1008 cm^{-1} are ascribed to nitro groups of nitro-celluloses, which indicates that the powder of the bullets is made of nitro-celluloses.

The structural information of a planar sample in a small area can be obtained by an IR microscope under conditions where the sample stage is moved along the x and the y axes successively. The structural information of a particular functional group in the area can be found by using special software. The imaging method, which illustrates the content of a particular functional group in a

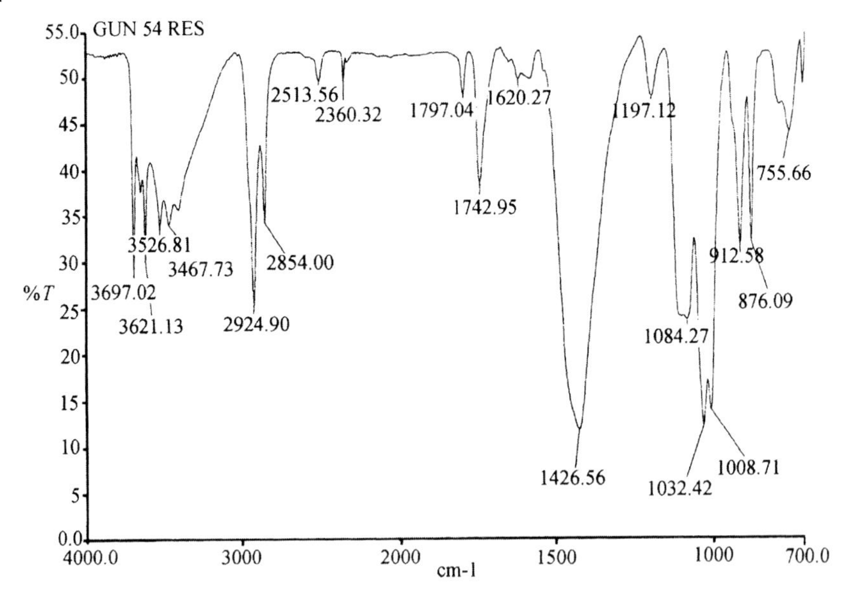

Fig. 7.13 The transmittance IR spectrum of gun shot residues.

sample, is known as chemical imaging. In general, chemical imaging is classified into vibrational spectroscopic imaging, magnetic resonance imaging, and so forth, according to the detection method.

Chemical imaging by scanning the sample successively is time-consuming. It takes several hours to record a set of data. The time duration can be significantly reduced by using a focal-plane array detector and an interferometer in the step scan mode. The detector consists of $64 \times 64 = 4096$ or $128 \times 128 = 16384$ detector elements. All detector elements record data simultaneously at every step. Because the step scan mode is in operation, the data acquired at the same step are accumulated to improve the signal-to-noise ratio for every step. In addition, acquired data are treated at every step so that the requirement for the computer memory is reduced greatly. If the number of the steps in scanning is the 256, the transformed spectrum has a resolution of 16 cm^{-1}. Note that 4096 or 16384 spectra are obtained simultaneously, which represent the distribution images of a selected functional group in a sample in detail.

7.5.6 GC-FT-IR

With knowledge of GC-MS, GC-FT-IR can be understood easily. Gas chromatography is a powerful method for separating a mixture that can be vaporized. However, the separated components are difficult to identify just by retention times. Although the mass spectra obtained from GC-MS provide structural information, which facilitates the identification of components, the identification

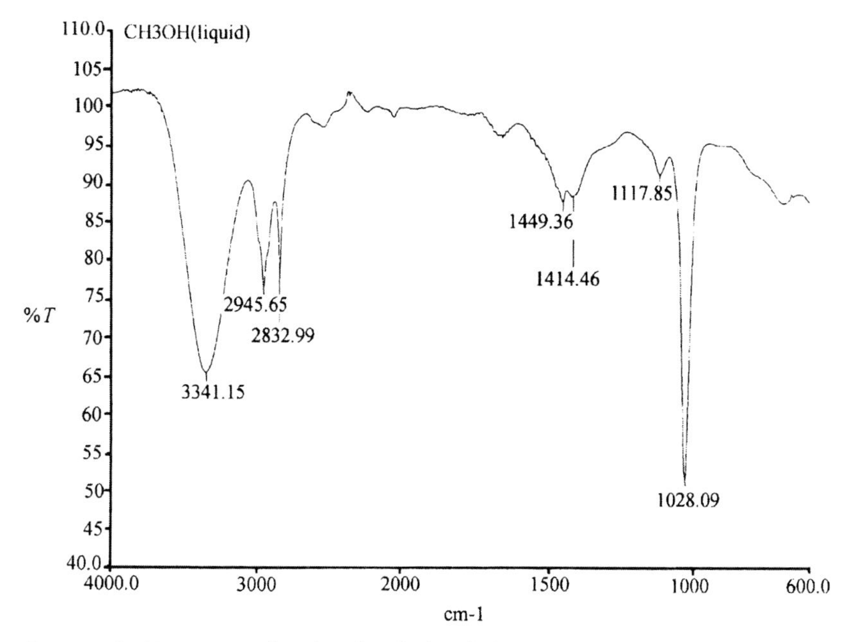

Fig. 7.14 The IR spectrum of methanol in the liquid phase.

can not always be accomplished. IR spectra offered by GC-IR provide valuable structural information, by which at least some functional groups can be determined. When the chromatographic conditions of GC-IR are identical to those of GC-MS, every IR spectrum from GC-IR can match the MS spectrum from GC-MS correctly.

Similar to the GC-MS system, the GC-IR system has a library search system, which compares every IR spectrum obtained from GC–IR with several IR spectra selected from the IR spectral library attached to the GC–IR system.

The advantage of GC-IR over GC-MS is the easy determination of some of the functional groups. However, GC–IR has a much lower sensitivity than GC-MS.

Because IR spectra of GC-IR are obtained in the vapor phase, they can be different from the IR spectra measured in the condensed phase. The IR spectrum of methanol in the liquid phase is shown in Fig. 7.14 and its IR spectrum measured in the vapor phase is shown in Fig. 7.15. Note that there are remarkable differences between the two spectra, especially in the absorption region of hydroxyl, because the association of hydroxyl is weakened considerably in the vapor phase, leading to an increase in the absorption wavenumbers of the hydroxyl group.

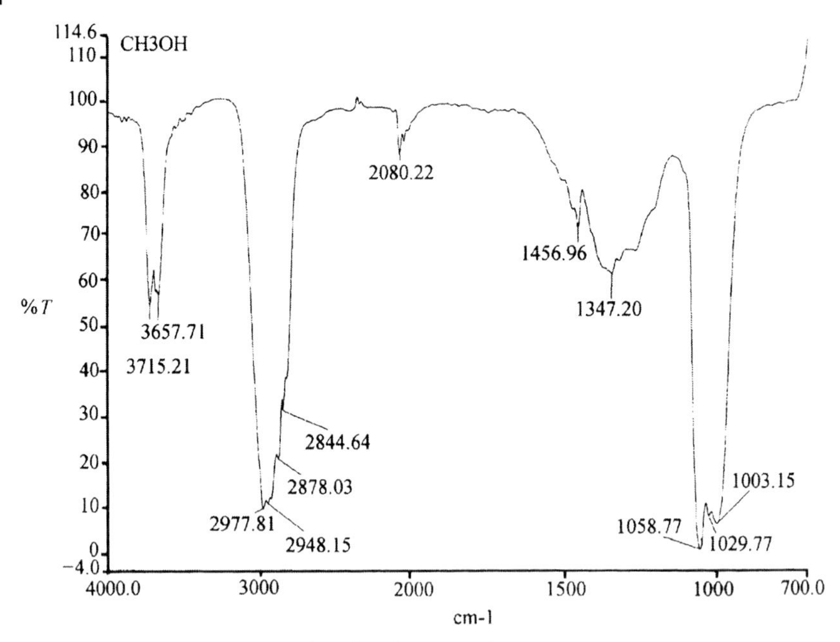

Fig. 7.15 The IR spectrum of methanol in the vapor phase.

7.6
Principle and Application of Raman Spectroscopy

Although both Raman and infrared spectroscopy are types of molecular vibrational spectroscopy, they are very different in their mechanisms. IR spectroscopy is an absorption spectroscopy while Raman spectroscopy is a scattering spectroscopy.

Because scattered light with much lower intensity than incident light is detected in Raman spectroscopy, the development of Raman spectroscopy was slow until the application of lasers, which have high intensity and good monochromaticity. The application of a laser as the light source in a Raman spectrometer is a milestone in the development of Raman spectroscopy.

7.6.1
Principle of Raman Spectroscopy

Raman scattering can be understood on the basis of the particle properties or wave properties of light. Firstly, it is necessary to understand the concept of light scattering.

1. Light scattering

Light scattering includes Raleigh scattering and Raman scattering.

- Raleigh scattering
  When a beam of light irradiates a sample, a great deal of light passes through the sample in the incident direction with a little absorption and scattering of the light. The scattered light diverges from the incident direction but possesses the same wavelength as the incident light. The intensity of the scattered light is in inverse proportion to the fourth power of the wavelength of the incident light. That is why the sky is blue because the blue light has the shortest wavelength among the components of visible light.
- Raman scattering
  In 1923 a German physicist, Smekal, predicted that scattered light with wavelengths different to that to the wavelength of the incident light should be present during scattering and that the shift in wavelength between the incident and scattered light would be characteristic of the molecular vibrations of the sample that scatters light. The prediction was verified experimentally by an Indian physicist, Raman in 1928. Because of the important contribution, he won the Nobel Prize in physics in 1930 and his name was used for this definition of the scattering.

2. Discussion of Raman scattering based on the particle properties of light

Light consists of photons, which possess energy E given by

$$E = h\nu \tag{7-39}$$

where h is Planck's constant; and ν is the frequency of light.

Raleigh scattering is elastic collisions between sample molecule and photons. Photons are scattered elastically so that their energies, and therefore their frequencies are unchanged.

Raman scattering is inelastic collisions between sample molecule and photons. Not only the motion direction of a photon but also the frequencies of a photon are changed after the collision, because the photon acquires some energy from or transfers some energy to the sample molecule during the collision.

Suppose that the sample molecule is situated at the ground state of the electronic energy levels and at the ground state or the first excited state of the vibrational energy levels. The incident photon possesses energy, which is much greater than the requirement for the transition between vibrational energy levels, but is less than the requirement for the transition between electronic energy levels. Through the interaction with the photon, the sample molecule is raised into a quasi-excited state, which is not related to the change in molecular electronic configuration and is not a fixed energy level. The quasi-excited molecules are unstable and they will return to the ground state of the electronic energy levels in one of the four ways, so as to emit photons. The four relaxation processes are shown at the top of Fig. 7.16.

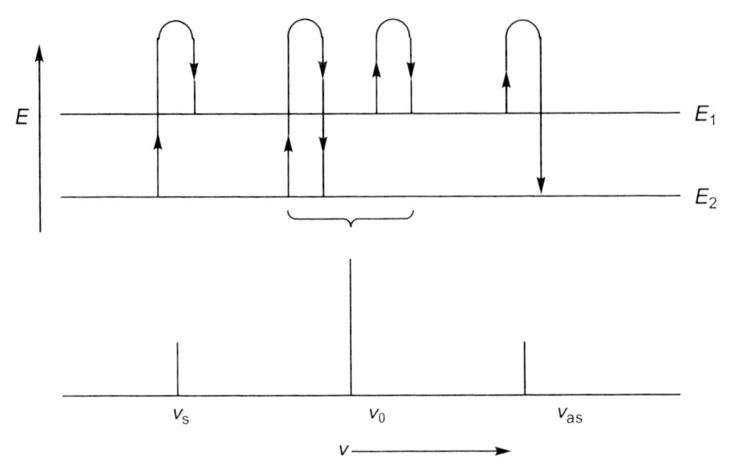

Fig. 7.16 Raleigh and Raman scattering.

If sample molecules are situated at the same energy level before the excitation and after the relaxation, they emit photons with an energy equal to that of the incident photons, which gives rise to Raleigh scattering. If sample molecules return to a higher vibrational energy level after the relaxation than they occupied before the excitation, they emit photons with a frequency given by

$$\nu_S = \nu_0 - \frac{E_1 - E_0}{h} \tag{7-40}$$

where ν_S is the Stokes frequency; ν_0 is the frequency of the incident photons; E_0 and E_1 are the energy at the ground state and the first excited state of the vibrational energy levels, respectively; and h is Planck's constant.

If sample molecules return to a lower vibrational energy level after the relaxation than that occupied before the excitation, they emit photons with frequency ν_{aS} given by

$$\nu_{aS} = \nu_0 + \frac{E_1 - E_0}{h} \tag{7-41}$$

where ν_{aS} is the anti-Stokes frequency.

Both the Stokes line and the anti-Stokes line are Raman scattering. According to Boltzmann's distribution law, the population at the ground state is greater than that at the first excited state. Therefore, the Stokes line is stronger than the anti-Stokes line. The frequency difference between the Rayleigh and the Stokes lines is equal to that between the Rayleigh and the anti-Stokes lines.

3. Discussion of Raman scattering on the basis of the wave properties of light
According to the wave properties of light, light is an electromagnetic wave, that is an alternating electromagnetic field traveling in one direction. The alternating electric and magnetic components fluctuate in planes at right angles to each other and to the path of the propagation of the wave. The alternating electric field intensity E is given by

$$E = E_0 \cos(2\pi\nu' t) \tag{7-42}$$

where E is the electric field intensity of the electric component at time t; E_0 is the amplitude of the electric component; and ν' is the frequency of the electromagnetic wave.

The electron cloud of the sample molecules can be distorted by the action of the alternating electric field. Consequently, an induced electric dipolar moment μ is produced,

$$\mu = \alpha E \tag{7-43}$$

where α is the polarizability of the simple molecule, which is a molecular property. Under the influence of an applied field, whether the deformation of the electron cloud is easy or difficult is measured by the polarizability.

Raman scattering can be produced when the polarizability is changed during molecular vibrations. Take a diatomic molecule as an example.

Suppose that the polarizability of the molecule changes with the molecular vibration. The polarizability can be developed according to Taylor's series. If the terms with higher powers are neglected, we have

$$\alpha = \alpha_0 + \left(\frac{d\alpha}{dq}\right)_0 q \tag{7-44}$$

where α_0 is the polarizability of the molecule at the equilibrium position; q is the vibrational coordinate, $q=r-r_e$ (see Section 7.2.1); and $\left(\frac{d\alpha}{dq}\right)_0$ is the derivative of α with respect to q at the equilibrium position.

The substitution of Eqs. (7-42), (7-44) and (7-14) into Eq. (7-43) gives

$$\begin{aligned}\mu &= \alpha E \\ &= \left[\alpha_0 + \left(\frac{d\alpha}{dq}\right)_0 q\right] E_0 \cos(2\pi\nu' t) \\ &= \left[\alpha_0 + \left(\frac{d\alpha}{dq}\right)_0 q_0 \cos 2\pi\nu t\right] E_0 \cos(2\pi\nu' t) \\ &= \alpha_0 E_0 \cos 2\pi\nu' t + q_0 E_0 \left(\frac{d\alpha}{dq}\right)_0 (\cos 2\pi\nu t)(\cos 2\pi\nu' t)\end{aligned} \tag{7-45}$$

However, as

$$\cos \alpha \cos \beta = \frac{1}{2}[\cos(\alpha + \beta) + \cos(\alpha - \beta)] \qquad (7\text{-}46)$$

Equation (7-45) can be rewritten as

$$\mu = \alpha_0 E_0 \cos 2\pi\nu' t + \frac{1}{2} q_0 E_0 \left(\frac{\mathrm{d}\alpha}{\mathrm{d}q}\right)_0 [\cos 2\pi(\nu' + \nu)t + \cos 2\pi(\nu' - \nu)t] \qquad (7\text{-}47)$$

The first term $\alpha_0 E_0 \cos 2\pi\nu' t$ represents an emission with the same frequency as the incident electromagnetic wave, that is the Raleigh scattering. The second term of the above equation represents two emissions with frequencies $\nu' + \nu$ and $\nu' - \nu$. This is Raman scattering, which consists of the anti-Stokes line (with the frequency $\nu' + \nu$) and the Stokes line (with the frequency $\nu' - \nu$). The displacements in the frequency are called the Raman shifts. The condition for producing Raman scattering is $\left(\frac{\mathrm{d}\alpha}{\mathrm{d}q}\right)_0 \neq 0$, that is, the polarizability changes with the vibration of the molecule. As for the incident light, it can be ultraviolet, visible or infrared light.

Both the absorption frequencies in infrared spectroscopy and the frequency displacements in Raman spectroscopy are molecular vibrational frequencies. However, infrared spectroscopy observes vibrations during which the dipolar nature of the sample molecule is changed, while Raman spectroscopy observes vibrations during which the polarizability of the sample molecule is changed. In general, infrared spectroscopy is suitable for measuring polar functional groups while Raman spectroscopy is suitable for measuring non-polar functional groups. Functional groups with high local symmetry, such as C=C, S–S and cycloalkyl groups, usually have high Raman activity. However, some polar groups for example, the nitro group, can also have high Raman activity.

For a molecule with a center of symmetry, vibrations that are infrared active are not Raman active, and vice versa. Compounds being studied do not always have a center of symmetry. Therefore, their functional groups may have both infrared activity and Raman activity. However, the intensities in an infrared spectrum and those in a Raman spectrum can be significantly different. There are also some functional groups that have neither infrared nor Raman activity.

7.6.2
Advantages and Applications of Raman Spectroscopy

Raman spectroscopy has the following advantages over infrared spectroscopy:

1. Some functional groups with weak absorption bands in infrared spectra may have strong peaks in Raman spectra. They are symmetrically substituted S–S, C=C, C≡C, N=N, C≡N, C=S, C=N=C, and so forth.

   Cyclic compounds have a strong Raman peak for their symmetrical stretching vibration, so they can be detected by Raman spectroscopy.

2. Molecules containing heavy atoms have rather low absorption frequencies. Their structural information can easily be obtained using Raman spectroscopy, because frequency displacements in Raman spectroscopy can be extended to a low wavenumber, for example, 40 cm^{-1}.
3. Raman spectroscopy is more sensitive than IR spectroscopy to structural changes. For example, although heroin, morphine and codeine have the same skeleton, they have distinct differences in their Raman spectra around 600–700 cm^{-1} [17].

   Raman spectra can show distinct differences in configurations and/or conformational variations (see Section 10.3).
4. Water is almost totally opaque to infrared radiation but transparent to visible light. Consequently, Raman spectroscopy with visible light as the source is especially suitable for measuring samples in an aqueous solution.
5. In general, peaks in Raman spectra are sharper than those in infrared spectra, so they can be resolved better and are seldom superimposed on top of each other. In addition, because of the absence of overtones and combination tones, the Raman spectrum is simpler than the IR spectrum of the same sample.
6. Solid samples can be measured directly without sample preparation, which is very useful and convenient for some areas of research, such as those of tacticity, crystallinity and degree of orientation for polymer samples.

Because glass is transparent to visible light, the sample solution in an ampoule can be measured directly using Raman spectroscopy with visible light as the source.

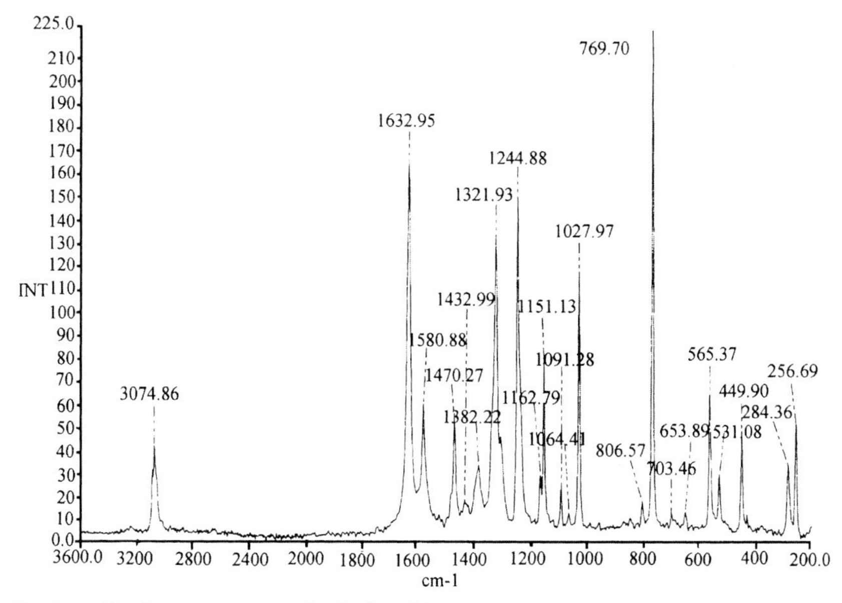

Fig. 7.17 The Raman spectrum of salicylic acid.

The Raman spectrum of salicylic acid (*ortho*-hydroxyl benzoic acid) is shown in Fig. 7.17. It is obvious that peaks are sharp and well resolved. The peak positions are almost the same as those in its infrared spectrum but the peak intensities are significantly different. For example, the strongest peak in Fig. 7.17 is situated at 769.7 cm^{-1} while the strongest band in its infrared spectrum is that for carbonyl.

Another advantage of Raman spectroscopy over infrared spectroscopy is better resolution in microanalysis because visible light, which has shorter wavelengths than those of infrared rays, can be used as the light source in Raman spectroscopy. The resolution in microanalysis is determined by the wavelength of the light to be used because of the restrictions due to diffraction. The shorter the wavelength of light, the better the resolution in microanalysis that can be obtained.

7.6.3 FT Raman Spectrometer

From what has been discussed in the preceding chapters it follows that all types of spectrometers, including NMR, MS and IR instruments, have developed from a non-Fourier transform mode to a Fourier transform mode, and this is also the case for a Raman spectrometer. This is a general trend.

A dispersion Raman spectrometer uses an Ar (514.5 nm) or He–Ne laser (632.8 nm) as its light source with a rotating grating and a fixed slit as the wavelength selection device to separate Rayleigh scattering, and a photomultiplier as the detector.

This has the following disadvantages [18]:

1. Because of the application of a visible laser, it is easy to produce fluorescence, which could come from the sample to be measured or from impurities in the sample. However, for an incident flux of 10^8 photons, on average, only one photon will be Raman scattered [19]. Therefore, the fluorescence could seriously interfere with the Raman scattering.
2. As wavelength selection by gratings has a low reproducibility, it is difficult to record difference spectra.
3. Visible light photons, which possess high energies, can photodecompose the sample being analyzed.
4. As the recording of a spectrum by a dispersion instrument is very slow, the rapidly changing system can not be studied using this type of instrument.

The FT Raman spectrometer also uses a Michelson interferometer with a Nd:YAG (yttrium aluminum garnet doped with neodymium) laser as the light source; and it uses an InGaAs or Ge detector, which possesses a good response to near infrared rays. A Fourier transform Raman spectrometer has the following advantages over a dispersion Raman spectrometer:

1. Because of the application of near infrared rays, which have a much lower energy than that of visible rays, as the light source, the problem of producing fluorescence can be greatly alleviated. Compounds that can not be measured

by a dispersion Raman spectrometer can be analyzed by an FT Raman spectrometer, such as:

- Compounds that produce fluorescence, for example, aromatic and hetero-aromatic compounds
- Polymers that produce strong fluorescence under the irradiation of visible light, such as synthetic rubber

Photons with low energy can also alleviate the photodecomposition problem.

2. Using a Michelson interferometer, the FT Raman spectrometer has good wavelength reproducibility, so that it can easily record difference spectra, which are particularly useful in some cases. For example, the spectrum of a medicine, which is in a tablet form and mixed with a filler, can be obtained easily by the subtraction of the filler spectrum from the tablet spectrum even if the concentration of the medicine is low. Another example is the measurement of a dye in a piece of textile.
3. Because of rapid acquisition, an FT Raman spectrometer can be used for the study of systems in which chemical reactions or physical variations occur rapidly.
4. A Michelson interferometer can be used both in an infrared spectrometer and in a Raman spectrometer so the cost will be more economical if users buy both types of spectrometers.
5. An FT Raman spectrometer has good prospects for telemetry because near infrared rays pass readily through the photo-conductive fiber.

Unfortunately, an FT Raman spectrometer can not improve the signal-to-noise ratio because Raman scattering is wavelength-dependent and the cross section for scattering falls off as $1/\lambda^4$. In the case of the substitution of a Nd:YAG laser for an Ar laser, the cross section for scattering decreases to 1/16, which counteracts the advantage of high flux offered by the Michelson interferometer. However, the signal-to-noise ratio can be improved quickly by means of rapid acquisition accumulations for the FT Raman spectrometer.

It must be pointed out that an FT Raman spectrometer can not surpass the dispersion Raman spectrometer in all aspects, which is different to the situation for FT-NMR, FT-MS and FT-IR spectrometers.

An FT-Raman spectrometer has the following shortcomings:

1. The lower limit of the measurable wavenumber is 50 cm^{-1} at least for the FT-Raman spectrometer while the limit to the wavenumber for a dispersion Raman spectrometer can be a few or even 1 cm^{-1} because the slit used in a dispersion instrument for wavelength selection is much more effective than the filter device used in an FT-Raman instrument.

   The upper limit of measurable wavenumber for an FT instrument is about 3600 cm^{-1}, which is lower than that for a dispersion instrument (about 7000 cm^{-1}).
2. Studying biological samples by an FT-Raman instrument is restricted because of the absorption of near infrared rays by water.

3. A dispersion Raman spectrometer can excite various systems by changing the light source while an FT-Raman spectrometer uses only a Nd:YAG source.
4. A strong background in a spectrum will be produced when the FT–Raman spectrometer records a sample with a dark color.
5. An FT–Raman spectrometer has worse resolution than a dispersion Raman instrument.
6. The detector of an FT-Raman spectrometer is less effective than the photomultiplier used for a dispersion Raman instrument. In addition, the detector of an FT Raman instrument has to be cooled by liquid nitrogen.

7.7 References

1 K. Nakanishi, P.H. Solomon, *Infrared Absorption Spectroscopy*, 2nd Edn., Holden-Day, San Francisco, 1977.
2 D. Debarre, A.C. Boccara et al., *Appl. Opt.*, **1981**, *20*, 4281–4285.
3 R.A. Palmer, J.L. Chao et al., *Appl. Spectrosc.*, **1993**, *47*, 1297–1310.
4 R.A. Crocombe, S.V. Compton, *FTS/IR Notes*, **1991**, No. 82, Bio-rad.
5 M.G. Sowa, H.H. Mantsch, *Appl. Spectrosc.*, **1994**, *48*, 316–319.
6 E.Y. Jiang, R.A. Palmer et al., *Appl. Spectrosc.*, **1997**, *51*, 1238–1244.
7 E.Y. Jiang, R.A. Palmer et al., *J. Appl. Phys.*, **1995**, *78*, 460–469.
8 K.L. Drapcho, R. Curbelo et al., *Appl. Spectrosc.*, **1997**, *51*, 453–460.
9 W.G. Fately, J.L. Koenig, *J. Polym. Sci.: Polym. Lett. Ed.*, **1982**, *20*, 445–452.
10 I. Noda, A.E. Dowrey et al., *Appl. Spectrosc.*, **1988**, *42*, 203–216.
11 I. Noda, *Appl. Spectrosc.*, **1990**, *44*, 550–561.
12 I. Noda, *J. Am. Chem. Soc.*, **1989**, *111*, 8116–8118.
13 I. Noda, A.E. Dowrey et al., *Appl. Spectrosc.*, **1993**, *47*, 1317–1328.
14 V.G. Gregorion, J.L. Chao et al., *Chem. Phys. Lett.*, **1991**, *179*, 491–496.
15 Su-Qin Sun et al., *Two-Dimensional IR Spectra to Identify Chinese Medicines* (in Chinese edition), Chemical Industry Publishing House, Beijing, 2003.
16 A. Nabet, M. Pezolet, *Appl. Spectrosc.*, **1997**, *51*, 466–469.
17 P. Hendra, C. Jones, G. Warnes, *Fourier Transform Raman Spectroscopy, Instrumentation and Chemical Applications*, Ellis Horwood, Chichester, 1991.
18 D.B. Chase, *Anal. Chem.*, **1987**, *59*, 881A–889A.
19 D.B. Chase, *J. Am. Chem. Soc.*, **1986**, *108*, 7485–7488.

8
Identification of an Unknown Compound through a Combination of Spectra

The preceding chapters discussed the correlation between structures and spectra, such as ^{1}H, ^{13}C, 2D NMR spectra, mass spectra and infrared spectra. In this chapter, on the basis of the combination of these spectra, the structure of an unknown compound will be deduced.

Although GC-MS, LC-MS and GC-IR can deal with a mixture, a component in the mixture can not be identified through the related data. Therefore, the separation of pure components from a mixture is necessary. The separation can be by various chromatographic methods such as column chromatography and thin-layer chromatography. Sometimes, re-crystallization and/or distillation, etc., can be applied as pretreatment methods. In any case, an unknown compound should be identified in a pure or approximately pure condition.

Although ultraviolet spectra can provide structural information on some types of compounds, a UV spectrum is not necessary in general for the identification of an unknown compound. Therefore, this monograph does not describe ultraviolet spectroscopy.

The role of NMR spectra in identifying an unknown compound is emphasized because NMR spectra, which includes ^{1}H, ^{13}C and many types of 2D NMR spectra, provide the most abundant structural information and compared with other spectra they are the easiest to interpret. Here the structural identification of organic compounds is based on NMR spectra. The method is straightforward and reliable.

This chapter deals with the structural identification of organic compounds in two steps: on the basis of one-dimensional and then on two-dimensional NMR spectra. Of course, MS and IR spectra should be used with NMR spectra. If an unknown compound has a fairly simple structure, it can be identified by the combination of its one-dimensional NMR spectra and MS and IR spectra. If an unknown compound has a complicated structure, its 2D NMR spectra and MS and IR spectra are necessary to deduce the structure.

When an unknown compound has a high molecular weight and a complicated structure, particularly if it belongs to a new structural type, it could be that its structure can not be identified by the combination of related 2D NMR spectra, MS and IR spectra. In this case, two different routes to follow can be considered.

Structural Identification of Organic Compounds with Spectroscopic Techniques. Yong-Cheng Ning
Copyright © 2005 WILEY-VCH Verlag GmbH & Co. KGaA, Weinheim
ISBN: 3-527-31240-4

1. Prepare a single crystal of the unknown compound and determine its structure by X-ray diffraction of the single crystal, through which all structural information: connections, bond lengths and bond angles, can be obtained. However, this method will fail when a single crystal of the unknown compound can not be prepared.
2. The unknown molecule can be "cut" into several segments by specific splitting decomposition reactions. The segments, which have fairly simple structures, can be identified and the structure of the unknown molecule is "assembled" from these segments.

References [1–7] are listed for the reader who wishes to consider more examples and/or to improve their skills.

8.1 Structural Identification of an Unknown Compound by Combination of One-dimensional NMR and Other Spectra

The steps proposed for identification are as follows:

1. Inspecting spectra and obtaining obvious conclusions from these spectra
Firstly, notice the working frequency of the NMR spectrometer with which the NMR spectra to be interpreted are recorded. The higher frequency the instrument has, the narrower intervals the coupling splitting will show. The frequency can be estimated from the ordinary ^{3}J coupling constant for an aliphatic chain group.

The ratio of hydrogen atoms of functional groups in a compound can be found from the integration curve of its ^{1}H spectrum. If a particular peak, such as a methyl group, or a *para*-substituted benzene ring, can be assigned, the numbers of hydrogen atoms in functional groups in the compound can be determined.

Carboxylic acid, aldehyde, aromatic ring, alkene, methoxy, normal aliphatic chain, isopropyl, *tert*-butyl groups, and so forth, can be determined solely on the basis of the ^{1}H spectrum because of their characteristic δ values and/or peak shapes.

In a ^{13}C spectrum with complete decoupling, the number of carbon atoms related to peaks can be estimated from the heights of the peaks although the heights of the peaks are not strictly proportional to these numbers. If a molecule is not symmetrical, the number of peaks in its ^{13}C spectrum is equal to the number of carbon atoms in the molecule. When a molecule is entirely or locally symmetrical, the number of peaks in its ^{13}C spectrum is less than the number of carbon atoms in the molecule.

In interpreting a ^{13}C spectrum, it is advisable to correlate the ^{13}C spectrum with its related ^{1}H spectrum. In most cases, the order of chemical shifts of functional groups in the ^{13}C spectrum coincides with that in the ^{1}H spectrum, which can be used for the correlation. However, some exceptions could exist.

Therefore, it is H,C-COSY which provides reliable correlations between directly connected carbon and hydrogen atoms.

On the basis of the correlation, the peaks representing chemically equivalent carbon atoms can be easily determined. For example, the fact that a peak in a ^{13}C spectrum correlates one peak with six hydrogen atoms in the related ^{1}H spectrum leads to the conclusion that the peak in the ^{13}C spectrum represents two carbon atoms. On the other hand, if the distribution of the numbers of hydrogen atoms of functional groups is not decisive from the integration curve in the ^{1}H spectrum, it can be determined by combining related ^{13}C spectral data.

The information about functional group types can be deduced on the basis of the three regions in the ^{13}C spectrum, namely the aliphatic, alkene and aromatic, and carbonyl regions.

Aliphatic and aromatic compounds can be distinguished from their mass spectrum characteristics, including the intensity of the molecular ion peak, the number of fragment ion peaks and the ion series in the light mass region. If a molecule contains S, Cl and Br, their numbers can be easily determined from the intensity ratio of its molecular ion peak cluster.

Functional groups can be deduced from the functional group region and verified from the fingerprint region in its IR spectrum.

2. Determination of the molecular formula

If high resolution MS data are available, the molecular formula can be obtained or at least several possible molecular formulae can be selected. Elemental compositions of important ions can be obtained at the same time. If only low resolution MS data are available, the molecular formula is deduced by the combination of its integral molecular weight with the data obtained from other spectra.

The number of carbon atoms in a molecule can be deduced from its ^{13}C spectrum. The number of hydrogen atoms attached to each carbon atom can be calculated from DEPT spectra. The total number of hydrogen atoms can be calculated from the integration curve in a ^{1}H spectrum. Some heteroatoms can be found from the ^{1}H and ^{13}C spectra on the basis of characteristic δ values and peak shapes of the functional groups. If the mass contributed from carbon, hydrogen and heteroatoms found equals the molecular weight, the molecular formula can be determined. If the added mass is less than the molecular weight, the missing elemental composition must be heteroatoms, which can also be determined from its mass spectrum.

The presence and numbers of S, Cl and Br can be determined from the peak intensities of the molecular ion peak cluster. The nitrogen rule provides information about the number of nitrogen atoms in the molecule. The presence of oxygen atoms in the molecule can be deduced from the fragment ion masses such as 31, 45, 93 and M–18. The presence of fluorine atoms can be deduced from the fragment ions M–19, M–20 (HF) and M–50 (CF_2). Similarly, the presence of iodine atoms can be deduced from M–127. Furthermore, the existence of iodine leads to a smaller value of $I(\mathrm{M}+1)/I(\mathrm{M})$.

If there is still a mass difference, such as 32 u, the molecule could contain one sulfur atom (with the molecular ion peak cluster) or two oxygen atoms.

After acquiring the molecular formula, the degree of unsaturation can be easily calculated. If the value is greater than 4, the unknown compound should contain an aromatic ring.

3. Determination of functional groups
This topic will be discussed in Section 8.2.

4. Extension of the structural unit through functional groups found
Coupling splitting and chemical shifts in a ^{1}H spectrum are important clues to finding neighboring functional groups. The mass difference between two fragment ions, metastable ions or important rearrangement ions also provides information on the connections, as do absorption band positions in an IR spectrum. For example, a carbonyl group has a fairly small wavenumber when it is conjugated with any other double bond group. Chemical shift values in a ^{13}C spectrum could also be useful for the determination of connections.

5. Construction of several possible structures of the unknown compound on the basis of the deduction of the structural units found
If the degree of unsaturation calculated from the functional groups found is equal to that calculated from the molecular formula, it is necessary to consider how these groups are connected. If the former is less than the latter, a connection with the formation of rings should be considered.

It is important to arrange the positions of heteroatoms and unsaturated bonds correctly, because their positions have an important influence on ^{1}H, ^{13}C, MS and IR spectra.

6. Establishing the most possible structure
On the basis of the assignments of all spectra of the unknown compound, establish its most possible structure. Sometimes a comparison between measured δ values and those calculated is necessary.

8.2 Determination of the Functional Groups (or Structural Units) of an Unknown Compound

Although the presence of a functional group can sometimes be determined just from one spectrum, it is advisable to determine the presence of a functional group by a combination of several spectra.

Several common functional groups will just be taken as examples to demonstrate this point.

8.2.1
Substituted Benzene Ring

The presence of a substituted benzene ring can be shown by the following spectra:

- ^{1}H spectrum: Peaks in the 6.5–8.0 ppm region with complex peak shapes, except for *para*-substituted benzene.
- ^{13}C spectrum: Peaks in the 110–165 ppm region.
- Mass spectrum: The ion series of *m/z* 39, 51, 65, 77 and 91 (sometimes including 92) shows the presence of a substituted benzene ring. The presence of a benzene ring increases the intensity of the molecular ion peak.
- IR spectrum: Absorption bands at about 3030, 1600 and 1500 cm^{-1} show the presence of a benzene ring in the unknown compound. The absorption bands reflecting the substitution positions on the benzene ring lie between 670 and 910 cm^{-1}.

1. The number of substituting functional groups on a benzene ring
   The number can be determined from the difference between the hydrogen atoms of unsubstituted aromatic rings and the remaining aromatic hydrogen atoms which are found from the integral curve of the ^{1}H spectrum or from the result of the DEPT spectra in the aromatic region. When the molecule is symmetrical, the overlaps of the related peaks should be considered.
2. The types of substituents
   The types of the substituents can be determined by the following spectra:
   - ^{1}H spectrum
     On the basis of the three patterns in ^{1}H spectra proposed in this monograph, the types of the substituting functional groups can be easily postulated. For example, the substitution of saturated heteroatoms decreases the δ values.
   - ^{13}C spectrum
     The substitution of a benzene ring by an oxygen or nitrogen atom strongly shifts the peak of the substituted carbon atom towards the lower field direction.
   - Mass spectrum
     The substituents on a benzene ring can be deduced from the lost neutral fragments from the benzene ring (see Tab. 6.4).
   - IR spectrum
     The absorption bands of some functional groups will be shifted when they are connected to a benzene ring. Therefore, the connection of these functional groups to a benzene ring can be postulated. These groups contain nitro, nitrile, carbonyl, hydroxyl, ether groups, etc.
3. The positions of substituents
   The positions of substituents on a benzene ring can be deduced from the following data:

- ^{13}C spectrum
  The δ values of the carbon atoms of a benzene ring are very sensitive with respect to their substituents. The agreement of these calculated δ values with the measured δ values can confirm the proposed substitution position. Because of the wide region of chemical shift in a ^{13}C spectrum for a substituted benzene, the conclusion is more reliable than those obtained from other spectra.
- ^{1}H spectrum
  Chemical shift values of the remaining hydrogen atoms in a substituted benzene ring can be used to postulate the positions of substituents on a benzene ring.
  The peak shapes in the aromatic region of the ^{1}H spectrum can also be used to postulate the positions of substituents on a benzene ring.
- IR spectrum
  Absorption bands in the 670–910 cm^{-1} range can be used to postulate the positions of substituents on a benzene ring.
- Mass spectrum
  Rearrangement ions produced by the *ortho*-substitution on a benzene ring could provide the information.

8.2.2 Normal Long-chain Alkyl Groups

The spectral characteristics of a normal long-chain alkyl group are as follows:

1. ^{1}H spectrum
   Except for the terminal methyl group, α-CH_2 and β-CH_2, all other methylene groups in a normal long-chain alkyl group have a δ value of approximately 1.25 ppm.
2. ^{13}C spectrum
   Similarly, most methylene groups in a normal long-chain alkyl have a δ value of about 28 ppm in the ^{13}C spectrum.
3. Mass spectrum
   A normal long-chain alkyl group produces the ion series m/z 29, 43, 57, 71 and so forth. The tops of these peaks form a smooth curve.
4. IR spectrum
   A normal long-chain alkyl group has absorption bands at about 2920, 2850 and 1470 cm^{-1} and a weak absorption band at about 723 cm^{-1}.

8.2.3 Alcohols and Phenols

1. ^{1}H spectrum
   Both alcohols and phenols have a hydroxyl peak, which can be removed by shaking the sample solution with several drops of D_2O.

2. ^{13}C spectrum
   The hydroxyl group of an alcohol or phenol can not be found directly in the ^{13}C spectra. Its presence can be estimated from the deshielding effect on their neighboring group.
3. Mass spectrum
   An alcohol compound has no molecular ion peak in general but the peak of M–18 usually exists. Alcohol compounds produce the ions m/z 31 and/or $31+n\times14$.
4. IR spectrum
   Both alcohols and phenols have a broad absorption band at about 3300 cm^{-1}. Their C–O vibration produce one strong absorption band in the region of 1050–1250 cm^{-1}.

8.2.4 Carbonyl Compounds

The presence of carbonyl groups can be shown distinctly in ^{13}C, IR and mass spectra. Some carbonyl compounds, such as aldehydes and carboxylic acids, can also be seen in their ^{1}H spectrum.

Aldehydes, ketones, carboxylic acids, esters and acid chlorides are taken as examples. Their spectral characteristics are shown in Tab. 8.1.

8.3 Deduction of the Structure of an Organic Compound on the Basis of 2D NMR Spectra

From the description in the Chapter 4, it is clear that 2D NMR spectra provide new methods for the structural identification of organic compounds. The 2D NMR method is more objective, reliable and flexible for structural identification than the 1D NMR method. More complex structures can be identified using 2D NMR spectra than by using conventional ^{1}H and ^{13}C spectra.

The method for deducing structures of organic compounds by using 2D NMR spectra can be described in three ways, which will be discussed in the following three sections.

It is helpful to take the following into consideration:

1. The application of 2D NMR spectra includes that of conventional 1D NMR spectra. Firstly, it is inevitable that 1D NMR spectra are recorded before measuring related 2D spectra, and hence some 2D NMR spectral parameters are determined from the measurement of 1D NMR spectra. Secondly, conventional 1D NMR spectra, which possesses a better numerical resolution than that of the projections of related 2D NMR spectra, are put above and/or beside 2D NMR spectra to replace the projections, so that the resolution of the projections can be improved. Furthermore, the spectral width of 2D NMR spectra is always narrower than that of related 1D NMR spectra. Therefore,

Tab. 8.1 Spectral characteristics of some carbonyl compounds

Spectra	Aldehyde	Ketone	Carboxylic acid	Ester	Acid chloride
^{1}H NMR	δ: 9.5–10.0 ppm		δ: 10–13 ppm	δ (OR): 3.3–4.5 ppm	
^{13}C NMR	δ >195 ppm	δ is near to or greater than 200 ppm	δ: 172–182 ppm	δ: 167–178 ppm	δ: 165–173 ppm
	δ >180 ppm (α-, β-unsaturated aldehyde)	δ>185 ppm (α-, β-unsaturated ketone)	δ: 165–175 ppm (α-, β-unsaturated carboxylic acid)	δ: 158–167 ppm (α-, β-unsaturated ester)	
MS	1. When a chained aldehyde has a γ-hydrogen atom, it produces ions m/z 44 + $n\times14$ by rearrangement. 2. Ion M-1, especially for aromatic aldehydes. 3. Ion m/z 29 (aliphatic aldehyde), M–29 (aromatic aldehyde).	When a chained ketone has a γ-hydrogen atom, it produces ions m/z 58 + $n\times14$ by rearrangement	1. When a chained carboxylic acid has a γ-hydrogen atom, it produces ions m/z 60 + $n\times14$ by rearrangement. 2. Ions m/z 45 + $n\times14$.	1. When a chained ester has a γ-hydrogen atom, it produces ions m/z 74 + $n\times14$ by rearrangement. 2. Two-hydrogen rearrangement produces ions with masses having two more units than masses of those produced by simple cleavages.	1. Isotopic peak cluster of molecular ion shows the presence of chlorine. 2. When a chained acid chloride has a γ-hydrogen atom, it produces ions m/z 78 + $n\times14$ with isotopic peak clusters.
I	$\approx$2820, $\approx$2720 cm^{-1}	$\approx$1715 cm^{-1}	3000–2500 cm^{-1}	1750–1735 cm^{-1}	1815–1770 cm^{-1}
R	$\approx$1720 cm^{-1}	$\approx$1675 cm^{-1}	1725–1700 cm^{-1}		1780–1750 cm^{-1}
	$\approx$1690 cm^{-1} (α-, β-unsaturated aldehyde)	(α-, β-unsaturated ketone)	1715 ~ 1690 cm^{-1} α-, β-unsaturated carboxylic acid		(α-, β-unsaturated acid chloride)

the application of 2D NMR spectra is associated with that of 1D NMR spectra.

2. Although the most abundant structural information is provided by NMR spectra, it is recommended that other types of spectra, such as MS and IR, are also applied.
3. The structural postulation of an unknown compound or satisfactory assignments of a known or a deduced structure can be reached just by 2D NMR spectra even without any related chemical knowledge (e.g., the correlation between chemical shifts and structures). However, it is better to use the knowledge obtained from 1D NMR spectra, so that the deduction process can be completed more rapidly.

8.3.1
Shift Correlation Spectra as the Key to Structural Postulation

At present this method is well-developed and is that most commonly used. It requires the following spectra: ^{1}H spectrum, ^{13}C spectrum and its DEPT, COSY, H,C-COSY (or HMQC, HSQC) and sometimes COLOC (or HMBC). The steps in the structural postulation are as follows:

1. Determining the functional groups containing carbon and hydrogen atoms in an unknown compound

Information on the presence of C–H functional groups in an unknown compound, that is, how much CH_3, CH_2, CH and C there is, can be obtained by combining the ^{1}H spectrum, ^{13}C spectrum and DEPT. A methylene group $-CH_2-$ and a substituted ethylene $-C{=}CH_2$ can be distinguished by its chemical shifts, so can a methane group –CH– and a –C=CH in a double bond or in an aromatic ring. A carbon atom in some cases, for example where a carbon atom connects to two oxygen atoms, has a δ value of over 100 ppm. This value might be confused with that of an unsaturated carbon atom, a fact that deserves to be noted.

The abundant information from H,C-COSY should be emphasized. H,C-COSY correlates carbon atoms with hydrogen atoms, which are carried by the carbon atoms. Consequently, all hydrogen atoms can be attached to their related carbon atoms (or, leaving some exchangeable protons on heteroatoms). This assignment of the H,C-COSY is distinct because the peaks in the ^{13}C spectrum are sharp and the cross-peaks in the H,C-COSY spectrum are apparent.

It is necessary to analyze the peak shape of every peak set in an ^{1}H spectrum carefully. Finding out the pairs of equally spaced peaks in a peak set and those in coupled two-peak sets is the key to the interpretation of a ^{1}H spectrum (see Section 2.9.2). This method does not work well when there are overlapped peak sets. However, the overlapped peak sets can be resolved by using H,C-COSY.

In the ^{1}H spectrum, four peak sets are overlapped at δ_b as shown in Fig. 8.1: (1) one H on a methane group (carbon A), (2) one H on a methane group (carbon B), (3) two H atoms on a methylene group (carbon C), (4) one H on a

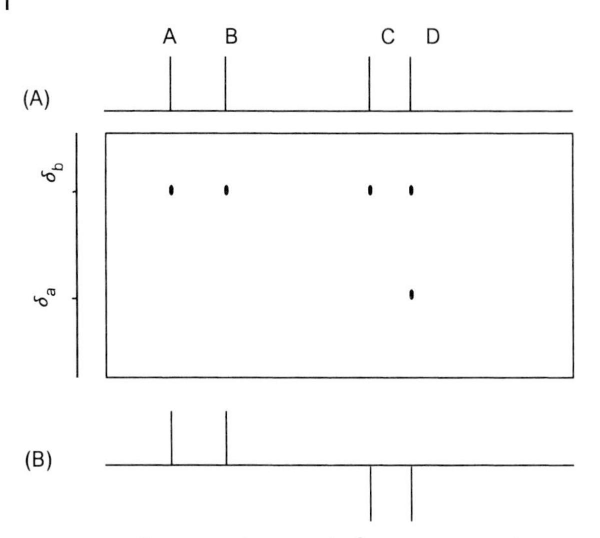

Fig. 8.1 Overlapping peak sets in the ^{1}H spectrum can be resolved. (A) H,C-COSY and (B) DEPT.

methylene group (carbon D). All the five protons in the four groups at δ_b are resolved and assigned by combining the H,C-COSY and DEPT.

A COSY reveals all ^3J coupling relationships in a compound. However, the interpretation of the COSY is troublesome if there are overlapped peak sets in a ^1H spectrum. Fortunately, such difficulty can often be overcome by using H,C-COSY.

In postulating an unknown structure, it is useful to correlate the peaks in its ^1H spectrum with those in its ^{13}C spectrum, so that functional groups containing carbon and hydrogen atoms as well as their chemical environments can be determined. For example, a CH_2 with δ_H of 5.0 ppm and δ_C of 74 ppm should be a methylene group connected to an oxygen atom and not a substituted ethylene; a CH with a small δ_H and a large δ_C should be a methane group connected to several large functional groups because δ_C is susceptible to the stereochemistry.

Now let us come back to the structural postulation of an unknown compound. So far all protons have been assigned to carbon atoms (and to heteroatoms, if they exist).

2. Determining the structural segments of an unknown compound

The carbon–carbon connections in a compound can be established through the COSY and H,C-COSY, and so can several structural segments that consist of carbon atoms and H atoms connected directly. These structural segments end at a quaternary carbon atom or a heteroatom.

Some cross peaks in COSY may be very weak, even in some particular situations with their intensities near to zero, for example, when the dihedral angle is close to 90°. On the other hand, a COSY may show some cross peaks from long-range couplings in addition to those from ^3J coupling when the long-range coupling constants have rather large values. These two facts should be kept in mind.

3. Establishing the connections around quaternary carbons

Let us establish the connections around a quaternary carbon atom through one of the spectra for a C–H long-range correlation, which include C–H long-range correlation, HMBC and COLOC. In general, the cross peaks from $^3J_{CH}$ appear frequently with those from $^2J_{CH}$ but occasionally with those from $^4J_{CH}$.

Through the long-range couplings from a quaternary carbon, a larger structural segment, which is made up of the adjacent structural segments across the quaternary carbon, can be constructed. Because a quaternary atom links up with four functional groups leading to many possibilities for connections, the correct connection can be achieved only with particular care.

4. Determining the existence of heteroatoms in the unknown compound and establishing their connections

i. Determining the existence of heteroatoms in the unknown compound
   The existence of heteroatoms can be deduced from the following spectral data:
   - ^1H spectrum: singlets with large δ values (–O–CH_3, –N–CH_3), peaks of exchangeable protons, first-order patterns in down field regions (heteroaromatic ring), etc.
   - ^{13}C spectrum: peaks of carbonyl group, cyano group, heteroaromatic ring, etc.

ii. Establishing connections of the heteroatoms with other functional groups
   There are two ways to establish the connections:
   - The C–H groups with large δ_C or δ_H values may be connected with a heteroatom.
   - The connection with a heteroatom can be found through ^{13}C–^1H long-range couplings, which can be detected by COLOC or HMBC. For example, the structural segment –C*–O–C–H*– in a saccharine can be determined through the long-range coupling between the carbon and hydrogen atoms separated by three chemical bonds.

   The structural postulation of an unknown compound is accomplished through the four steps above. It is still necessary to assign the spectra to increase the reliability of the postulated structure.

5. Assigning the spectra and checking the deduced structure

i. ^1H spectrum and ^{13}C spectrum
   It is necessary to inspect all the δ_H and δ_C values of the functional groups of the unknown compound to see whether they are reasonable or not. Chemical shifts vary with the substitution of electronegative groups, and the conjunction with double bands. In addition, δ_C varies with the substitution by large groups.

   Splitting patterns in an ^1H spectrum, which are clear when the spectrum is recorded with a high frequency spectrometer, can be used to assist in the judgment of the structure of the unknown compound.

ii. 2D NMR spectra
The cross peaks in H,C-COSY, HSQC or HMQC are generally discernible, so assignments can be easily accomplished. The assignment of the cross peaks of COSY should be done carefully. The interpretation of COLOC or HMBC must be carried out prudently because of the diverse and complex coupling relationships.
For complex compounds, it is appropriate to record NOESY-like spectra in which some functional groups adjacent in space can be shown, so that the structure can be further confirmed.

8.3.2 Deduction of the Structure of an Unknown Compound by Using Mainly HMQC-TOCSY

The method mentioned above will not work when there are serious overlapping peak sets in the ^1H spectrum. The combination of an HMQC with TOCSY is an improvement upon the former method.

Suppose that an unknown compound has the structural segment **101**.

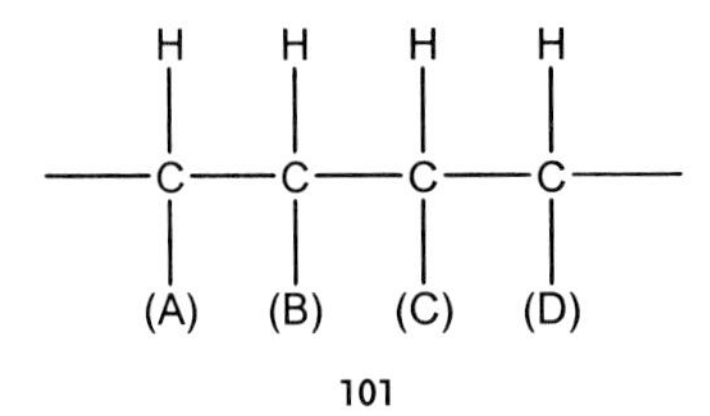

The connections of C(A)–H(A), C(B)–H(B), C(C)–H(C) and C(D)–H(D) can be drawn from HMQC. Then several HMQC-TOCSY spectra are recorded, respectively, with different mixing times. By choosing C(A) as the beginning of the observation, the correlated peaks of C(A)–H(A), C(A)–H(A)–H(B), C(A)–H(A)–H(B)–H(C) and C(A)–H(A)–H(B)–H(C)–H(D) will appear one after another as the mixing time is increased. In other words, the coupling sequence is obtained. By using the IDR (inverted direct response)-HMQC-TOCSY spectrum [8], the correlated peaks from $^1J_{CH}$ are negative and those from the couplings between indirectly connected carbon and hydrogen atoms are positive, so that the cross peak of C(A)–H(A) can be clearly differentiated from those of the couplings between indirectly connected carbon and hydrogen atoms.

The process starting at H(A) is similar to that mentioned above.

In the HMQC-TOCSY spectrum, the four cross peaks form a rectangle, as shown in Fig. 8.2, i.e., at: δ_C(A), δ_H(A); δ_C(A), δ_H(B); δ_C(B), δ_H(A); and δ_C(B), δ_H(B). The rectangle illustrates a structural unit of $-C_A(H_A)-C_B(B_B)-$.

From these rectangles the connections of an unknown compound can be found.

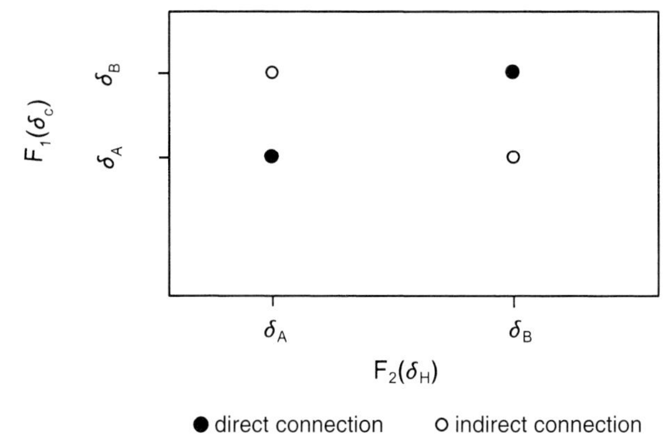

Fig. 8.2 The rectangle of the HMQC-TOCSY spectrum reveals a structural unit of $-C_A(H_A)-C_B(B_B)-$.

It is worthwhile emphasizing that the HMQC-TOCSY spectrum reveals the connections of carbon atoms, which in some cases can not be obtained by the combination of the COSY and H,C-COSY. Suppose that an unknown compound contains the following two structural units: $-C_A(H_A)-C_B(H_B)-$ and $-C_X(H_X)-C_Y(H_Y)-$ and that $\delta_H(A)$ and $\delta_H(X)$ are close to each other. In its COSY spectrum, the cross peak of H_A may be confused with that of H_X. However, the connections between C_A and C_B and between C_X and C_Y are clearly illustrated in its HMQC-TOCSY spectrum as shown in Fig. 8.3.

For the connections with heteroatoms, the reader is referred to Section 8.3.1.

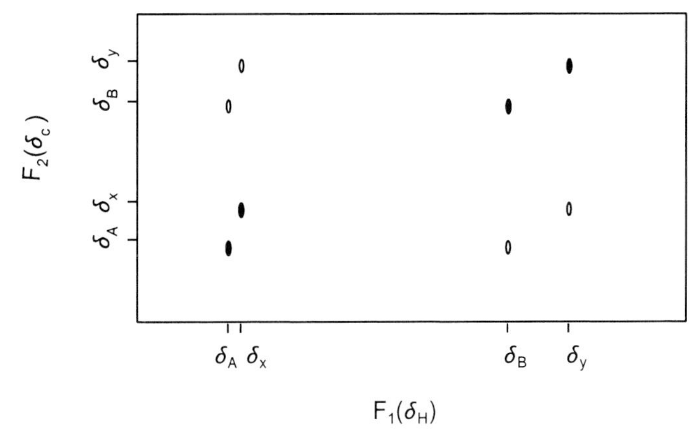

Fig. 8.3 The HMQC-TOCSY spectrum reveals exactly the connections between carbon atoms.

8.3.3
Postulating an Unknown Structure by 2D INADEQUATE

Compared with the two methods described above, this method is quite different, and can directly determine the connections between carbon atoms in an unknown compound. It is the most reliable of all methods for the structural identification of unknown compounds.

The main disadvantage of the method is a very low sensitivity, so that a considerable amount of sample and a long acquisition time are required for measurements. These problems have been gradually improved with the development of NMR spectroscopy.

1. Increasing the spectrometer frequency
   The NMR sensitivity will be enhanced considerably with an increase in the spectrometer frequency. Commercial spectrometer frequencies have reached 920 MHz, so that the accumulation time can be greatly reduced.
2. New probe development
   A new probe, of less than 50 μL in volume, has been developed. The sample to be measured concentrates around the axis, and hence its measurement efficiency is greatly enhanced.
3. Magic angle spinning (MAS)
   Dedicated to solid NMR, MAS is a method in which a sample tube rotates along the axis at an angle of 54.7° with respect to $\boldsymbol{B}_0$. The broadened linewidth at the bottom of a peak, caused by insufficient height of a sample solution, can be decreased using MAS.
4. Using special software
   Special software is used to effectively improve the sensitivity of 2D INADEQUATE. The software can identify INADEQUATE signals buried in the noise of the spectrum on the basis of the following facts:
   i. The signals of INADEQUATE are paired. The two signals of a pair are situated on a horizontal line parallel to the F_2 axis.
   ii. All paired signals are symmetrically located about the quasi-diagonal.
   iii. The signals agree with the positions of the ^{13}C peaks on the F_2 axis.
   iv. Every signal consists of a doublet with opposite intensities and with an interval equal to $^1J_{CC}$.

The software selects the signals from the considerable noise, lists the parameters of the signals in a table and then constructs the structural segment, which contains all of the carbon atoms of the unknown compound.

So far the connections between all carbon atoms in an unknown compound have been found by use of the 2D INADEQUATE spectrum. To complete the structural formula, it is necessary to:

i. Add related hydrogen atoms to the skeleton obtained according to the DEPT.
ii. Add the heteroatoms of the molecular formula to the unfinished structural formula (see Section 8.3.1).

8.4 Examples of Structural Identification or Assignment

Several examples of the structural identification of unknown compounds or the assignment of known compounds are presented in this section. The types of spectra used in the examples are not the same because the degrees of difficulty of the examples are different. The examples do not cover the preceding three sections and they are used just for explanation.

Example 1

An unknown compound needs to be identified. Its ^{13}C NMR data are shown in Tab. 8.2. Its mass spectrum, ^{1}H NMR spectrum and IR spectrum are shown in Figs. 8.4, 8.5 and 8.6, respectively.

Tab. 8.2 The ^{13}C NMR data of the unknown compound in Example 1

Ordinal number	*δ_C (ppm)*	*Carbon atom number*	*Ordinal number*	*δ_C (ppm)*	*Carbon atom number*
1	171.45	1	8	27.22	1
2	46.98	1	9	26.35	1
3	45.68	1	10	25.05	1
4	35.02	1	11	21.76	1
5	29.99	2	12	20.72	1
6	27.74	4	13	11.96	1
7	27.48	2			

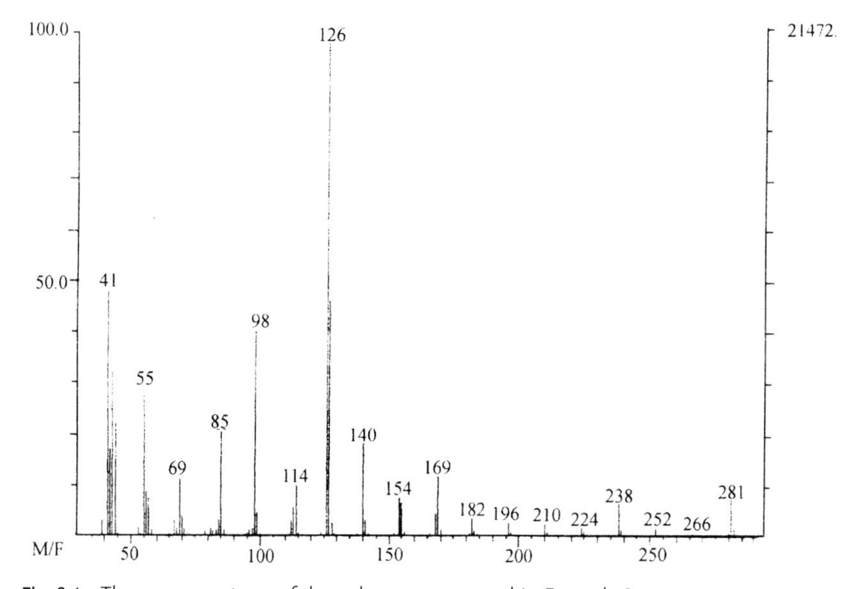

Fig. 8.4 The mass spectrum of the unknown compound in Example 1.

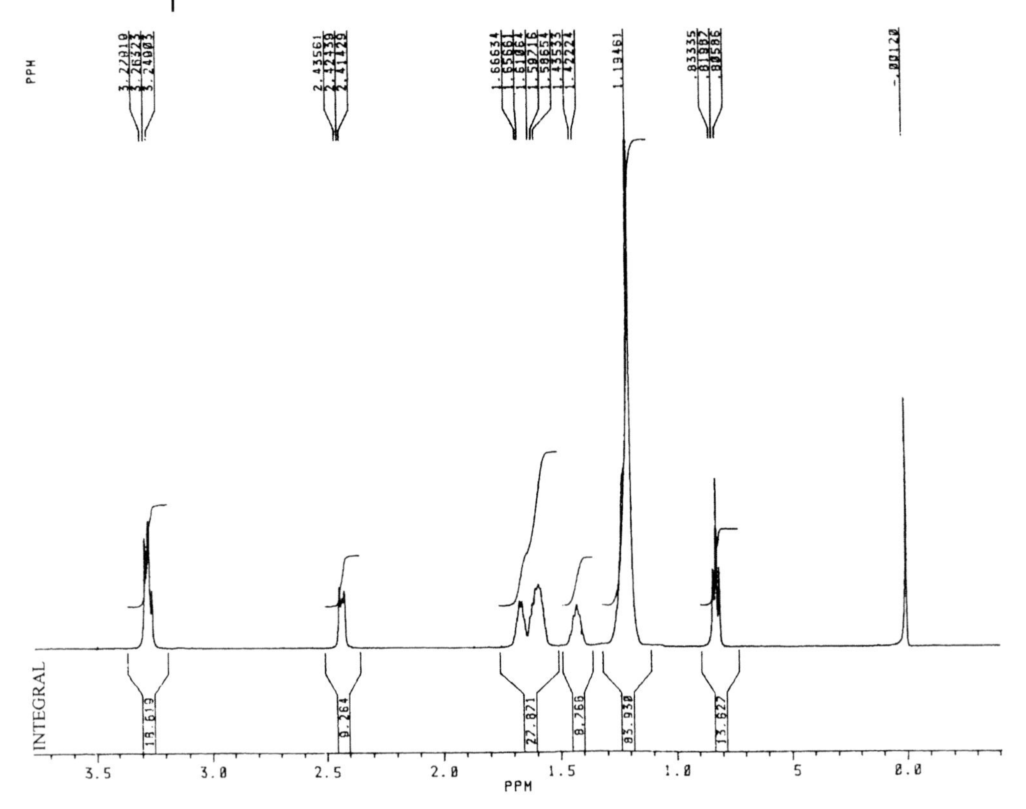

Fig. 8.5 The ^{1}H NMR spectrum of the unknown compound in Example 1.

Solution

The elemental composition of the unknown compound should be determined first.

From Tab. 8.2 it can be found that the unknown compound contains 18 carbon atoms.

The peak set at 0.819 ppm in the ^{1}H spectrum can be assigned to a methyl group as a quantitative point of reference. Consequently, it can be calculated that the unknown compound has 35 hydrogen atoms.

The peak at m/z 281 agrees with the criterion for a molecular ion peak. Therefore, this peak can be assumed to be the molecular ion peak, which accounts for the fact that the unknown compound contains an odd number of nitrogen atoms.

The fact that the unknown compound has a carbonyl group can be easily deduced from the peak at 171.45 ppm in the ^{13}C spectrum or from the absorption band at 1649.1 cm^{-1} in the IR spectrum.

By combining the conclusions above, the molecular formula of the unknown compound can be written as $C_{18}H_{35}ON$.

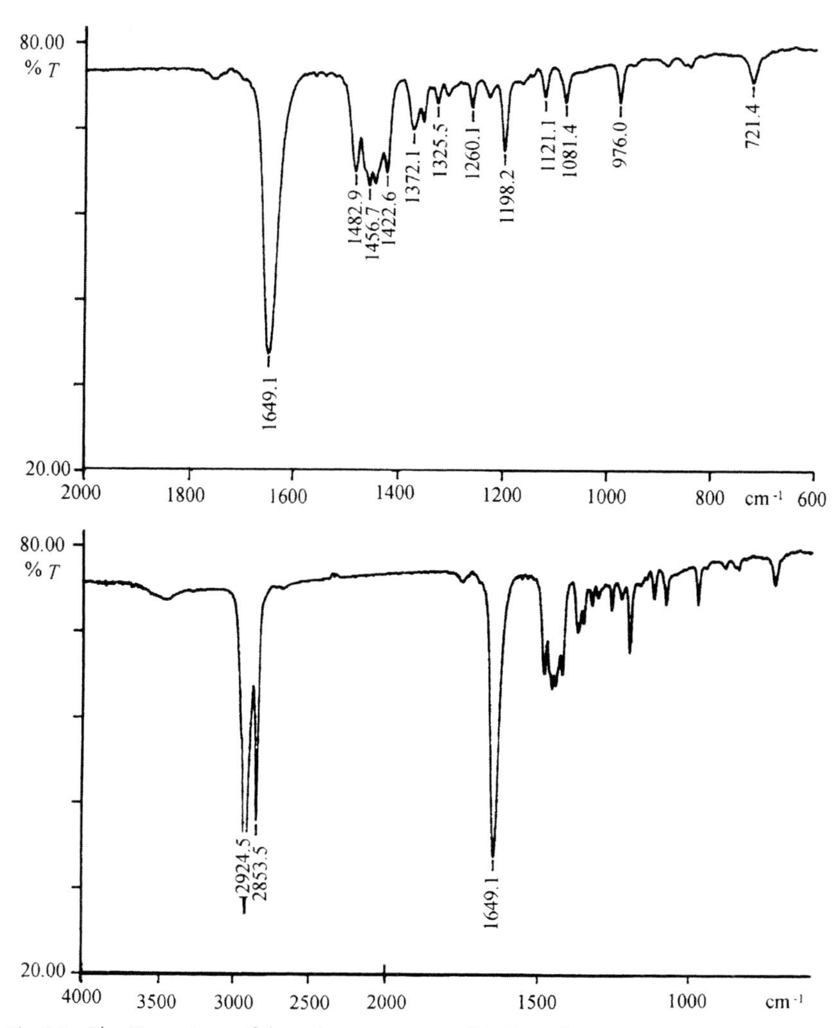

Fig. 8.6 The IR spectrum of the unknown compound in Example 1.

Now let us deduce the functional groups in the unknown compound.

1. It is easy to determine that the unknown compound contains an amide group from the following data:
   - The peak at 171.45 ppm in the ^{13}C spectrum reveals that the carbonyl group is connected to a heteroatom, which must be a nitrogen atom according to the molecular formula.
   - The absorption band at 1649.1 cm^{-1} is in accord with the typical value of an amide absorption.
2. The unknown compound contains a normal alkyl group, whose characteristics are very obviously as follows:

- The peaks at 11, 20, 21, 25, 26 and 27 ppm in the ^{13}C spectrum.
- The strong peak at 1.195 ppm and the triplet at 0.819 ppm in the ^{1}H spectrum.
- The strong absorption bands at 2924.5 and 2853.5 cm^{-1}.
- The peak clusters from 238 to 98 u spaced by 14 u in the mass spectrum.

3. The unknown compound should contain a ring to form a lactam group because of the following reasons.
   - The unknown compound contains only an unsaturated functional group (a carbonyl group), but its unsaturation degree is 2.
   - The two peaks at 46.98 and 45.68 ppm imply that the two carbon atoms should be connected by a nitrogen atom, but they are situated in different chemical environments.
   - The four absorption bands in the 1422.6–1482.9 cm^{-1} region show the presence of methylene groups in different chemical environments.

On the basis of the conclusions above the structure of the unknown compound can be written as structure **102**.

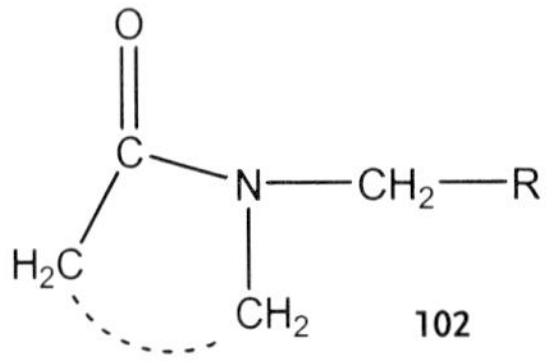

O=C(–H₂C⋯CH₂–)N–CH₂–R

102

Now the only uncertainty that remains about the structural formula is the distribution of the methylene groups between the ring and the side-chain.

The most important information relevant to the distribution is the base peak at m/z 126 in the mass spectrum, which should be as in structure **103**.

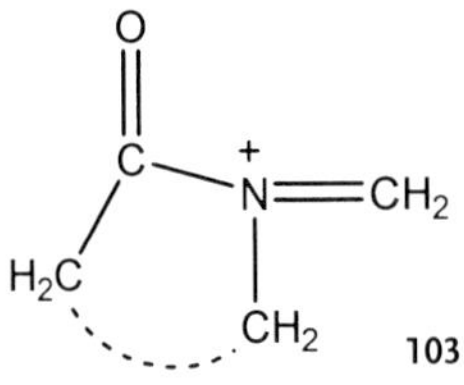

O=C(–H₂C⋯CH₂–)N⁺=CH₂

103

Thus, from m/z 126 the structural formula of the ring is structure **104**.

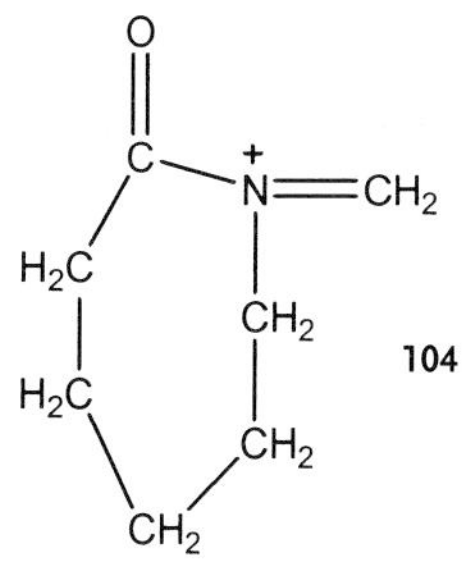

104

Therefore, the normal alkyl is $-C_{12}H_{23}$ and the unknown structure is **105**.

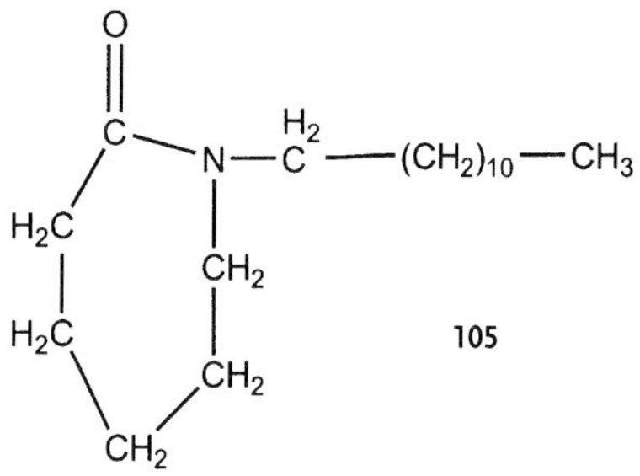

105

Example 2
The ^{13}C spectrum, ^{1}H spectrum, mass spectrum and IR spectrum of an unknown compound are shown in Figs. 8.7 to 8.10. Try to identify its structure.

Solution
The ^{13}C specimen indicates that the unknown compound contains 29 carbon atoms (notice that two peaks are close at 18 ppm, and likewise at 26 and 28 ppm).

The ^{1}H spectrum shows that the unknown compound has 39 hydrogen atoms.

The peak at m/z 461.3 agrees with the criterion for a molecular ion peak. Therefore, the unknown compound should contain an odd number of nitrogen atoms.

There are two strong absorption bands near to 1674 cm^{-1} in the IR spectrum, which should be two amide groups.

The conclusions above allow us to write the molecular formula for the unknown compound as $C_{29}H_{39}O_2N_3$. Consequently, its degree of unsaturation can be calculated as 12, which means that the unknown compound has a complex structure.

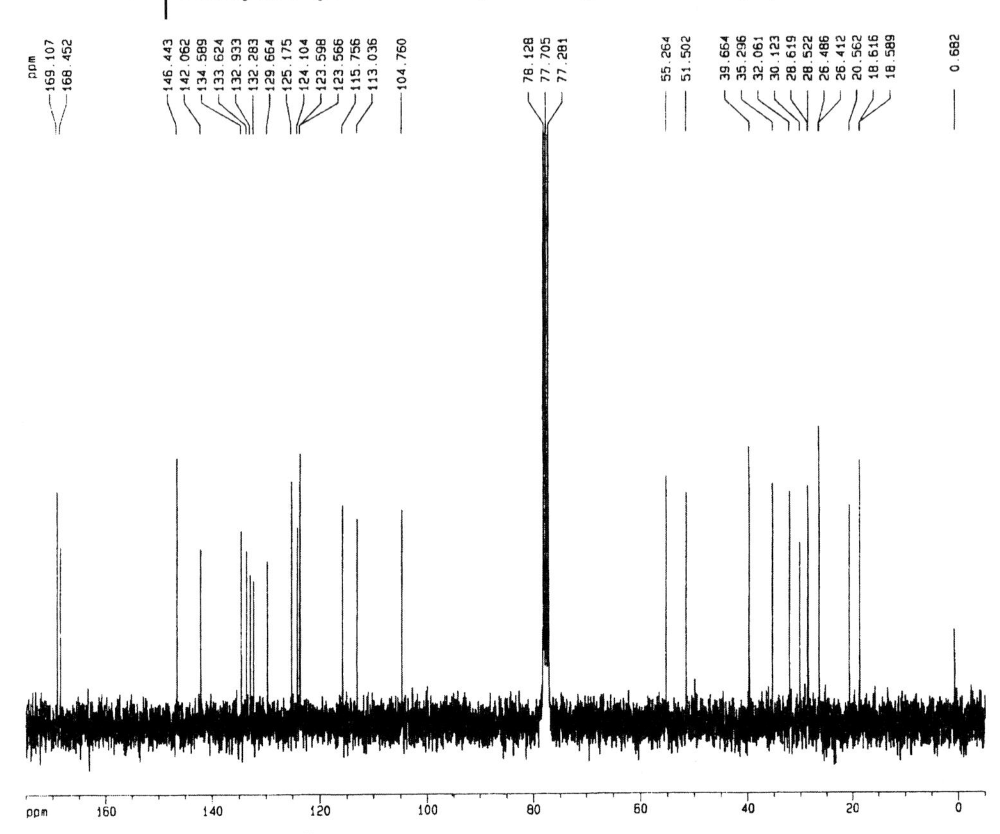

Fig. 8.7 The ^{13}C spectrum of the unknown compound in Example 2.

Fortunately, there is only one natural product in the data bank of natural products, whose structure is that of **106**.

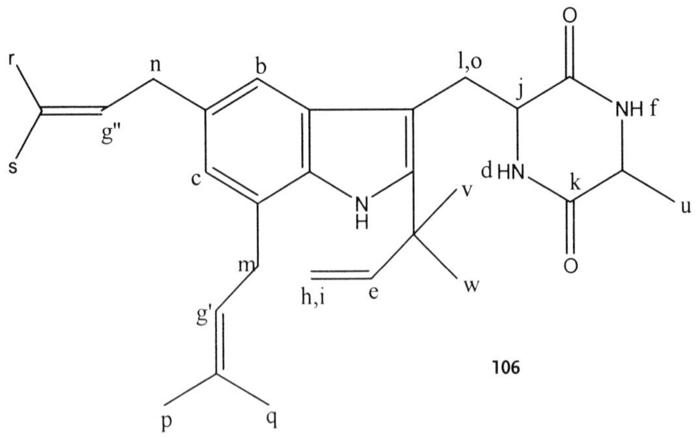

106

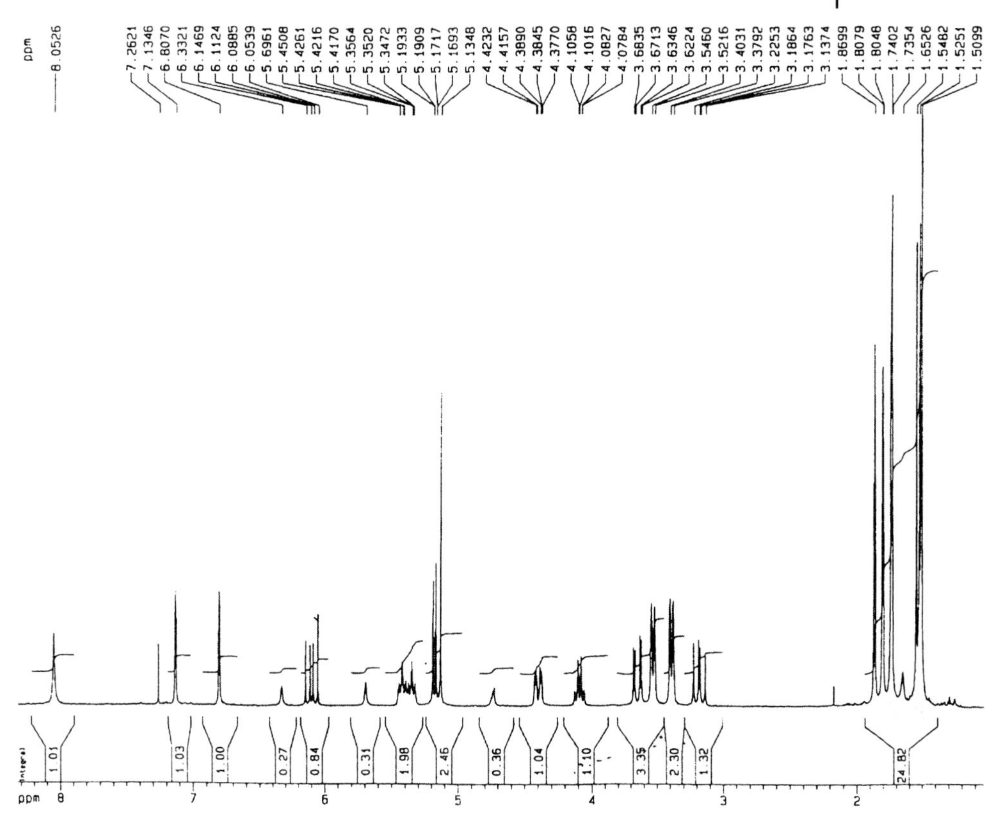

Fig. 8.8 The ^{1}H spectrum of the unknown compound in Example 2.

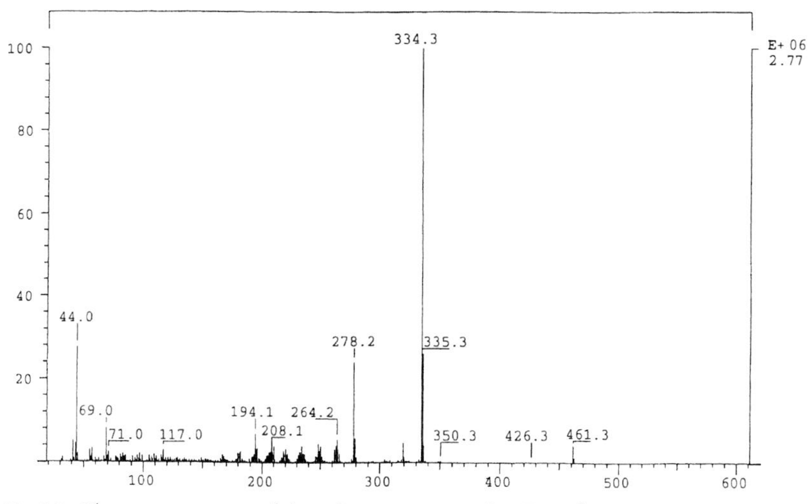

Fig. 8.9 The mass spectrum of the unknown compound in Example 2.

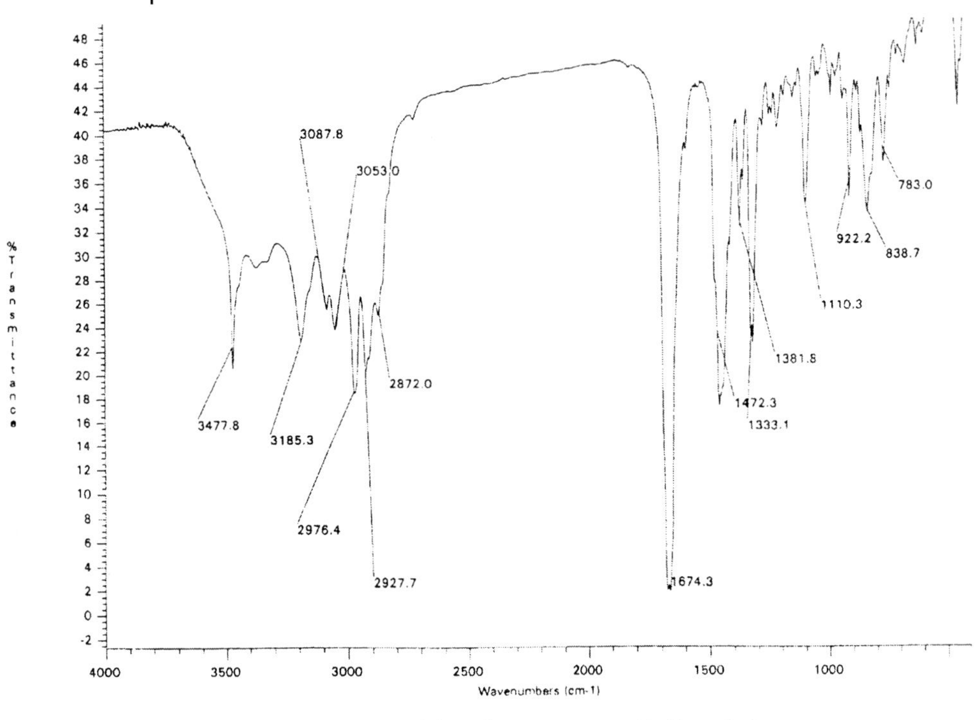

Fig. 8.10 The IR spectrum of the unknown compound in Example 2.

This structure is in accordance with all of the above spectra. It can be further confirmed by the COSY. The assignment of the ^{1}H spectrum in the decrease of δ values is denoted by small letters from a to w in Fig. 8.11.

Example 3

The data of the ^{13}C spectrum and the ^{1}H spectrum of an unknown compound are shown in Tabs. 8.3 and 8.4, respectively. Its molecular formula is determined as $C_{20}H_{22}O_6$ by mass spectrometry with a high resolution. Try to identify the unknown compound.

Solution

The degree of unsaturation of the unknown compound is calculated from its molecular formula to be 10, which is a rather large number.

Tab. 8.3 The data of the ^{13}C spectrum of the unknown compound in Example 3

δ (ppm)	*54.1*	*55.9*	*71.7*	*85.9*	*108.6*	*114.3*	*119.0*	*132.9*	*145.2*	*146.7*
Multiplicity	d	q	t	d	d	d	d	s	s	s

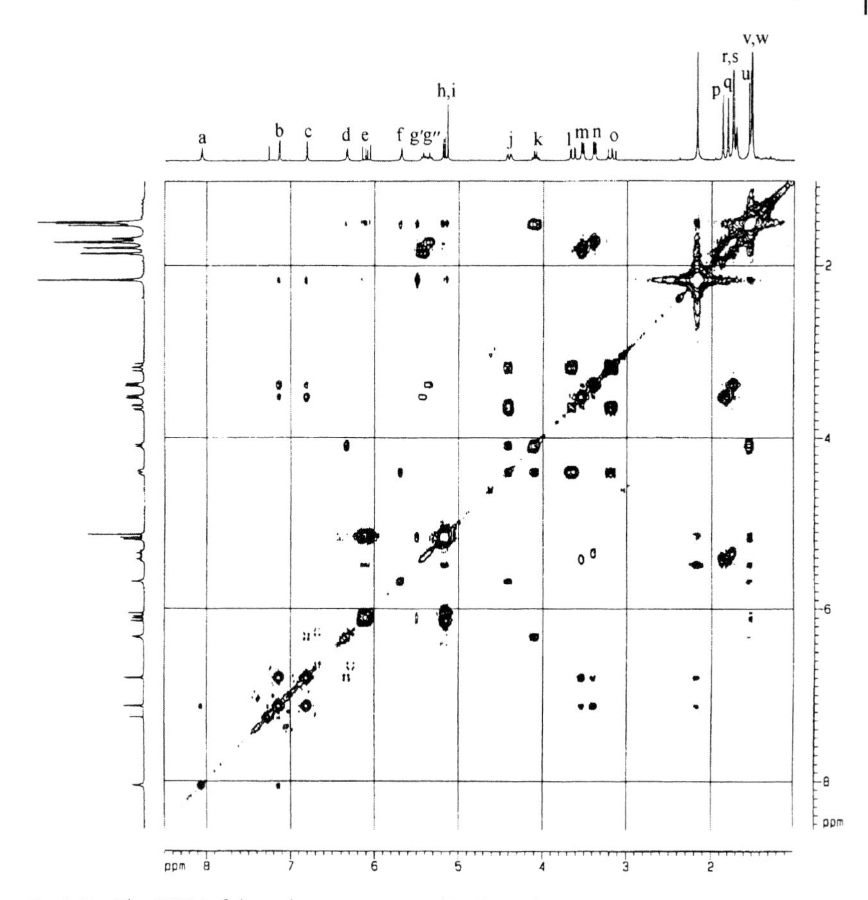

Fig. 8.11 The COSY of the unknown compound in Example 2.

Tab. 8.4 The data of the ^{1}H spectrum of the unknown compound in Example 3

δ (ppm)	*6.90*	*6.88*	*6.83*	*5.72*	*4.74*	*4.24*	*3.89*	*3.87*	*3.10*
Peak shape	d	d	d, d	blunt	d	d, d	s	m [b)]	m
J (Hz)	2.0	7.8	7.9, 2.0		4.4	9.3, 6.8			
Hydrogen atom number [a)]	1	1	1	1	1	1	3	1	1

a) This number should be multiplied by 2.
b) Partial peaks are covered by the methoxy group peak.

From Tab. 8.3 it can be seen that this unknown compound has a symmetrical structure because 10 peaks represent 20 carbon atoms. The number of hydrogen atoms connected to carbon atoms, as calculated from the multiplicity, is 20. Therefore, the two other hydrogen atoms should connect to the heteroatoms, that is, they belong to two hydroxyl groups.

On the basis of the ^{13}C spectrum data, two tri-substituted benzene rings and two methoxy groups can be deduced. By comparing the deduced structural units with the molecular formula, it is clear that the unknown compound should contain another two oxygen atoms.

Except for a lignin skeleton it is rare that the remaining six carbon atoms and two oxygen atoms would make up two symmetrical rings (structure **107**).

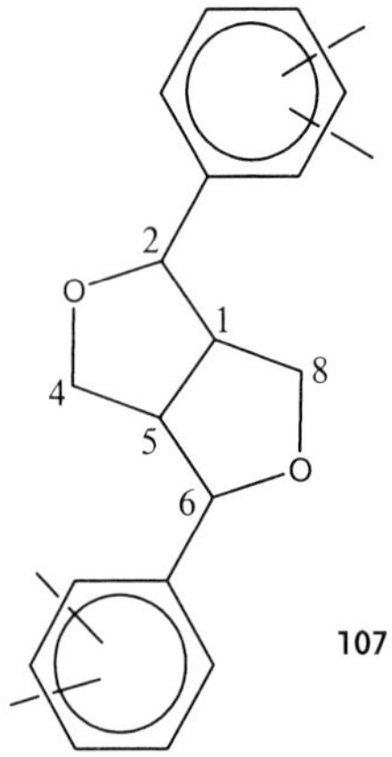

107

This structure agrees with all δ values and the multiplicity shown in Tab. 8.3: C-1/C-5, 54.1 ppm, d; C-4/C-8, 71.7 ppm, t; C-2/C-6, 85.9 ppm, d.

It is known that two substituents in a benzene ring are a hydroxyl group and a methoxy group. From the δ values of the two substituted carbon atoms, 145.2 and 146.7 ppm, the two substituents should be adjacent to each other, otherwise the values would be much greater than 150 ppm. On the other hand, the third substituted carbon atom can not be adjacent to either one of the two substituents.

All the information available allows us to confine the unknown to two possible structures (**108**).

The ^{1}H spectral data agree with the two structures. The data in the range 6.83–6.90 ppm shows the presence of a substituted benzene ring at the 1-, 2- and 4-positions.

The most probable structure should be chosen from (a) and (b). As a methoxy group is less effective than a hydroxyl group in shifting its adjacent hydrogen atom in the direction of the high field and the hydrogen atom without ^{3}J coupling has a fairly large δ value, (a) would be a reasonable structure. The COLOC is necessary to confirm the structure definitively.

a OCH3 OH HO OCH3 O O

b OH O(H3CO OH O O

108

Example 4

The ^{1}H spectrum, ^{13}C spectrum, DEPT, COSY and H,C-COSY of an unknown compound are shown in Figs. 8.12 to 8.16, respectively. The mass spectrum shows the molecular peak at m/z 372. Try to identify the structure.

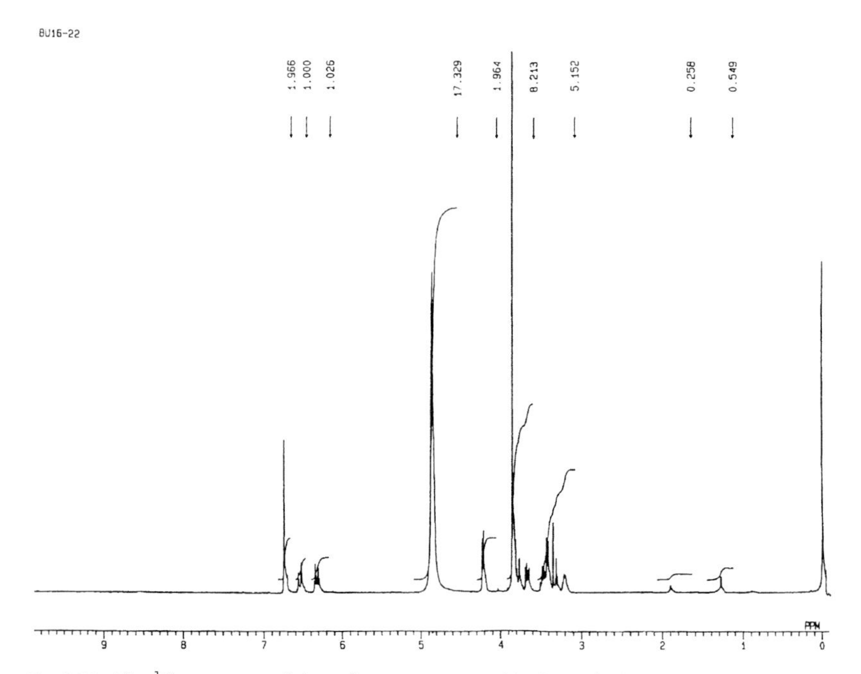

Fig. 8.12 The ^{1}H spectrum of the unknown compound in Example 4.

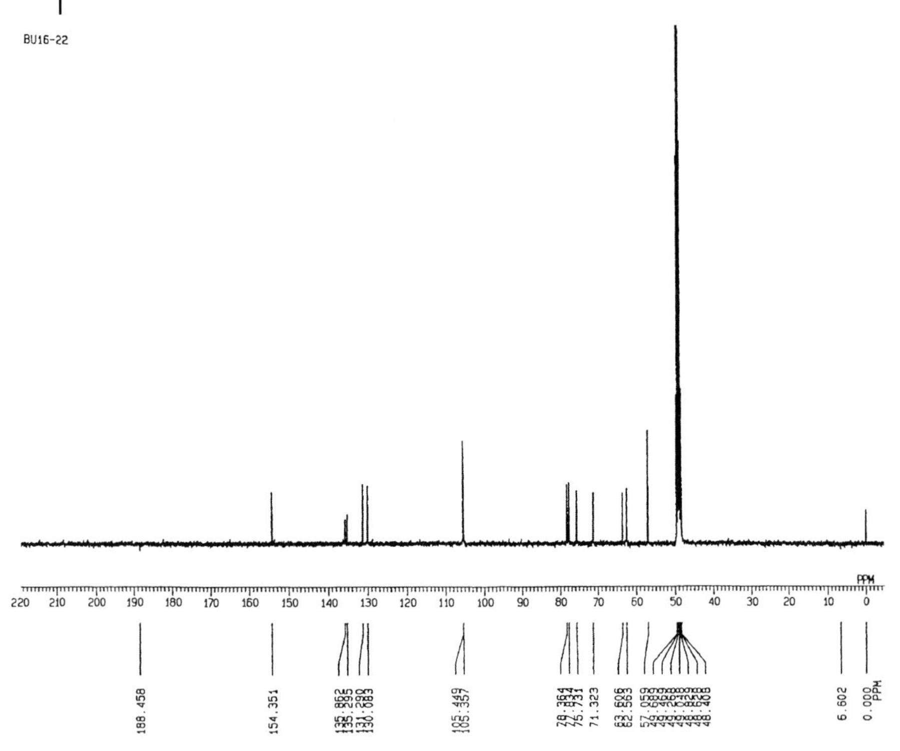

Fig. 8.13 The ^{13}C spectrum of the unknown compound in Example 4.

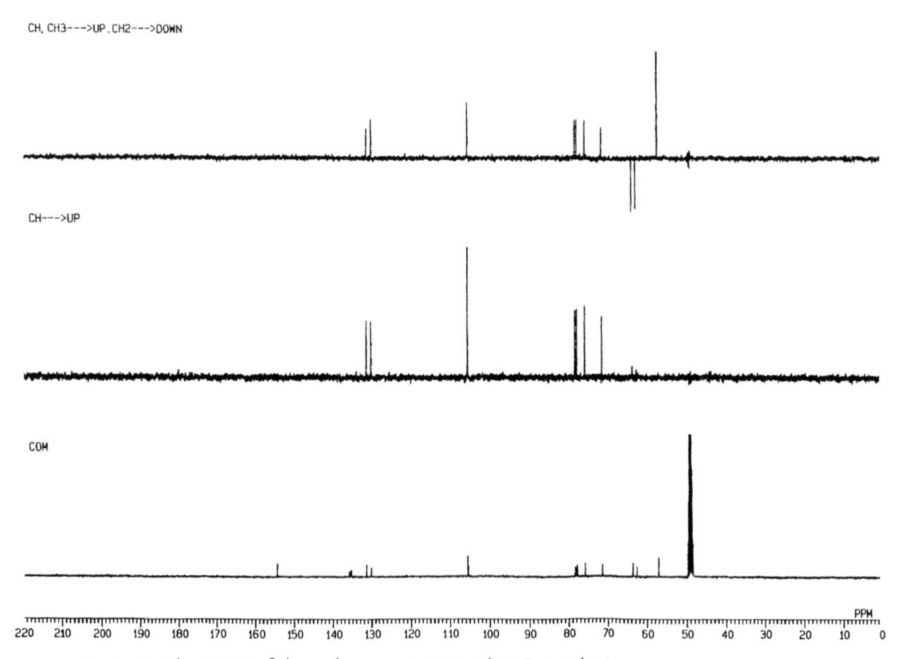

Fig. 8.14 The DEPT of the unknown compound in Example 4.

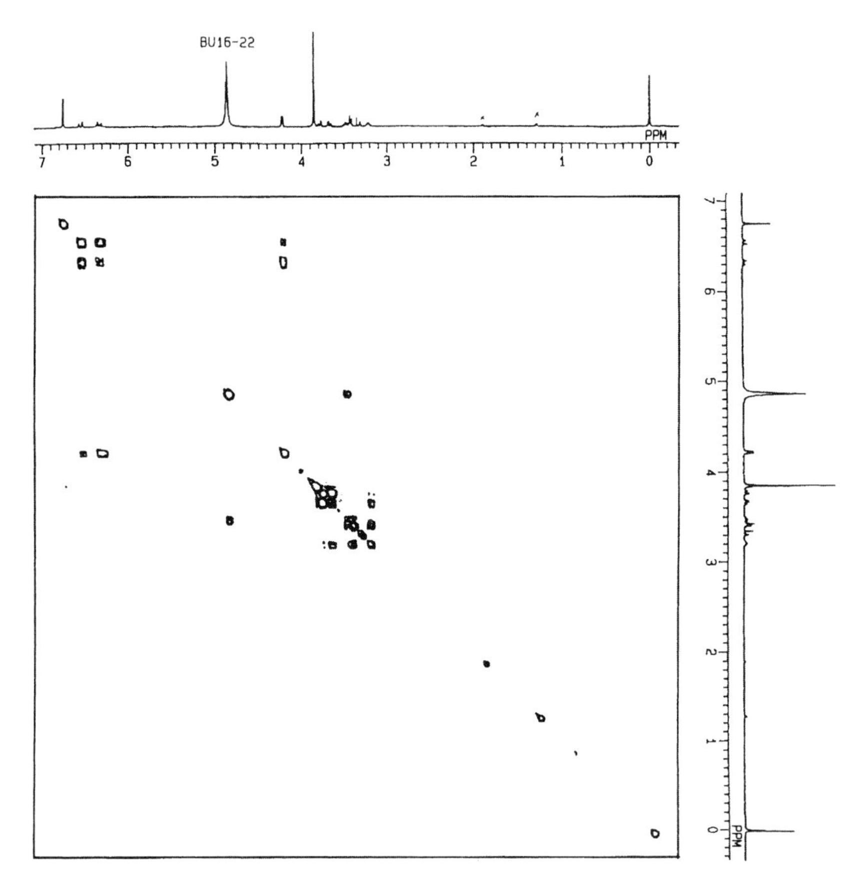

Fig. 8.15 The COSY of the unknown compound in Example 4.

Solution

The ^{1}H spectrum clearly shows a singlet at 3.86 ppm with six hydrogen atoms determined from the integral curve. Therefore, the unknown compound contains two chemically equivalent methoxy groups.

From the peak shape at 6.25–6.60 ppm and the COSY, it is known that the unknown compound contains the structure unit, **109**.

6.5 (d, J=16Hz) 4.2 (d, J=5Hz)

H CH_2—

C=C **109**

H

6.3 (d×t, J=16Hz, 5Hz)

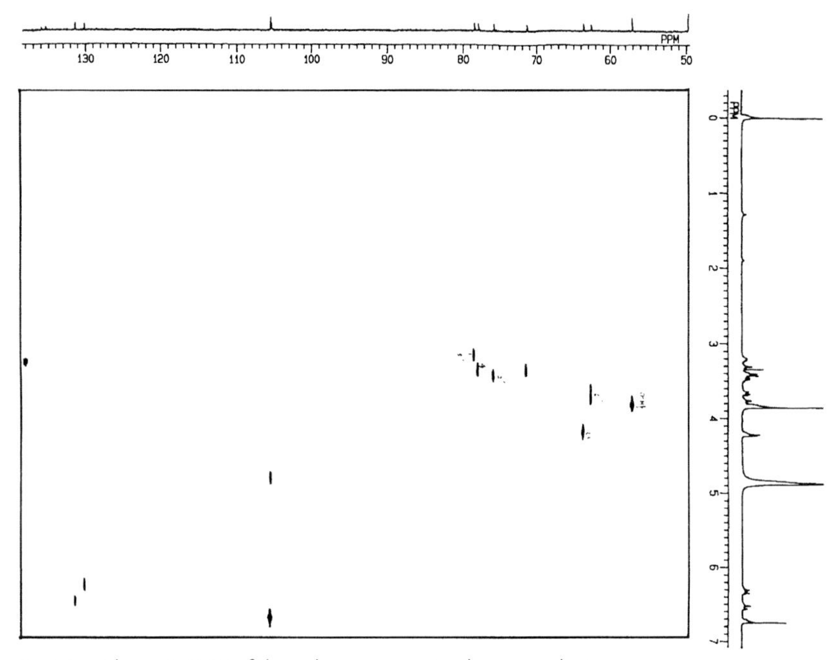

Fig. 8.16 The H,C-COSY of the unknown compound in Example 4.

Using the H,C-COSY, the structure unit can be assigned as **110**.

—CH=CH—CH$_2$—
131.3 130.1 63.6

110

The two methoxy groups can be determined at 57.1 ppm in the ^{13}C spectrum.

It appears that the unknown compound is associated with a sugar because of several peaks beyond 70 ppm in the ^{13}C spectrum and the peaks around 3.1–4.8 ppm in the ^{1}H spectrum. The huge peak is attributed to HDO. However, the cross peak at 4.8 ppm in the COSY reveals the presence of a hydrogen atom in the sample. The COSY further confirms the existence of a sugar unit and illustrates the following coupling relationships:

4.84–3.47–3.32–3.21–3.67, 3.77 (ppm)

In addition to H,C-COSY, a glycoside is determined, structure **111**.

So far the remaining peaks to be determined are only one at 6.75 ppm with two hydrogen atoms in the ^{1}H spectrum and four peaks at 154.4, 135.9, 135.3

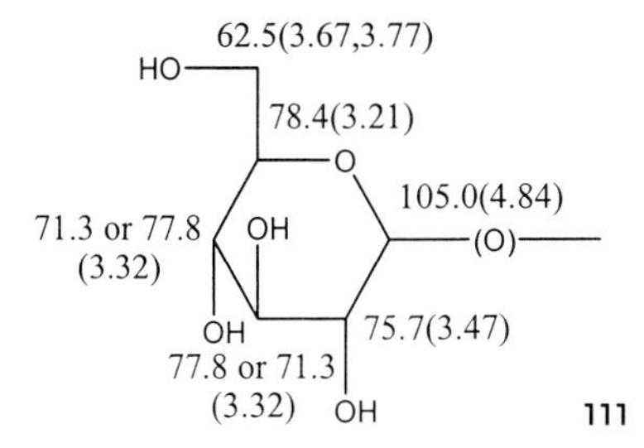

111

and 105 ppm, respectively, in the ^{13}C spectrum. On the basis of the above data and the related hydrogen atom number, structural unit **112** can be constructed.

112

As the unknown compound contains two methoxy groups, they must be the two X atoms in the above structural formula. The substituent containing the double bond should be Z because the δ value of the isolated hydrogen atom in the benzene ring is 6.75 ppm. If Z is the glycoside group, the δ value of the isolated hydrogen atom will be about 6.0 ppm.

All the deductions lead to structure **113** for the unknown compound.

113

The molecular weight calculated from the structural formula agrees with that obtained from the mass spectrum, which further confirms the structure.

Example 5

The ^{1}H spectrum, ^{13}C spectrum, COSY, H,C-COSY and COLOC of an unknown compound are shown in Figs. 8.17 to 8.21. Its high resolution CI mass spectrum shows the M+1 at 383.1327 u. Try to identify the structure [9].

Solution

The molecular formula is determined to be $C_{18}H_{22}O_9$ by combination of the high resolution MS and the ^{13}C spectrum. Consequently, the degree of unsaturation of the unknown compound can be calculated to be 8 from the molecular formula.

The ^{13}C spectrum shows four carbonyl groups and two double bonds. As these six functional groups only have a degree of unsaturation of 6, the unknown compound should contain two rings.

The ^{1}H spectrum shows the hydrogen atom ratio of 1:1:2:2:1:1:1:1:3:3:3:3 from left to right. This ratio agrees with the molecular formula.

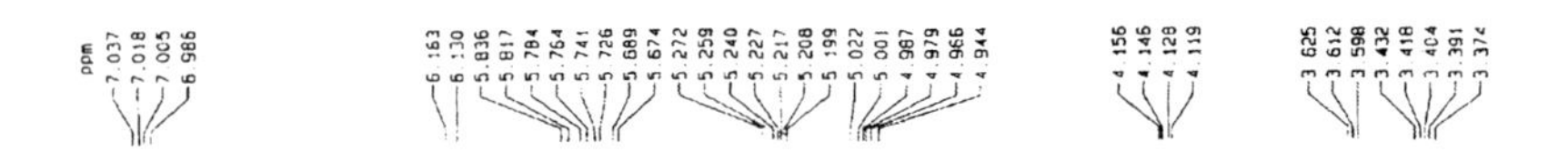

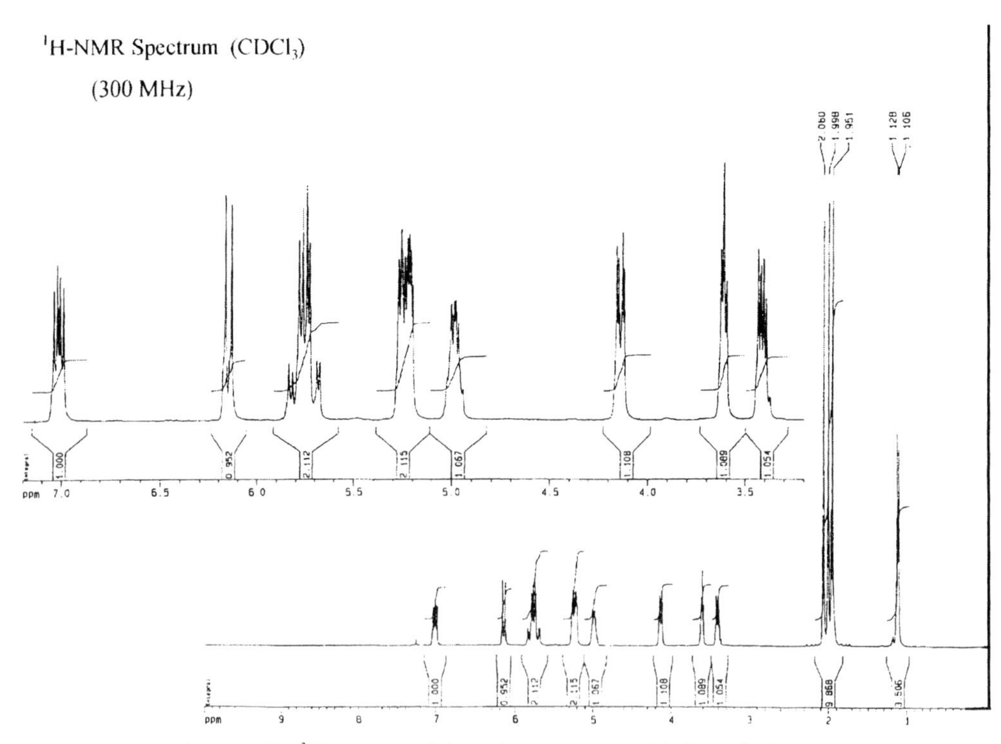

Fig. 8.17 The ^{1}H spectrum of the unknown compound in Example 5.

Proton Decoupled C NMR Spectrum
Showing multiplicity: Q = quartet
D = doublet
S = singlet

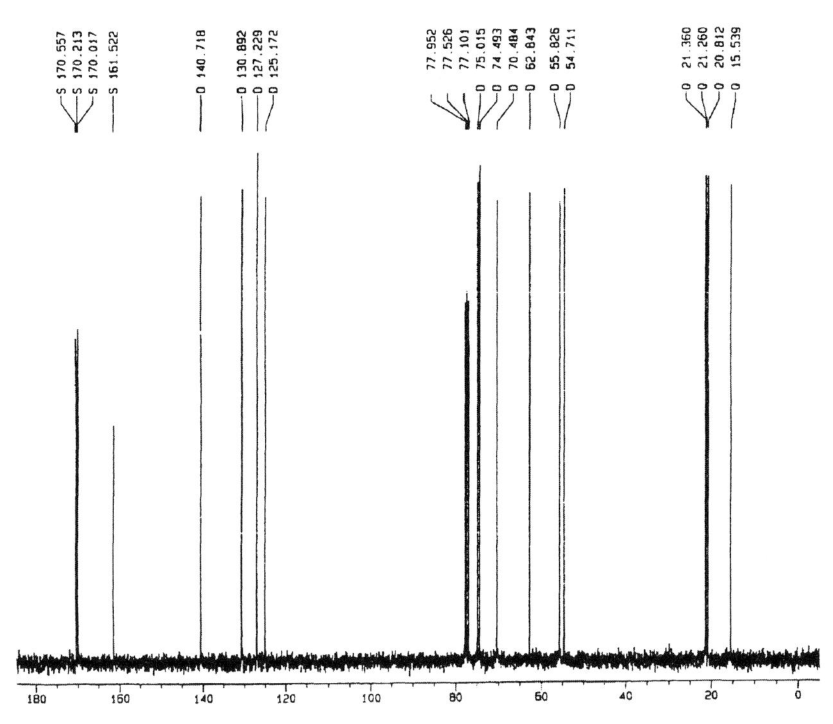

Fig. 8.18 The ^{13}C spectrum of the unknown compound in Example 5.

The cross peaks in the COSY provide the connection information for the unknown compound. From downfield to upfield the coupling relationship can be found to be as follows:

δ (ppm) 1.04–4.98–5.26–5.80–5.71–3.61–3.41–4.14–5.21 7.01–6.15

Such connections together with the H,C-COSY and the values of δ_H and δ_C lead to the structure segment **114**.

(O) (O) (O)

CH_3—CH—CH—CH=CH—CH—CH—CH—CH—CH=CH—

(O) (O) (O)

114

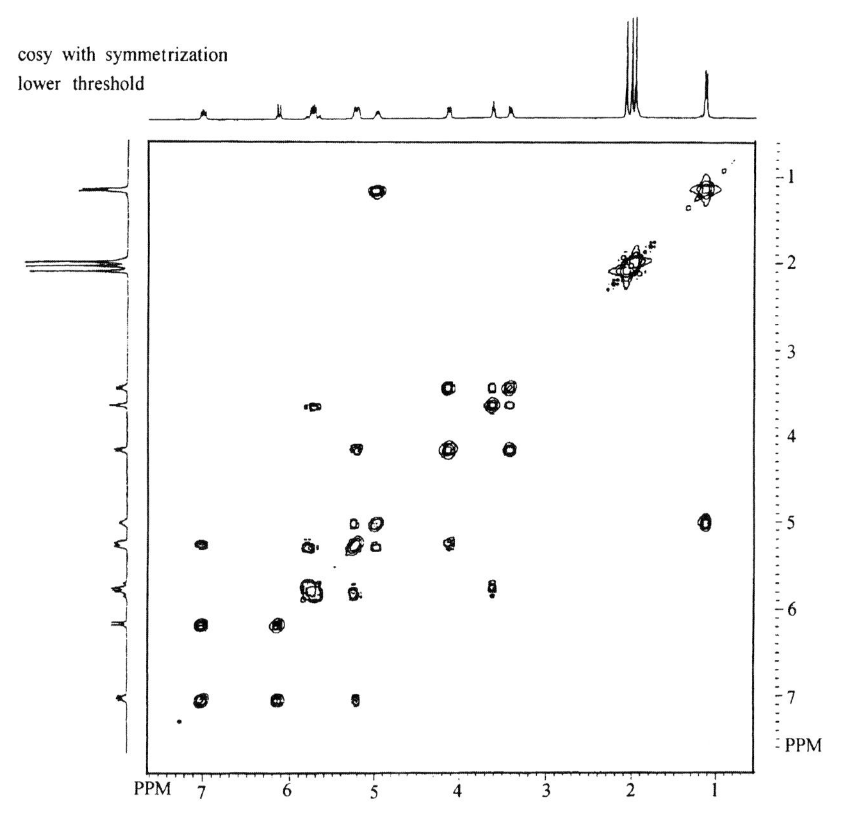

Fig. 8.19 The COSY of the unknown compound in Example 5.

Besides the structure unit, there are four carbonyl groups and three methyl groups without any couplings in the unknown compound. From their δ_H and δ_C, it is clear that they should be three acetyl groups and a carbonyl group, which is connected to a heteroatom and conjugated with a double bond. Therefore, the skeleton of the unknown compound can be extended to structure **115**.

```
   (O)              (O)     (O)                   O
    |                |       |                    ||
CH3—CH—CH—CH=CH—CH—CH—CH—CH—CH=CH—C—O—
        |                |       |
       (O)              (O)     (O)
```

115

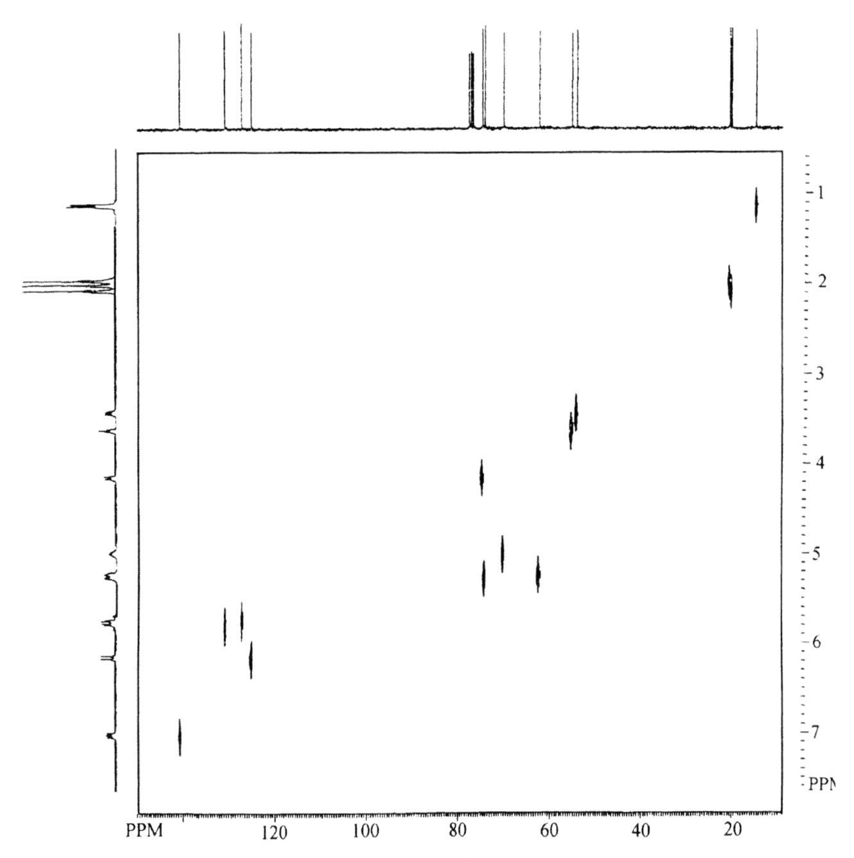

Fig. 8.20 The H,C-COSY of the unknown compound in Example 5.

In addition to the fact that the unknown compound must contain two rings, and that one oxygen atom remains, the structure of the unknown compound can be further extended to **116**.

H_3C—CH—CH—HC=HC—HC—C(O)—CH—CH— (atoms numbered 12, 11, 10, 9, 8, 7, 6, 5, 4, 3, 2, 1)

116

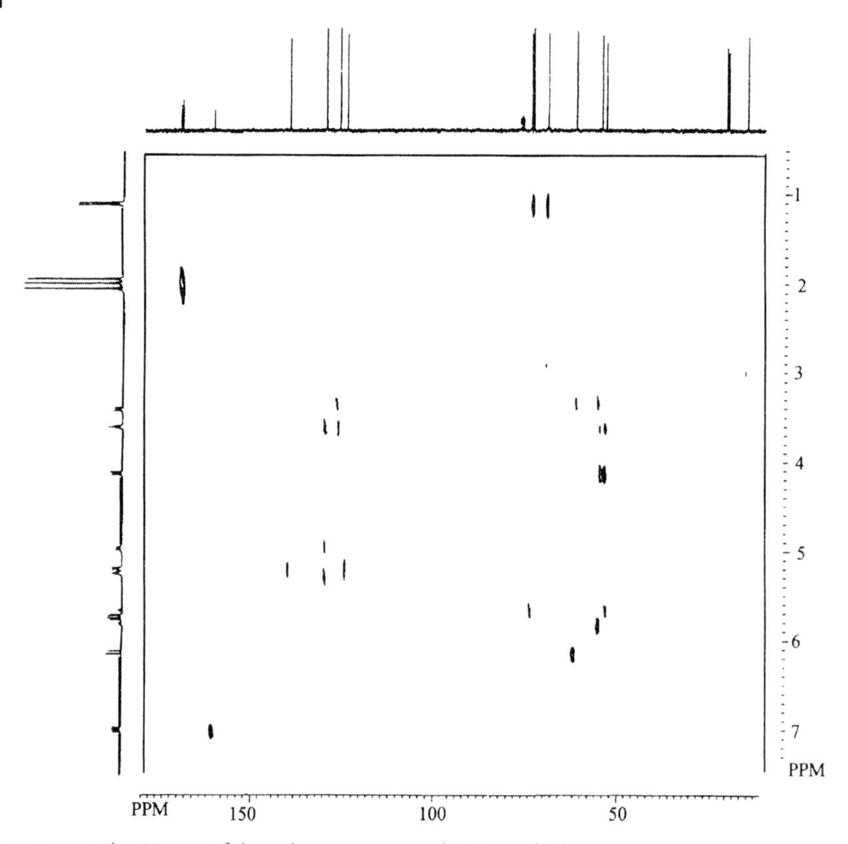

Fig. 8.21 The COLOC of the unknown compound in Example 5.

The following points should be noted:

1. 55.8 and 54.7 ppm are less than the δ values of other –CH–O– groups.
2. The carbonyl group conjugated with a double bond has a δ value of 161.5 ppm, which agrees with that of a lactone group in a six-membered ring.
3. The three CH moieties with δ values of 70.4, 74.4 and 75.0 ppm should be connected to the three acetyl groups, respectively.

All the deductions lead to the complete structural formula for the unknown compound as that of **117**.

From the ^3J coupling constant, it is known that H-8 and H-9 are in the *trans*-configuration.

The assignments of δ_H and δ_C for the unknown compound are listed in Tab. 8.5.

The following correlations are shown by the COLOC.

- C-1 H-3
- C-2 H-4
- C-3 H-4

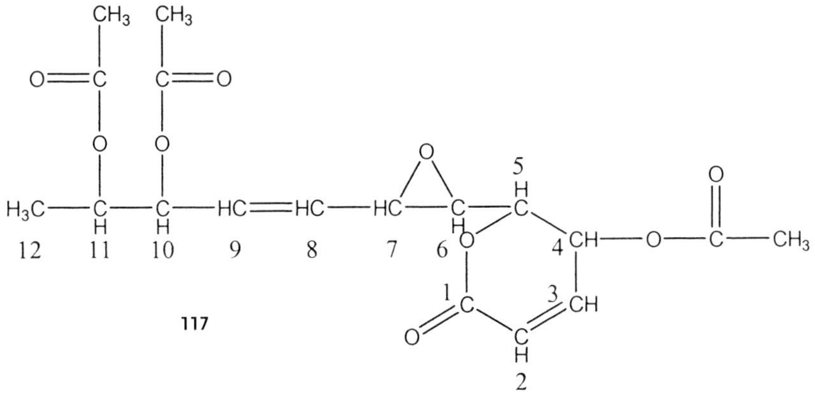

117

Tab. 8.5 The assignments of δ_H and δ_C for the unknown compound in Example 5

Position	δ_C	δ_H
1	161.5	6.15
2	125.1	7.01
3	140.1	5.21
4	62.8	4.14
5	75.0	3.41
6	54.7	3.61
7	55.8	5.71
8	127.2	5.80
9	130.8	5.26
10	74.4	4.98
11	70.4	4.95
12	15.5	1.04
$OC^{*}OCH_3$	170.0	
	170.2	
	170.5	
$OCOC^{*}H_3$	20.8	1.93
	21.2	2.01
	21.3	2.04

- C-4 H-2, H-6
- C-6 H-5, H-7, H-8
- C-7 H-5, H-6, H-9
- C-8 H-6, H-7
- C-9 H-7, H-10, H-11
- C-10 H-8, H-12
- C-11 H-12

These correlations further confirm the structural formula.

Example 6

Clivorine is a pyrolizidine alkaloid. Its structural formula is shown as compound **C9-1**. Its molecular formula is $C_{21}H_{27}NO_7$. Its ^{13}C spectrum and DEPT, COSY, H,C-COSY and COLOC are shown in Figs. 8.22 to 8.25, respectively. Try to assign the ^1H spectrum and ^{13}C spectrum.

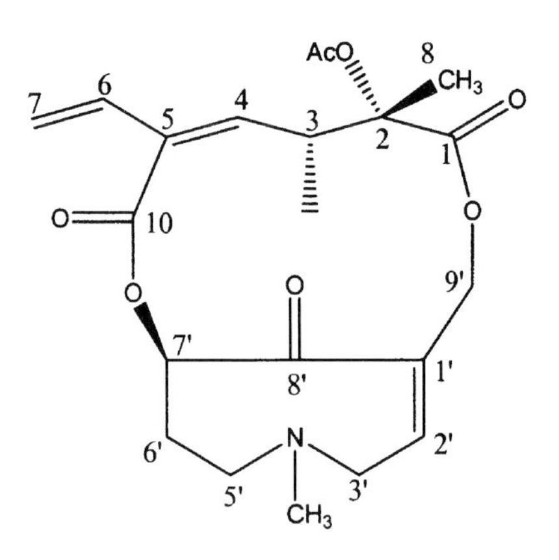

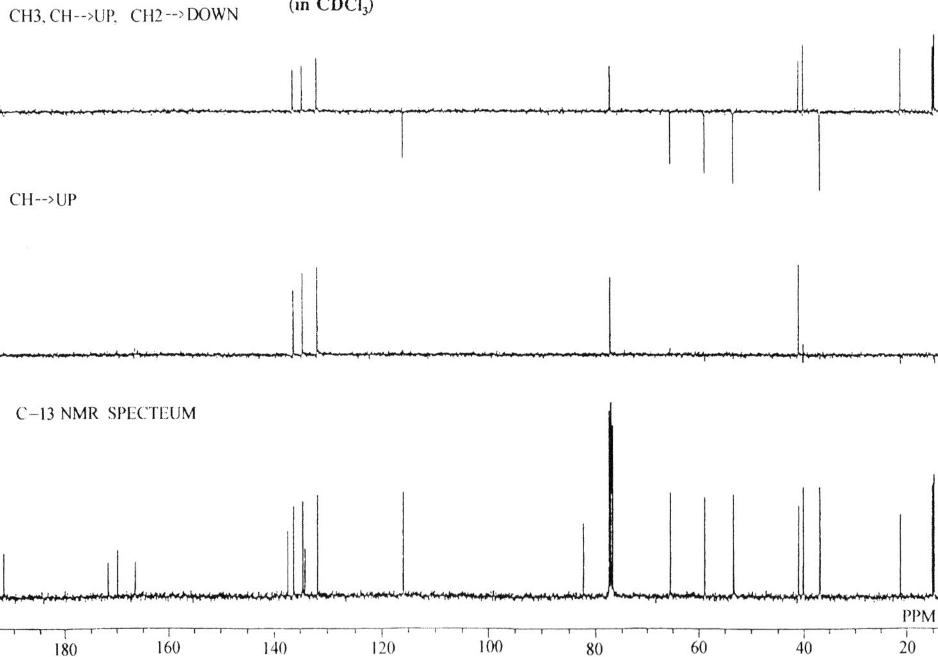

Fig. 8.22 The ^{13}C spectrum and DEPT of clivorine.

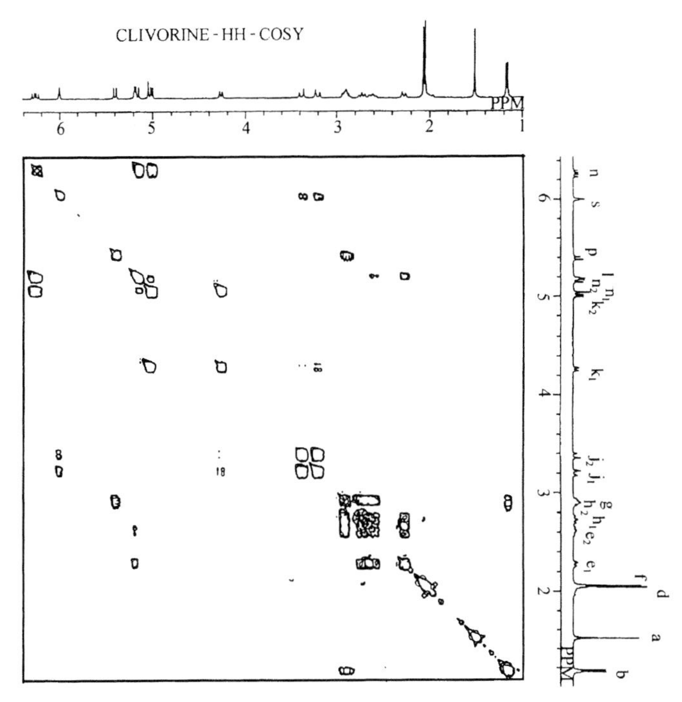

Fig. 8.23 The COSY of clivorine.

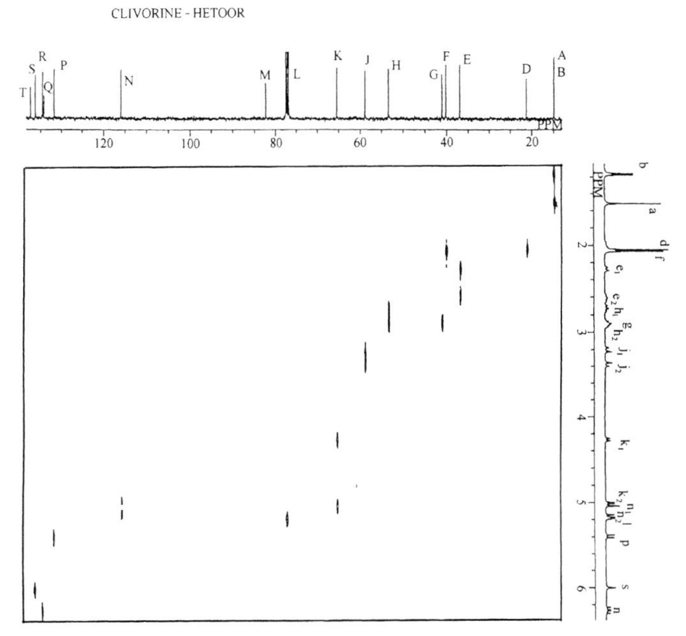

Fig. 8.24 The H,C-COSY of clivorine.

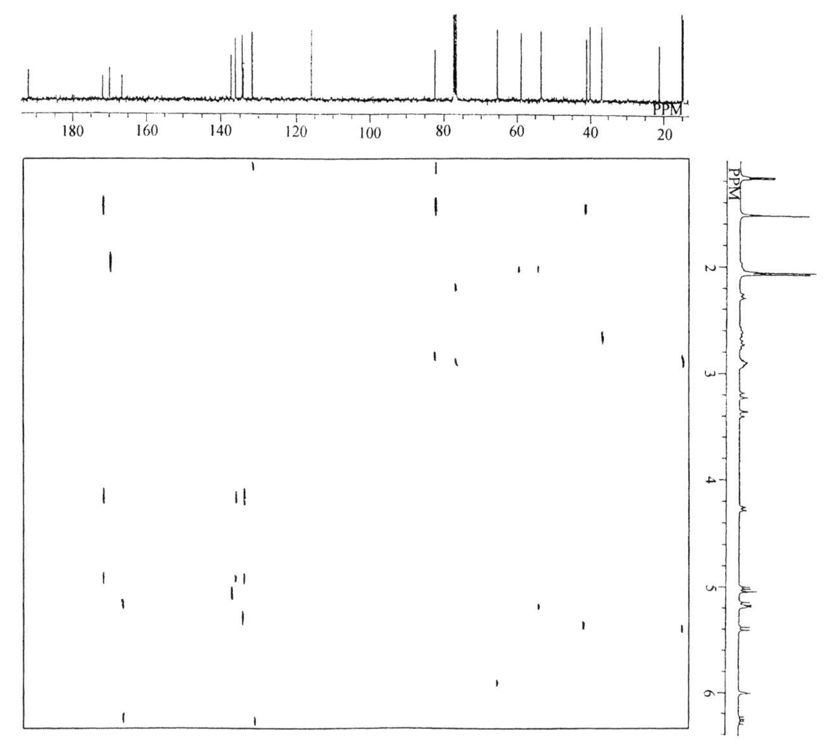

Fig. 8.25 The COLOC of clivorine.

Solution

There are 20 peaks in the ^{13}C spectrum. However, another peak, hidden in the solvent peak, is shown in the DEPT. The 21 peaks are marked with capital letters from A to X (except for C, I and O), respectively, in the order of decreasing δ values.

Using the H,C-COSY, we mark related hydrogen atoms with small letters according to the markings in the ^{13}C spectrum.

Using these marks, the NMR data are summarized in Tab. 8.6.

Because clivorine possesses a cyclic structure, every CH_2 group has two chemically non-equivalent hydrogen atoms.

On the basis of the COSY, structural units **118** can be found and assigned by referring to the structure formula.

The related δ_H and δ_C values agree with these assignments.

Hence, only three methyl group signals without splitting at 1.52, 2.05 and 2.07 ppm in the ^1H spectrum remain to be assigned. The signal at 1.52 ppm can be assigned as 8-CH_3 from its δ_H value. The other two methyl groups, which have two close δ_H values, can be assigned by their δ_C values. The δ_C value of N–CH_3 is greater than that of CH_3–C=O. Therefore, CH_3 (δ_C 40.12, δ_H 2.07) is assigned as N–CH_3 and CH_3 (δ_C 21.16, δ_H 2.05) is assigned as CH_3–C=O (in the acetyl group).

Tab. 8.6 Summary of NMR data for clivorine

δ_c (ppm)	C mark	DEPT	H mark	δ_H (ppm)	Peak shape	J (Hz)
14.78	A, B	CH_3	a	1.52	s	
14.83	B	CH_3	b	1.17	d	7
21.16	D	CH_3	d	2.05	s	
36.83	E	CH_2	e_1	2.27	d	14
			e_2	2.61	m	
40.12	F	CH_3	f	2.07	s	
41.06	G	CH	g	2.90	m	
53.31	H	CH_2	h_1	2.73	m	
			h_2	2.90	m	
58.85	J	CH_2	j_1	3.20	d	18
			j_2	3.38	d	18
65.43	K	CH_2	k_1	4.27	d	12
			k_2	5.03	d	12
76.76	L	CH	l	5.18	m	
112.26	M	C				
115.91	N	CH_2	n_1	5.02	d	18
			n_2	5.16	d	11
131.71	P	CH	p	5.39	d	12
134.03	Q	C				
134.44	R	CH	r	6.26	d, d	11, 18
136.16	S	CH	s	6.00	broad	
137.27	T	C				
166.51	U	C				
169.86	V	C				
171.71	W	C				
191.90	X	C				

b ⟶ g ⟶ p 9 3 4 H_3C—CH_2—CH=

e_1,e_2—h_1,h_2 (| 1) 6' 5' H_2C—CH_2 (| 7' CH)

j_1,j_2 —s 3' 2' CH_2—CH=

k_1,k_2 9' CH_2

n_1,n_2—r 7 6 CH_2—CH=

118

Tab. 8.7 Summary of assignment of remaining quaternary carbon atoms for Example 6 using COLOC

Carbon atom	*Correlated H with long range coupling*	*Assigned position*
A	H-3	8
B	H-4	9
E	H-5′	6′
G	8-CH_3	3
K	H-2′	9′
L	H-6′, H-5′	7′
M	9-CH_3, 8-CH_3, H-3	2
P	H-6	4
Q	9′-CH_2	1′
R	H-4	6
S	9′-CH_2	2′
T	7-CH_2	5
U	H-7′, H-6	10
V	Methyl in acetyl	Acetyl
W	8-CH_3, 9′-CH_3	1

The assignment of the remaining quaternary carbon atoms is completed by using COLOC, with the results summarized in Tab. 8.7.

The quaternary carbon atom X (191.9 ppm) surely belongs to 8′. As the carbonyl group does not connect any heteroatom, it has the greatest δ_C value.

Example 7

The data of the ^{13}C spectrum and DEPT ($\theta = 135°$) of an unknown compound are shown in Tab. 8.8. Its ^{1}H spectrum, IR spectrum and MS spectrum are shown in Figs. 8.26 to 8.28, respectively. In fact, the enlarged aromatic region of the ^{1}H spectrum was shown in Fig. 2.27 as Example 6 in Section 2.9. Try to postulate its structure.

Solution

From the ^{13}C spectrum and DEPT ($\theta = 135°$) it can be shown that the unknown compound contains 17 carbon atoms all of whose orders can be determined. The peak at 13.6 ppm can be established as a terminal methyl group from its particular chemical shift value. The peak at 55.8 ppm can be determined as a

Tab. 8.8 Data for the ^{13}C spectrum and DEPT ($\theta = 135°$) of the unknown compound in Example 7

δ (ppm)	*163.8*	*163.2*	*160.0*	*132.6*	*130.7*	*128.5*	*127.8*	*125.3*	*122.7*	*121.7*	*114.4*	*104.6*	*55.8*	*39.7*	*29.9*	*20.1*	*13.6*
Type of C	C	C	C	CH	CH	C	CH	CH	C	C	CH	CH	CH_3	CH_2	CH_2	CH_2	CH_3

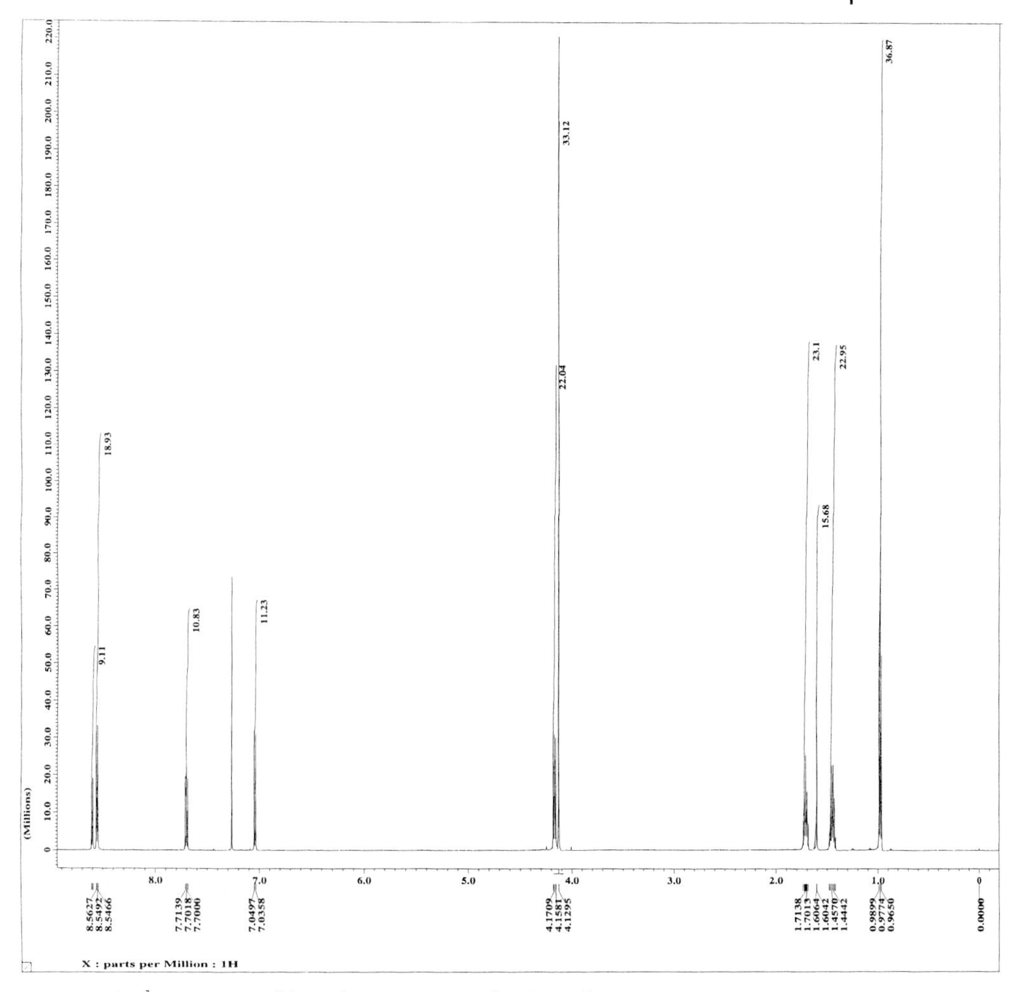

Fig. 8.26 The ^{1}H spectrum of the unknown compound in Example 7.

methoxy group from its particular chemical shift value and its sharp peak in the ^{1}H spectrum. The result is shown in Tab. 8.8.

Fig. 8.26 shows the ^{1}H spectrum of the unknown compound. The peak at 7.26 ppm is the solvent peak of $CDCl_3$. The peak at 1.60 ppm is the peak of the water remaining in the sample when deuterated chloroform is used as the solvent. The peak at 4.13 ppm corresponding to three hydrogen atoms shows the existence of a methoxy group. The ^{1}H spectrum shows that the unknown compound contains 17 hydrogen atoms, which coincides with the result obtained from DEPT. Therefore, the compound does not contain active hydrogen atoms (see Section 3.5.2).

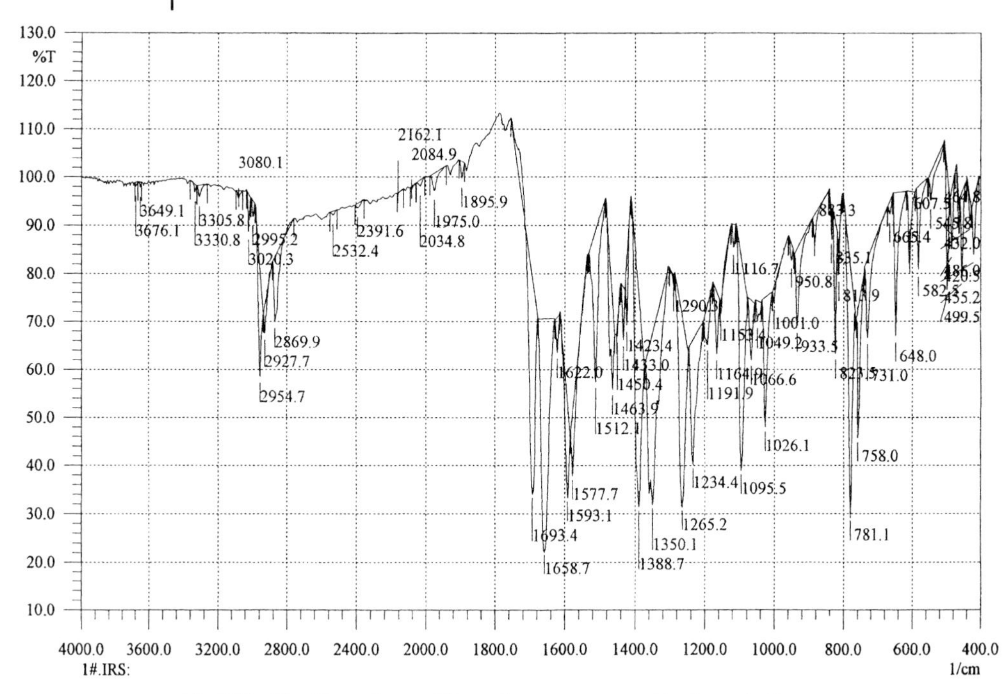

Fig. 8.27 The IR spectrum of the unknown compound in Example 7.

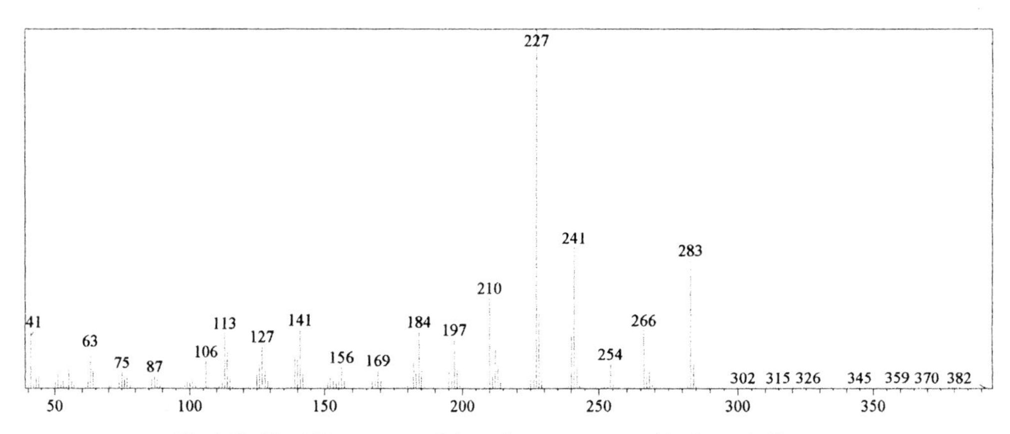

Fig. 8.28 The MS spectrum of the unknown compound in Example 7.

The IR spectrum clearly shows two absorption bands of carbonyl groups (at 1693.4 and 1658.7 cm^{-1}). Several absorption bands (at 3020, 1593, 1577 and 1512 cm^{-1}, etc.) illustrate the presence of aromatic rings.

The molecular ion peak can be postulated as the peak at m/z 283 in Fig. 8.28. Therefore, the compound should contain odd-numbered nitrogen atoms.

Let us now summarize the conclusions mentioned above. The compound contains 17 carbon atoms, 17 hydrogen atoms, 3 oxygen atoms (two carbonyl groups and one methoxy group) and one nitrogen atom, as the sum of the masses is equal to 283. Its unsaturation number can be calculated as being 10 from its molecular formula of $C_{17}H_{17}O_3N$.

A butyl group can be deduced from the ^{13}C spectrum, DEPT and ^{1}H spectrum.

Because the two carbonyl groups have rather small δ values, they should be connected to heteroatoms. In such a case, the two carbonyl groups are connected to the unique nitrogen atom.

Consider also that the chemical shift of the first CH_2 of the butyl group and the CH_2 should be connected to the nitrogen atom.

So far we have two structural units: structure **119** and a methoxy group.

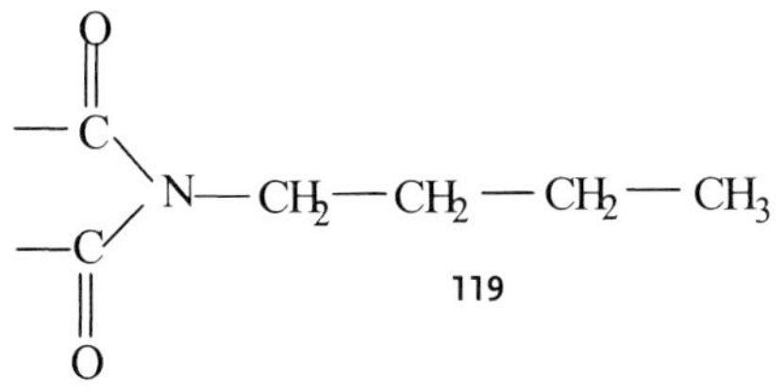

O
‖
—C
N—CH_2—CH_2—CH_2—CH_3
—C
‖
O

119

By subtracting the elemental composition of the two structural units, $C_7H_{12}O_3N$, from the molecular formula $C_{17}H_{17}O_3N$, the remaining elemental composition of the unknown compound is $C_{10}H_5$. As the compound has a rather high unsaturation number of 10 and just 10 carbon atoms, the remaining structural unit should be a substituted naphthalene ring.

So all that remains to be done in order to complete the structure is the determination of the substituted positions on the naphthalene ring. As has been emphasized many times in this monograph, the analysis of peak shapes in the ^{1}H spectrum is what is most important and reliable for a determination. This example also illustrates this point of view. Please read Example 6 in Section 2.9 to complete the postulation of the unknown structure.

The assignment for the ^{1}H spectrum and the assignment of the carbon atoms, except for some of the quaternary carbon atoms for the ^{13}C spectrum, confirm the structure as shown in **120** and **121**.

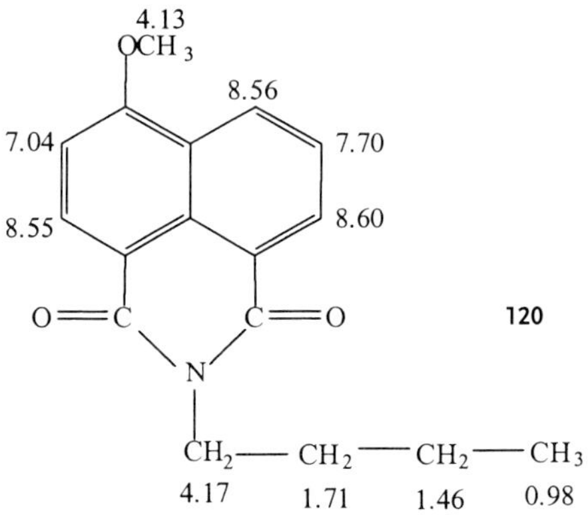

4.13
OCH_3
8.56
7.04
7.70
8.55
8.60
O=C C=O
N
CH_2— CH_2—CH_2—CH_3
4.17 1.71 1.46 0.98

120

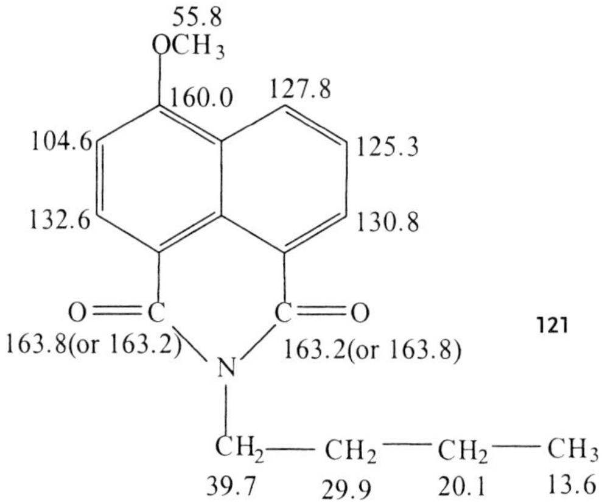

55.8
OCH_3
160.0 127.8
104.6
125.3
132.6
130.8
O=C C=O
163.8(or 163.2) N 163.2(or 163.8)
CH_2— CH_2—CH_2—CH_3
39.7 29.9 20.1 13.6

121

8.5 References

1 P.L. Fuches, L.A. Bunnell, *Carbon-13 NMR Based Organic Spectral Problems,* John Wiley, Chichester, 1979.
2 S. Sternhell, J.R. Kalman, *Organic Structures from Spectra,* John Wiley, Chichester, 1986.
3 R.M. Silverstein, G.C. Bassler, *Spectrometric Identification of Organic Compounds,* 6th Edn., John Wiley, Chichester, 1998.
4 S. Tanaka, *Methods of Structural Determination of Organic Compounds* (Japanese Version), Industrial Books, 1979.
5 G.E. Martin, A.S. Zektzer, *Two-Dimensional Methods for Establishing Molecular Connectivity,* VCH, Weinheim, 1988.
6 H. Duddeck, D. Dietrich, *Structure Elucidation by Modern NMR,* A Workbook, 3rd Edn., Springer, Berlin, 1998.
7 W.R. Croasmun, R.M.K. Carlson, *Two Dimensional NMR Spectroscopy: Applications for Chemists and Biochemists,* 2nd Edn., VCH, Weinheim, 1994.
8 T. Domke, *J. Magn. Reson.*, **1991**, *95*, 174.
9 G.H. Lu, C.T. Che et al., *J. Nat. Prod.*, **1997**, *60*, 425–427.

9
Determination of Configuration and Conformation of Organic Compounds by Spectroscopic Methods

As early as 1874, virtually at the same time, Van't Hoff and Le Bel independently proposed the theory that all molecules exist in a three-dimensional space, a milestone in organic chemistry. The term "conformation" was first proposed by Haworth in 1929. The development and application of the conformational theory by Hassel and Barton led to them being awarded the Nobel Prize in chemistry in 1969. In 1975, Prelog and Cornforth also received the Nobel Prize in chemistry for their contribution to stereochemistry. It is now well established that the configuration and conformation of an organic compound can affect its physical, chemical, as well as biological properties. Research on various aspects of stereochemistry has received extensive attention in recent years.

Although the method of diffraction of a single crystal by X-rays is excellent for the determination of the absolute configuration of a molecule, its prerequisite is the preparation of a single crystal, which prohibits the application of this technique to the determination of substances that do not form crystals or hardly form a single crystal. Obviously, this method cannot be used to study the conformation of a sample in its solution. Therefore, spectroscopic methods provide useful alternatives for stereochemical determinations without consumption of the sample to be measured (NMR, IR, etc.) and do not consume a large amount of the sample (MS). Nevertheless, spectroscopic methods for configuration and conformation determinations have scarcely been touched upon in some books on stereochemistry [1, 2] or only one method is mentioned [3–7]. To provide the reader with a comprehensive discussion on this area, the author has made a challenging attempt to dedicate a whole chapter to this topic from the point view of methodology.

Chapter 8 dealt with the connections between the atoms of an unknown compound (the constituents) and simple stereochemical determinations, such as *trans*- and *cis*-orientations on a double bond. This chapter discusses configuration and conformation measurements one by one according to spectroscopic techniques.

As the life sciences are developing very rapidly and the measurement methods are similar to those used in organic chemistry, the discussion in this chapter also covers biological molecules.

Structural Identification of Organic Compounds with Spectroscopic Techniques. Yong-Cheng Ning
Copyright © 2005 WILEY-VCH Verlag GmbH & Co. KGaA, Weinheim
ISBN: 3-527-31240-4

9.1 NMR

From the preceding chapters, it is known that NMR plays an important role in the structural identification of organic compounds. It can be estimated that NMR is also an excellent method for stereochemical determinations. Firstly, chemical shifts (δ_H and δ_C), homonuclear and heteronuclear coupling constants and NOE provide abundant stereochemical information about the configurations and conformations of organic compounds. Moreover, NMR has an appropriate time scale that can be used to study interchangeable processes of conformations and to distinguish different conformations at a lower temperature. Consequently, NMR is the most frequently used and effective method for the determination of configurations and conformations.

With the development of 2D NMR and multi-dimensional NMR, the use of NMR for stereochemistry will find wider and wider applications.

9.1.1 Chemical Shift

In an NMR spectrum, the chemical shift is the most fundamental parameter, which refers to the chemical and stereochemical environment for the given atoms. Even in the early 1970s, the δ values of ^{13}C spectra had been adopted in research on stereochemistry [8].

^{13}C chemical shifts

^{13}C chemical shift values provide more stereochemical information than those of ^{1}H because δ_C is more sensitive to the stereochemical environment than δ_H and it covers a wider range. In general, δ_C is affected by steric hindrance, shielding and deshielding effects, which can be described as follows:

1. The γ-gauche effect
   A substituent produces the upfield shift for carbon atoms, which are three bonds away from it (see Section 3.2.2).
2. Some functional groups, especially cyclically conjugated functional groups, produce an anisotropic shielding effect on neighboring functional groups. Steric relationships among these functional groups can be deduced from this effect.
3. The carbon atom substituted by a large functional group possesses a rather large δ_C value.
4. In a cyclic molecule, the bonds along different orientations (e.g., equatorial or axial bonds in a six-membered ring) have different shielding (or deshielding) effects on the other carbon atoms of the ring.

These effects can be illustrated by the following examples.

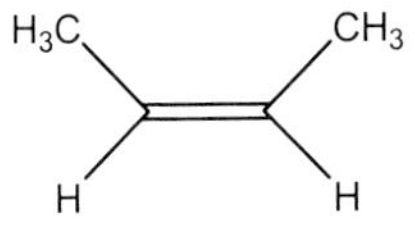

C9-1

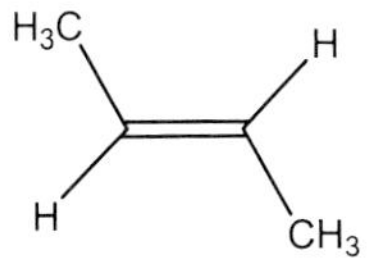

C9-2

Because of the γ-gauche effect of the *cis*-butene-2 (**C9-1**), two methyl groups extrude together so that they have smaller δ values than the two methyl groups of the *trans*-butene-2 (**C9-2**) by 5 ppm.

Moreover, let us compare the two stereoisomers **C9-3** and **C9-4**, in which the C-7 in the *exo*-configuration has a much smaller δ value than that in the *endo*-configuration because the C-7 of **C9-3** is affected by the γ-gauche effect from C-1 and C-3 while the C-7 in the *endo*-configuration is not.

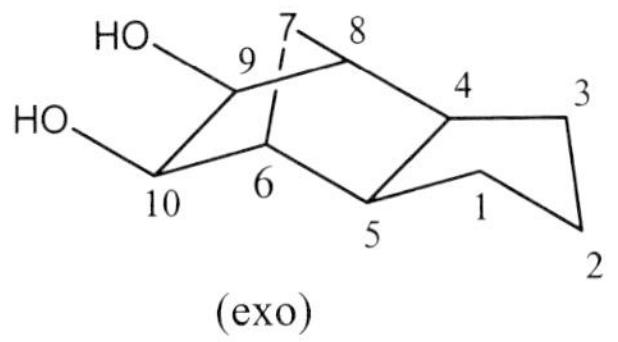

(exo)

C9-3

(endo)

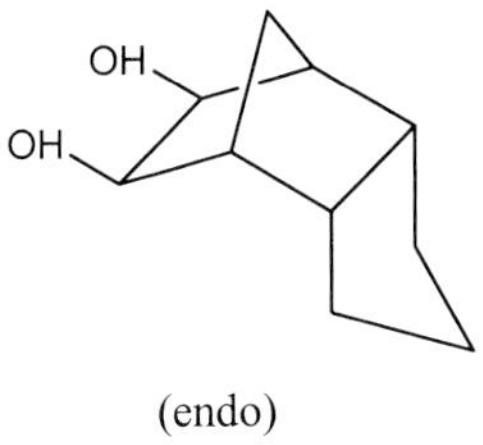

C9-4

On the other hand, the C-1 and C-3 in the *endo*-configuration resonate at higher field positions than the C-1 and C-3 in the *exo*-configuration because of the γ-gauche effect from the C-9 and C-10, respectively, in the *endo*-configuration. However, the C-1 and C-3 in the *exo*-configuration are affected by the γ-gauche effect just by the C-7 so that they have rather large δ values. The chemical shift values of C-9 and C-10 can be analyzed in a similar way [9].

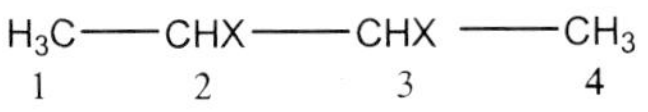

H_3C—CHX—CHX—CH_3
1 2 3 4

C9-5

An *erythro*-form and a *threo*-form of a compound can be distinguished by their chemical shifts. Compound **C9-5** can be taken as an example [10].

As the functional groups can rotate around the C–C axis, both the *erythro*- and *threo*-isomers of **C9-5** have conformers. However, when the three functional groups connecting to the C-2 are superimposed on those connecting to the C-3, the eclipsed conformations have a fairly high internal energy. Therefore, both the *erythro*- and *threo*-isomers of **C9-5** have three conformers respectively, as shown in Fig. 9.1.

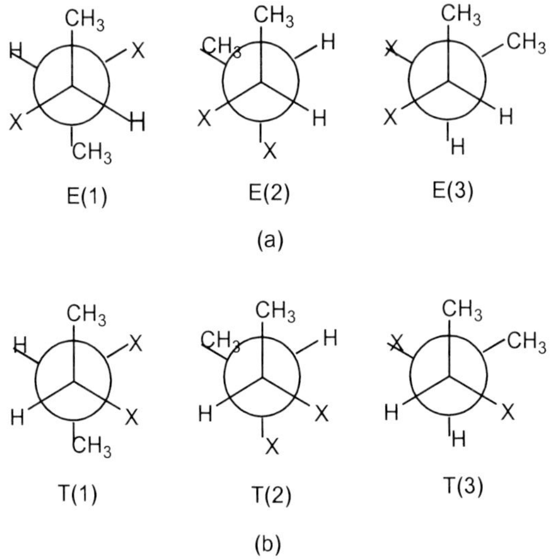

Fig. 9.1 Both the *erythro*- and *threo*-isomers have three conformers, respectively.

Being large in size, the two halogen atoms produce electron repulsion easily, thus making conformations E(1) and T(3) significantly preferable.

In all these six conformers, the peaks of the terminal methyl groups are displaced upfield by the γ-gauche effect. The methyl groups of E(1) have a smaller upfield shift than those of E(2) or E(3) in which four functional groups connecting to C-2 and C-3, respectively, are close to each other, leading to a strong γ-gauche effect. On the other hand, the methyl groups of T(3) have a large upfield shift because of a strong γ-gauche effect. In brief, the methyl groups of the preferred E(1) have a smaller shift towards the high field while the methyl groups of the preferred T(3) have a greater shift towards the high field. Thus the methyl groups of the *erythro*-isomer have a greater δ value than those of the *threo*-isomer, which agrees with the experimental results. $\delta_{erythro}-\delta_{threo}$ equals 3.7, 5.8 and 9.5 ppm when X in **C9-5** is chlorine, bromine or iodine, respectively.

The conformational equilibrium is always an important area in stereochemistry. In fact, Barton's related paper led him to win the Nobel Prize. It has been proved that the mono-substituted cyclohexane has a preferable conformer, in which the substituent is in the equatorial direction, as shown in Fig. 9.2e.

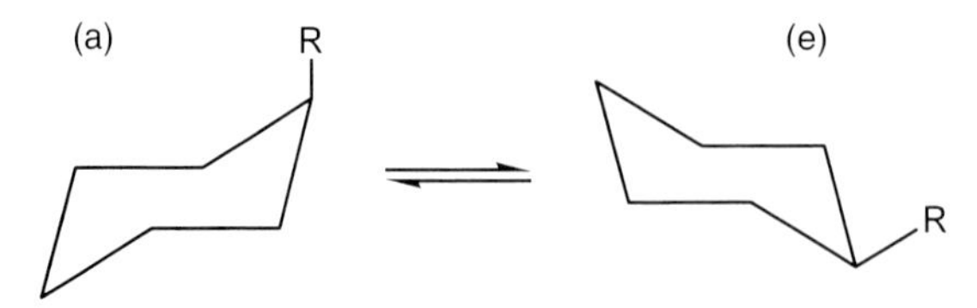

Fig. 9.2 The mono-substituted cyclohexane has a preferred conformer.

As expected, the γ-effects from R on C-3 and C-5 in conformation (a) are stronger than those in conformer (e), which leads to the fact that C-3 and C-5 in conformation (a) have smaller chemical shifts than those in conformer (e). When R is $-CH_3$, $-CH_2-CH_3$, or $-CH-(CH_3)_2$, $\Delta\delta$ is 6.0, 5.6 or 5.3 ppm, respectively [11]. Because the two conformations are exchangeable, it is necessary to measure them at a sufficiently low temperature so that the NMR signals of the two conformers can be differentiated.

Anomeric configurations of glycosides can be distinguished by the ^{13}C chemical shift values of C-1 in a glycoside. For example, the δ_C of C-1 of an α-glycopyranoside is larger than that of a β-anomer by about 4 ppm. However, the δ values of C-3 and C-5 of the α-anomer are smaller than those of the β-anomer by about 6 ppm.

After crown ethers and cyclodextrins, a new type of compound in host-guest chemistry was found, that is, the calixarene, which consists of units of a *para*-substituted phenolic ring and methylene. According to the number of substituted phenolic ring units, calixarenes are denoted as calix[ ]arenes, for example, calix[4]arenes, which is the smallest of the known calixarenes. Other numbers can be 6, 8, and so forth.

Calix[4]arenes have four conformations: cone, partial cone, 1,2-alternate and 1,3-alternate as shown in Fig. 9.3.

With respect to the conformation, calixarenes are flexible compounds. Conformational exchanges will definitely lead to variations of δ_C and δ_H of the methylene groups [12, 13]. On the basis of the study of 24 calix[4]arenes (**C9-6**), the conclusion has been drawn that the δ values of about 31 ppm are those of the methylene groups that are located between two substituted phenolic rings, $-Ar-CH_2-Ar-$, in the *syn*-orientations of the phenol rings, and the δ values of about 37 ppm in the *anti*-orientations [13]. Therefore, if a calix[4]arene has a cone conformation, its methylene groups have δ values of about 31 ppm while those of about 31 and 37 ppm are for the partial cone conformation.

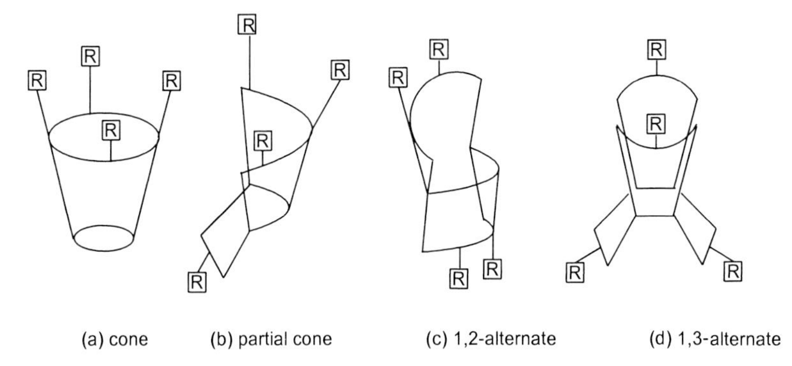

Fig. 9.3 The four conformations of calix[4]arenes.

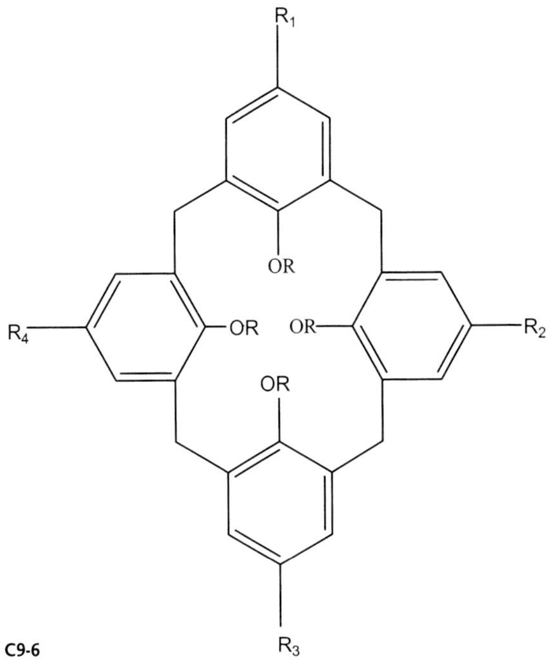

C9-6

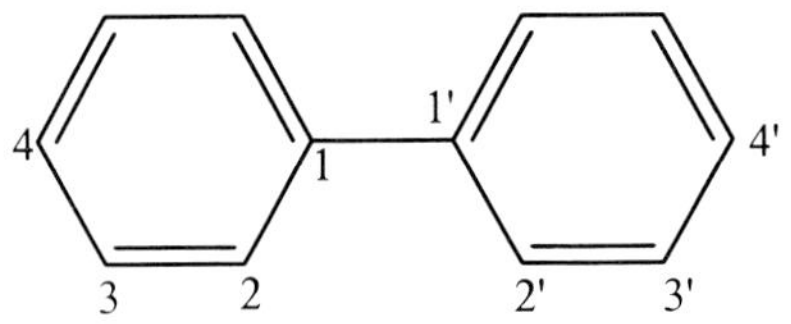

C9-7

^{13}C chemical shifts can also be used to estimate inter-planar angles (θ) for biphenyl derivatives (**C9-7**). When a biphenyl is substituted at 2- and 2′-, θ can be found by calculating the difference between the δ(C-1′) and the δ(C-4′) [14].

Now a sensitive example for configuration determination by ^{13}C chemical shifts will be given [15].

Prostacyclin (**C9-8**) is a potent vasodilator and an inhibitor of blood-platelet aggregation. In view of the stability of the synthesized product, the oxygen atom in the cyclic enol-ether of **C9-8** was replaced by an ethylene in the synthesis process so as to produce carbacyclin (**C9-9**). For further studies iloprost (**C9-10**) was adopted, which has practically the same biological function as prostacyclin.

In synthesis, isoiloprost (**C9-11**) was mixed with iloprost. These two configuration isomers showed quite different biological effects, so they should be differentiated.

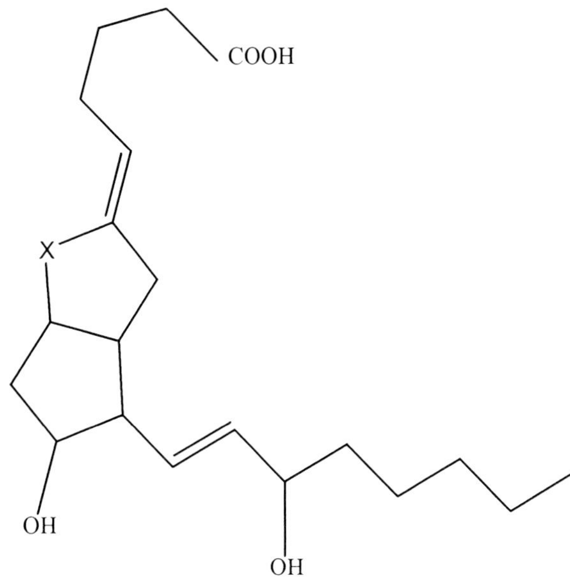

C9-8 X=O (prostacyclin)
C9-9 $X=CH_2$ (carbacyclin)

C9-8 and C9-9

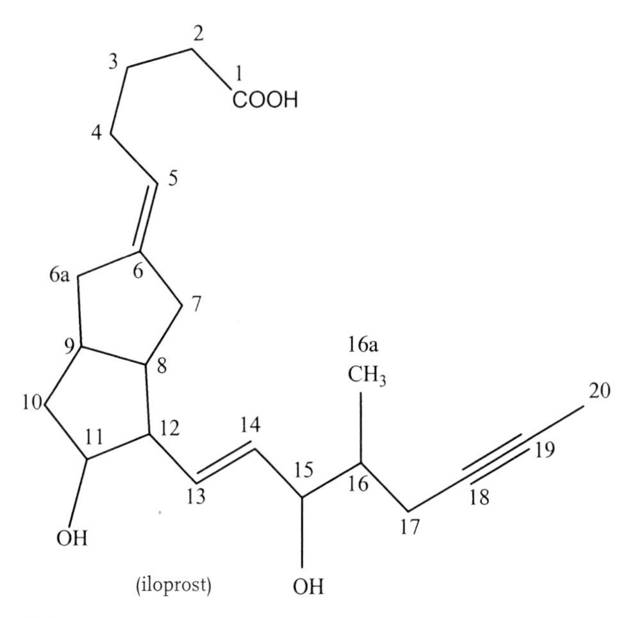

C9-10

HOOC

CH_3

OH

(isoiloprost) OH

C9-11

After the two configurational isomers were isolated, their ^{13}C spectra were recorded.

The strategy for the determination of the two isomers is that the configuration change will alter the chemical shifts of some carbon atoms owing to the steric effects. Consequently, the complete assignments of the two ^{13}C spectra will be a prerequisite. However, as C-16 in **C9-10** or **C9-11** is a chiral carbon atom, both iloprost and isoiloprost have two diastereoisomers. Thus some carbon atoms in **C9-10** or **C9-11** have a pair of peaks, which makes the assignments more difficult.

The ^{13}C spectrum assignments were accomplished by the combination of the DEPT, COSY and relayed coherence transfer spectra.

Finally, the result was as given in Tab. 9.1.

Isomer 1 should be iloprost on account of its smaller $\delta(6\,a)$ in this configuration through the γ-gauche effect, which was produced by C-4. Likewise isomer 2 should be isoiloprost because of its smaller $\delta(7)$ resulting from the γ-gauche effect, which was produced by C-4 in its configuration.

Furthermore, ^{13}C chemical shifts can even reveal the secondary structure of proteins. On the basis of the statistics of more than 100 amino acid residues, the results obtained are shown in the Tab. 9.2 [16].

Tab. 9.1 ^{13}C assignments for isomers

δ (ppm)	*6 a*	*7*
Isomer 1	35.9	38.1
Isomer 2	41.2	32.5

Tab. 9.2 Averaged shifts (ppm) for 100 amino acids

	C-α	C-β
In α-helix	3.09 ± 1.06	–0.38 ± 0.85
In β-sheet	1.48 ± 1.23	2.16 ± 1.91

^{1}H Chemical shifts

In many cases, ^{1}H chemical shifts are not so effective in the application of stereochemistry as ^{13}C chemical shifts for the following reasons: (1) ^{1}H chemical shifts cover a much narrower range than those of ^{13}C. (2) ^{1}H chemical shifts are less sensitive to steric factors than those of ^{13}C. (3) Sometimes coupling splittings in ^{1}H spectra cause difficulties in the assignment.

However, abundant stereochemical information can be deduced from ^{1}H chemical shifts, for example, the configurational determination of substituted double bonds, the distinction of the equatorial or axial orientation of a substituent in a six-membered ring, and so forth.

When a molecule contains anisotropic functional groups, stereochemical information can be obtained from ^{1}H chemical shifts.

It is worthwhile recommending that secondary structures of proteins be determined by ^{1}H chemical shifts. On the basis of the statistics for a large number of protein molecules with known secondary structures, the chemical shift index (CSI) method was proposed by Wishart et al. [17, 18]. The secondary structures of proteins will be reflected in the δ_H of α-CH moieties of amino acids. The δ_H of the α-CH of an amino acid in a random coil is set as the reference of the index, (0). When an amino acid is in an α-helix, the δ_H of its α-CH will be shifted upfield. In this case the index is set as –1. When an amino acid is in a β-sheet, the δ_H of its α-CH will be shifted downfield. In this case the index is set as 1. If four (or more) amino acid residues have the index –1 without being interrupted by an index of 1, this segment of the amino acid residues can be determined as being in an α-helix. If three (or more) amino acid residues have an index of 1 without being interrupted by an index of –1, this segment of the amino acid residues can be determined as being in a β-sheet. This method is simple but has an accuracy approximating that of the traditional NOE method.

9.1.2
Coupling Constants

It is easy to understand that stereochemical information can be obtained by the measurement of coupling constants as their values are affected by steric factors. Comprehensive summaries of the relationships between coupling constants and stereochemistry are given in references [19–21].

Homonuclear coupling constants

Although coupling constants between ^{13}C–^{13}C also reveal stereochemical information [19], its application is greatly limited because of the difficulties with their measurement. Therefore, only $J_{H\text{–}H}$ will be discussed in this section.

1. *Geminal* coupling constants ^2J
   ^2J values are related to molecular stereochemical environments. The ^2J values in isomers **C9-12** and **C9-13** change with the orientation of the *ortho*-substituent containing the heteroatom.

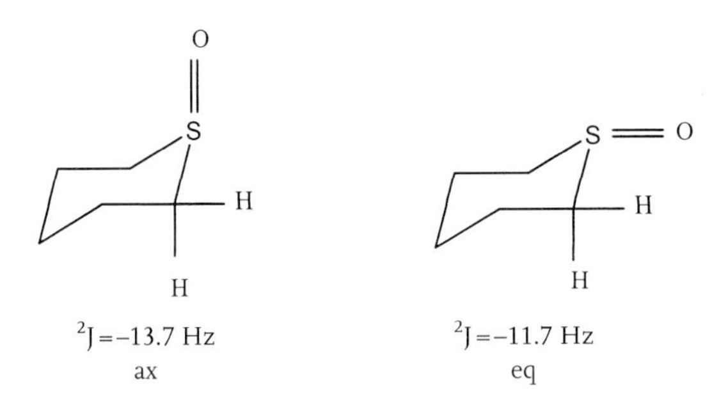

C9-12

In $(\mathbf{C9\text{-}12})_{eq}$, the lonely electron pair of the heteroatom is parallel to the axial bond at the adjacent carbon atom so that the ^2J is more "positive" (a smaller absolute value) than that of $(\mathbf{C9\text{-}12})_{ax}$.

The two ^2J values of $(\mathbf{C9\text{-}13})_{ax}$ and $(\mathbf{C9\text{-}13})_{eq}$ can be explained similarly.

H-2a
H-2e
S
S
N
Cl
^{2}J = −13.6 Hz
ax

H
H
S
S
N
Cl
^{2}J = −12.6 Hz
eq

C9-13

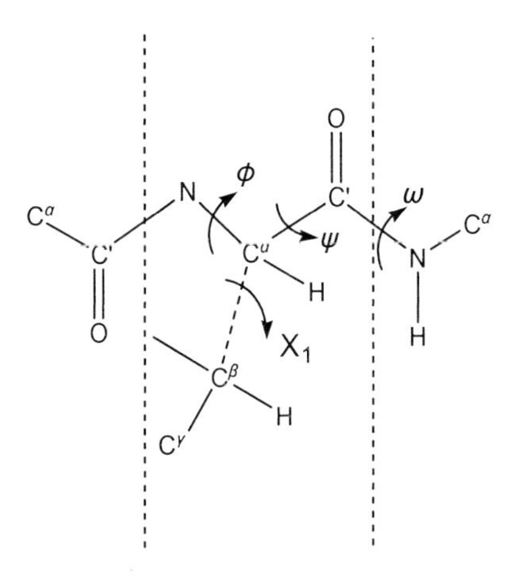

Fig. 9.4 The conformation of an amino acid residue in a peptide.

The conformation of an amino acid residue in a peptide is determined by four torsional angles: φ, ψ, χ_1 and ω, as shown in Fig. 9.4.

The torsional ψ is related to the ^2J of the two H atoms born by the C-α (if there are two hydrogen atoms connected to the carbon atom).

The H–C_α–H *geminal* proton coupling constants in glycine and sarcosine residues have been summarized [20]. When ψ equals 90°, ^2J has a maximum algebraic value (a minimum absolute value because of the negative sign of ^2J). When ψ equals 0° or 180°, ^2J has a minimum algebraic value. The difference between the maximum and the minimum was about 6 Hz.

2. *Vicinal* coupling constants ^3J

The Karplus equation describes the relationship between ^3J values and its related dihedral angles, which is used frequently not only in organic compounds but also in biological molecules. Although the application of the Karplus equation in stereochemistry has been discussed in Section 2.2.3, more examples are given as follows.

As $J_{aa}>J_{ae}\geq J_{ee}$ exists for vicinal couplings in a six-membered ring, the orientation of the substituent on C-1 can be postulated by the coupling constants between H-1 and H-2.

An unknown compound has either structure **C9-14** or **C9-15**.

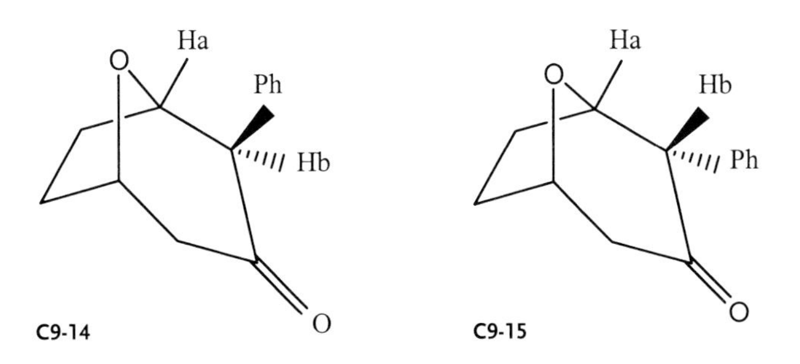

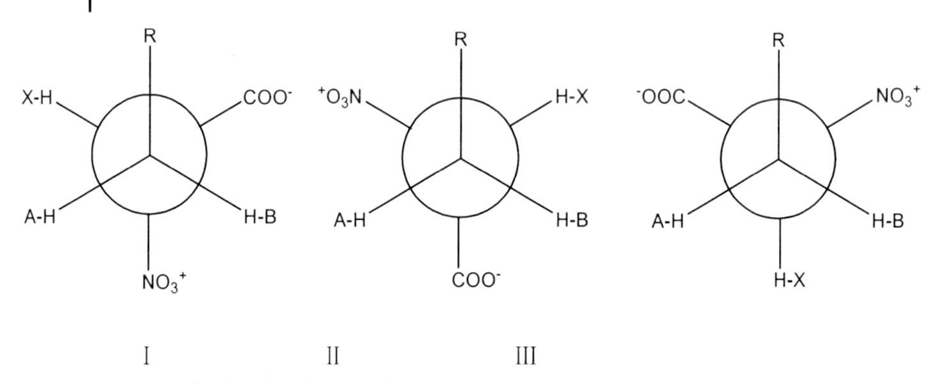

Fig. 9.5 The three conformers of an amino acid molecule.

It is concluded that the unknown compound is **C9-15** based on the fact that the dihedral angle formed from H_a–C–C–H_b in **C9-15** is about 45°, which agrees with the measured coupling constant of 4 Hz. On the other hand, if the unknown compound is **C9-14**, the dihedral angle of nearly 90° leads to a coupling constant close to zero Hz.

An amino acid molecule has three conformers: I, II and III, as shown in Fig. 9.5.

The notations of H-A, H-B and H-X in Fig. 9.5 are used to distinguish the three H atoms. In a phenylalanine molecule, the conformational populations can be calculated from the measured *vicinal* coupling constants: ^3J (H-X, H-A) and ^3J (H-X, H-B) on account of the fact that ^3J (H, H)$_g$ equals 2.6 Hz and ^3J (H, H)$_t$ 13.6 Hz, where *g* means *gauche* and *t trans*. ^3J (HH)$_g$ corresponds to a dihedral angle of H–C–C–H of about 60° and ^3J (HH)$_t$ about 180°. It is necessary to deuterate stereospecifically for the assignment of H-A and H-B [22].

From Fig. 9.4, it can be seen that φ can be calculated from the measured coupling constant about H–N–C$^\alpha$–H and the χ_1 can be calculated from the measured coupling constant about H–C$^\alpha$–C$^\beta$–H.

It should be noted that the electronegativity of a substituent and the orientation of an electronegative substituent affect the value of ^3J.

Ha
X
He
X=Cl, OAc, OH

C9-16

X
He
Ha
X=Br, OAc, OH

C9-17

When the substituent X is situated at the equatorial bond as in **C9-16**, J_{ae} is about 5.5 Hz, which is greater than that (about 2.5 Hz) in **C9-17**.

3. Long-range coupling constants
   Although long-range coupling constants are difficult to use to obtain stereochemical information because of their small values, there are some effective examples for specific cases [7].

Heteronuclear coupling constants

Heteronuclear coupling constants include coupling constants between C and H, N and H, P and H, and so forth, of which the coupling constants between C and H are the most important.

1. $^1J_{CH}$
   It is easy to measure $^1J_{CH}$ due to its large value.
   A good example of obtaining stereochemical information by the measurement of $^1J_{CH}$ is the determination of anomeric isomers. The anomeric configurations of glycosides can be differentiated on the basis of the fact that the coupling constants $^1J_{C\text{-}1,\,H\text{-}eq}$ of α anomers are generally larger than those of $^1J_{C\text{-}1,\,H\text{-}ax}$ of β anomers, the former being about 170 Hz and the latter being about 160 Hz. Often this difference of 10 Hz could be due to the presence of other substituents in the glycosides. Similar results from other six-membered heterocycles have been reported at great length in the literature, such as reference [23].
   $^1J_{CH}$ can also be used for the study of protein conformations. An empirical correlation between $^1J_{C\alpha H\alpha}$ and protein backbone conformations exists: residues in an extended β-sheet conformation have a relatively small average $^1J_{C\alpha H\alpha}$ value (140.5 ± 1.8 Hz), whereas larger values (146.5 ± 1.8 Hz) are found for α-helical residues [16].
2. $^2J_{CH}$
   $^2J_{CH}$ is also known as the heteronuclear *geminal* coupling constant because the coupled carbon and hydrogen atoms are separated by two chemical bonds.
   Heteronuclear *geminal* coupling constants are also used for the differentiation of glycoside anomers. When an oxygen atom is antiperiplanar to a C–C–H linkage (Fig. 9.6 a), $^2J_{C-C-H}$ is positive; when the oxygen atom is of opposite orientation [Fig. 9.6(b)], $^2J_{C-C-H}$ is negative (but with a comparable absolute value) [19].

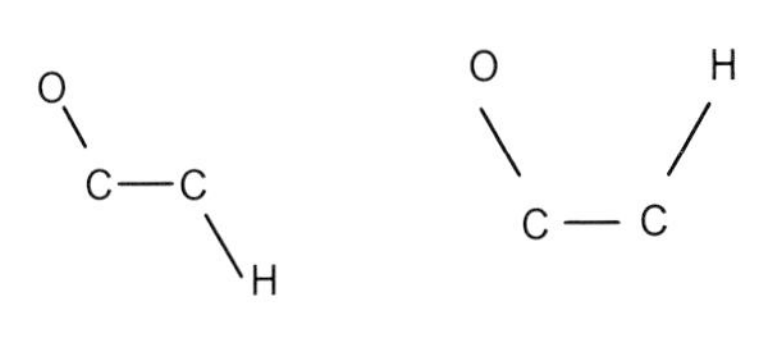

Fig. 9.6 $^2J_{C-C-H}$ has a positive sign (a) or a negative sign (b).

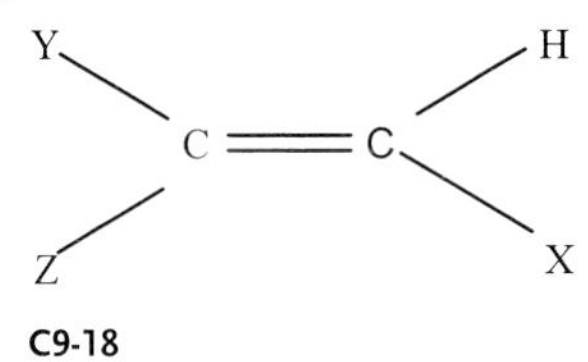

C9-18

On the basis of the statistics for a vast amount of data, it is known that the $^{2}J_{C\text{-}1,\ H}$ in an olefin substituted by three functional groups: X, Y and Z (**C9-18**) can be calculated with a set of parameters, which means the $^{2}J_{C\text{-}1,\ H}$ value is additive [19]. These parameters have been summarized to calculate the related $^{2}J_{CH}$. Calculated values are close to the measured values, for example, **C9-19** and **C9-20**.

measured $^{2}J_{C\text{-}C\text{-}H}$=15.4Hz
calculated $^{2}J_{C\text{-}C\text{-}H}$=16.3Hz

C9-19

measured $^{2}J_{C\text{-}C\text{-}H}$<0.3Hz
calculated $^{2}J_{C\text{-}C\text{-}H}$=0.9Hz

C9-20

Therefore, the configurations of substituted olefins can be determined by measuring related $^{2}J_{CH}$ values.

$^{2}J_{CH}$ for an acyclic structure in which a methine group is substituted by a methyl group, a hydroxyl group or an –OR group, which are commonly found in natural products, has been studied [24]. As shown in Fig. 9.1, three staggered rotamers exist in a *threo*- or an *erythro*-isomer. A rotamer which has a rather large ^{1}H–^{1}H coupling constant can be differentiated because the dihedral angle is 180°. Having rather small ^{1}H–^{1}H coupling constants, the two other rotamers can not be differentiated because both are of *vicinal* H/H–*gauch* conformation. When the methine group is substituted by a hydroxyl or an –OR group, the two rotamers in a *threo*- or an *erythro*-isomer can be differentiated because they have different $^{2}J_{CH}$. For the substitution of a methyl group, when the methyl group is *gauch* to its *vicinal* proton, the $^{2}J_{CH}$ is small while when it is *anti*, the value is large. For the substitution of a hydroxyl group or an –OR group, when the group is *gauch* to its *vicinal* proton, the $^{2}J_{CH}$ is large while when it is *anti*, the value is small. Therefore, the two rotamers can be differentiated by their different $^{2}J_{CH}$ values. On the basis of $^{2}J_{CH}$, $^{3}J_{CH}$ and $^{3}J_{HH}$, a method, called "*J*-based configuration analysis" was proposed [24].

3. $^3J_{CH}$

Because the Karplus equation holds for H–H ^3J couplings, it is reasonable to try to establish an analogous correlation for carbon–proton ^3J couplings. A similar correlation between the dihedral angle constituted by C–C–C–H and $^3J_{CH}$ has been established based on both theoretical calculation and experimental measurements (dihedral angles were measured by X-ray diffraction and $^3J_{CH}$ by NMR). For example, in a theoretical study of propane, the following equation was derived:

$$^3J_{CH} = 4.26 - 1.00 \cos \theta + 3.56 \cos 2\theta$$

In spite of the difficulties in measuring $^3J_{CH}$, the application of $^3J_{CH}$ in stereochemistry is extensive when no $^3J_{HH}$ exists. The two isomers **C9-21** and **C9-22** are examples of the determination of the configuration of the substituted double bond, where "*" is used to denote the related C and H, which are coupled.

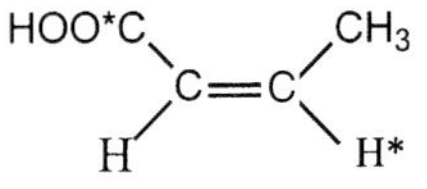

HOO*C(H)C=C(CH3)H* $^3J=14.5Hz$

C9-21

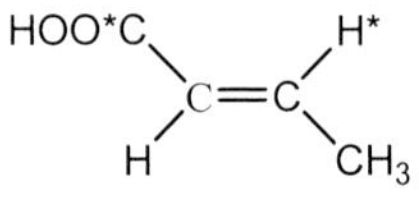

HOO*C(H)C=C(H*)CH3 $^3J=6.78Hz$

C9-22

Now we come back to Fig. 9.5. As in the discussion on the application of $^3J_{HH}$, the conformational populations of an amino acid can also be calculated by measuring $^3J(C,H)_g$ and $^3J(C,H)_t$ on account of the fact that $^3J(C,H)_g = 1.3$ Hz and $^3J(C, H)_t = 9.8$ Hz. The result of the calculation from $^3J_{CH}$ agrees with that from $^3J_{HH}$ [22].

4. The other heteronuclear ^3J

In addition to $^3J_{CH}$, $^3J_{NH}$ and $^3J_{NC}$ are used for stereochemical studies of proteins when samples of enriched ^{15}N are used [25].

Measurement of coupling constants

There are hundreds of papers describing the methods of measuring coupling constants, the simplest of which is to read them out directly from the conventional (1D) NMR spectra.

Two-dimensional NMR spectra provide diverse methods for the measurement of coupling constants. As these methods are developing very rapidly and are difficult to summarize, only some of them are presented in what follows.

Accurate J values can be drawn from J-resolved spectra for the cases of weak coupling systems.

The phase sensitive COSY is used to obtain J values together with their signs. However, this method is less effective when digital resolution is insufficient or positive peaks and negative peaks partially overlap. In this case the disadvantage can be overcome by E.COSY [26] (exclusive COSY), which is a weighted addition of DQF-COSY and TQF-COSY. J values can be easily read out from the simplified fine structures of cross peaks of E.COSY.

The presentations above are the methods of reading out J values from peak shapes of conventional ^{1}H spectra or fine structures of cross peaks of 2D NMR spectra.

J values can also be obtained from cross peak intensities as J values are related to the intensities of the cross peaks (see Appendix 1). One example is the measurement of J values from the cross peaks of TOCSY by iterative back-calculation [27].

When measuring heteronuclear coupling constants, the acquisition in the conventional mode (detecting carbon atoms or heteroatoms) has very low sensitivity. Therefore, the inverse mode is recommended. Reference [28] is one of the many papers on using the inverse mode to obtain heteronuclear coupling constants.

Many new pulse sequences have been developed to measure heteronuclear coupling constants. An example is given of measuring long-range ^{1}H–^{13}C coupling constants by using the selective excitation of ^{13}C [29].

Several methods for the measurement of heteronuclear coupling constants, such as HETLOC (heteronuclear long-range coupling), HSQC–TOCSY (heteronuclear single-quantum correlation–total correlation spectroscopy), phase-sensitive HMBC, and so forth, have been discussed and compared [30]. Methods to obtain heteronuclear coupling constants are presented. For example, the value of $^{2}J_{CH}$ or $^{3}J_{CH}$ is obtained from the shift of the second slice with respect to the first slice in the F_2 dimension in a HETLOC spectrum. These two slices pass through two coupled partners of a cross peak in a HETLOC spectrum.

Isotopically enriched samples are used to measure coupling constants between ^{13}C and heteroatoms, for example $^{3}J_{CN}$.

9.1.3 NOE

Although NOE was discovered in the middle of the 1950s, it was first applied in chemistry in 1965 by Anet and Boum [31]. Six years later, the first monograph on NOE was published [32]. The second monograph on NOE was published in 1989 [4]. There have been many published papers on NOE.

Configuration and conformation determinations of organic compounds by using NOE are always more effective than by using chemical shifts or coupling constants. The study by Brakta et al. is a good example [33], which will be discussed later.

The following points should be noted:

1. The prerequisite for studying NOE is the assignment of related NMR spectra.
2. There are two methods to measure NOE: one-dimensional spectra (NOE difference spectra) and two-dimensional spectra (NOESY or ROESY). The NOE

difference spectrum has the advantage of high sensitivity. However, to irradiate many peak sets in a ^{1}H spectrum is time-consuming.

Some "one-dimensional spectra", obtained by using 2D NMR pulse sequences together with selective excitations, such as GOESY (see Section 4.1.9), are very sensitive to NOE without a distorted baseline.

3. NOE enhancements are related to the molecules being studied, as well as to the NMR instruments (frequencies) and other experimental conditions.
4. The information from NOE is related to the number of chemical bonds, across which the NOE is shown. The more bonds, through which two nuclei show the NOE, the NOE spans, the more valuable the steric information is. Under such conditions, configurations or conformations can be restricted to a few possibilities.
5. Rigid molecules are appropriate for NOE measurement.
6. For flexible molecules, in which conformational exchanges occur, their NOE can not be directly measured. Three methods are feasible:
   i. Reduce the sample temperature so that conformational exchanges become slow compared with the NMR time scale.
   ii. Add a specific substance that makes the sample solution viscous, so that the speed of the conformational exchange process decreases, which leads to a preferential conformation in the solution.
   iii. The sample molecules can be modified chemically to decrease the speed of conformational exchange process.

Some examples to show the applications of NOE in stereochemistry are presented as follows. Brakta's study [33] is convincing for this application.

There are α- and β-anomers of glycopyranoside: the α-anomer **C9-23** and the β-anomer **C9-24**.

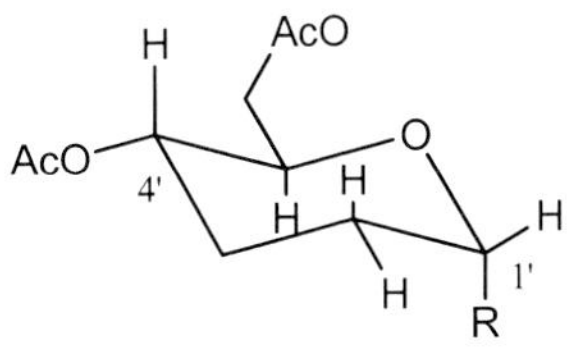

C9-23

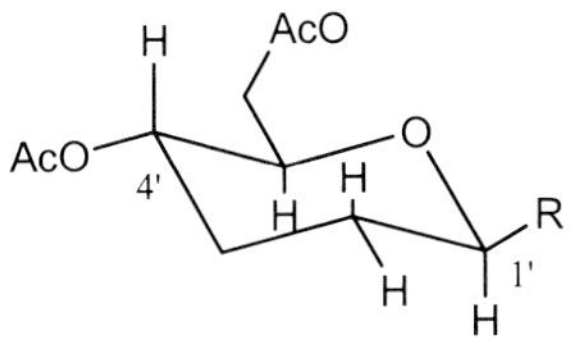

C9-24

As the H-1 of the β-anomer is situated at the axial bond position, its δ_H value is smaller than that of the α-anomer. On the basis of the consideration above, these two anomers can be differentiated. However, sometimes the differences between the δ_H values are small, even if the order of the δ_H values is reversed so that the judgment is not reliable.

In most cases, the value of $^{1}J(\text{C-1}', \text{H-1}'_{eq})$ is larger than that of $^{1}J(\text{C-1}', \text{H-1}'_{ax})$. However, sometimes these two values are close, which leads to difficulty in the differentiation of these two anomers.

In the β-anomer, because H-1′ is situated at the axial bond, its $^3J(\text{H-}1'_{ax}, \text{H-}2'_{ax})$ is greater than the $^3J(\text{H-}1'_{eq}, \text{H-}2'_{ax})$ of the α-anomer, it appears that these two anomers can be differentiated by the $^3J_{HH}$. However, because the substituent is situated at the axial bond, the conformation of the α-anomer is not stable, which leads to conformational exchanges between these two anomers. Therefore, the measured $^3J_{HH}$ value of the α-anomer is always greater than its real value, which results in the difficulty in differentiating between the two anomers.

Unambiguous conclusions are drawn by using NOE difference spectra. The peak intensity of H-5′ in the β-anomer varies when the peak of the H-1′ is irradiated because the two hydrogen atoms are situated on the same side and they are close in space. Thus the NOE shows that it is the β-anomer. On the contrary, this phenomenon does not exist in the α-anomer. Therefore, the most reliable method for the differentiation of these two anomers is NOE.

Another example is as follows [34]. There is a pair of isomers, **C9-25** and **C9-26**.

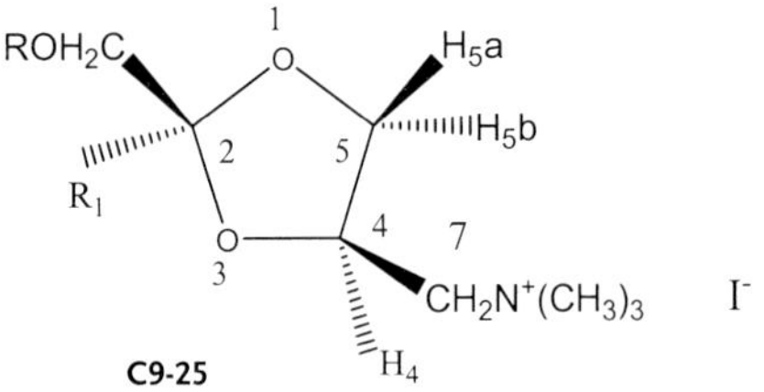

C9-25 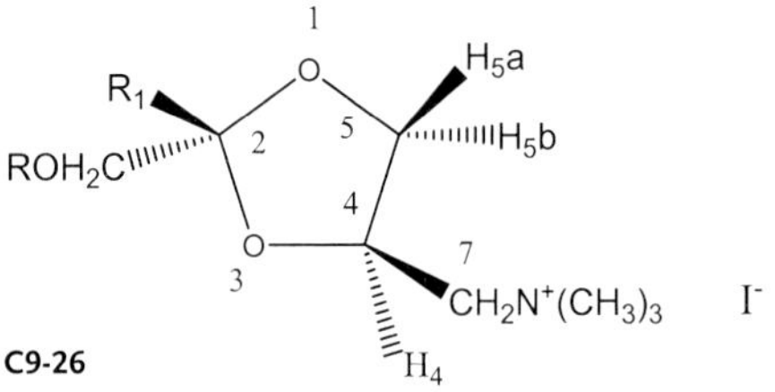 **C9-26**

When $ROCH_2^*$– is irradiated, the NOE for H-4 and H-5b is observed so that this compound can be assigned as being **C9-26**. With the same irradiation, the NOE for H-5a and H-7b is observed so that the compound can be assigned as being **C9-25**.

Heteronuclear NOE is preferable sometimes, for example, in the case of differentiating the isomers **C9-27** and **C9-28**.

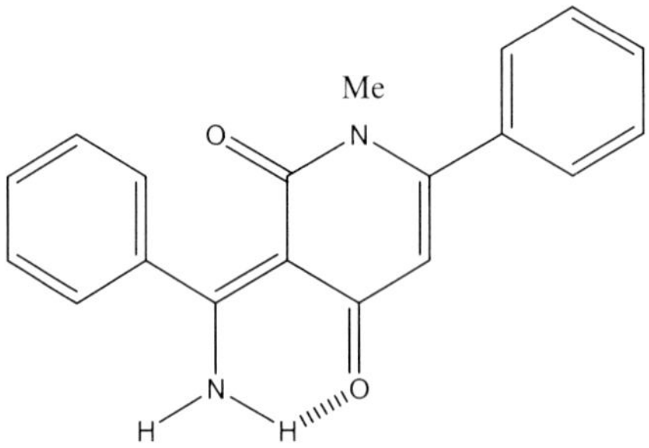

C9-27

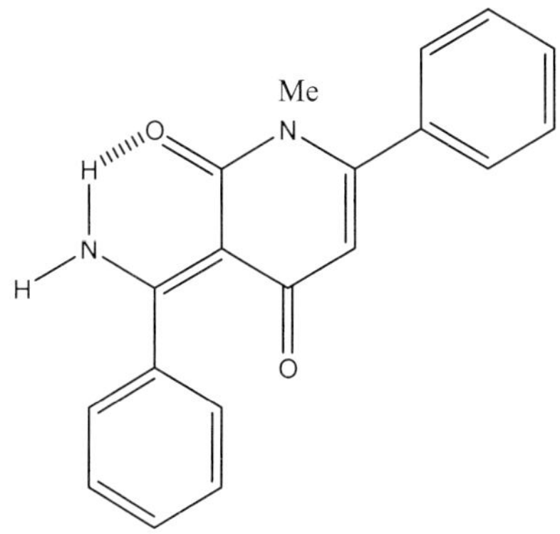

Me
N
O
H
N
H
O

C9-28

From these two structural formulae, it is known that the NH_2 group is close to one of the two carbonyl groups of a compound in space. When the NH_2 group is irradiated, one of the two carbonyl groups of a compound will show heteronuclear NOE. Consequently, a differentiation between **C9-27** and **C9-28** can be achieved.

The preceding examples are the qualitative application of NOE, which is used mainly for the differentiation of isomers. On the other hand, NOE can be applied quantitatively. Distances between some pairs of nuclei can be calculated from related NOE data.

This type of calculation is performed by using NOESY-like spectra. In fact, a 2D NMR spectrum is the presentation of a three-dimensional figure. The calculation of NOE intensities is the integration of the related cross peak volume. There are simple methods for the calculation, as described in reference [35]. Today the calculations are performed by specific software. Two methods are frequently used [36–40].

The distances of pairs of nuclei of a molecule can be used as an important limit to the calculation of molecular dynamics, MD, by which the conformations of a compound in its solution can be obtained [36, 40].

Wüthrich has made important contributions to the conformational determination of proteins and nucleic acids, especially of proteins [41], which led to him sharing the Nobel Prize in chemistry in 2002.

9.2 Mass Spectrometry

Since the 1960s, organic mass spectroscopists have been devoting themselves to stereochemical studies by mass spectrometry. Although the MS method is less effective and less versatile than NMR in general for stereochemical studies, MS is the only feasible stereochemical method if the amount of a sample available is very small. In addition, mass spectrometry is effective for the stereochemical study of many types of organic compounds, which is described in reference [6].

As this chapter deals with the configuration and conformation studies by spectroscopic methods, the discussion is arranged according to ionization methods: EI, soft ionizations and reaction mass spectrometry, RMS.

9.2.1
Utilizing Electron Impact Ionization

As EI was earliest method to be developed and is the widest used for organic mass spectrometry, this section starts with it.

Using EI as the ion source, some organic compounds with different configurations can be differentiated because of their various fragmentation processes, and consequently different abundances of some ions. Maleic acid (**C9-29**) and fumaric acid (**C9-30**) as well as their derivatives are good examples.

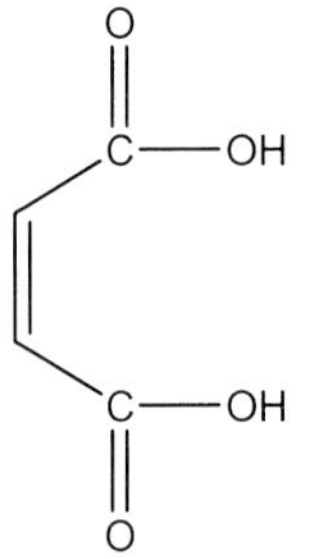

C9-29 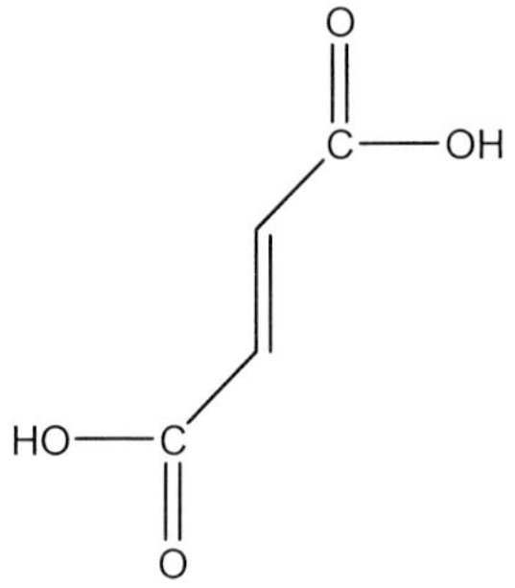 **C9-30**

In a maleic acid molecule, two carboxyl groups are close to each other. Consequently, it is easy for the molecule to lose a CO_2 molecule through the replacement of a hydrogen atom for the carboxyl groups. Therefore, $M^{+\bullet}$–44 (*m/z* 72) is the base peak in the MS spectrum and thus, the intensity of the molecular ion peak is low. On the contrary, the mass spectrum of fumaric acid has a strong molecular peak, a weak peak of $M^{+\bullet}$–44 and the fumaric acid is characterized by the peak of $M^{+\bullet}$–18 (with the loss of water) [42].

Similarly, a greater difference exists between the mass spectrum of esters of the maleic acid and that for the fumaric acid. The mechanism of the related mass spectrometry reactions are postulated in reactions **122** and **123** by isotopically labeled compounds [43].

O, C, OC_2H_5, C, OC_2H_5, O (molecular ion, ⌝+·)

$-OC_2H_5$

$^+OC_2H_5$, C, O, C, O

m/z 127

122

$-C_2H_4$

^+OH, C, O, C, O

m/z 99

O, C, OC_2H_5, C_2H_5O, C, O (molecular ion, ⌝+·)

$-OC_2H_5$

$C\equiv O^+$, C_2H_5O, C, O

m/z 127

$-C_2H_4$

$C\equiv O^+$, HO, C, O

m/z 99

$-CO$

$C_2H_5O-CH=CH-C\equiv O^+$

m/z 99

H_3C, CH, H, O, C, O, $C\equiv O^+$

123

$-(CH_3CHO+CO)$

$H_2C=CH-C\equiv O^+$

m/z 55

Stereochemical factors may play an important role in RDA fragmentation. As an example, **C9-31***cis* with a *cis*-junction of a cyclohexene and an adjacent ring gives rise to very abundant diene ions through RDA fragmentation. This ion is of negligible abundance in the mass spectrum of the *trans*-isomer. Therefore, the mass spectrum of **C9-31***cis* shows a weak molecular ion peak while that of **C9-31***trans* shows a strong molecular ion peak because its RDA fragmentation does not occur [6].

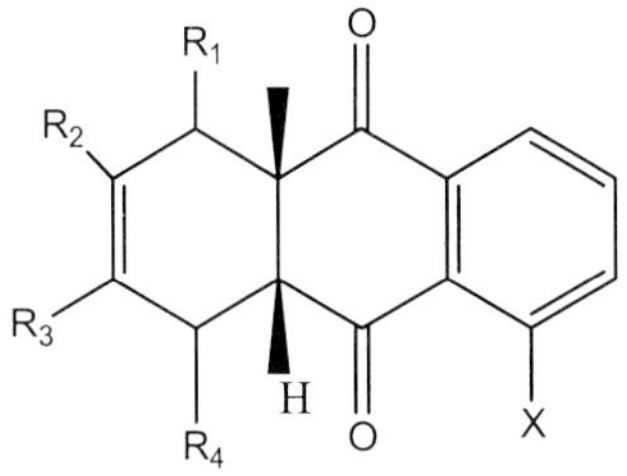

C9-31 cis 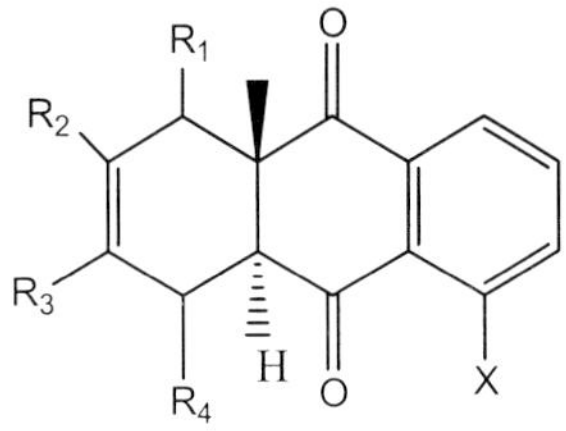 **C9-31 trans**

Sometimes monoterpenes with different skeleton structures may show similar mass spectra. However, it is possible to distinguish between different configurational isomers by EI, especially with a fairly low electron beam energy [44].

9.2.2 Utilizing Soft Ionization

On the basis of the presentation above, the mass spectrum differences among stereochemical isomers generally increase when the electron beam energy in EI decreases, for example, from 70 eV to less than 20 eV. It can be estimated that more stereochemical information will be obtained if EI is replaced by soft ionization. This estimation has been proved by related experiments. As CI was earliest of all soft ionization techniques to be used and now it is that most widely applied for stereochemistry, references to CI dominate this field. Therefore, only some examples of CI can be discussed in this section. An example for FAB is cited in reference [45].

The differentiation of stereo-isomers of monoterpenes is taken as an example.

The four stereo-isomers **C9-32**, **C9-33**, **C9-34** and **C9-35** can be differentiated into two groups using NICI–CID (negative ion chemical ionization associated with collision-induced dissociation) [46]. The isomers with the axial hydroxyl groups (**C9-32** and **C9-33**) are characterized by m/z 135 and m/z 83, while the isomers with the equatorial hydroxyl groups (**C9-34** and **C9-35**) are characterized by m/z 133, 109 and 41. This result comes from the strong effect of the hydroxyl group on the fragmentation of related compounds. On the other hand, the mass spectra for **C9-32** and **C9-33** are similar, as are those for **C9-34** and **C9-35**, because there is little influence of the methyl group upon fragmentation.

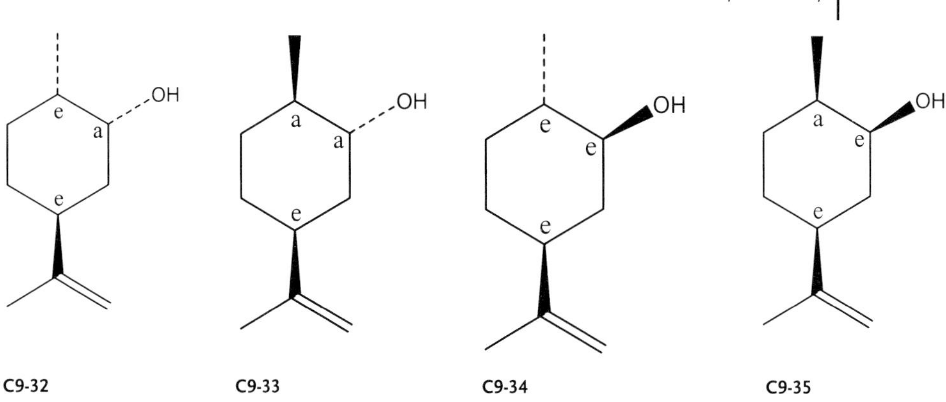

Similar to the fragmentation of maleate and fumarate, *cis*-1,4-cyclohexane dicarboxylate undergoes alcohol elimination using CI while *trans*-1,4-cyclohexane dicarboxylate does not undergo this reaction because the two substituents of the former are close in space [47].

The differentiation of four isomers of adamantane derivatives is similar to that discussed above [48].

9.2.3
Reaction Mass Spectrometry

Since the middle of the 1980s, several groups, led by Chen, Winkler and Nikolaev respectively, have made great progress in the application of mass spectrometry in stereochemistry. The concept of reaction mass spectrometry, RMS or stereoselective reaction mass spectrometry has been proposed.

However, there are some differences in the definitions of reaction mass spectrometry. Here that proposed by Chen is cited [49]: "RMS is a technique in which a reagent is introduced into the ion source EI, CI or FAB, where it reacts stereoselectively with the sample through ion–molecule reactions to form some characteristic ions. From the relative abundances of these ions, the stereochemical information about the sample molecule may be obtained".

Although RMS is associated with soft ionization, it uses a reagent for the reaction and important progress has been made, so it is presented here.

The study of RMS is based on the idea that stereochemical reactions, which take place in a condensed phase, may occur in the gaseous phase and it is possible to detect the ions produced using mass spectrometry.

Some chiral compounds with known absolute configurations, are used as reagents of the reaction, and sometimes prochiral compounds are also used [50].

A typical example is the absolute configuration determination of asymmetric secondary alcohols [51].

The asymmetric secondary alcohols include cinchonine, cinchonidine, (–)-ephedrine, (+)-pseudoephedrine, quinine and (+)-methylmandelate. *R*- and *S*-1-phenylbutyric anhydrides are used as reagents for RMS. Isobutane is used as the reagent gas for CI. Every secondary alcohol reacts with a pair of reagents of RMS and produces the characteristic ester ion $[M_S+M_r+H–C_6H_5CHEtCO_2H]^+$ where M_s and M_r are the sample and the reagent molecules, respectively. The ion has a higher abundance when a sample and a reagent have the same configuration and a lower abundance for different configurations. The ratio $r=B/A$, where B and A are the abundances of the characteristic ester ion and $[M_S+H]^+$, respectively, shows the capacity for producing the characteristic ester ions. Subscripts *R* and *S* are used for the ratio *r*. The parameter r_R shows the ratio when the reagent is in the *R*-configuration, and similarly, the r_S for the *S*-configuration. It is found that for the sample with the *R*-configuration, $r_R:r_S$ is greater than 1, and for the compound with *S*-configuration, the ratio is less than 1.

Some monosaccharide configurations are determined in the same way.

9.3 Infrared and Raman Spectroscopy

Molecular vibration spectra, that is, infrared or Raman spectra, are also used in the configuration and conformation studies. However, the amount of related literature is much less than that on NMR or MS. In addition, it appears that there is no systematic summary on this topic.

Cis- and *trans*-double bond isomers can be distinguished by IR, Tab. 9.3.

They can also be distinguished by Raman spectroscopy, Tab. 9.4.

Because both infrared and Raman spectra show molecular vibration frequencies for a given vibration, the absorption frequencies in infrared spectra and Raman shifts in Raman spectra are generally close. However, the intensities of the corresponding peaks in the two types of spectra may differ greatly.

The absorption position of the carbonyl group of a steroid ketone is related to the position of the carbonyl group in the steroid as well as to the direction of the hydrogen atom situated at the common point of two rings, denoted as *α*- and *β*-.

Tab. 9.3 Distinguishing *cis*- and *trans*-double bond isomers by IR

Bond	Vibration	cis-	trans-
C=C	stretching	≈1650 cm^{-1}	≈1675 cm^{-1}
C–H	bending	750–675 cm^{-1}	965 cm^{-1}

Tab. 9.4 Distinguishing *cis*- and *trans*-double bond isomers by Raman spectroscopy

Bond	Vibration	cis-	trans-
C=C	stretching	≈1650 cm^{-1}	≈1670 cm^{-1}
C–H	bending	≈690 cm^{-1}	≈970 cm^{-1}

The skeleton of a steroid is numbered as shown in compound **C9-36**. The absorption position of the 6-carbonyl changes with the direction of 5-hydrogen atom (5α- or 5β-).

For the 6-CO (5α-) and 6-CO (5β-) isomers the absorption bands are at 1712–1714 and 1706–1708 cm^{-1}, respectively.

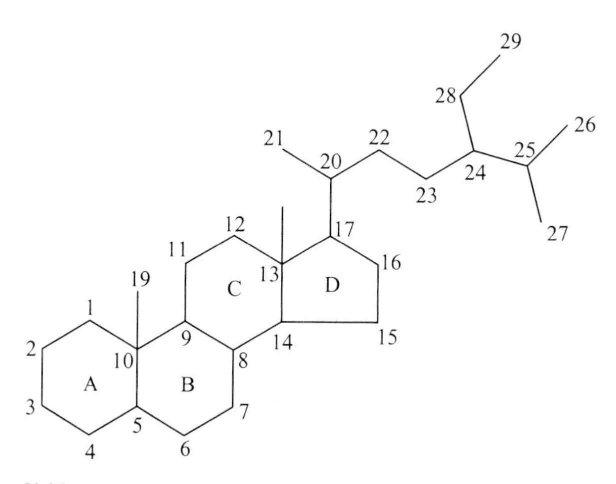

C9-36

The conformations of six-membered rings can be determined by IR spectra. For example, in fluorocyclohexane the equatorial C–F bond has an absorption band at 1062 cm^{-1} while the axial C–F bond has an absorption band at 1129 cm^{-1}. In chlorocyclohexane the equatorial C–Cl bond has an absorption band at 742 cm^{-1} while the axial C–Cl bond has an absorption band at 688 cm^{-1}. The coexistence of these two peaks in the same IR spectrum shows the coexistence of these two conformers. The equilibrium constant for these two conformers can be calculated through variable-temperature experiments [52].

The formation of hydrogen bonds decreases related absorption frequencies, which may be used for isomer differentiation in some cases, for example compounds **C9-37** and **C9-38** as shown in Tab. 9.5.

Tab. 9.5 Isomer differentiation using related absorption frequencies

Bond	*C9-37*	*C9-38*
O–H	3520 cm^{-1}	3610 cm^{-1}
C–OH	$\approx$964 cm^{-1}	$\approx$1060 cm^{-1}

C9-37 **C9-38**

Infrared spectroscopy is suitable for the study of hydrogen bonds because the interchange between the hydrogen bonded state and the non-hydrogen bonded state is a rapid process. If this process is studied using NMR, which has a rather slow time scale, only an average result will be obtained. However, this process can be observed perfectly by using IR, whose time scale is faster than that of NMR by several orders of magnitude. Therefore, both the hydrogen bonded state and the non-hydrogen bonded state, which are in equilibrium, can be “observed” clearly using IR.

Gellman et al. selected a series of diamides as a simplified model for the study of non-covalent interaction of protein molecules. The selected diamides have two conformers, with or without hydrogen bonds, compounds **C9-39** and **C9-40**.

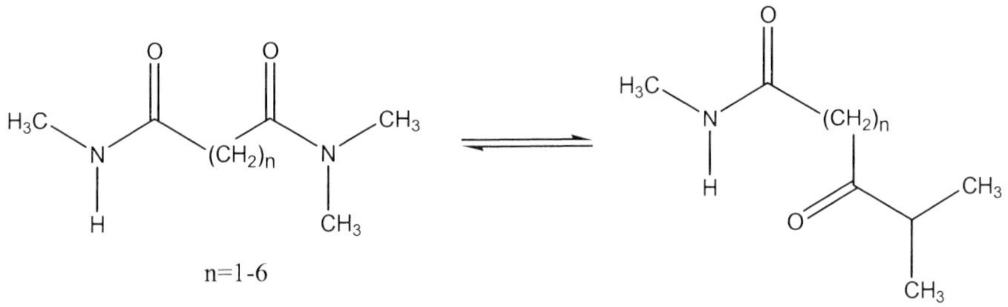

n=1-6

C9-39 **C9-40**

Two absorption bands are observed. A sharp peak at 3450–3460 cm^{-1} is assigned as the N–H stretching vibrational absorption of the isomer without hydrogen bonds. Another broadened band about 3300–3330 cm^{-1} is assigned as the N–H stretching vibrational absorption of the isomer with hydrogen bonds. The former dominates when $n=2$–5 and the latter dominates when $n=1$. The number of conformers with hydrogen bonds increases when the temperature decreases. The equilibrium constant can be obtained [53].

Protein or peptide molecules contain amide units, and thus are suitable for study by IR or Raman spectroscopy. The amide units of proteins have nine absorption bands in their vibrational spectra, namely the bands of A, B and I to VII in the order of decreasing wavenumbers. The amide I, II and III bands are important for the study of protein conformations. The most important absorption band, the amide I band, resulting from the stretching vibration of the amide, which is situated at 1680–1600 cm^{-1}, is especially useful for the determination of the secondary structures of proteins. The amide II band situated between 1580 and 1480 cm^{-1} is ascribed mainly to the bend vibration of N–H mixed with the stretching vibration of C–N. The amide III band situated between 1300 and 1230 cm^{-1} is ascribed mainly to the stretching vibration of C–N mixed with the bending vibration of N–H. These absorption bands are sensitive to the conformations of proteins. Therefore, the secondary structures of proteins, i.e., α-helix, parallel or anti-parallel β-sheet, β-turn and random coil, etc., can be postulated from these absorption bands [54, 55].

Nabet and Pezolet studied the secondary structure of muscle myoglobin by using 2D IR spectra and H–D chemical exchange. The decomposed peaks were shown clearly in synchronous and asynchronous 2D IR spectra [56]. They also succeeded in using non-cyclic perturbation to replace cyclic perturbation, so 2D IR spectra can be used more widely.

Raman spectra can show the slight changes in the structure of the sample studied, even in the configuration and/or conformation of an organic compound [57]. Sometimes Raman spectra of isomers can differ greatly from each other, for example, between α-D-glucose and β-D-glucose.

Stereo-isomers can be well differentiated using Fourier transform vibrational circular dichroism, FT–VCD [58].

9.4 References

1 E. Juarsti, *Introduction to Stereochemistry and Conformational Analysis*, John Wiley, New York, 1991.

2 H. Dodziuk, *Modern Conformational Analysis, Elucidating Novel Exciting Molecular Structures*, VCH, Weinheim, 1995.

3 G. E. Martin, A. S. Zektzer, *Two-Dimensional NMR Methods for Establishing Molecular Connectivity: A Chemist's Guide to Experiment Selection, Performance, and Interpretation*, VCH, Weinheim, 1998.

4 D. Neuhaus, M. Williamson, *The Overhauser Effect in Structural and Conformational Analysis*, VCH, Weinheim, 1989.

5 W. R. Croasmun, R. M. K. Carlson, *Two-Dimensional NMR Spectroscopy: Applications for Chemists and Biochemists*, 2nd Edn., VCH, Weinheim, 1995.

6 J. S. Splitter, F. Turecek, *Applications of Mass Spectrometry to Organic Stereochemistry*, VCH, Weinheim, 1994.

7 K. Pihlaja, E. Kleinpeter, *Carbon-13 Chemical Shifts in Structural and Stereochemical Analysis*, VCH, Weinheim, 1994.

8 J. W. D. Haan, L. J. M. van de Ven, *Org. Magn. Reson.*, **1973**, *5*, 147–153.

9 E. Kleinpeter, H. Kuhn, *Org. Magn. Reson.*, **1976**, *8*, 279.

10 H. J. Schneider, M. Lonsdorfer, *Org. Magn. Reson.*, **1981**, *16*, 133.

11 M. E. Squillacote, J. M. Neth, *Magn. Reson. Chem.*, **1987**, *25*, 53–56.

12 C. D. Gutsche, B. Dhawan et al., *Tetrahedron*, **1983**, *39*, 409–426.

13 C. Jaime, J. D. Mondoza et al., *J. Org. Chem.*, **1991**, *56*, 3372–3376.

14 R. M. G. Roberts, *Magn. Reson. Chem.*, **1985**, *23*, 52–54.

15 K. V. Schenker, W. V. Philipsborn, *Helv. Chim. Acta*, **1986**, *69*, 1718–1723.

16 S. Spera, A. Bax, *J. Am. Chem. Soc.*, **1991**, *113*, 5490–5492.

17 D. S. Wishart, B. D. Sykes et al., *Biochemistry*, **1992**, *31*, 1647–1651.

18 D. S. Wishart, B. D. Sykes et al., *J. Mol. Biol.*, **1991**, *222*, 311–333.

19 J. L. Marshall, *Carbon-Carbon and Carbon-Proton NMR Couplings: Applications to Organic Stereochemistry and Conformational Analysis*, Verlag Chemie, Weinheim, 1983
20 V. F. Bystrov, *Prog. Nucl. Magn. Reson. Spectrosc.*, **1976**, *10*, 41–81.
21 P. E. Hansen, *Prog. Nucl. Magn. Reson. Spectrosc.*, **1981**, *14*, 175.
22 B. P. Mikhova, M. F. Simeonov et al., *Mag. Reson. Chem.*, **1985**, *23*, 474–477.
23 G. W. Vuister, F. Delaglio, A. Bax, *J. Am. Chem. Soc.*, **1992**, *114*, 9674–9675.
24 N. Matsumori, N. Kaneno et al., *J. Org. Chem.*, **1999**, *64*, 866–876.
25 G. W. Vuister, A. C. Wang et al., *J. Am. Chem. Soc.*, **1993**, *115*, 5334–5335.
26 C. Griesinger, O. W. Sorensen, R. R. Ernst, *J. Chem. Phys.*, **1986**, *85*, 6837–6852.
27 J. P. M. van Duynhoven, J. Gougriaan et al., *J. Am. Chem. Soc.*, **1992**, *114*, 10055–10056.
28 U. Wollborn, W. Willker et al., *J. Mag. Reson., A*, **1993**, *103*, 86–89.
29 B. Adams, L. Lerner, *J. Mag. Reson., A*, **1993**, *103*, 97–102.
30 B. L. Marquez, W. H. Gerwick, T. Williamson, *Mag. Reson. Chem.*, **2001**, *39*, 499–530.
31 F. A. L. Anet, A. Boarn, *J. Am. Chem. Soc.*, **1965**, *87*, 5250.
32 T. H. Noggle, R. E. Schimer, *The Overhanser Effect*, Academic Press, New York, 1971.
33 M. Brakta, R. N. Farr et al., *J. Org. Chem.*, **1993**, *58*, 2292–2298.
34 A. Mucci, L. Schenetti et al., *Magn. Reson. Chem.*, **1995**, *33*, 167–173.
35 T. A. Holak, J. N. Scarsoale et al., *J. Magn. Reson.*, **1987**, *74*, 546.
36 M. Reggelin, H. Hoffmann et al., *J. Am. Chem. Soc.*, **1992**, *114*, 3272–3277.
37 C. Yu, T. H. Yang et al., *Biochem. Biophys. Acta*, **1991**, *1075*, 141–145.
38 T. Cohno, J. Kim et al., *Biochemistry*, **1995**, *34*, 10 256–10 265.
39 R. Boelens, T. M. G. Koning et al., *J. Mol. Struct.*, **1988**, *173*, 299.
40 S. Mronga, G. Muller et al., *J. Am. Chem. Soc.*, **1993**, *115*, 8414–8420.
41 K. Wüthrich, *NMR of Proteins and Nucleic Acids*, John Wiley, Chichester, 1986.
42 F. Benoit, J. L. Holmes et al., *Org. Mass Spectrom.*, **1969**, *2*, 591.
43 A. G. Harrison, A. Mandelbaum et al., *Org. Mass Spectrom.*, **1987**, *22*, 283–288.
44 Y. Yugawa et al., *Collection of the Spectra of Terpines* (Japanese Version), 1974.
45 M. Sawada, M. Shizuma et al., *J. Am. Chem. Soc.*, **1992**, *114*, 4405–4406.
46 M. Decouzon, J. F. Gal et al., *Org. Mass Spectrom.*, **1990**, *25*, 312–316.
47 A. Mandelbaum, in *Advances in Mass Spectrometry*, Vol. 13, 227–240,
48 B. Munson, B. L. Jelus et al., *Org. Mass Spectrom.*, **1980**, *15*, 161–165.
49 J. R. Chapman, *Practical Organic Mass Spectrometry*, 2nd Edn., 1993.
50 Y. Z. Chen, H. Li et al., *Org. Mass Spectrom.*, **1988**, *23*, 821–824.
51 Y. P. Tu, G. Y. Yang et al., *Org. Mass Spectrom.*, **1991**, *26*, 645–648.
52 B. A. Marpies, *Elementary Organic Stereochemistry and Conformational Analysis*, The Royal Society of Chemistry, London, 1981.
53 S. H. Gellman, G. P. Dado et al., *J. Am. Chem. Soc.*, **1991**, *113*, 1164–1173.
54 A. Dong, P. Huang et al., *Biochemistry*, **1990**, *19*, 3303–3308.
55 D. M. Byler, H. Susi, *Biopolymers*, **1986**, *25*, 469–487.
56 A. Nabet, M. Pezolet, *Appl. Spectrosc.*, **1997**, *51*, 466–469.
57 P. Hendra, C. Jones, G. Warnes, *Fourier Transform Raman Spectroscopy*, Ellis Horwood Limited, 1991.
58 F. Long, T. B. Freedman et al., *Appl. Spectrosc.*, **1997**, *51*, 504–507.

Appendix 1
Product Operator Formalism for Pulse Sequences

The magnetization vector model illustrates intuitively the production of NMR signals, the function of pulse sequences and thus the principles of 2D NMR. However, application of the magnetization vector model is limited. The function of pulses other than 90° or 180° cannot be discussed using this model, two sequential 90° pulses are even difficult to analyze. Therefore, pulse sequences that have been discussed previously, such as DEPT, COSY, INADEQUATE, cannot be treated with the magnetization vector model.

Quantum mechanics provides a general method, that is the density matrix formalism (particularly, product operator formalism) for analyzing pulse sequences, which is an integrated and strict theoretical system. However, chemists have great difficulty in understanding the theory to gain an overall comprehension of this formalism. It is more difficult for chemists to apply the formalism to analyze pulse sequences. In fact, introduction to the complete theoretical system is complicated but its rigid calculations are quite simple. There is a great deal of literature on the formalism. However, many references present the formalism in fragments or just cite some of the conclusions of the formalism without a complete description. Therefore, it is necessary to include an Appendix here, in order to avoid the difficult aspects of quantum mechanism. It is not a rigorous treatment but it is systematic and easy for chemists to understand. After reading this Appendix, the reader should be able to analyze pulse sequences using the formalism.

Two considerations will be taken into account as follows:

1. Quantum mechanics methods for the analysis of pulse sequences are limited to product operator formalism, which is applied in weakly coupled systems.
   For simplicity, only the system consisting of two spins will be described and on this basis, the main part of the pulse sequences can be analyzed.
2. Great efforts will be made to correlate the formalism with our familiar magnetization vector model.

Structural Identification of Organic Compounds with Spectroscopic Techniques. Yong-Cheng Ning
Copyright © 2005 WILEY-VCH Verlag GmbH & Co. KGaA, Weinheim
ISBN: 3-527-31240-4

A1.1
Spin States and Effects in NMR are Characterized by Operators

It should first of all be noted that bold-faced majuscules are used to describe operators as well as physical quantities.

Because all our discussions in this Appendix are concerned with a rotating frame, the sign "," used to denote the rotating frame is omitted.

In Chapter 4, $\boldsymbol{M}_x$, $\boldsymbol{M}_y$ and $\boldsymbol{M}_z$ were used to denote the components of the magnetization vector along the x, y and z axe, respectively. Here they are replaced correspondingly by $\boldsymbol{I}_x$, $\boldsymbol{I}_y$ and $\boldsymbol{I}_z$ with respect to the quantum mechanics. The rotation of the magnetization vector about the x or the y axis by a pulse is considered to be the result of fact that the magnetization vector is affected by $\boldsymbol{I}_x$ or $\boldsymbol{I}_y$.

The precession of the transverse magnetization vector in the x–y plane about the z axis is described as the action of $2\pi\delta t\boldsymbol{I}_z$.

For a system with two spins, their coupling is described as the action of πJtI_zS_z, where $\boldsymbol{I}$ and $\boldsymbol{S}$ represent the operators of the two nuclei of the system.

In brief, spin states, actions of radiofrequency pulses and evolutions by chemical shift or coupling are expressed completely as the actions of operators.

A1.2
Operation Rules of Operators

As discussed above, the analysis of the action of a pulse sequence is an operation of operators, that is, an operator is acted on by other operators to become another operator (sometimes unchanged).

In fact, there are only two cases between any two operators: commutation and anticommutation.

For any two operators $\boldsymbol{A}$ and $\boldsymbol{B}$, the following relationship holds:

$$[\boldsymbol{A}, \boldsymbol{B}] \equiv \boldsymbol{AB} - \boldsymbol{BA} \tag{A1}$$

$\boldsymbol{A}$ and $\boldsymbol{B}$ commute, if

$$[\boldsymbol{A}, \boldsymbol{B}] = 0 \tag{A2}$$

that is

$$\boldsymbol{AB} = \boldsymbol{BA} \tag{A3}$$

$\boldsymbol{A}$ and $\boldsymbol{B}$ anticommute, if

$$\boldsymbol{AB} = -\boldsymbol{BA} \tag{A4}$$

If the two operators commute, one operator will not be changed after the action of the other operator, which is expressed as:

$$\begin{array}{c} A \\ B \Downarrow \\ A \end{array} \tag{A5}$$

If two operators $\boldsymbol{B}_1$ and $\boldsymbol{B}_2$ commute, the action resulting from $\boldsymbol{B}_1$ and $\boldsymbol{B}_2$ for $\boldsymbol{A}$ does not change with the action order of $\boldsymbol{B}_1$ and $\boldsymbol{B}_2$. For example, because $\boldsymbol{I}_z$ describing the evolution of δ and $\boldsymbol{I}_z\boldsymbol{S}_z$ describing the spin coupling commute, the effective order of either $\boldsymbol{I}_z$, $\boldsymbol{I}_z\boldsymbol{S}_z$ or $\boldsymbol{I}_z\boldsymbol{S}_z$, $\boldsymbol{I}_z$ will lead to the same result.

If $\boldsymbol{A}$ and $\boldsymbol{B}$ anticommute, under the action of $\boldsymbol{B}$, $\boldsymbol{A}$ becomes two operators: one is the unchanged $\boldsymbol{A}$ with a coefficient of $\cos b$, and the other is a new operator $\mathbf{C}$ with a coefficient of $\sin b$, which is denoted as:

$$\begin{array}{ccc} & \boldsymbol{A} & \\ b\boldsymbol{B} \swarrow & & \searrow \\ \cos b\boldsymbol{A} & & \sin b\mathbf{C} \end{array} \tag{A6}$$

where b is the coefficient of $\boldsymbol{B}$, and

$$\mathbf{C} = \boldsymbol{AB} \tag{A7}$$

For simplicity, this operation is expressed as:

(A8)

We will of course add $\cos b$ and $\sin b$ to the final result. As for $\mathbf{C}$, it is obtained from a table.

Under the condition of $\cos b = 0$, the operation leads only to $\mathbf{C}$, which is expressed as:

(A9)

where S stands for the coefficient of $\sin b$.

Likewise, under the condition of sin $b=0$, the operation leads only to $\boldsymbol{A}$, which is expressed as:

(A9′)

where C stands for the coefficient of cos b.

A1.3
Pauli Matrices

Pauli matrices are the representations of spin operators in matrices. For the system of a single spin with $I = \frac{1}{2}$, we have

$$\sigma_x = \frac{1}{2}\begin{pmatrix} 0 & 1 \\ 1 & 0 \end{pmatrix} \tag{A10}$$

$$\sigma_y = \frac{1}{2}\begin{pmatrix} 0 & -i \\ i & 0 \end{pmatrix} \tag{A11}$$

$$\sigma_z = \frac{1}{2}\begin{pmatrix} 1 & 0 \\ 0 & -1 \end{pmatrix} \tag{A12}$$

They are related to $\boldsymbol{I}_x$, $\boldsymbol{I}_y$ and $\boldsymbol{I}_z$, respectively, and

$$\sigma_0 = \frac{1}{2}\begin{pmatrix} 1 & 0 \\ 0 & 1 \end{pmatrix} \tag{A13}$$

corresponds to the unit matrix.

Pauli matrices possess the following properties:

$$\text{i. } \sigma_u^2 \equiv \frac{1}{2}\sigma_0 \qquad u = 0, x, y \text{ and } z \tag{A14}$$

$$\text{ii. } \sigma_u\sigma_v = \frac{1}{2}\sigma_w \qquad \text{(in the order } x, y \text{ and } z\text{)} \tag{A15}$$

For example,

$$\sigma_x\sigma_y = \frac{1}{2}\sigma_z$$

or

$$\sigma_u \sigma_v = -\frac{i}{2}\sigma_w \quad \text{(in the order } z, y \text{ and } x\text{)} \tag{A16}$$

For example,

$$\sigma_y \sigma_x = -\frac{i}{2}\sigma_z$$

$$\text{iii. } \sigma_0 \cdot \sigma_u = \frac{1}{2}\sigma_u \quad (u = x, y \text{ and } z) \tag{A17}$$

These four equations can be easily proved by matrix multiplication.

A1.4 Product Operators of a System with Two Spins

A system with two spins, which may be homonuclear or heteronuclear, is denoted as IS.

The product operators of this system are formed by the direct product of those for a single spin system. This is why product operators are so named. As the system with a single spin has four product operators, the system with two spins has 4×4=16 product operators, which are classified into five groups.

i. I_0S_0
ii. I_xS_0, I_yS_0, I_zS_0, I_0S_x, I_0S_y, I_0S_z
iii. I_zS_z
iv. I_xS_z, I_yS_z, I_zS_x, I_zS_y
v. I_xS_x, I_xS_y, I_yS_x, I_yS_y

Now the product operators will be discussed according to the five groups.

i. I_0S_0 corresponds to the unit operator.
ii. The six product operators are actually I_x, I_y, I_z, S_x, S_y and S_z. As I_0 or S_0 corresponds to a unit operator, they are associated with the components of transverse magnetization vectors along the three coordinate axes. I_x, I_y, S_x and S_y are detectable signals.
iii. I_zS_z represents spin longitudinal ordered distribution.
iv. Each operator in the fourth group corresponds to two transverse magnetization vectors with opposite directions. For example, I_xS_z corresponds to two transverse magnetization vectors of I along the $\pm x$ axes under the condition of two magnetization vectors of S along the $\pm z$ axes, respectively.
v. These operators are the combinations of zero-quantum coherence and multi-quantum coherence.

These product operators are formed as follows:

$$I_uS_v = 2I_u \otimes S_v \tag{A18}$$

where $\otimes$ represents the direct product.

For example,

$$I_xS_z = 2I_x \otimes S_z = 2 \times \frac{1}{2}\begin{pmatrix} 0 & 1 \\ 1 & 0 \end{pmatrix} \otimes S_z$$

$$= \begin{bmatrix} 0 & 0 & & S_z \\ 0 & 0 & & \\ & & 0 & 0 \\ S_z & & 0 & 0 \end{bmatrix} = \frac{1}{2}\begin{bmatrix} 0 & 0 & 1 & 0 \\ 0 & 0 & 0 & -1 \\ 1 & 0 & 0 & 0 \\ 0 & -1 & 0 & 0 \end{bmatrix}$$

The 16 product operators possess the following properties, which are similar to Eqs. (A14) to (A17).

i. $(I_uS_v)^2 = \frac{1}{2}I_0S_0$ (A19)

$u = 0, x, y$ and z; and $v = 0, x, y$ and z.

ii. Any two of the 16 operators are commuted or anticommuted, that is

$$[I_uS_v, I_nS_m] = 0 \quad \text{(A20)}$$

$$I_uS_v \cdot I_nS_m = I_nS_m \cdot I_uS_v \quad \text{(A21)}$$

or

$$I_uS_v \cdot I_nS_m = -I_nS_m \cdot I_uS_v \quad \text{(A22)}$$

iii. $\boldsymbol{I}_u\boldsymbol{S}_v \cdot \boldsymbol{I}_n\boldsymbol{S}_m = 4(\boldsymbol{I}_u \cdot \boldsymbol{S}_v) \otimes (\boldsymbol{I}_n \cdot \boldsymbol{S}_m)$

$$= 4(\boldsymbol{I}_u \cdot \boldsymbol{I}_n) \otimes (\boldsymbol{S}_v \cdot \boldsymbol{S}_m)$$

$$= \boldsymbol{I}_a \cdot \boldsymbol{S}_b \quad \text{(A23)}$$

where

$$I_a = 2I_u \cdot I_n \quad \text{(A24)}$$

$$S_b = 2S_v \cdot S_m \quad \text{(A25)}$$

Equations (A23) to (A25) play an important role in simplifying the product operator calculations, because they decompose the operations of a two spin system into the operations of the single spin systems. On the basis of these equations, Table A1, which is frequently used in operations, is constructed.

A1.5
Operation Table of the Product Operators of a System with Two Spins

It is known that all states of a system with two spins are characterized by 16 product operators. The actions of pulses and the evolution by δ or J are the operations of related product operators. Any result of an operation is still one of the 16 product operators. Thus we can construct a table from which the result of operations can be found immediately. The table is formed on the basis of Eqs. (23) to (25).

Seven product operators belong to acting operators. They are as follows:

- I_x, I_y, S_x and S_y: for the actions of pulses
- I_z, S_z: for the evolution of δ
- I_zS_z: for the evolution of coupling

These seven product operators are placed in the first row of Table A1.

The 16 product operators that characterize the states of the system with two spins can be simplified further. As I_0S_0 is the unit operator, it can be removed. I_0S_x, I_0S_y and I_0S_z can be simplified as S_x, S_y and S_z, respectively. Likewise I_xS_0, I_yS_0 and I_zS_0 can be simplified as I_x, I_y and I_z, respectively. These simplified operators are placed in the left column of Table A1.

All of the results of any spin state by an action of an operator are listed in Table A1, in which "E" expresses the fact that original operator remains unchanged.

It should be noted that two systems with different rotation directions correspond to two sign systems for operator operations: one is towards the right-hand and the other to the left-hand. Therefore, the signs in Table A1 can be reversed. We have adopted these signs, which are in accordance with the concept

Tab. A1 The operation table of product operators of a system with two spins

	I_x	I_y	I_z	S_x	S_y	S_z	I_zS_z
I_x	E	I_z	$-I_y$	E	E	E	$-I_yS_z$
I_y	$-I_z$	E	I_x	E	E	E	I_xS_z
I_z	I_y	$-I_x$	E	E	E	E	E
S_x	E	E	E	E	S_z	$-S_y$	$-I_zS_y$
S_y	E	E	E	$-S_z$	E	S_x	I_zS_x
S_z	E	E	E	S_y	$-S_x$	E	E
I_zS_z	I_yS_z	$-I_xS_z$	E	I_zS_y	$-I_zS_x$	E	E
I_xS_z	E	I_zS_z	$-I_yS_z$	I_xS_y	$-I_xS_x$	E	$-I_y$
I_yS_z	$-I_zS_z$	E	I_xS_z	I_yS_y	$-I_yS_x$	E	I_x
I_zS_x	I_yS_x	$-I_xS_x$	E	E	I_zS_z	$-I_zS_y$	$-S_y$
I_zS_y	I_yS_y	$-I_xS_y$	E	$-I_zS_z$	E	I_zS_x	S_x
I_xS_x	E	I_zS_x	$-I_yS_x$	E	I_xS_z	$-I_xS_y$	E
I_xS_y	E	I_zS_y	$-I_xS_z$	$-I_xS_z$	E	I_xS_x	E
I_yS_x	$-I_zS_x$	E	I_xS_x	E	I_yS_z	$-I_yS_y$	E
I_yS_y	$-I_zS_y$	E	I_xS_y	$-I_yS_z$	E	I_yS_x	E

of the magnetization vector model. For example, I_z becomes I_y (not $-I_y$) after the action of a 90°_X pulse.

A1.6 Operations Around Pulse Sequences

The information necessary to analyze pulse sequences by product operators has thus been introduced. The operation is summarized as the following steps:

1. Determine the initial state. In general, it is I_z, which corresponds to M_0 in the equilibrium state.
2. The action of a pulse is initially considered. It is an operation of I_x or I_y (or S_x, S_y) according to the axis on which the pulse is applied.
3. The evolutions from δ and J are then considered. Because I_zS_z commutes with I_z (or S_z), their order of operation does not influence the final result.
4. The above operations are drawn according to Eqs. (A5), (A8) and (A9) or (A9′) as in an inverse tree. If the arrow points to the left, the unchanged operator is recorded. If the arrow points to the right, a new operator found according to the related operators from Tab. A1 is recorded.
5. Repeat steps 2 to 4 according to the pulse sequence.
6. Find the detectable signals (I_x, I_y, S_x and S_y) from the final result.
7. Write down the coefficients of the detectable signals.

The pulse sequence of COSY is used as an example, as follows (Fig. A1).

The operation starts from I_z (or from S_z) (Fig. A2).

From the above operations, it is known that the operations are simple but the process is tedious. For two pulses, an operation with ten steps must be performed.

Finally, 13 terms are obtained, in which I_x, $-I_y$, S_x and $-S_y$ are detectable signals. Their coefficients should be in continued multiplication.

I_x is taken as an example. Its coefficient is

$$\sin(2\pi\delta_I t_1)\cos(\pi J t_1)\cos(2\pi\delta_I t_2)\cos(\pi J t_2)$$

From the coefficient, it is known that this signal is a diagonal peak because δ_I is correlated with both t_1 and t_2, which means that the peak is situated at $\omega_1 = \omega_2 = \delta_I$.

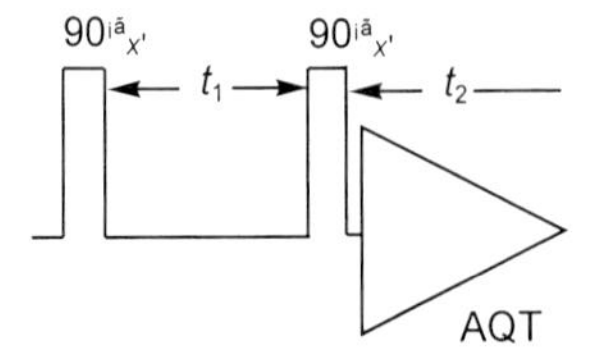

Fig. A1 Pulse sequence of COSY.

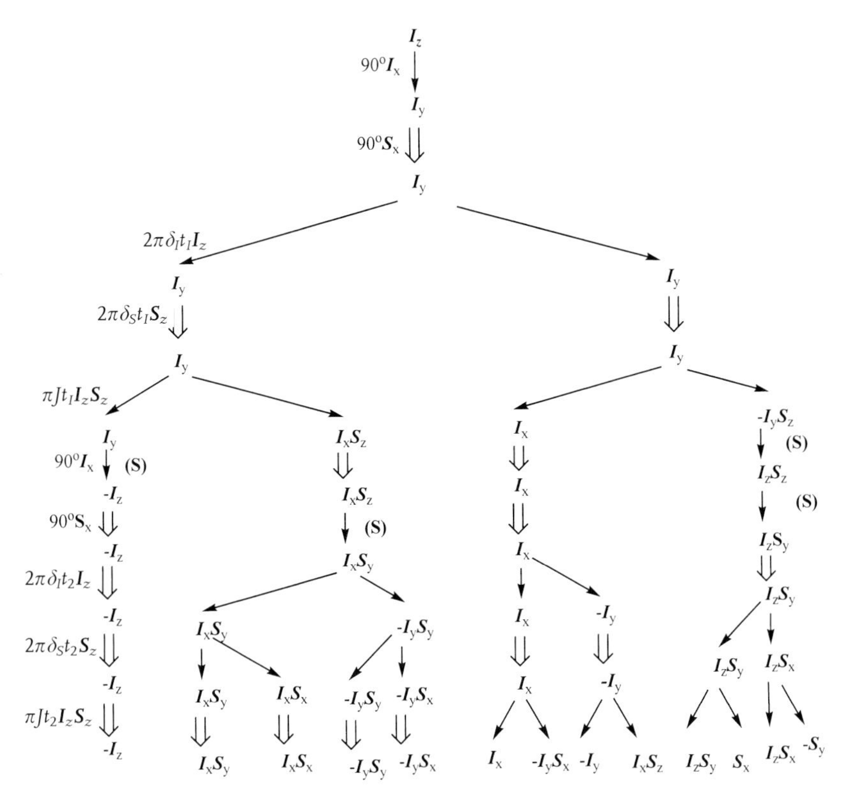

Fig. A2 Pulse sequence operation starting from I_z.

From the two factors of $\cos(\pi J t_1)$ and $\cos(\pi J t_2)$ in the coefficient, it is also known that the peak shows coupling splitting in both ω_1 and ω_2 because J is correlated with t_1 and t_2.

This result can be understood another way. Because both the original and final states are concerned with $\boldsymbol{I}$ without a transfer from $\boldsymbol{I}$ into $\mathbf{S}$, the peak should be a diagonal peak.

The term of $-\boldsymbol{I}_y$ is similar to $\boldsymbol{I}_x$.

The coefficient of $\boldsymbol{S}_x$ is

$$\sin(2\pi\delta_I t_1)\sin(\pi J t_1)\cos(2\pi\delta_S t_2)\sin(\pi J t_2)$$

From the coefficient, it is known that the signal is a cross peak because δ_I is correlated with t_1 and δ_S with t_2, which means that the peak is situated at $\omega_1 = \delta_I$ and $\omega_2 = \delta_S$.

From the two factors of $\sin(\pi J t_1)$ and $\sin(\pi J t_2)$ in the coefficient, it is also known that the peak shows coupling splitting in both ω_1 and ω_2.

This result can be understood another way. Because the original state is $\boldsymbol{I}_z$ but the final state is $\boldsymbol{S}_x$, that is, there is a transfer from $\boldsymbol{I}$ into $\boldsymbol{S}$, the peak should be a cross peak.

The term of $-\boldsymbol{S}_y$ is similar to $\boldsymbol{S}_x$.

Because the cross peak intensity is proportional to $\sin(\pi J t_1)\sin(\pi J t_2)$, increasing t_1 and t_2 will increase the intensity when J is small. That is why a time duration Δ is added before and after the second 90° pulse in the COSYLR sequence.

References

1 R. Freeman, *A Handbook of Nuclear Magnetic Resonance*, Longman Scientific & Technical, London, **1988**, 164–173.

2 J. Y. Lallemand, *Seminars of 2D NMR*, Beijing, 1985.

Appendix 2
Characteristic Frequencies of Common Functional Groups

Compound	*Group*	*Absorption frequency* (cm^{-1})				
		4000–2500	*2500–2000*	*2000–1500*	*1500–900*	*<900*
Alkyls	$-CH_3$	2960, sharp [70] 2870, sharp [30] (1)			1460 [<15] 1380 [15] (2)	
	$-CH_2$	2925, sharp [75] 2850, sharp [45] (3)			1470 [8]	725–720 [3] (4)
	Δ (cyclopropane)	3000–3080 (5) [variable]				
Unsaturated hydrocarbons	$=CH_2$	3080 [30] 2975 [medium]				
	$=CH-$	3020 [medium]				
	C=C			1675–1600 [medium-weak] (6)		
	$-CH=CH_2$				990, sharp [50] 910, sharp [110]	
	$-C=CH_2$					895, sharp [100–150]
	trans-				965, sharp [100]	
	cis-					800–650 (7) [40–100]

Structural Identification of Organic Compounds with Spectroscopic Techniques. Yong-Cheng Ning
Copyright © 2005 WILEY-VCH Verlag GmbH & Co. KGaA, Weinheim
ISBN: 3-527-31240-4

Compound	**Group**	**Absorption frequency (cm^{-1})**				
		4000–2500	**2500–2000**	**2000–1500**	**1500–900**	**<900**
Unsaturated hydrocarbons	tri-substituted					840–800 sharp [40]
	≡CH	3300, sharp [100]				
	–C≡C–		2140–2100 [5] (8) 2260–2190 [1] (9)			
Phenyl and condensed aromatic rings	C=C			1600, sharp [<100] 1500, sharp [<100]	1450 [medium]	
	=CH	3030 [<60]				
				2000–1600 [5] (10)		900–850 [medium] (11) 860–800 sharp [strong] (12) 800–750 sharp [strong] (13) 770–730 sharp [strong] (14) 710–690 sharp [strong] (15)

Compound	*Group*	*Absorption frequency (cm⁻¹)*				
		4000–2500	***2500–2000***	***2000–1500***	***1500–900***	***<900***
Heteroaromatics	pyridine	3075–3020 sharp [strong]		1620–1590 [medium] 1500 [medium]		920–720 sharp [strong] (16)
	furan	3165–3125 [medium, weak]		≈1600 ≈1500	≈1400	
	pyrrole	3490, sharp [strong] (17) 3125–3100 [weak]		1600–1500 (18) [variable] (two bands)		
	thiophene	3125–3050		≈1520	≈1410	750–690 [strong]
Alcohols and phenols	isolated (19):					
	primary	3640, sharp [70]			1050, sharp [60–200]	
	secondary	3630, sharp [55]			1100, sharp [60–200]	
	tertiary	3620, sharp [45]			1150, sharp [60–200]	
	phenolic	3610, sharp [medium]			1200, sharp [60–200]	
	inter-molecular: H-bond				1200, sharp [60–200]	
	dimeric	3300–3500 (20)				
	polymeric	3600, broad [strong]				
	intra-molecular: H-bond					
	polyol	3600–3500 [50–100]				
	π-H-bond	3600–3500				
	chelated	3200–2500, broad [weak]				
Ethers	C–O–C				1150–1070 [strong]	
	=C–O–C				1275–1200 [strong]	

Compound	***Group***	***Absorption frequency (cm^{-1})***				
		4000–2500	***2500–2000***	***2000–1500***	***1500–900***	***<900***
Ketones	saturated			1725–1705 sharp [300–600]		
	cyclic					
	seven membered and larger			1720–1700 sharp [extremely strong]		
	six membered			1725–1705 sharp [extremely strong]		
	five membered			1750–1740 sharp [extremely strong]		
	four membered			1775 sharp [extremely strong]		
	three membered			1850 sharp [extremely strong]		
	unsaturated					
	α,β-unsaturated			1685–1665 sharp [extremely strong] (21)		
				1650–1600 sharp [extremely strong] (22)		
	Ar–CO–			1700–1680, sharp [extremely strong] (23)		
	α,β-, α',β'-unsaturated Ar–CO–Ar			1670–1660, sharp [extremely strong] (24)		

Compound	*Group*	*Absorption frequency (cm^{-1})*				
		4000–2500	*2500–2000*	*2000–1500*	*1500–900*	*<900*
Ketones	α-substituted α-halogenated			1745–1725, sharp [extremely strong]		
	α-di-halogenated			1765–1745, sharp [extremely strong]		
	diketone			1730–1710, sharp [extremely strong]		
	quinone 1,2-quinone 1,4-quinone			1690–1660, sharp [extremely strong]		
	tropone			1650, sharp [extremely strong]		
Aldehydes	saturated	2820 [weak] 2720 [weak]		1740–1720, sharp [extremely strong]		
	unsaturated α,β-unsaturated			1700–1680, sharp [extremely strong]		
	$\alpha,\beta,\gamma,\delta$-unsaturated			1680–1660, sharp [extremely strong]		
	Ar–CHO			1715–1695, sharp [extremely strong]		
Carboxylic acids	saturated	3000–2500, broad		1760 [1500] (25)	1440–1395 [medium, strong]	
				1725–1700 [1500] (26)	1320–1210 [strong] 920, broad [medium]	

Compound	Group	Absorption frequency (cm^{-1})				
		4000–2500	2500–2000	2000–1500	1500–900	<900
Carboxylic acids	α,β-unsaturated			1720 [extremely strong] (27) 1715–1690 [extremely strong] (28)		
	Ar–COOH			1700–1680 [extremely strong]		
	α-halogenated			1740–1720 [extremely strong]		
Acid anhydrides	chain saturated			1820 [extremely strong] 1760 [extremely strong]	1170–1045 [extremely strong]	
	α,β-unsaturated			1775 [extremely strong] 1720 [extremely strong]		
	six membered			1800 [extremely strong] 1750 [extremely strong]	1300–1175 [extremely strong]	
	five membered			1865 [extremely strong] 1785 [extremely strong]	1300–1200 [extremely strong]	
Esters	chain saturated			1750–1730 sharp [500–1000]	1300–1050 two bands [extremely strong]	

Compound	*Group*	*Absorption frequency* (cm^{-1})				
		4000–2500	*2500–2000*	*2000–1500*	*1500–900*	*<900*
Esters	α,β-unsaturated			1730–1715 [extremely strong]	1300–1250 [extremely strong] 1200–1050 [extremely strong]	
	α-halogenated			1770–1745 [extremely strong]		
	Ar–COOR			1730–1715 [extremely strong]	1300–1250 [extremely strong] 1180–1100 [extremely strong]	
	CO–O–C=C–			1770–1745 [extremely strong]		
	CO–O–Ar			1740 [extremely strong]		
Carboxylates	$-COO^-$			1610–1550 [strong, broad]	1420–1300 [strong]	
Acid halides (29)	saturated			1815–1770 sharp [extremely strong]		
	α,β-unsaturated			1780–1750 sharp [extremely strong]		
Amides (30) (31) (38)	primary $-CONH_2$	3500, 3400 two bands [strong] (32) (3350–3200) two bands (32)				
	primary $-CONH_2$			1690 (1650), sharp [extremely strong] (33) 1600 (1640) [strong] (34)		

Compound	Group	Absorption frequency (cm^{-1})				
		4000–2500	2500–2000	2000–1500	1500–900	<900
Amides	secondary –CONH-	3440, sharp (3300, 3070) (35)		1680 (1665) sharp [extremely strong] (33)		
				1530 (1550) [variable] (36)		
					1260 (1300) [medium, strong] (37)	
	tertiary –CON<			1650 (1650)		
Amines (38)	primary	3500 (3400) [medium, strong]		1640–1560 [strong, medium]		
		3400 (3300) [medium strong]				
	secondary	3350–3310 [weak]				
	Ar–NHR	3450 [medium]				
	Ar–NHAr′	3490 [medium]				
Amines	heterocyclic NH	3490 [strong]				
	tertiary*				1350–1260 [medium]	
Ammonium salts	$-NH_3^+$	3000–2000 [strong] (39)		1600–1575 [strong]		
				1550–1500 [strong]		
	$-NH_2^+$	3000–2250 [strong] (39)		1620–1560 [medium]		
	$-NH^+$	2700–2250 [strong] (39)				
Cyanide	R–CN		2260–2240, sharp [variable]			
	α,β-un-saturated		2240–2215, sharp [variable]			
	Ar–CN		2240–2215, sharp [variable]			

Compound	*Group*	*Absorption frequency (cm^{-1})*				
		4000–2500	***2500–2000***	***2000–1500***	***1500–900***	***<900***
Thiocyanates	R–S–C≡N		2140, sharp [extremely strong]			
	Ar–S–C≡N		2175–2160, sharp [extremely strong]			
Iso-thiocyanates	R–N=C=S		2140–1990, sharp [extremely strong]			
	Ar–N=C=S		2130–2040, sharp [extremely strong]			
Imines	⟩C=N–			1690–1630 [medium] (40)		
Oximes	⟩C=N–OH	3650–3500 broad [strong] (41)		1680–1630 [variable]	960–930	
Diazo-compounds	–N=N			1630–1575 [variable]		
Nitros	R–NO_2			1550 sharp [extremely strong]	1370 sharp [extremely strong]	
	Ar–NO_2			1535 sharp [extremely strong]	1345 sharp [extremely strong]	
Nitrates	–O–NO_2			1650–1600 [strong]	1300–1250 [strong]	
Nitrosos	–NO			1600–1500 [strong]		
Nitrites	–ONO			1680–1650 [variable] 1625–1610 [variable]		

Compound	***Group***	***Absorption frequency (cm^{-1})***				
		4000–2500	***2500–2000***	***2000–1500***	***1500–900***	***<900***
Sulfur-containing compounds	thiols, –SH	2600–2550				
	>C=S				1200–1050, [strong]	
	sulfoxides >S=O				1060–1040 sharp [300]	
	sulfones $>S\langle^{O}_{O}$				1350–1310 sharp [250–600] 1160–1120 sharp [500–900]	
	sulfonates $R–SO_3^-M^+$ (42)				1200, broad [extremely strong] 1050 [strong]	
	sulfon-amides $R–SO_2–N\langle$				1370–1330 [extremely strong] 1180–1160 [extremely strong]	
Halides	C–F				1400–1000 [extremely strong]	
	C–Cl					800–600 [strong]
	C–Br					600–500 [strong]
	C–I					500 [strong]
Phosphorus groups	P–H		2440–2280 [medium, weak]			
	P–C					750–650
	P=O				1300–1250 [strong]	
	P–O–R				1050–1030 [strong]	
	P–O–Ar				1190 [strong]	

Remarks:

(1) Moves to a lower frequency in $–O–CH_3$ or $–N–CH_3$.

(2) Doublet in *geminal* methyl groups.

(3) Shifts to a lower frequency in $–O–CH_2$ or $–N–CH_2$.

(4) Exists in $–(CH_2)_n$ – when $n>4$. Moves to a higher frequency when $n<4$.

(5) When H is situated in the ring.

(6) Moves to a lower frequency when it conjugates with any other alkene.
(7) Frequently in 730–675 cm^{-1}.
(8) Terminal alkyne.
(9) Alkyne in a chain.
(10) Several weak bands. It can be obscured by other bands in this region.
(11) Isolated H.
(12) Two adjacent H atoms.
(13) Three adjacent H atoms.
(14) Four or five adjacent H atoms.
(15) Appears in mono, 1,3-, 1,3,5-, 1,2,3-substituted phenyls.
(16) Similar to the consideration of (11)–(15).
(17) Absorption bands of NH.
(18) Absorption of C=C.
(19) In non-polar diluted solutions.
(20) Usually obscured by the band of the polymer.
(21) Absorption band of carbonyl.
(22) Absorption band of alkene.
(23) Absorption band of carbonyl.
(24) Absorption band of carbonyl.
(25) Absorption band of the monomer.
(26) Absorption band of the dimer.
(27) Absorption band of the monomer.
(28) Absorption band of the dimer.
(29) Frequencies of these absorption bands decrease in the order of –F, –Br, –I.
(30) The frequency noted in parentheses is that of associated molecules.
(31) The absorption frequency of lactam is higher when the ring is smaller.
(32) Absorption bands of N–H.
(33) Absorption band of carbonyl. Amide I band.
(34) Amide II band. Two bands in solid phase.
(35) Absorption band of NH.
(36) Amide II band.
(37) Amide III band.
(38) The frequency noted in parentheses is that of associated molecules.
(39) Several bands on a broad band.
(40) Moves to a lower frequency when conjugated.
(41) Moves to a lower frequency when associated.
(42) Metallic ion.

Notes:

A. Indicated frequencies in the table are characteristic absorption bands of common functional groups.
B. "Sharp" and "broad" are used for special band shapes.
C. Absorption intensities are expressed as follows:
  - Extremely strong: the apparent molecular absorption coefficient >200
  - Strong: the apparent molecular absorption coefficient is 75–200
  - Medium: the apparent molecular absorption coefficient is 25–75
  - Weak: the apparent molecular absorption coefficient <25
  - or values are denoted in the parentheses.

References

1. K. Nakanishi et al., *Infrared Absorption Spectroscopy*, 2nd Edn., Holden-Day, San Francisco, 1977.
2. N. B. Colthup, H. D. Lawrence, E. W. Stephen, *Introduction to Infrared and Raman Spectroscopy*, 3rd Edn., Academic Press, New York, 1990.
3. R. M. Silverstein, G. C. Bassler, T. C. Morill, *Spectrometric Identification of Organic Compounds*, 5th Edn., John Wiley, Chichester, 1991.

Index

Structural Identification of Organic Compounds with Spectroscopic Techniques. Yong-Cheng Ning
Copyright © 2005 WILEY-VCH Verlag GmbH & Co. KGaA, Weinheim
ISBN: 3-527-31240-4